Date Due

Keywords — brazing processes, brazing atmospheres, brazing fluxes, brazing fundamentals, silver brazing, nickel brazing, copper brazing, brazing of stainless steel, brazing of refractory metals, brazing of ceramics, brazing of low expansion alloys, cleaning of brazements, brazing honey-comb.

Brazing Handbook

Fourth Edition

Prepared by
AWS Committee on Brazing and Soldering

Under the Direction of
AWS Technical Activities Committee

Approved by
AWS Board of Directors

American Welding Society

550 N.W. LeJeune Road, P.O. Box 351040, Miami, Florida 33135

Library of Congress Number: 91-70805
International Standard Book Number: 0-87171-359-4

American Welding Society, 550 N.W. LeJeune Road,
P.O. Box 351040, Miami, Florida 33135

THE BRAZING HANDBOOK is a collective effort of many volunteer technical specialists to provide information to assist with the design and application of brazing processes.

Reasonable care is taken in the compilation and publication of the Brazing Handbook to insure authenticity of the contents. No representation or warranty is made as to the accuracy or reliability of this information.

The information contained in the Brazing Handbook shall not be construed as a grant of any right of manufacture, sale, use, or reproduction in connection with any method, process, apparatus, product, composition or system, which is covered by patent, copyright, or trademark. Also, it shall not be construed as a defense against any liability for such infringement. No effort has been made to determine whether any information in the Handbook is covered by any patent, copyright, or trademark, and no research has been conducted to determine whether an infringement would occur.

Printed in the United States of America

Contents

Personnel

AWS Committee on Brazing and Soldering

W. R. Frick, Chairman	Varian
W. G. Bader, Chairman (1986-1989)	Consultant
N. C. Cole, Chairperson (1983-1986)	Oak Ridge National Laboratories
J. R. Terrill, Chairman (1980-1983)	Molycorp, Incorporated
D. D. Peter, 1st Vice Chairman	Consultant
M. J. Lucas, Jr., 2nd Vice Chairman	General Electric Company
H. F. Reid, Secretary	American Welding Society
G. A. Andreano*	Gana & Associates, Incorporated
W. G. Bader	Consultant
R. E. Ballentine*	Consultant
T. S. Bannos	Engelhard Corporation
R. E. Beal	Amalgamated Technologies
C. R. Behringer+	Consultant
S. S. Bhargava	Allison Gas Turbine Division General Motors Corporation
D. W. Bucholz	IBM Corporation, FSD
A. B. Cedilote	Westinghouse Electric Corporation, E.M.D.
N. C. Cole	Oak Ridge National Laboratories
R. E. Cook	The Wilkinson Company
G. D. Cremer*	Consultant
P. H. Dauve+	Hughes Aircraft Company
R. G. Fairbanks	Scarrott Metallurgical Company
E. B. Gempler*	Consultant
P. F. Gerbosi	Lepel Corporation
I. S. Goodman+	Consultant
P. K. Gupta	General Electric Company
K. L. Gustafson	Consultant
M. J. Higgins	Techni Braze, Incorporated
T. Hikido	Pyromet Industries
T. P. Hirthe	Lucas-Milhaupt, Incorporated
F. M. Hosking	Sandia National Laboratories
J. R. Jachna	Modine Manufacturing Company
J. J. Kozelski*	Consultant
S. Liu	Colorado School of Mines
E. Lugscheider*	University of AAchen
D. J. Manente	Vac-Aero International
M. M. McDonald	Rockwell International
J. A. Miller	Essex Consultants
H. Mizuhara	GTE WESGO Division GTE Products Corporation

A. J. Moorhead*	Oak Ridge National Laboratories
E. P. Patrick*	Aluminum Company of America
D. W. Patterson*	Harrison Radiator Division General Motors Corporation
R. L. Peaslee	Wall Colmonoy Corporation
C. W. Philp	Consultant
M. Prager*	Welding Research Council
A. Rabinkin	Metglas Products Division Allied Signal, Incorporated
M. L. Santella*	Oak Ridge National Laboratories
J. L. Schuster	E.G. & G., Incorporated
A. Severin	Lucas Milhaupt, Incorporated
J. F. Smith*	Lead Industries Association
R. Squillacioti*	Army Materials Technology Laboratory
J. R. Terrill	Molycorp, Incorporated
P. Vianco	Sandia National Laboratories
R. W. Walls	Princeton University
C. E. T. White*	Indium Corporation of America
W. L. Winterbottom	Ford Motor Company

+Deceased
*Advisor

AWS Subcommittee on Brazing Education and Information

A. Severin, Chairman	Lucas-Milhaupt, Incorporated
J. R. Jachna, Vice Chairman	Modine Manufacturing Company
H. F. Reid, Secretary	American Welding Society
N. C. Cole	Oak Ridge National Laboratories
R. E. Cook	The Wilkenson Company
W. R. Frick	Varian
E. B. Gempler*	Consultant
J. J. Kozelski*	Consultant
S. B. Kulkarni	Multi Metals
M. J. Lucas, Jr.	General Electric Corporation
J. A. Miller	Essex Consultants
D. E. Paris	Parfuse Corpration
R. L. Peaslee	Wall Colmonoy Corporation
M. L. Santella	Oak Ridge National Laboratories
J. L. Schuster	EG&G Incorporated

*Advisor

AWS Subcommittee on Brazing Applications

A. B. Cedilote, Chairman	Westinghouse Electric Corporation, E.M.D.
H. F. Reid, Secretary	American Welding Society
R. E. Beal	Amalgamated Technologies
N. C. Cole	Oak Ridge National Laboratories
R. G. Fairbanks	Scarratt Metallurgical Company
C. E. Fuerstenau	Lucas-Milhaupt, Incorporated
*E. B. Gempler**	Consultant
P. F. Gerbosi	Lepel Corporation
P. K. Gupta	General Electric Corporation
T. Hikido	Pyromet Industries
J. R. Jachna	Modine Manufacturing Company
*J. J. Kozelski**	Consultant
S. B. Kulkarni	Multi Metals
M. K. Megerle	Naval Air Engineering Center
J. A. Miller	Essex Consultants
D. D. Peter	Consultant
J. L. Schuster	EG&G Incorporated
M. K. Tarby	Wall Colmonoy Corporation
R. W. Walls	Princeton University
D. L. Warburton	Varian
W. L. Winterbottom	Ford Motor Company

*Advisor

AWS Subcommittee on Brazing Safety

S. S. Bhargava, Chairman	Allison Gas Turbine Division of General Motors Corporation
H. F. Reid, Secretary	American Welding Society
A. B Cedilote	Westinghouse Electric Corporation, E.M.D.
*E. B. Gempler**	Consultant
R. Hensen	J. W. Harris Company
T. P. Hirthe	Lucas-Milhaupt, Incorporated
L. Liu	Colorado School of Mines
C. W. Philp	Handy and Harman

* Advisor

Contributors - Not Members of the Brazing and Soldering Committee

C. H. Barken	Unitrol Electronics
C. D. Budzko	Taylor-Winfield Consultant
J. Hawkins	Textron-Lycoming
W. T. Hooven, III	Hooven Metal Treating Incorporated
J. S. Libsch	Consultant
L. F. Lockwood	Dow Chemical Company
R. Oliva DDS	Consultant
R. D. Sulton	Lepal Corporation
W. Sisolak	Consultant
M. Vallaro	University of Connecticut
C. VanDyke	Lucas Milhaupt, Incorporated
D. Zambrosky	Consultant
AWS Safety and Health Committee	

Foreword

This Fourth Edition of the Brazing Handbook is the culmination of several years of effort by more than 100 brazing experts. Material from the Third Edition of the Brazing Manual has been completely reorganized and updated. Extensive chapters on furnace brazing, brazing safety, and the corrosion of brazements have been added.

Thermal expansion data for a number of brazing materials are included in the Appendix.

Preface

Knowledge of the ancient art of brazing is continuously being supplemented by an ever-increasing amount of technical information about metals and their behavior, so that today brazing must be considered both an art and a science. This fourth edition of the Brazing Handbook (formerly the Brazing Manual) includes the fundamental concepts of brazing and incorporates the many advances made since the Manual was first published.

The American Welding Society defines brazing as "a group of welding processes that produces coalescence of materials by heating them to the brazing temperature in the presence of a filler metal having a liquidus above 840°F (450°C) and below the solidus of the base metal. The filler metal is distributed between the closely fitted faying surfaces of the joint by capillary action."

Brazing then must meet each of three criteria:

(1) The parts must be joined without melting the base metals.

(2) The filler metal must have a liquidus temperature above 840°F (450°C).

(3) The filler metal must wet the base metal surfaces and be drawn into or held in the joint by capillary action.

To achieve a good joint using any of the various brazing processes described in this Handbook, the parts must be properly cleaned and must be protected, either by fluxing or atmosphere during the heating process, to prevent excessive oxidation. The parts must be designed to afford a capillary for the filler metal when properly aligned, and a heating process must be selected that will provide the proper brazing temperature and heat distribution.

No analysis of a subject that is continuously being improved can hope to be complete, nor can the subject be covered with a thoroughness that would satisfy the specialist. For this reason, most chapters provide a list of references that give additional and more detailed information on the subject. Yet even after the additional research, trial and error may be required to successfully complete unusual applications. It is hoped, however, that the trials and errors will be fewer for having this Handbook as a guide.

Comments, inquiries, and suggestions for future revisions of the Handbook are welcome. They should be sent to the Director, Technical Standards and Publications, American Welding Society, 550 N.W. LeJeune Road, Miami, Florida 33135.

Keywords — brazing, brazing process, brazing fundamentals, brazed joints

Chapter 1

FUNDAMENTALS

INTRODUCTION

The process of brazing that we know today began as an ancient art. What began as art, however, evolved through our increased understanding of the nature and behavior of materials into art, technology, and science. In a very general sense, brazing is a process for joining materials which relies on the melting, flow, and solidification of a filler metal to form a leaktight seal, a strong structural bond, or both. The uniqueness of the process is that metallurgical bonds are formed during brazing by melting only the filler metal and not the parts being joined. Brazing is a well established commercial process, and is widely used in industry, in large part, because almost every metallic and ceramic material can be joined by brazing. Generally, brazing can easily be performed by manual techniques, but, in many cases, it can just as easily be automated if necessary.

The American Welding Society defines brazing as "a group of welding processes which produces coalescence of materials by heating them to a suitable temperature and by using a filler metal having a liquidus temperature above 840°F (450°C) and below the solidus temperature of the base materials. The filler metal is distributed between the closely fitted surfaces of the joint by capillary attraction." This definition serves to distinguish brazing from other joining processes of soldering and welding. Brazing and soldering share many important features, but the term "brazing" is used when the joining process is performed above 840°F (450°C), while "soldering" is used below that temperature. Brazing differs from welding in that in braze processing the intention is to melt only the braze filler metal and not the base materials. In welding, both filler metals and base metals are melted during the process.

To achieve a good joint by any variation of the brazing process, the parts must be properly cleaned and must be protected from excessive oxidation by fluxing, or by use of a controlled atmosphere. In addition, the parts must be designed so that when they are properly aligned a capillary is formed in which the molten filler metal can flow. Also, a heating process must be selected that will produce the proper brazing temperature and heat distribution. The various brazing processes, joint design, cleaning and heating methods, and details specific to particular materials are outlined in this *Brazing Handbook*. The purpose of this chapter is to provide a basic understanding of the brazing process through a review of the factors fundamental to the process itself. To assist the reader with unfamiliar terms, a glossary of terms commonly used in the field is provided in Appendix A.

HISTORICAL PERSPECTIVE

Brazing is perhaps the oldest technique for joining metals other than mechanical fastening, and its historical development parallels that of the development of materials. Earliest evidence indicates that mixtures of metal salts and organic reducing agents (e.g., copper hydrate and organic gum) were used as filler metals to braze gold and silver base metals. Figure 1.1 shows a wall painting from a tomb at Thebes in Egypt dating back to 1475 B.C. It depicts a slave engaged in gold brazing with a charcoal fire in a clay bowl, a reed for a blowpipe, and tongs. Later advances led to the use of copper and brass alloys as braze filler metals. As the development of the base materials progressed from the noble metals to stronger copper-based and iron-based alloys, to the present range of metallic alloys and ceramics, parallel developments in the sophistication of braze filler metals have occurred. Advances in processing and quality control methods have led to a similar sophistication of brazing techniques.

PRINCIPLES AND DEFINITIONS

In the process of brazing, the temperature of the assembly to be brazed (brazement) is raised to a point where the filler metal be-

Figure 1.1 — Wall Painting From a Tomb in Thebes, Egypt, Made in Approximately 1475 B.C. Which Shows a Slave Engaged in Gold Soldering. (This Figure was Taken From the October 1979 Issue of *Welding Journal*)

comes molten and fills the joint gap between base materials as illustrated in Figure 1.2. In most cases the interaction between the molten filler metal and the base materials results in the establishment of a metallurgical bond when the filler metal solidifies. Although the representation of brazing depicted by Figure 1.2 is relatively simple, basic and often complex metallurgical and chemical processes take place within the joint and on the surfaces of the materials involved. An appreciation of the complexity of the process is necessary in order to design and produce braze joints with closely controlled physical and chemical properties.

The phenomena of wetting and flow of a liquid on the surface of a solid are basic to most models developed to describe the formation of a braze joint. Wetting of the base materials by the filler metal is required to provide intimate contact between them and to develop the bonding needed. The driving forces for these phenomena are characterized through the thermodynamic concepts of surface free energy, and the free energy of formation of phases which may result from chemical reactions that occur during brazing. After wetting conditions are established, capillary forces produce flow of the liquid filler metal, and act to fill the joint gap with the molten metal. Both wetting and flow are strongly influenced by the chemical reactions occurring at the interfaces and within the filler metal itself, as well as by the geometry of the joint. The quality of wetting and flow strongly influences the final properties of the joint.

A full description of the theories of wetting and flow as they apply to brazing is beyond the scope of this book. However, the introduction of these basic ideas will be helpful in outlining the general characteristics of solids and liquids that influence the process.

Wetting

Factors important for determining the extent of wetting can be illustrated by the familiar problem of a liquid droplet in contact with a flat solid surface. In the idealized case where there are no chemical reactions between solid, liquid, and vapor phases, and where gravitational forces can be ignored (i.e., for relatively small drop-

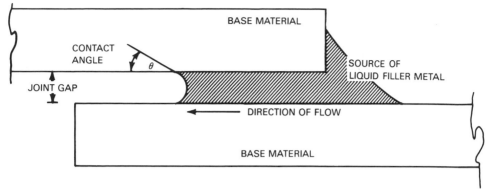

Figure 1.2 — Schematic Representation of a Braze Joint

lets), the liquid droplet will assume an equilibrium configuration dictated by surface free energy considerations. The shape of the liquid droplet is uniquely characterized by θ, its contact angle with the solid as shown in Figure 1.3. The relationship between contact angle and the surface free energies at the liquid-vapor γLV, solid-vapor γSV, and solid-liquid SL interfaces is expressed by:

$$cos\theta = (\gamma SV - \gamma SL)/\gamma LV \qquad \text{(Eq. 1)}$$

The boundary between wetting and non-wetting conditions is generally taken as θ = 90°. For θ < 90°, wetting occurs, while θ > 90° represents a condition of non-wetting. Spreading may be defined as the condition where the liquid completely covers the solid surfaces. This condition occurs when θ approaches the value of 0°. For most brazing systems the optimum value of θ is in the range of 10-45°, and is determined by joint gap or thickness, i.e. small θ for very thin joints. It should be remembered that when appreciable chemical reactions occur during brazing, the equation relating contact angle to surface energies is of qualitative value only.

It is also important to understand that the liquid and solid surface free energies can be markedly lowered by the absorption of surface active impurities at any of the three interfaces shown in Figure 1.3. All real surfaces of liquids and solids are modified to some extent by absorption of surface active elements and particularly oxidation.

Indeed, the presence of oxide on a solid metal surface suppresses wetting and inhibits its spreading of liquid metal over the surface. Therefore, much of the technology of brazing is directed to eliminate possible detrimental effects of oxide presence on wetting.

Capillary Flow

There is no simple treatment for describing filler metal flow in brazing. Although the field of fluid dynamics provides a basis for quantitative insight, its complexities are beyond the scope of this introductory discussion and only a qualitative treatment of flow will be discussed here.

Experience shows that filler metal flow is a function of capillary driving force, the viscosity and density of the molten metal, and the geometry of the joint. Viscosity, η, which is the resistance to flow of liquid, is found empirically to be a moderate exponential function of temperature T, of the form

$$ln\eta = A + B/T + C/T^2 \qquad \text{(Eq. 2)}$$

where A, B, and C are constants characterizing the liquid. Figure 1.4 shows the measured viscosity of iron, nickel, and copper as a function of temperature, until erosion of the base metal begins. This behavior is typical of other metals and alloys. The nearly linear dependence indicates that temperature has a strong influence upon viscosity and that filler metal flow can be markedly enhanced by an increase in braz-

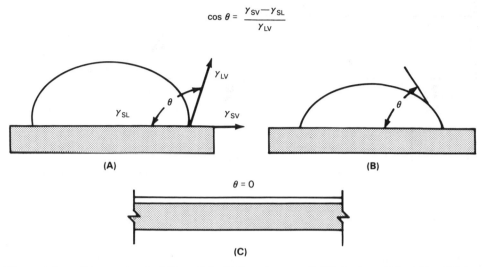

$$\cos \theta = \frac{\gamma_{SV} - \gamma_{SL}}{\gamma_{LV}}$$

Figure 1.3 — Contact Angle θ For a Liquid Droplet on a Solid Surface: (A) < 90°, (B) θ > 90°, (C) θ + 0°

ing temperature, until erosion of the base metal begins.

Explicit flow rate expressions for even simple geometries are complex and not easy to verify by experiment. However, analyses and experiments show that flow rates can be high (of the order of cm/s), and tend to increase with the magnitude of the relationship:

$$(\gamma LV \cdot \cos \theta)/\eta \qquad \text{(Eq. 3)}$$

Unless impeded by other factors such as base metal surface roughness and chemical reactions, flow is expected to be rapid and flow rate is not of concern in the time frame of most braze processing.

FACTORS CONTROLLING THE PROPERTIES OF THE BRAZEMENT

Anticipated service conditions will determine how important factors such as the selection of joint designs, filler metals, and processing parameters are for obtaining the desired properties. These factors have major effects on the geometries and micro-structures of braze joints, and eventually determine the properties of joints. Factors that influence joint properties are briefly discussed below; more thorough treatment can be found in the following chapters.

Joint Design

Effect of joint clearance on braze joint tensile strength dramatically illustrates the influence of one variable of joint design on braze joint properties. Figure 1.5 shows the variation of tensile strength with joint clearance for butt joints of stainless steel brazed with BAg-1a filler metal taken from an early study of this phenomena. These data show that at small joint clearances [< 0.006 in (150 μm)] joint tensile strength is quite high and even approaches that of the stainless steel. This occurs even though the intrinsic strength of the BAg-1a is only in the range of 40 ksi (275 MPa). The reason the joint strength can be so much higher than the braze filler metal strength is that necking of the thin filler metal layer is suppressed. During tensile testing the filler metal layer approaches a stress state of very high triaxial tension which effectively increases its tensile strength. As joint clearance increases, suppression of necking is

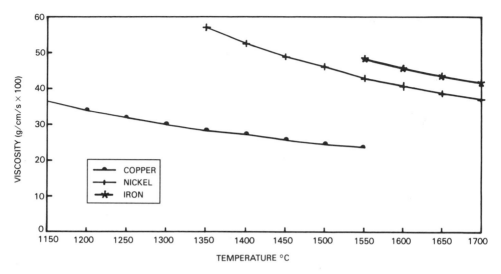

Figure 1.4 — The Variation With Temperature of the Viscosity of Pure Iron, Nickel, and Copper (After Cavalier)

reduced or eliminated, and joint strength approaches the intrinsic strength of the filler metal.

Filler Metal

Filler metals can be fairly complex alloys and their melting can take place over a range of temperatures. The implications of such behavior can be deduced from the phase diagram for the commercially important silver-copper system shown in Fig. 1.6. Except for the eutectic composition (72 Ag-28 Cu wt percent), melting of Ag-Cu alloys occurs over a range of temperatures as illustrated for the 50 Ag-50 Cu wt percent composition. Melting, which begins at solidus temperature of 1435°F (780°C), is not complete until a temperature in excess of 1560°F (850°C) is reached. Within that temperature range there is a "mushy" mixture of liquid and solid which wets and flows in a manner distinctly different from the completely liquid alloy. Flow is reduced when the filler metal is in the partially melted state, and the wetting and spreading behavior of the low melting liquid phase in the mixture leads to the tendency of the low melting phase to separate from the solid constituents. This behavior, in turn, can re-

sult in the formation of defective joints because of insufficient or nonuniform joint filling. Figure 1.7 provides a visual illustration of the differences in flow between an alloy with a narrow melting range and one with a wide melting range. The rate of heating through the melting temperature range, the choice of brazing temperature, and the brazing time are all important control parameters for avoiding these types of problems.

In addition to problems associated with the melting characteristics of braze filler metals, alloying can occur between the liquid filler metal and the base metal during brazing. The wetting and flow of a filler metal can be markedly influenced by alloying depending on the extent to which the melting points of the filler metal and base material are affected, and the tendency for formation of new phases. Liquid filler metal can cause excessive dissolution (erosion) of the base material, and significant changes of the composition of the filler metal and its melting characteristics may result. Also, significant changes in the base material composition near the joint surfaces may be produced by diffusion of elements from the filler metal. The extent of the effect of al-

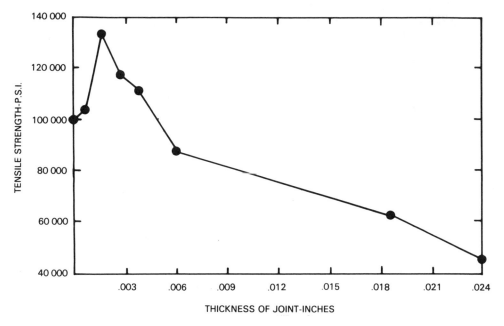

Figure 1.5 — Relationship of Tensile Strength to Joint Clearance of Stainless Steel Brazed with BAg-1a Filler Metal (From Work of Leach and Edelson as Described by Bredzs)

loying on the base material depends upon factors such as solubility of filler metal elements in the base material, time and temperature, kinetics of solid state diffusion, grain size in the base material, and its composition. Problems with wetting or mechanical properties can result from these interactions.

Alloying occurs to some extent in almost all braze processing; therefore, it is generally desirable to control alloying in order to avoid deleterious effects on joint microstructure and properties. Under certain circumstances, however, alloying can have a beneficial effect by either increasing the solidus temperature of the braze layer or by improving the intrinsic mechanical properties of the filler metal.

Residual Stresses

When two different base materials (e.g., a ferritic steel and an austenitic steel, or an austenitic steel and a ceramic) must be joined by brazing, large residual stresses may form in the final assembly because of the difference in thermal expansion coeffi-

cients between the two materials. The residual stresses are produced during cooling from the brazing temperature as one component of the assembly shrinks at a different rate than the other. When the thermal expansion coefficients of the materials being joined are very different, these residual stresses can be large enough to cause localized deformation or cracking in the materials or warping of brazed assemblies. Residual stresses can be controlled to some extent by programmed cooling from the brazing temperature to promote stress relaxation.

Assembly, Fixturing and Filler Metal Placement

Fixturing methods for aligning joint components must provide the correct relative positioning throughout the braze cycle to achieve a successful joint. Also, placement of the filler metal must provide intimate contact for adequate flow to ensure complete filling of the joint. For assemblies containing joints of dissimilar materials, differences in thermal expansion coeffi-

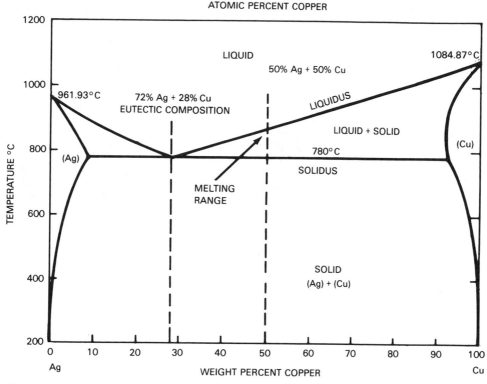

Figure 1.6 — Ag-Cu Phase Diagram. This Figure Is Taken From Binary Phase Diagrams, ASM

cients must also be considered for proper selection of component tolerances, fixturing and filler metal placement.

BIBLIOGRAPHY

Adamson, A. W. *Physical chemistry of surfaces.* New York: John Wiley & Sons, 1976.

Aksay, I. A., Hoge, C. E., and Pask, J. A. "Wetting under chemical equilibrium and nonequilibrium conditions." *Journal of Physical Chemistry* 78(12): 1178-1183; 1974.

Bredzs, N. "Investigation of factors determining the tensile strength of brazed joints." *Welding Journal* 33(11): 545s-563s; 1954.

Cavalier, G. "Measurement of viscosity of iron, cobalt and nickel." *c.r. Acad. Sci.* 256(6): 1308; 1963.

Colbus, J., Keel, C. G., and Blanc, G. M. "Notes on the strength of brazed joints." *Welding Journal* 41(9): 413s-419s; 1962.

Funk, E. R. and Undin, H. "Brazing hydrodynamics." *Welding Journal* 31(6): 310s-316s; 1952.

Gibbs, J. W. *Collected works,* Volume 1. New Haven: Yale University Press, 1948.

Kingery, W. D. "Role of surface energies and wetting in metal-ceramic sealing." *Ceramic Bulletin* 35(3): 108-112; 1956.

Milner, D. R. "A Survey of the scientific principles related to wetting and spreading." *British Welding Journal* 5: 90-105; 1958.

Roberts, P. M. "Gold brazing in antiquity." *Gold Bulletin* 6(4): 112-119; 1973.

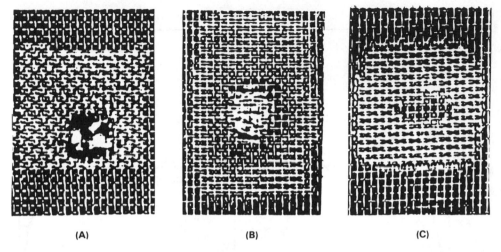

(A)   (B)   (C)

Figure 1.7 — Illustration of the Difference in Flow Properties Between Eutectic and Noneutectic Braze Filler Metals: (A) Slow Furnace Heating of BAg-2 Filler Produced Nonuniform Flow Because of Liquation. BAg-2 Has a Melting Range of About 170°F (95°C). (B) Rapid Heating of BAg-2 Results in Better Flow. (C) Slow Furnace Heating of BAg-1 Results in Uniform Flow Because no Liquation Takes Place in Filler Metals With Eutectic Compositions. (This Figure Is Taken From Handy & Harmon Technical Bulletin No. T-2)

Slaughter, G. M. "Adams Lecture: The technology of brazing and soldering is broad-based and vital to the industrial economy." *Welding Journal* 58(10): 17-28; 1979.

Swaney, O. W. Trace, D. E., and Winterbottom, W. L. "Brazing automotive heat exchangers in vacuum: process and materials." *Welding Journal* 65(5): 49-57; 1986.

Keywords — brazement design, brazed joints, joint design, joint selection, joint clearance, fit-up, joint length, properties of joint, design data, drafting room practice, design variables

Chapter 2

BRAZEMENT DESIGN AND DRAFTING ROOM PRACTICE

Many variables are to be considered in designing and manufacturing a reliable mechanical assembly. From the mechanical standpoint the design of a brazement is no different from the design of any other part. The rules applying to concentrated loads, stress risers, stress concentration, static loading, dynamic loading, etc., that apply to machined or other fabricated parts also apply to the brazement.

The design of a brazed joint does have specific requirements that must be met when adequate operating characteristics are to be achieved. Some of the more important factors are:

(1) Composition of the base metal and filler metal (members to be brazed may be of similar or dissimilar materials).

(2) Type and design of joint

(3) Service requirements such as mechanical performance, electrical conductivity, pressure tightness, corrosion resistance, and service temperature.

Reference should be made to the appropriate chapter for information on filler metals and base metals.

SELECTION OF BASE METAL

In addition to the normal mechanical requirements of the base metal in the brazement, the effect of the brazing cycle on the base metal and the final joint strength must be considered. Cold work strengthened base metals will be stress relieved and the overall strength reduced by the brazing process temperature and time when in the stress relieving range of the base metal being processed. "Hot-cold worked" heat resistant base metals can also be brazed; however, only the stress relieved physical properties will be available in the final brazement. The brazing cycle by its very nature will usually stress relieve the cold worked base metal unless the brazing temperature is below the stress relief time/temperature of the base metal. It is not practical to cold work the base metal after the brazing operation.

When it is essential to design a brazement having strength above the annealed properties of the base metal after the brazing operation, it will be necessary to specify a heat treatable base metal. The base metals can be of the oil quench type, the air quench type that can be brazed and hardened in the same or separate operation, or the precipitation hardening type in which the brazing cycle and solution treatment cycle may be combined; or the parts may be hardened and then brazed with a low temperature filler metal using short times at temperature in order to maintain the mechanical properties.

Some brazing operations must be carried out above the optimum austenitizing temperature for steels or the optimum solution treating temperature for age hardening base metals. This will result in enlarged grain size in the base metal and less than optimum mechanical properties in the base metal. In some applications, the lower mechanical properties will be adequate for the

9

service requirements to which it will be exposed. In more critical applications where the maximum mechanical properties are required, the desired physical properties and grain refinement can be obtained by: reaustenitizing, quenching and tempering steels, or resolution treating and aging of age hardenable base metals. The grain refinement cannot be accomplished in austenitic stainless steels by any heat treatment procedure. Another problem with grain growth is associated with fabricating of all base metals. The cold work reduction of base metal thicknesses of approximately one to 10 percent reduces the recrystallization temperature by as much as 500°F, and can cause large grains to grow at normal annealing temperatures of the base metal.

The strength of the base metal has a profound effect on the strength of the brazed joint; thus, this property must be clearly kept in mind when designing the joint for specific properties. Some base metals are easier to braze than others, particularly by specific brazing processes. For example, nickel base metal containing high titanium or aluminum additions or both, will present special problems in furnace brazing. For this reason vacuum furnace brazing is usually selected because of its excellent control of available oxygen that combines with the aluminum and titanium and impairs braze flow. An alternate procedure, nickel plating, is sometimes used as a barrier coating to prevent the oxidation of the titanium or aluminum and presents a readily wettable surface to the brazing filler metal.

SELECTION OF FILLER METAL WITH RESPECT TO JOINT DESIGN

When selecting a filler metal, it first must be compatible with the base metal, joint clearance, and brazing procedure to be used. Then, when designing a brazed joint for a specific service application, it is important to consider the properties and compatibility of the base metal and filler metal in the brazing operation as well as the final brazed joint in the environment in which they will operate. There are a great many filler metals on the market today; however, there are relatively few filler metals that find extensive usage. These filler metals are listed in ANSI/AWS A5.8, *Specification for Filler Metals for Brazing*. Also see Chapter 3 on Filler Metals.

Brazing filler metals and their properties are covered more fully in the filler metal section.

The properties of the filler metal in the joint after brazing and heat treating, if required, must be taken into consideration when designing a joint for a given service temperature. Some of the more important factors are: strength at service temperature, fracture toughness, galvanic corrosion under service conditions, corrosive media at service conditions, fatigue properties, electrical properties, and heat transfer properties.

TYPES OF JOINTS

Several factors influence the selection of the type of joint to be used. These factors include brazing process to be used, fabrication techniques prior to brazing, number of items to be brazed, method of applying filler metal, and the ultimate service requirement of the joint.

The unit strength of the brazing filler metal may occasionally be higher than that of the base metal. In general, however, the strength of the filler metal is considerably lower. The joint strength will vary according to the joint clearance, degree of filler metal-base metal interaction (diffusion and solution of base metal), the presence of defects in the brazement, and to a greater extent, the design of the specific joint.

There are basically two types of joints used in brazing operations — the lap joint and the butt joint. These joints are illustrated in Figure 2.1.

In the lap joint, the area of overlap may be varied so that the joint will be as strong as the weaker member, even when using a lower strength filler metal or with the presence of small defects in the joint. An overlap at least three to four times the thickness of the thinner member will usually yield maximum joint efficiency where the joint is stronger than the base metal or the joint is strengthened by diffusion brazing. Lap joints are usually used because they offer

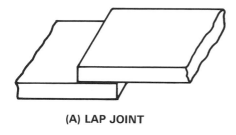

(A) LAP JOINT

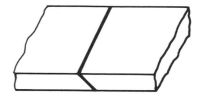

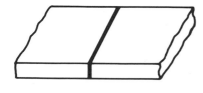

(B) BUTT JOINT

Figure 2.1 — Basic Lap and Butt Joints for Brazing

Figure 2.2 — Typical Scarf Joint Design

the greatest possibility of obtaining joints with maximum efficiency and the greatest ease of fabrication. They do, however, have the disadvantage of increasing the metal thickness at the joint and of creating a stress concentration at each end of the joint where there is an abrupt change in cross section.

A variation of the butt joint is the scarf joint (Figure 2.2). In this joint an attempt is made to increase the cross-sectional area of the joint without an increase in thickness. The disadvantages to this joint are: the sections are difficult to align and the joint is difficult to prepare, particularly in thin members. Since the joint is not perpendicular to the axis of tensile loading, the load-carrying capacity is similar to the lap joint rather than to the butt tensile joint. The standard design consideration of the percent of tensile plus the percent of shear value is not valid in angular brazed joints as represented by the scarf joint.

Butt joints may be used where the thickness of the lap joint would be objectionable and where the strength of the completed joint will meet the requirements of the brazement in service. The strength of a properly executed butt joint may be sufficiently high in laboratory tests, but in service will almost always fall far below the base metal strength due to stresses applied in service. The butt joint strength will depend on the stress application in service, the filler metal strength in the joint as compared to the base metal strength, the clearance (joint gap), the degree of brazing filler metal-base metal interaction during the brazing cycle, and the service requirements. Maximum efficiency may not be obtained with the butt joint when the filler metal in the joint is much weaker than the base metal. To obtain the greatest efficiency, the brazed joint should contain a minimum number of defects (i.e., flux inclusions, voids, unbrazed areas, pores or porosity) sufficient to meet the service requirements. The diffusion brazing process can markedly increase the strength of the joint by diffusion of the filler metal into the base metal(s). The joint after brazing and heat treatment (if required) should exhibit sufficient strength to meet the intended service of the part.

For our purposes T-joints and corner joints are regarded as variations of the butt joint.

JOINT CLEARANCE OR "FIT-UP"

Clearance is the distance between the faying surfaces (see Appendix A for definition) of the joint. The clearance between members of similar materials is easily maintained in assemblies where parts are pressed or require a shrink fit. When designing with dissimilar materials, the clearance must be calculated for the brazing temperature to be used.

Joint clearance has a significant effect on the mechanical performance of a brazed joint. This applies to all types of loading, including static, fatigue, impact, etc., and is applicable to all joint designs. Joint clearance has several effects on mechanical performance and includes (1) the purely mechanical effect of restraint to plastic flow of the filler metal offered by the greater strength of the base metal, (2) the possibility of flux entrapment, (3) the possibility of voids, and (4) the relationship between joint clearance and capillary force which accounts for filler metal distribution.

To assure the proper clearance gap for optimum flow into the joint, it may be necessary to use spacer wires, shims, prick punch marks, grit blasting, etc. Clearance between the brazed parts must be considered in terms of conditions at one specific instant, i.e. room temperature or brazing temperature. With similar materials of about equal mass, the room temperature clearance (before brazing) is a satisfactory guide. In brazing dissimilar materials, the one with the higher coefficient of thermal expansion may tend to increase or decrease the clearance, depending on the relative positions and configurations of the material. Thus, when brazing dissimilar materials (or greatly differing masses of similar materials), consideration must be given to the clearance at brazing temperature and adjustments made in the room temperature clearance to achieve the desired clearance at brazing temperature. See further information under Effect of Dissimilar Materials.

The influence of joint clearance is illustrated in Figure 2.3, which indicates joint shear strength for various joint clearances. Table 2.1 may be used as a guide for clearances when designing brazed joints for maximum strength.

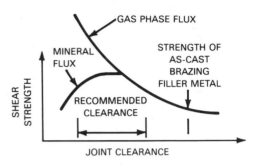

Figure 2.3 — Shear Strength Versus Joint Clearance (Brazing Low Carbon Steel With Silver Alloy Filler Metal)

Some specific clearance versus strength data for a given base material-filler metal combination and brazing condition are contained in Figure 2.4 and Figure 2.5.[1] Figure 2.4 shows the optimum shear values obtained on a 0.5 in. (12.7 mm) round drill rod using pure silver. The rods were butt brazed by an induction heating method in a dry 10% hydrogen-90% nitrogen atmosphere. Testing was accomplished in a specially designed guillotine-type fixture. Figure 2.5 shows the curve produced with a butt brazed joint tested in tension. Although this curve shows the normal ascending joint strength with decreasing clearance at practical production clearance, it is of interest to note that the strength drops off when extremely small clearances are obtained. It must be pointed out that each individual design of test specimens will vary the indicated "joint strength" obtained; thus tests must be conducted on a standard test specimen or on the final brazement in order to obtain load-carrying values for a specific design or service condition.

Variations in the process used, clearance, brazing temperature, fluxing agent (mineral flux or protective atmosphere), etc. will produce variations in the strength-versus-clearance curve. For example, a free-flowing filler metal melted in a high-quality protective atmosphere will adequately flow through a joint having a very low clearance

---

1.. Note: These data were obtained with nonstandard test specimens but do provide specific laboratory data only for the specific specimen design and test method as noted.

Table 2.1
Recommended Joint Clearance at Brazing Temperature

Filler Metal AWS Classification	Joint Clearance,' in.
BAlSi group	0.000-0.002 for furnace brazing in a vacuum atmosphere and clad brazing sheet in salt bath
	0.002-0.008 for length of lap less than 0.25 in.
	0.008-0.010 for length of lap greater than 0.25 in.
BCuP group	0.001-0.005 no flux and for flux brazing for joint length under 1 in.
	0.007-0.015 no flux and for flux brazing for joint length greater than 1 in.
BAg group	0.002-0.005 flux brazing
	0.000-0.002 **atmosphere brazing
BAu group	0.002-0.005 flux brazing
	0.000-0.002 **atmosphere brazing
BCu group	0.000-0.002 **atmosphere brazing
BCuZn group	0.002-0.005 flux brazing
BMg	0.004-0.010 flux brazing
BNi group	0.002-0.005 general applications (flux/atmosphere)
	0.000-0.002 free-flowing types, atmosphere brazing

*Clearance is measured on the radius when rings, plugs, or tubular members are involved. On some applications it may be necessary to use the recommended clearance on the diameter to assure not having excessive clearance when all the clearance is on one side. An excessive clearance will produce voids. This is particularly true when brazing is accomplished in a high quality atmosphere
**For maximum strength a press fit of 0.001 in. per in. of diameter should be used.

or press fit. However, when the atmosphere deteriorates, it is often necessary to use larger clearances in order to obtain adequate flow; thus, identical clearance data and curves would not be expected from sources having a different quality protective atmosphere even though all other variables were identical. Figures 2.4 and 2.5 show the optimum strength values for given clearances when employing a specific set of materials and brazing variables. Thus, it will be necessary to evaluate the brazing variables of the materials combination along with the specific production brazing facility. The current method for joint evaluation (shear joints) may be found in ANSI/AWS C3.2, *Standard Method for Evaluating the Strength of Brazed Joints in Shear.*

Preplaced Filler Material

Preplaced filler is foil, wire or powdered filler metal put into the joint, such as between two plates or into grooves machined in the joint area. When filler metal is applied between two plates the clearances noted in Table 2.1 may not apply. In applications where filler metal is preplaced, the two members of the joint may be weighted or have force applied to them so that the joint clearance is reduced during the brazing operation. This reduction in clearance enables the filler metal to fill the voids in the normal roughness of the faying surfaces (note Figure 2.22). In general, use of preplaced filler metal is not recommended in a fixed joint, such as a tube to tube lap joint or tack welded flat surfaces. The problem is that there is not an adequate volume of filler metal to fill the joint as the surface finish roughness and surface waviness may require as much as 0.001 to 0.002 in. of additional filler metal. Thicker filler metal cannot be used, as the maximum clearance may be exceeded, while the surface finish roughness and waviness voids are still present and require still more filler metal.

In some cases, very thin filler metal foils can be used to preset the clearances when the major quantity of filler metal is applied at the end of the joint. With joints having small clearances, preplacing filler metal in a tube-to-tube joint is difficult. In some applications additional filler metal is added by extending the filler metal shim out beyond the joint.

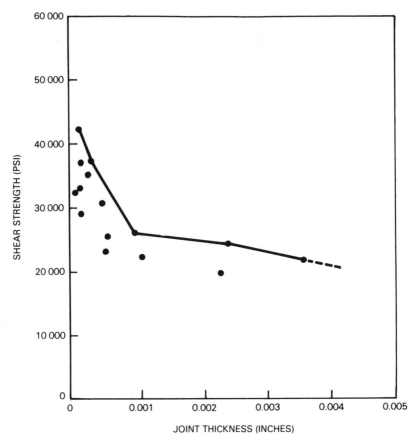

Figure 2.4 — Relationship of Shear Strength to Joint Thickness for Pure Silver Joints in 0.5 in. (12.7 mm) Diameter Drill Rod. Induction Brazed in a Dry Atmosphere of 10% H$_2$-90%N$_2$ (no Flux Used)

Face fed filler metal is filler metal that is applied to the exterior surfaces adjacent to the joint. This method is usually accomplished by hand during torch brazing or by using semi-automated torch brazing machines.

The joint strength of all joints, and particularly butt joints, can be improved by utilizing joint clearances as small as practical. Although the minimum joint clearance has been demonstrated to produce optimum joint strength, producing such clearances is sometimes considered impractical economically. However, with the current sophisticated metal working techniques, maintaining proper clearances is not a major problem.

It is important to point out that if high quality and highly reliable brazements are to be produced it is imperative that clearances be controlled.

Effect of Fluxes and Protective Atmospheres on Clearance

Flux is usually used in brazing. The American Welding Society document A3.0, *Standard Welding Terms and Definitions*, defines a flux as follows: "A material used to hinder or prevent the formation of oxides and other undesirable substances in molten metal and on solid metal surfaces, and to dissolve or otherwise facilitate the removal of such substances."

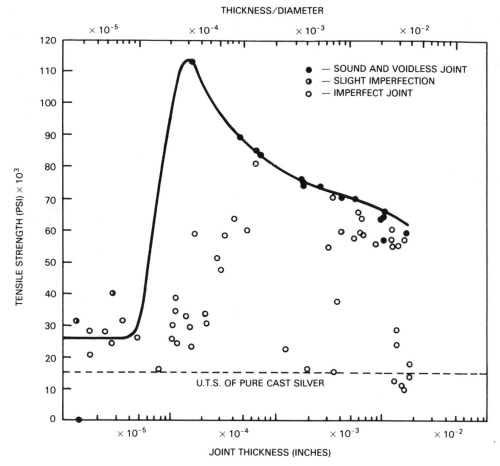

Figure 2.5 — Tensile Strength of 0.5 in (12.7 mm) Diameter Silver Brazed Butt Joints in 4340 Steel as a Function of Joint Thickness

As you will note, this definition fits both flux as well as protective atmospheres (including vacuum). Fluxes melt at a temperature below the brazing filler metal and must flow into the joint to prepare it for the filler metal. The filler metal, on melting, must force out the flux as the filler metal is drawn into the joint by capillary action. When the clearance is too small, the flux will be held in the joint so firmly that displacement by the liquid brazing filler metal may be difficult or impossible. Thus joint defects may be produced. When the clearance is too large, the liquid filler metal will travel around pockets of flux, thus giving rise to excessive flux inclusions.

Protective atmospheres have the same function as the fluxes, and various types are: combusted fuel gas, hydrogen, nitrogen base, vacuum, argon, etc. (see Chapter 4). Protective atmospheres also have their effect on the design of clearances for a specific brazement; it is essential to specify the brazing process on the blueprint when the clearance is determined so that they are compatible. The protective atmospheres require lower clearances to obtain optimum strength, so the load carrying capacity of the joint can be higher. In general, when protective atmosphere furnace brazing a joint in a vertical position, a free flowing filler metal will flow out of a joint having a clearance in excess of 0.003 in. (0.08

mm). Larger clearances can be filled by using filler metals that have a wide liquidus to solidus range and using a brazing temperature between the liquidus and solidus. This technique is often helpful in experimental applications, but it is not recommended for production applications because it is difficult to control and the joints have lower load carrying capacity with the larger clearances. The "purity" of a protective atmosphere, as measured by the ability to dissociate the oxides of a specific base metal in the brazement, will also vary the clearance requirement. When a high-quality protective atmosphere (low partial pressure of oxygen) rapidly and thoroughly dissociates the metal oxides of the base metal and the free-flowing filler metal, the filler metal may be drawn out of a 0.002 in. (0.05 mm) clearance joint and onto the base metal surface. In this case, a lower quality atmosphere (higher partial pressure of oxygen) or a lower clearance can be used. Lower clearance is recommended because it is easier to control in production operations, i.e., copper brazing filler metal on a low carbon steel in a -60°F (-51°C) dew point (or lower) pure dry hydrogen atmosphere and other high quality atmospheres where the flow of copper is excessive. Similarly, when brazing carbon steel with copper in an exothermic atmosphere the normally very fluid copper can be held in larger than normal clearances and holes by adequately increasing the oxygen partial pressure of the atmosphere as measured by dew point (water content), by increasing the carbon dioxide content, or both.

Effect of Surface Finish On Clearance

In general, the filler metal is drawn into the joint by capillary action. When the surface finish of the base metal is too smooth, the surfaces come together with little or no spacing between them and the filler metal may not distribute itself throughout the entire joint. The resulting voids will lower the strength of the joint. When atmosphere furnace brazing carbon steels with a fluid filler metal such as copper, there is adequate surface roughness and waviness on tubing, machined parts, stamped parts, etc. to allow flow of the copper through joints with a press fit-up to .002 in. (0.05 mm) on the diameter.

Other filler metals, not so fluid, need the surface roughness to help pull the filler metal into the joint and distribute it throughout the joint. Formed sheet metal parts are not normally perfectly round; thus, when brazing a sheet metal part to a machined part some points will have less than the required clearance.

To assure adequate flow of the filler metal throughout the joint, particularly when the clearance is zero or a press fit, the faying surfaces of the joint should be roughened, preferably with a clean metallic grit compatible with the base metal. Care must be taken not to contaminate the surfaces to be brazed with nonmetallic materials (ceramics, nonmetallic blasting materials, minerals from a water rinse bath, etc.) or other detrimental contaminants as this will seriously lower the joint strength. While grit blasting roughens the surface, it is not the roughness that assures flow through the joint. Tests on jet engine components have shown that it is the stress induced in the surface by blasting that improves the flow.

Surfaces that are too rough (250 in. RMS) will also lower the joint strength because only the high points may braze; in effect the average clearance is too large. A surface roughness of 30 to 80 in. (0.7 to 2.0 m) RMS is generally acceptable, but is not to be considered as optimum in all base material-filler metal combinations, and tests must be conducted to assure optimum conditions for a specific brazement.

Effect of Base Material-Filler Metal Interaction on Clearance

The mutual solubility of many base materials and filler metals causes interaction to take place through solution of the base metal with the liquid filler metal and diffusion both in the liquid and solid states. This interaction has its effect on the clearance for a specific brazement. If the interaction is low, the clearances will in general be smaller. However, when the interaction is high, the clearances required will have to be larger. This also affects the flow through the joint; thus, when the joint is long and interaction is large, the clearance should be increased. Since the aluminum base metal is brazed with a lower melting alloy of aluminum at a temperature close to the base metal melting point, interaction can be expected.

Note that in Table 2.1 there are two clearances listed for aluminum: one clearance for lap joints less than 0.25 in. (6 mm) and a larger clearance for lap joints above 0.25 in.

(6 mm). The larger clearance is to prevent the filler metal from freezing in the joint as it picks up more base metal.

Aluminum is by no means the only filler metal that exhibits interaction, as all filler metals show varying degrees of interaction to different base materials, i.e., silver filler metals with copper base metal, copper filler metal with nickel and copper-nickel base metals, gold filler metals with nickel base metals, nickel filler metals with nickel and cobalt base metals, etc.

The interaction (solubility) of silver filler metals with iron base metal is so low that it has no effect on the clearance. Copper filler metals, likewise, have a very low interaction with steel and that is one reason why they flow so well through tight fitting joints.

Effect of Base Material on Clearance

The chemistry of base materials often includes one or more elements in varying quantities whose oxide is not easily dissociated in a specific atmosphere or by a specific mineral flux. Since the metal-metal oxide dissociation of a given base metal by a specific flux or protective atmosphere is dependent on many factors (see Chapter 4), it is important to match the proper clearance with the brazing process and flux or atmosphere. For example, with aluminum additions, the clearance can be small with extremely small amounts of aluminum present. When a slightly larger aluminum addition is incorporated, however, the clearance may have to be increased. When still larger aluminum additions are incorporated, brazing may not take place and the protective atmosphere must be made more active (low dew point, lower vacuum pressures, a cleaner vacuum atmosphere or addition of activators), or the surface must be protected with a barrier coating, such as electrolytic nickel plating. The clearances would then require reappraisal to obtain optimum braze joint properties.

Thus the degree of mutual solubility of the base material and filler metal will also have an effect on the optimum clearance. When the solubility is high and the clearance low, the pick up of base metal as the filler metal flows through the joint will progressively increase the melting temperature of the filler metal in the joint. Thus, the filler metal can solidify in the joint without completely filling the joint. For additional information, see Chapter 3 on Filler Metals.

The coefficient of thermal expansion will, of course, have an effect on the joint clearance at brazing temperature when the base materials are dissimilar. Refer to the section on Brazing Filler Metal and Base Metal Interaction in Chapter 3 for additional information.

Effect of Brazing Filler Metal on Clearance

Filler metals have a broad spectrum of flowability and viscosity at various brazing temperatures and conditions. Some flow freely and others are sluggish. The free-flowing filler metals generally require a lower clearance then the sluggish filler metals. Filler metals that have a single melting point, such as pure metals (i.e. copper), or eutectic melting filler metals and self-fluxing filler metals, will usually be free flowing, particularly when there is no or very little interaction with the base metal.

Variations in the quality of fluxes and atmospheres can enhance the free flowing qualities of the filler metals or can result in no flow at all as fluxes may be a poor choice for a given application. For example, standard low temperature fluxes will not be suitable for aluminum bronze base metal, or become oxidized, or atmospheres may be of low relative purity. Thus, the clearances may appear to be improper for a given set of brazing conditions when in fact the flux or atmosphere is not proper for the given application.

Effect of Joint Length and Geometry on Clearance

As the joint length increases, the clearance must also increase, particularly when there is interaction between the filler metal and base material. As the filler metal is drawn into a long joint, it may pick up enough base metal to freeze before it reaches the other end. The more interaction that exists with a specific base material-filler metal combination under given brazing conditions, the larger the clearance must be as the joint becomes longer. This is only one of a number of reasons why it is important to make the joint length as short as possible, consistent with an optimum joint strength.

Effect of Dissimilar Base Materials

When designing a joint where dissimilar base materials are involved, it will be necessary to calculate the joint clearance at the brazing temperature. Figure 2.6 shows the expansion with temperature of a few base metals; see the appendix for additional data. The thermal coefficient of expansion for each member and the brazing temperature must be taken into consideration to assure the desired mechanical strength in the final joint (Figure 2.7). With high differential thermal expansion between two parts, the brazing filler metal must be strong enough to resist fracture and at least one of the base materials, filler metal, or an added layer of ductile base metal must yield during cooling. It is important to remember that some residual stress will remain in the final brazement as a result of joining at the brazing temperature and subsequently cooling to room temperature. Thermal cycling of such a brazement during its service life will also stress the joint area which may or may not shorten the service life of the brazement. Whenever possible, the brazement should be designed so that any residual stresses do not add to the stress imposed during service. One example of differential expansion is the brazing of a 20 in. (508 mm) outside diameter carbon steel ring with BCu filler metal to the inside of a 300 series stainless steel ring. If the filler metal rings were machined for 0.000 in. clearance at room temperature, at brazing temperature the joint clearance would open up to 0.100 to 0.150 in. (2.5 to 3.8 mm) on the diameter or 0.050 to 0.075 in. (1.3 to 1.9 mm) clearance per side. With this clearance, no filler metal would flow into the joint by capillary action, and filler metal

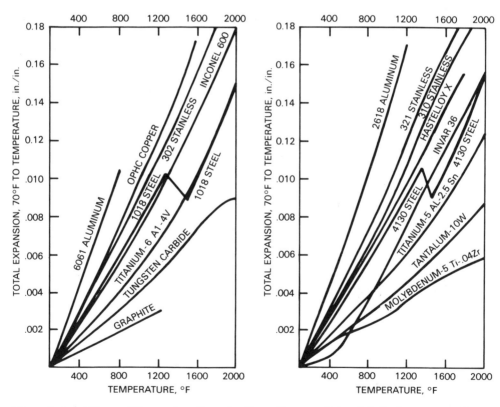

Figure 2.6 — Thermal Expansion Curve for Some Common Materials Refer to Appendix for Additional Data

flowing into the joint by gravity would flow out the other side. If such a joint were made, it would be of poor quality. One solution of this problem would be to employ a 300 series stainless steel backup ring inside the carbon steel ring to stretch the carbon steel ring at the same rate as the outer stainless ring expands, thus maintaining the clearance preset at room temperature.

In a few specific cases the base metals, such as carbides, do not possess sufficient ductility to yield upon cooling, and it is nec-

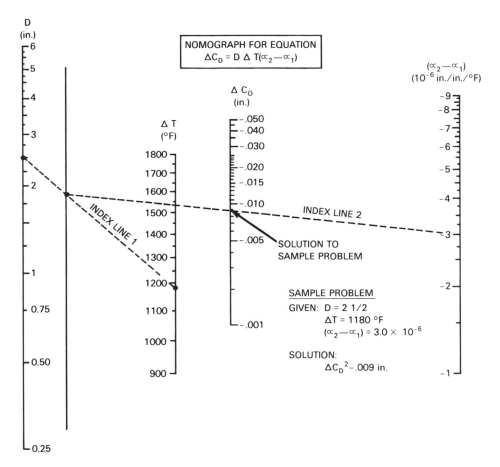

NOTES:
1. THIS NOMOGRAPH GIVES CHANGE IN DIAMETER CAUSED BY HEATING. CLEARANCE TO PROMOTE BRAZING FILLER METAL FLOW MUST BE PROVIDED AT BRAZING TEMPERATURE.
2. D = NOMINAL DIAMETER OF JOINT, in.
ΔC_D = CHANGE IN CLEARANCE, in.
ΔT = BRAZING TEMPERATURE MINUS ROOM TEMPERATURE, °F
α_1 = MEAN COEFFICIENT OF THERMAL EXPANSION, MALE MEMBER, in./in./°F
α_2 = MEAN COEFFICIENT OF THERMAL EXPANSION, FEMALE MEMBER, in./in./°F
3. THIS NOMOGRAPH ASSUMES A CASE WHERE α_1 EXCEEDS α_2; THUS, SCALE VALUE FOR $(\alpha_2 - \alpha_1)$ IS NEGATIVE. RESULTANT VALUES FOR ΔC_D ARE THEREFORE ALSO NEGATIVE, SIGNIFYING THAT THE JOINT GAP REDUCES UPON HEATING. WHERE $(\alpha_2 - \alpha_1)$ IS POSITIVE, VALUES OF ΔC_D ARE READ AS POSITIVE, SIGNIFYING ENLARGEMENT OF THE JOINT GAP UPON HEATING.

Figure 2.7 — Nomograph for Finding the Change in Diametrical Clearance in Joints of Dissimilar Metals for a Variety of Brazing Situations

essary to use a soft ductile base metal spacer in the joint, such as nickel or copper, which will yield during the cooling cycle to prevent high stress between two base materials, such as tungsten carbide brazed to a nickel base or cobalt base metal. Problems can also be expected when brazing low ductility materials, such as carbides and ceramics, to heat treatable base metals requiring a fast to moderate quenching rate and base metals that have a volume increase as a result of transformation. This problem is particularly acute when the volume change is at the lower temperature ranges [600°F (316°C) and below]. These problems result in either shearing the brazed joint on cooling or fracture of the ceramic, carbide, or similar material.

STRESS DISTRIBUTION

A properly designed and executed high strength brazement may fail in the base metal. Lightly loaded joints may be more economically fabricated using simplified joint designs which should function adequately for the intended service.

In general, loading of a brazed joint demands the same design consideration given any other joint or change in base material cross section. Brazed joints have a few specific requirements that are usually only apparent under conditions of high dynamic stresses or of high static stresses. It is well known that the base metal itself can best withstand high stresses and dynamic loading; thus, a good brazement design will incorporate a joint design that will remove any high stress concentration from the edges of the braze joint and will distribute the stresses into the base metal. Numerous examples of this will be noted in the joint design sketches, Figures 2.8 through 2.17.

A fillet of brazing filler metal is often erroneously incorporated by the designer as a method of eliminating stress concentrations. This very poor brazing design method is imprecise and seldom allows the

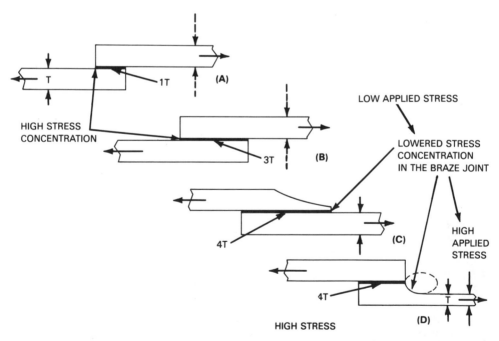

Figure 2.8 — Lap Joint Designs to be Fused at Low and High Stress. Flexure of Right Member in (C) and (D) Will Distribute Load Through Base Metal Thus Reducing the Stress Concentration and Improving the Fatigue Life of the Brazement

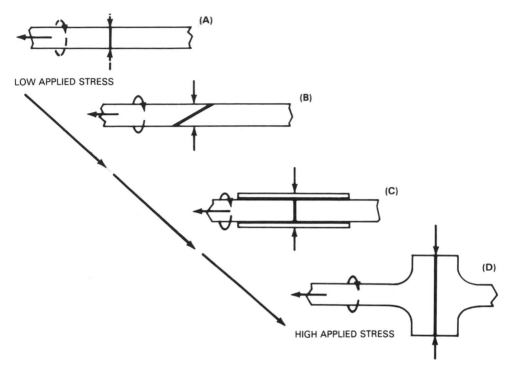

Figure 2.9 — Butt Joint Designs to Increases Load-Carrying Capacity of Joint Under High Stress and Dynamic Loading. High Stress Applied to Joints Made With Design (A). Lower Stress Applied Joints Require Joints Designed Similar to (C) and (D)

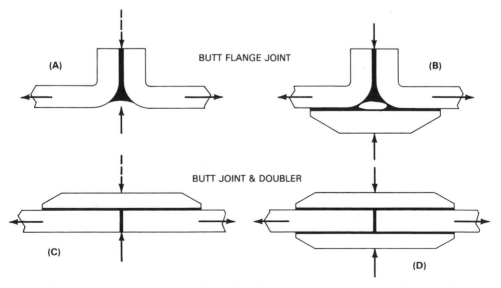

Figure 2.10 — Joint Designs Useful With Sheet Metal Brazements. Note that Loading in Sketch (A) Cannot be Symmetrical. Load Carrying Capability Increase From (A) to (D)

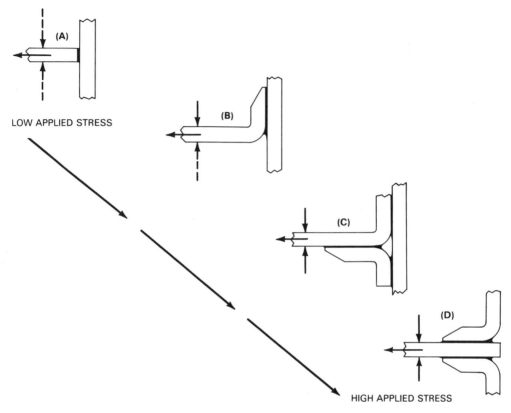

LOW APPLIED STRESS

HIGH APPLIED STRESS

Figure 2.11 — Joint Designs for Sheet Metal Brazements, Increasing Ability to Carry Higher Static and Dynamic Loading From (A) to (D)

brazing filler metal to form the desired fillet size and contour. Large fillets of brazing filler metal are of a coarse structure although the filler metal in a properly brazed joint will generally be a finer grain and high density joint structure. When the fillets become too large, shrinkage will occur thus leaving porosity which will act as a stress riser rather than the fillet acting as a stress reducer. Fillets of brazing filler metal, while they can be beneficial under some limited conditions, should not be used to replace a base metal fillet required to produce an optimum joint design.

A common pitfall in the use of a brazed joint operating with high static stresses is to disregard the flexibility of the design and assume the brazement to be a rigid member for purposes of stress analysis. Thus, a brazed joint that is calculated to satisfacto-

rily operate with a load of 130 000 psi (910 MPa) in the base metal may fail when loaded to 5000 psi (35 MPa) in the base metal due to the excessive stress concentration that is imposed by the bending of one member [Figure 2.16(A)]. Here again, it is imperative to move the stress into the base metal by using a base metal fillet, or else to increase the rigidity of the assembly in the vicinity of the joint. See Figure 2.16(B), (C), (D), and (E).

PLACEMENT OF BRAZING FILLER METAL

Before a brazed joint is designed, it is necessary to determine the brazing process to be used and the manner in which the filler

LOW APPLIED STRESS

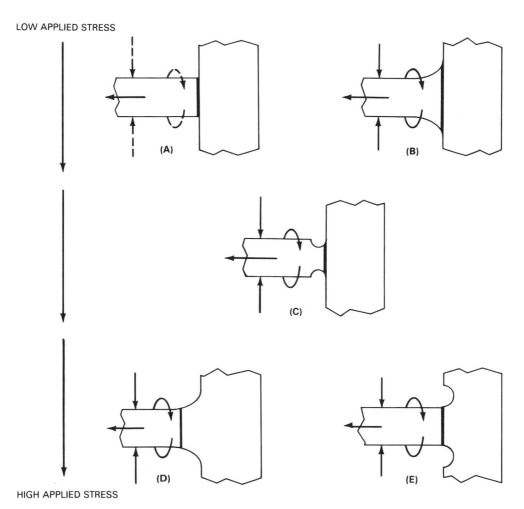

HIGH APPLIED STRESS

Figure 2.12 — Plate or Hub and Shaft Joint Designs for Rotating Members or Wheels. The Joints are in Tension Due to the Centrifugal Loading of a Rotating Wheel

metal will be introduced to the joint. In most manually brazed joints, the filler metal is simply face fed.

Furnace brazing and high production brazing operations, however, employ preplacement of the filler metal and may also incorporate the use of automatic dispensing equipment. Figures 2.18 through 2.22 illustrate a number of ways in which the filler metal may be preplaced. Brazing filler metal may take the form of wire, shims, strip, powder, or paste. It may be sprayed, plated, or vacuum deposited. Preplaced filler metal is most commonly

used in the form of wire, strips, or powder pastes. When the base metal is grooved to accept the preplaced filler metal, the groove should be cut in the heavier section (Figures 2.18 and 2.19). When designing for strength, it is essential that the groove area be subtracted from the joint area. This step is necessary since the brazing filler metal flows from the groove to the joint interfaces.

Powdered filler metals can be applied in any of the locations indicated in Figure 2.18, and are available in suspension mixtures with neutral binders or binders con-

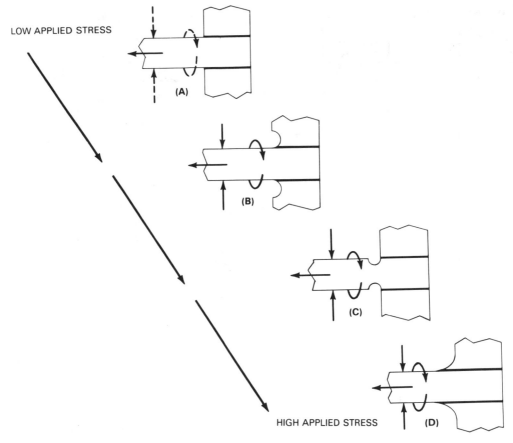

Figure 2.13 — Sheet and Plate or Hub and Shaft Joint Designs Showing Methods of Removing Stress Concentration From Edge of Brazed Joint. Note That Though Grooves Indicated in (B) and (C) May Appear to Weaken Member, They actually Increase Overall Brazement Strength

taining fluxes. Powdered filler metals are also applied to the joint area and wet down with binders or can be premixed with the binder and applied to the joint. Note that the density of the powder is less than the solid wire or shim, and therefore, the groove volume must be larger since the powder is 50 to 70 percent of the solid metal volume.

In applications where there is to be a large difference in cross section of the members to be brazed, and where there is a large degree of mutual solubility between the filler metal and the base metal, diffusion with and erosion of the thin member can be an-

ticipated. This usually takes place because the filler metal and the thinner joint member reach the filler metal melting temperature well before the heavy member of the joint; thus the filler metal and the thinner member continue to increase in temperature. With this larger volume of liquid filler metal in contact with the thinner member and continuing to increase in temperature, erosion and undercutting may occur. Figure 2.19 shows two methods of preventing the liquid filler metal from reaching the tube wall until the heavier member reaches the brazing temperature. If erosion does occur in the trepanned section (bottom two

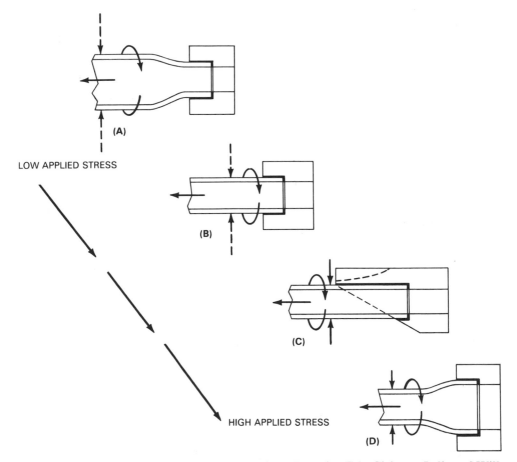

Figure 2.14 — Joint Designs for Tube and Fittings. Reducing Tube Sizing as Indicated Will Produce a Serviceable Brazement for Low Stress Applications; However, Multiple Stress Concentrations Preclude its Use in High Stressed Dynamic Loading. Expanding the Tube as Indicated in (D) Will Produce Best Overall Strength. Exact Increase in Diameter, Contour of Radii, and Angle of Expanded Tube (as in D), Will Have to be Developed for Optimum Service Life With a Specific Design Requirement, While (C) is Better Than (A) or (B) it Can be Significantly Improved by Reducing the Stress Concentration at the Left End of the Joint by Using the Dashed Line Design

sketches, Figure 2.19) or on the shield ferrule (top sketch, Figure 2.19) it is of little concern because the properties of the tube are unaffected.

Erosion and undercutting are controllable reactions and should not occur on a properly designed part or in a furnace cycle that has been programmed to prevent wide temperature differences between thin and heavy sections of the part. Placement of the filler metal within the joint or at the bottom of the joint will prevent its melting before the entire joint members reach a temperature where the filler metal will flow into the joint. Once the filler metal is in the joint, further heating may result in solid state diffusion, but will not cause solution of the base metal or erosion, unless a very large excess of filler metal is left in the fillet or at some location on the part.

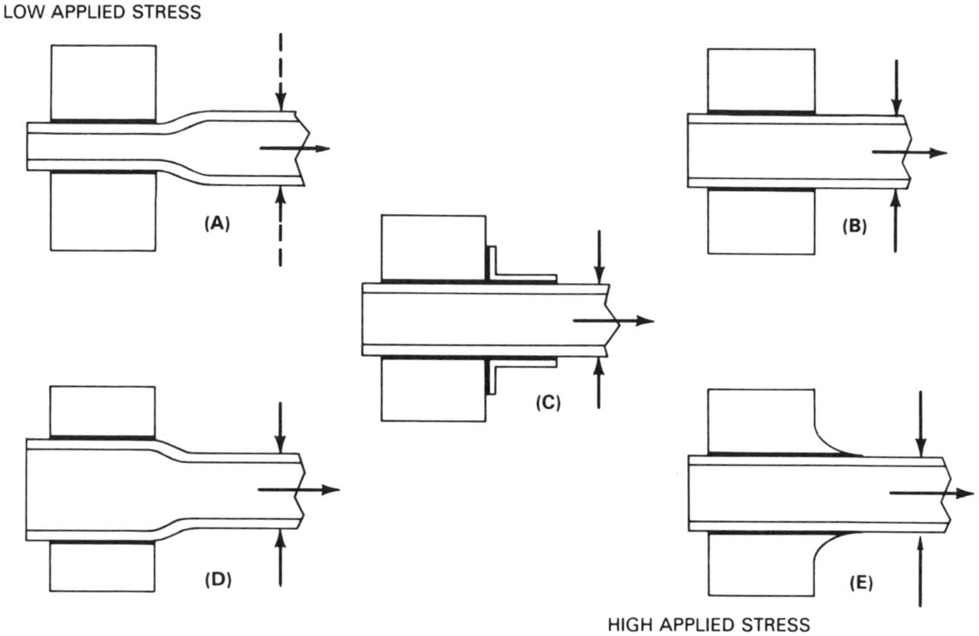

LOW APPLIED STRESS

Figure 2.15 — Designs for Tube and Header Joints in a Heat-Exchanger. Pulsations, Sonic Stresses, and High Velocity Air Across Tubes Can Vibrate Tubes to Destruction. In (A) and (B) Maximum Stress is at the Header; Tube Will Thus Fail in Fatigue at This Point. Designs (C), (D), and (E) Distribute Stresses, Thus Increasing Allowable Loading and Service Life

Where preplaced shims are used, the sections being brazed should be free to move with respect to one another so that the excess filler metal and flux will be forced out of the joint by the application of pressure, either deliberately or by the expansion of the joint elements during the brazing cycle. The reason for requiring pressure to close the space between the two faying surfaces is that the surfaces have an inherent roughness and waviness. It will take .001 in. (0.003 mm) or more of filler metal just to fill the roughness and waviness. Thus, if two flat surfaces with a .001 in. (0.03 mm) shim between them are tack welded, there is insufficient quantity of filler metal to fill the joint. So tack welding should not be used and the joint and appropriate loading should reduce the clearance between the parts as the shim melts.

The designer should also consider subsequent inspection needs when designing a brazed joint. The designer can help in this matter, for example, by providing for the placement of filler metal inside the assembly so that it will flow outward and thus provide visible proof of filler metal flow and complete joint penetration. In many joints the filler metal cannot be placed in the joint. To assure adequate flow through the joint with proper provision for inspection, the filler metal should be placed only on one side of the joint and allowed to flow through the joint. Inspection for flow can then be made on the side of the joint opposite the filler metal placement.

In some applications, perforations or witness holes provide an excellent method of inspection where fillets can be observed to assure that braze flow has reached the area where witness holes are placed. Inspection (witness) holes are particularly desirable in long sheet metal joints and some tube and fitting joints. On large doublers, larger holes can be used to reduce the weight of the brazement and filler metal can be applied in this area. If filler metal is applied in these holes they cannot be used for inspection.

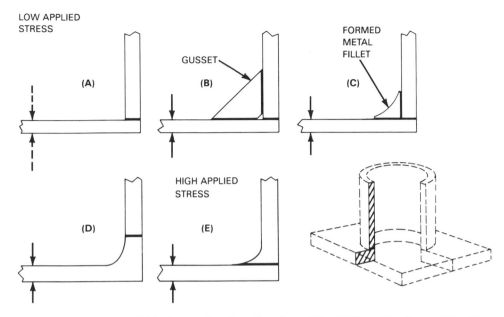

Figure 2.16 — Tube and Diaphragm Section of an Actual Part Where Diaphragm Was Hydraulically Loaded. When Designed as in (A), Premature Failure Resulted; the Redesign Stiffened the Assembly to Remove High Stresses at One Side of the Joint. Fillets, Such as in (D) and (E), Will Also Improve Serviceability

When witness holes are the only inspection process used, the first parts must be inspected using nondestructive (X-ray or ultrasonic, if applicable) or destructive (cross section) evaluation to prove that the procedure and design are producing the quality of internal joint required. After this evaluation, strict process control of parameters, such as ramp of temperature, braze temperature, time at temperature, cleanliness, and braze gap are required to assure quality brazements.

PROPERTIES OF JOINT AS INFLUENCED BY DESIGN

The mechanical properties of the joint will depend to a large extent on design of the joint. Although unlimited variations of joint designs are possible, a few of the more commonly used designs are shown in Figure 2.8 through 2.17. It will be noted that a given joint design, with low joint loading, can be made to handle higher loading by changing joint cross section and base metal contour. A good example of this is seen in Figure 2.8.

Mechanical property data for specific joint applications are not included in this chapter since such data are dependent on joint geometry. Test methods may vary considerably for each joint design. It is recommended that primary data be obtained using ANSI/AWS C3.2. Testing of simulated joints and the final brazement should be carried out under service conditions to assure adequate service life of the joint design.

Laboratory testing of brazing filler metals and base metal combination as well as other brazing variables should also be accomplished using ANSI/AWS C3.2.

In specific cases it will be necessary to use other test specimens to obtain additional data for structural members. Data on fatigue (using nonstandard specimens) for silver brazed butt joint specimens is shown in Figure 2.23; for nickel brazed lap joints see Figure 2.24.

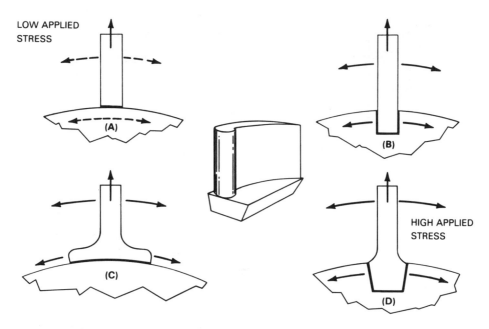

Figure 2.17 — Brazed Blade Attachments for Thin Blading. Long, Thin Blades Should Not be Attached as Indicated in A; However, Short, Squat Blades Have Been Adequately Brazed and Operated Using This Design. Design D Has Proven Serviceable for High Vibrating and Centrifugal Stresses

ELECTRICAL CONDUCTIVITY

The principal factor to be considered in the design of an electrical joint is electrical conductivity. The joint, when properly designed, should not add appreciable electrical resistance to the circuit.

Brazing filler metals in general have very low electrical conductivity compared to copper. As an example, BCuP-5 filler metal has approximately 10 percent of the electrical conductivity of copper; and BAg-6 filler metal 24.4 percent of the conductivity of copper. However, a brazed joint will not add any appreciable resistance to the circuit. The shorter path through the brazing filler metal, as compared to the longer path through the conductor, results in only a negligible increase in the total resistance in the circuit.

As an illustration, consider the extreme case shown in Figure 2.25. This figure shows a copper conductor of one square inch cross section containing a joint brazed with BCuP-5 brazing filler metal using an 0.005 in. (0.127 mm) joint clearance.

R_{Cu} = Resistance of copper
= 6.79 x 10^{-7} ohms/in./in.^{2} IACS at 207C
R_{FM} = Resistance of filler metal
$$= \frac{1}{C_{FM}} R_{Cu}$$

where
C_{FM} = conductivity of filler metal
= approximately 10 percent for BCuP-5.

Therefore,

$$^R FM = \frac{1}{0.1} \times 6.79 \times 10^{-7}$$

$$= 67.9 \times 10^{-7} \; ohms/in./in.^2$$

l = Joint clearance = Electrical path
= 0.005 in. for BCuP-5 through joint

Let L = Length of copper conductor + length brazed butt joint
(assume L = 1.0 in.)

Then total resistance = R_T

$$= (L - l) R_{Cu} + l R_{FM}$$

$$R_T = (0.995) 6.79 \times 10^{-7} + (0.005) 67.9 \times 10^{-7}$$

$$R_T = 7.10 \times 10^{-7} \; ohms/in./in.^2$$

Percent resistance raised by butt joint in 1 in. of

$$conductor = \frac{R_T - R_{Cu}}{R_{Cu}} \times 100 = 4 \; 1/2\%$$

As the length of the conductor increases, the influence of the brazed joint decreases. Thus, if the length of conductor were 2 in. instead of 1 in. the increase in electrical resistance would be 2.25%.

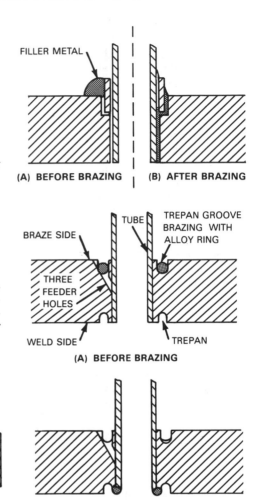

(A) BEFORE BRAZING (B) AFTER BRAZING

(A) BEFORE BRAZING

(B) AFTER BRAZING

Figure 2.19 — Braze Joint Design and Filler Metal Feeding for Tube and header in Nuclear Application. Note That the Filler Metal Cannot Flow Into the Joint Until the Entire Joint is up to Heat and Will Accept the Filler Metal. This is Very Helpful Where Base Metal Filler Metal Interaction Occurs

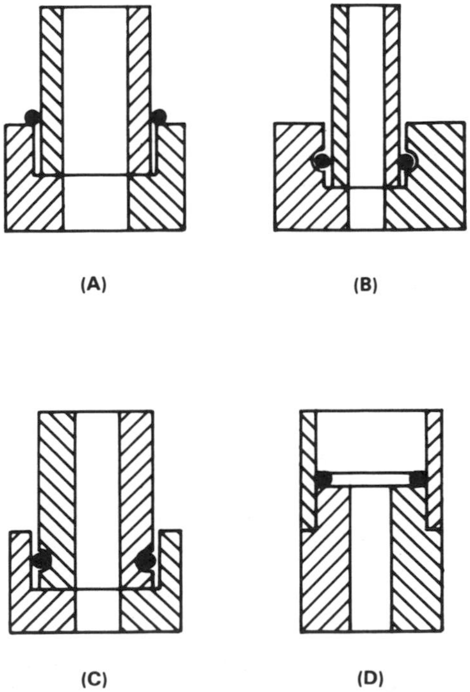

(A)   (B)

(C)   (D)

Figure 2.18 — Methods of Preplacing Brazing Filler Metal in Wire form

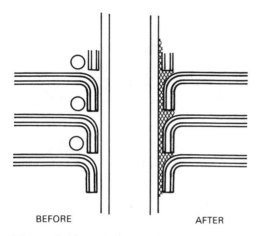

Figure 2.20 — Joint Design for Bonding Fins to Tubes for Heat Exchanger. Fins Can be Solid Metal or Clad, i.e., Copper-Stainless Clad for High Temperature Applications

BEFORE   AFTER

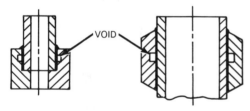

Figure 2.21 — Brazed Joints With Grooves for Preplacement of Filler Metal

From a practical standpoint it must be recognized that there will be a certain number of voids which will cut the effective area of the electrical path. For this reason lap joints are recommended where design will permit.

A lap length of 1.5 times the thickness of the thinner member of the joint will give a resistance approximately equal to the same length in solid copper (Figure 2.26). This is a rule of thumb which has been used extensively with good results. A longer lap may be used as necessary for convenience or for some special reason. Also, with increasing number and size of voids it may be necessary to increase the overlap.

Other joint designs for electrical applications are shown in Figure 2.27. In applications where leads must have flexibility to prevent breakage, designs such as shown in Figure 2.28 may be used.

PRESSURE TIGHTNESS (PRESSURE OR VACUUM)

Joints in brazements required to be pressure tight should be of the lap type wherever possible. This type of joint provides the large braze area (interface) with less chance of leakage through the joint (Figure 2.29).

Joint clearance is an important factor in obtaining the best quality pressure-tight joint. It is generally best to use the lower side of the clearance range (Table 2.1), since a joint having fewer inclusions and open passageways can be obtained. Most problems result from excessive clearances, poor flowing filler metal-base material combinations, and poor brazing technique.

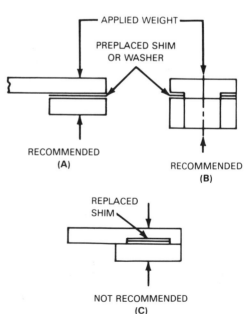

Figure 2.22 — Preplacement of Brazing Filler Metal in Shim Form

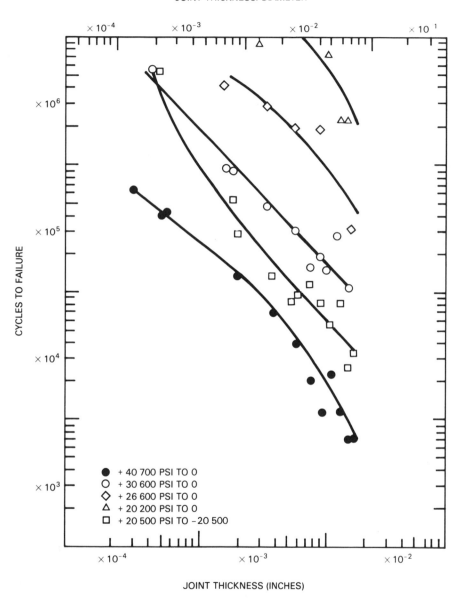

Figure 2.23 — Fatigue Properties of 0.25 in. (6.35 mm) Diameter Silver Brazed Butt Joints in 4340 Steel

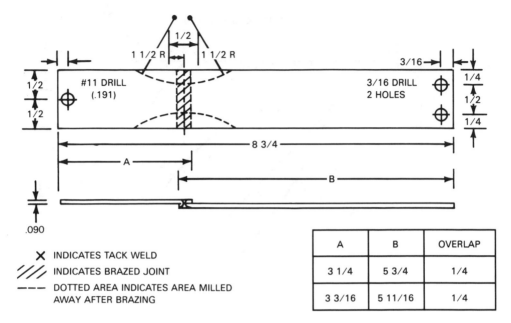

Figure 2.24A — Design of Single-Lap-Joint Fatigue Specimen of AISI 347 Sheet Metal Used to Obtain Fatigue Strength of Brazed Joints

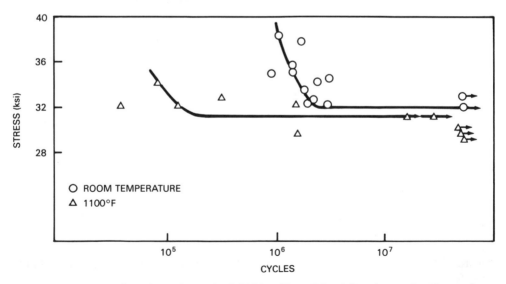

Figure 2.24B — Fatigue Data for AISI 347 Flat Sheet Metal Specimens for Comparison With the Brazed Specimens Showing the Effect of Temperature on the Fatigue Strength of AISI 347 Stainless Steel Sheet

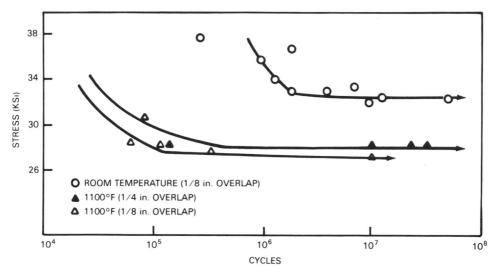

Figure 2.24C — Fatigue Data for AISI 347 Sheet Metal Brazed With BNi-1 and for the Base Metal Processed Through the Brazing Cycle Showing the Effect of Temperature and Joint Overlap on the Fatigue Strength of Single Lap Joints

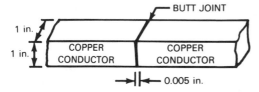

Figure 2.25 — Brazed Joint in an Electrical Conductor

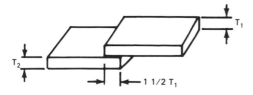

Figure 2.26 — Optimum Lap for Joints in Electrical Conductors

High pressure brazements are currently being fabricated satisfactorily when proper attention is paid to joint design and properties of the filler metals and base metals. The principal factor other than strength to be considered in the design of pressure-or vac-uum-tight assemblies is the importance of proper venting. Heat from the brazing operation expands the air or gases within the closed assembly so rapidly that unless the assembly is well vented (open to the atmosphere), it is likely to be forced apart. At the same time, forces may act on the filler metal entering and flowing through the joint that tend to minimize the effect of capillary action (Figure 2.30). If the unvented brazement survives expanding gases as the part is heated to brazing temperature, cooling of the brazement to solidus temperature of the filler metal (where filler metal is completely solid) will usually provide sufficient forces on liquid filler metal in the joint to pull the fillets all to the inside. As the gases bubble through the solidifying filler metal, air passages will be left through the joint from the outside directly to the inside of the unvented brazement.

A similar problem can exist when brazing in a vacuum atmosphere where desorption (out-gassing) on increasing temperature or absorption on decreasing temperature can result in leakage through the joint if venting is not provided. This problem is independent of the ones caused by pumping down to a vacuum when no venting is provided.

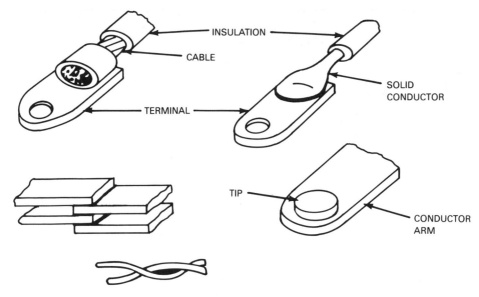

Figure 2.27 — Typical Joints for Brazing

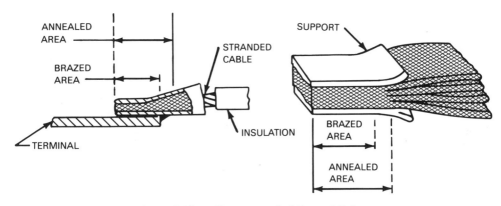

Figure 2.28 — Recommended Brazed Joints

Figure 2.29 — Typical Brazed Joints for Pressure-Tight Containers

However, when vacuum brazing a clean stainless steel base metal that has a minimum of outgassing, vacuum-tight sealed containers can be obtained. One application of sealing a container during brazing to obtain a vacuum inside is the large volume of vacuum bottles produced that are used to keep foods hot or cold. As the bottles are heated in the vacuum atmosphere brazing furnace, the metals are outgassed through a hole approximately 1.5 in. (38 mm) in diameter covered by a circular piece of stainless sheet. BNi-7 filler metal is used to seal the circular joint and excellent results are obtained.

VENT TO RELIEVE PRESSURE
IN CONTAINER DURING
BRAZING CYCLE

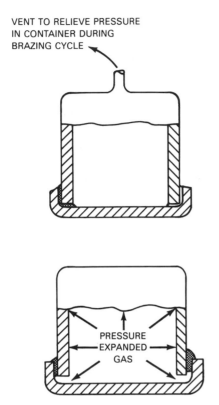

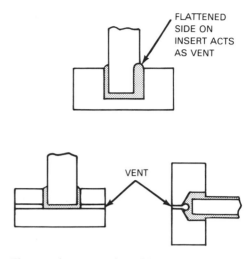

Figure 2.31 — Venting of Solid Joints During Brazing

PRESSURE
EXPANDED
GAS

Figure 2.30 — Venting of Solid Joints During Brazing

Dead-end holes may be considered as small pressure containers. Figure 2.31 illustrates a means of designing joints with vents to eliminate buildup of pressure during the brazing cycle. Similarly, any brazement that is designed with closed segments or areas may be subject to problems just mentioned. Thus, it is essential to have adequate venting of all enclosed containers or passages to assure good quality joints.

TESTING OF BRAZED JOINTS FOR DESIGN DATA

The importance of a standard method for evaluating the strength of brazed joints cannot be over emphasized. Different designs of test specimens will result in different sets of data. It is readily apparent in Figure 2.32 that

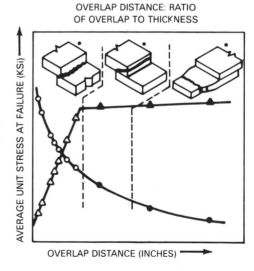

Figure 2.32 — Average Unit Shear Stress (Circles) and Base Metal Tensile Strength (Triangles) as Functions of Overlap Distance. (Open Symbols Represent Failure in the Filler Metal; Filled Symbols Represent Failure in the Base Metal)

the "apparent joint strength" at a very short overlap distance (where the overlap T = overlap distance in inches ÷ thickness of thinner sheet member in inches) is much higher in comparison to the long overlap. Thus, two laboratories that each test a single overlap specimen distance may be testing at opposite ends of the curve, and as a result come up with different conclusions. It is therefore highly desirable to use a wide enough overlap range of the curve to obtain adequate data.

The most sensitive portion of the curve is the short overlap portion where the apparent joint strength (average unit stress) is the highest curve (starting at the upper left corner of Figure 2.32). At this same portion of the curve the most scatter (or variation in results) will occur, as the joint is very sensitive to all the factors affecting the brazing of the joint.

The designer, on the other hand, is looking for the load-carrying capacity of the joint and thus is most interested in the right-hand portion of the base metal curve, the lines starting at the lower left of Figure 2.32 and continuing to the upper right. Where design constraints permit, the brazement should be designed, so that the failure occurs in the base metal and without an excessive overlap. Figures 2.33, 2.34, 2.35, and 2.36 give data on specific materials obtained using the standard test method. The two curves for each combination show the highest and lowest results obtained by two different laboratories in a ten laboratory testing program.

For further information refer to ANSI/AWS C3.2.

DRAFTING ROOM PRACTICES

As with any other industrial fabricating process, it is important that brazed joints be properly engineered, processed and inspected. The engineer must design the joint so that it can be fabricated, adequately inspected, and will withstand the service conditions to which the finished fabrication will be exposed.

At the same time, the joint should be so designed that it can be made and inspected without inconvenience and, as far as possible, with existing shop facilities. For this reason, close cooperation between shop and engineering personnel cannot be over emphasized.

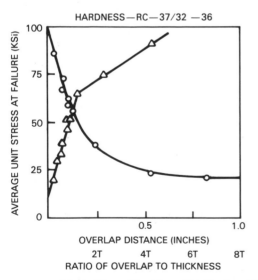
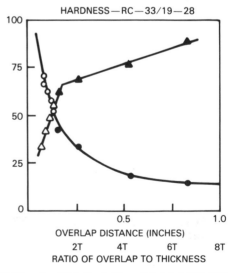

Figure 2.33 — Shear Test Data Employing AWS Standard Method. Base Metal AISI 410 Stainless Steel, Filler Metal — BNi-1, Brazing Method — Furnace at 2150⁰F (1177⁰Ca) for 30 Min. in Pure Dry Hydrogen. The Left-Hand Data Is 'as Hardened' With No Stress Relief. The Right-Hand Data Is Hardened and Draw to Rc 28 Avg.

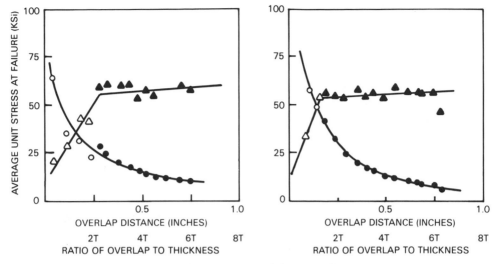

Figure 2.34 — Shear Test Data Employing AWS Standard Method. Base Metal — SAE 1010 Mild Steel, Filler Metal — BAg-1, Brazing Method — Torch

The designer should convey his design to the fabricating department by using drawings of the latest edition of ANSI/AWS A2.4, *Standard Symbols for Welding, Brazing and Nondestructive Examination.* These symbols provide a means for identifying the brazing procedure, as shown in Figures 2.37 through 2.41. The type of brazing process to be used is indicated by the letter in the tail of the symbol, in accordance with Table 2.2. If necessary, supplementary sketches and notes should be included on the design drawings to show details of the joint preparation and configuration.

It is strongly recommended that the designer familiarize himself with the other

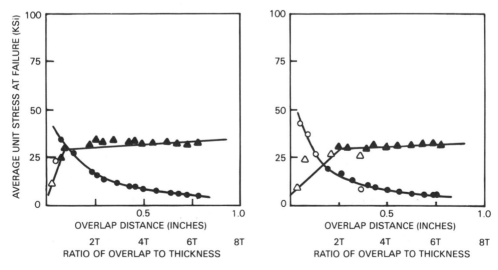

Figure 2.35 — Shear Test Data Employing AWS Standard Method. Base Metal — Copper, Filler Metal — BAg-1, Brazing Method — Torch

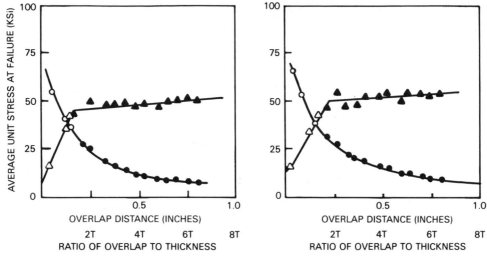

Figure 2.36 — Shear Test Data Employing AWS Standard Method. Base Metal — SAE 1010 Mild Steel — Filler Metal — BCu-1, Brazing Method — Furnace at 20507F (11217C) for 30 Min. in Pure Dry Hydrogen

chapters in this manual to fully appreciate and give due consideration to such matters as assembly, filler metal feed, and the selection of the brazing process and postbrazing processes.

DESIGN VARIABLES

The designer and draftsman must define as many of the design variables of the brazed joint as practical in order to assure obtaining the desired service properties and life of the joint and the completed brazement.

The following items list the specific variables of the brazement that should be described. This listing does not contain all of the possible variables and thus, for more complex brazements, it will be necessary to develop and define additional details.

(1) Joint design: (a) joint clearance at room temperature and at brazing temperature; (b) physical shape of members, i.e. stress concentration points, base metal fillets

(2) Base metal(s), i.e. specifications, chemistry, properties, etc.

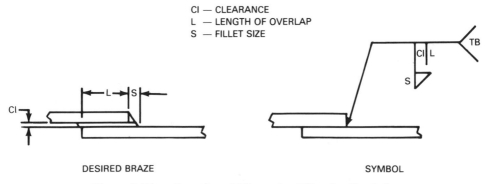

Figure 2.37 — Location of Elements of Brazing Symbols

Table 2.2
Brazing Process and Braze Welding Letter Designation Symbols

Process	Symbol
BRAZING PROCESSES	
Block Brazing*	BB
Diffusion Brazing	DFB
Dip Brazing	DB
Electron Beam Brazing	EBB
Exothermic Brazing	EXB
Flow Brazing*	FLB
Furnace Brazing	FB
Induction Brazing	IB
Infrared Brazing	IRB
Laser Beam Brazing	LBB
Resistance Brazing	RB
Torch Brazing	TB
Twin Carbon Arc Brazing*	TCAB
BRAZE WELDING	
Arc Braze Welding**	ABW
Braze Welding**	BW
Carbon Arc Braze Welding	CABW
Torch Braze Welding**	TBW

*These processes are seldom used but do have specific applications and thus are included.
**These letter designations are recommended but have not yet been added to the AWS Terms & Definitions.

(3) Brazing filler metals (see ANSI/AWS A5.8 *Specification for Brazing Filler Metals*) type, joint strength, melting characteristics, vapor pressure characteristics, and method of placement.
(4) Brazing atmosphere or flux (Tables 4.1 and 4.2)
(5) Brazing process and process variables:
 (a) brazing processes (Table 2.2)
(b) brazing process variables, i.e. temperature, atmosphere, time at temperature, heating, and cooling slopes, if required
(6) Prebraze cleaning, i.e. grease and oil removal, oxide removal, prebraze clean-up cycle (out gassing) in furnace with the appropriate atmosphere of gas or vacuum, etc.
(7) Postbraze cleaning, i.e. flux or oxide removal, stop-off removal, etc.
(8) Postbraze heat treatment, i.e. tempering, annealing, hardening heat treatment, solution treatment, aging, etc.
(9) Inspection method, i.e. type of test(s), test requirements, frequency, test limits, qualification requirements, etc.
(10) Special precautions, techniques, and other special information and requirements, i.e. rebraze if required, etc.
(11) Inspection criteria

All the brazement variables and instructions should be included or referenced on the assembly drawing or brazing procedure specification used during brazing so that adequate information is available to the brazing engineer and shop.

INSPECTION

Not only is it important for the design engineer to consider the inspection requirements of each joint and the ease with which it may be inspected, but the inspection method and limits must be indicated or referenced on the drawing to assure that the brazement quality meets its design requirements.

ANSI/AWS A2.4, should be referred to for the nondestructive testing symbols (NDT) to be used on the drawing.

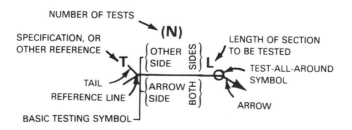

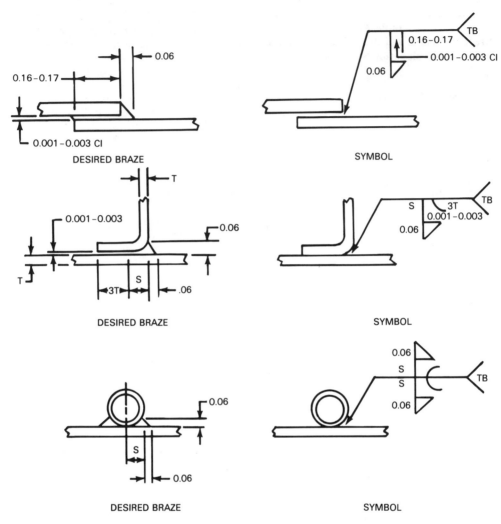

Figure 2.38 — Application of Torch Brazing Symbols. Note: The Above Symbols With Filler Size Should be Used for Brazements Made With Local Heating Methods Only, e.g. Torch, Induction. Where High-Strength Joints are Desired and Where Furnace Brazing is Used, Fillets (When Required) Should be Fabricated in the Base Metal. In General, Fillet Sizes Cannot be Controlled in Furnace Brazing. In Particular a Filler Metal That is Fluid Will Flow Off of Vertical Joints, Leaving Small Capillary Fillets

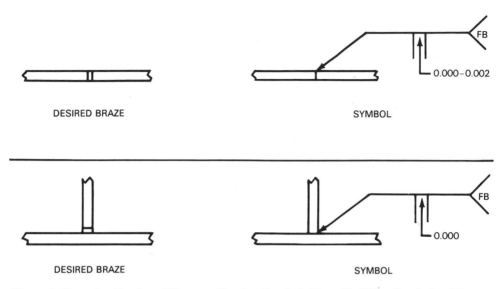

DESIRED BRAZE — SYMBOL

DESIRED BRAZE — SYMBOL

Figure 2.39 — Application of Furnace Brazing Symbols Note: No Fillet Symbol or Dimensions Should be Used When Furnace Brazing Unless for a Specific Application in a Flat Position or When Special Filler Metals or Techniques are Used

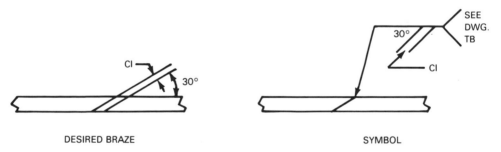

DESIRED BRAZE — SYMBOL

Figure 2.40 — Scarf Joint

QUALIFICATION OF THE BRAZING PROCESS AND/OR BRAZING OPERATOR (BRAZER)

In applications where critical quality control is required, such as nuclear brazements, aerospace brazements, boiler and pressure vessel brazements, the design engineer may wish to incorporate on the drawing a reference to qualify the brazing process, the brazing operator (brazer), or both, to the latest revision of ANSI/AWS B2.2, *Standard for Brazing Procedure and Performance Qualification*, or to the latest revision of the *ASME Boiler and Pressure Vessel Code*, Section IX, Welding and Brazing Qualifications or other similar specification.

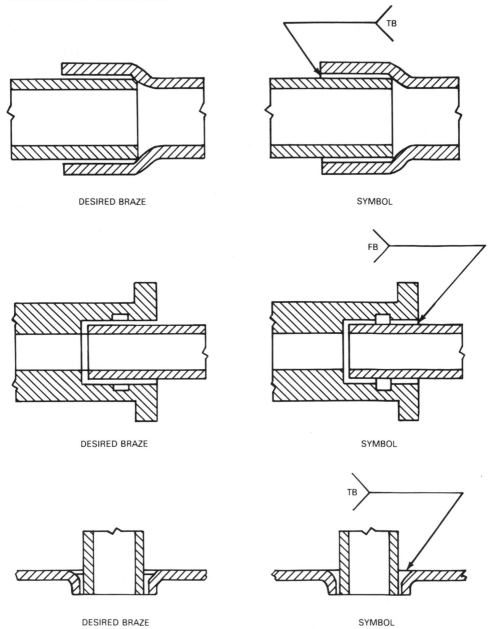

DESIRED BRAZE   SYMBOL

DESIRED BRAZE   SYMBOL

DESIRED BRAZE   SYMBOL

Figure 2.41 — Desired Braze Joint and Application of Brazing Symbols

Keywords — brazing filler metal, hard solder, silver solder, gold solder, brazing alloy, liquation, base metal interaction, melting ranges, brazing temperature, form of filler metal, filler metal classifications, preformed filler metal shapes

Chapter 3

BRAZING FILLER METALS

DEFINITION AND GENERAL CHARACTERISTICS

Brazing filler metal, as defined by the American Welding Society, is a metal to be added when making a braze. Brazing filler metals are metals or alloys that have liquidus temperatures above 840°F (450°C) but below the solidus temperatures of the metals being joined. The clearly defined term "brazing filler metal" has replaced some of the terms formerly used, i.e., hard solder, silver solder, gold solder, and brazing alloy.

For satisfactory use as a brazing filler metal, a metal or alloy should have the following characteristics:

(1) The ability to wet the base materials on which it is used.

(2) Suitable melting point or melting range and fluidity to permit its distribution by capillary attraction into properly prepared joints.

(3) A composition of sufficient homogeneity and stability to minimize separation of constituents by liquation under the brazing conditions to be encountered.

(4) The ability to form brazed joints possessing suitable mechanical and physical properties for the intended service application.

(5) Depending on requirements, the ability to produce or avoid certain base metal/filler metal interactions. Brittle intermetallic compounds or excessive erosion may be undesirable, while higher joint remelt temperature might be an attribute.

MELTING OF FILLER METALS

Melting and Fluidity

Pure metals melt at a constant temperature and are generally very fluid. For example, pure silver melts at 1761°F (961°C) and pure copper melts at 1981°F (1083°C). Combinations of copper and silver have very different melting characteristics than when they are in their pure forms, depending on the relative contents of the two metals.

Figure 3.1 is the equilibrium diagram for the silver-copper binary system. The solidus temperature line, ADCEB, represents the start of melting for all alloy combinations of silver and copper in the system. The liquidus temperature line, ACB, represents the temperatures above which each of these alloys in the system are completely liquid. At point C the liquidus and solidus temperature lines meet, indicating that a particular alloy melts at a constant temperature instead of melting over a range of temperatures. This temperature is known as the eutectic temperature and the alloy, in this case 72% silver and 28% copper, is known as the eutectic composition. This alloy is essentially as fluid as a pure metal.

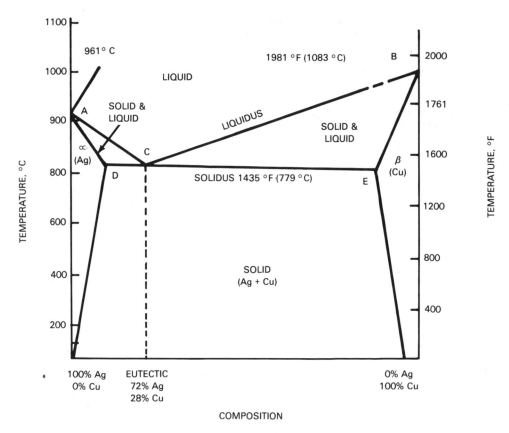

Figure 3.1 — Equilibrium Diagram for the Silver-Copper Binary System

The other non-eutectic alloy compositions are mushy between the solidus and liquidus temperatures. In general, alloys with melting ranges are more sluggish than eutectics with respect to flow in a capillary joint. For instance, the alloy of 50% silver and 50% copper will have a solidus temperature of 1435°F (779°C) and a liquidus temperature of 1580°F (860°C). It begins melting at the solidus and is not completely molten until it reaches the liquidus. In this zone some constituents of the alloy are in the liquid phase and others are in the solid phase. Until this alloy becomes completely liquid it has a diminished ability to flow.

Two terms which had previously been commonly used are melt point and flow point. Melt point was usually used to describe the point where melting was complete and flow point used to describe the temperature necessary to achieve proper capillary flow. They have not always been used with consistent meaning and therefore the terms solidus and liquidus, which can be more clearly defined, are now used.

Liquation

Brazing filler metals with narrow melting ranges and eutectic compositions, both of which behave during melting like a pure metal, are widely used for brazing in applications where a joint clearance of 0.001 to 0.003 in. (0.026 to 0.080mm) is maintained. They are suitable for processes with either a slow or rapid heating rate. When brazing with compositions having a wide melting range, some precautions may be necessary to avoid separation of the solid and liquid phases. This separation, called liquation, occurs when the filler metal is heated slowly through its melting range.

Liquation is more often encountered when preplaced filler metal is used. For example, when a filler metal with a wide melting range is preplaced at the joint and gradually heated within this melting range, there may be time to allow the low melting, fluid portion to flow into the joint capillary or plate onto the outside of the components being joined. This leaves behind a "skull". (See Figure 3.2) The tendency for a filler metal to liquate depends on the relative amounts of liquid and solid phases present as the filler metal is heated through its melting range.

Temperature Required For Brazing

Generally, a brazing filler metal is completely molten before it flows into a joint and is distributed by capillary attraction. Therefore, the liquidus normally may be considered the lowest temperature which should be used for brazing, and all sections of the joint should be heated to this temperature or higher. However, a few alloys that are used as brazing filler metals become sufficiently fluid below the actual liquidus to form good joints. With these filler metals successful brazing can be accomplished even though the liquidus is not attained.

In practice, it is impossible to uniformly heat both internal and external sections of a joint to the minimum brazing temperature. The location and rate of heat input, the mass of the parts, and the thermal conductivity of the base metal are factors that influence the distribution of heat within the work. With localized, rapid heating methods, such as induction, temperature may not be uniform. The temperature at the outer surfaces may be greater than in inner areas where the joint surfaces are located. To minimize this temperature differential, heat should be concentrated at larger masses or at points where heat can be conducted readily to the braze area. With slower heating methods, such as a furnace, the assembly heating will be more uniform.

Table 3.1 lists the solidus, liquidus and recommended brazing temperature range for filler metals in specification ANSI/AWS A5.8, *Specification for Filler Metals for Brazing*. The table indicates brazing temperature ranges begin at, or below, the liquidus. It is good practice to use the lowest temperature within the recommended range that will assure complete flow of filler metal through the joint. Brazing temperature selection will depend upon (1) the materials being joined, (2) the filler metal, (3) the brazing process, and (4) the joint design, including clearance between the parts.

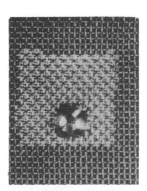

BAg-2 HEATED SLOWLY IN FURNACE. NOTE RATHER LARGE SKULL REMAINING BECAUSE OF LIQUATION OF THE NON-EUTECTIC TYPE FILLER METAL.

BAg-2 HEATED RAPIDLY. NOTE PRESENCE OF MUCH SMALLER SKULL.

BAg-1 HEATED SLOWLY IN FURNACE. NOTE ABSENCE OF SKULL BECAUSE NO LIQUATION TAKES PLACE WITH EUTECTIC TYPE FILLER METALS.

Figure 3.2 — Examples of Heating BAg-1 AND BAg-2 Filler Metals and Liquation (Photos Courtesy of Handy & Harman)

Table 3.1
Solidus, Liquidus, and Brazing Temperature Ranges*

AWS Classification	Solidus °F	Solidus °C	Liquidus °F	Liquidus °C	Brazing Temperature Range °F	Brazing Temperature Range °C
			SILVER			
BAg-1	1125	607	1145	618	1145-1400	618-760
BAg-1a	1160	627	1175	635	1175-1400	635-760
BAg-2	1125	607	1295	702	1295-1550	702-843
BAg-2a	1125	607	1310	710	1310-1550	710-843
BAg-3	1170	632	1270	688	1270-1500	688-816
BAg-4	1240	671	1435	779	1435-1650	779-899
BAg-5	1225	663	1370	743	1370-1550	743-843
BAg-6	1270	688	1425	774	1425-1600	774-871
BAg-7	1145	618	1205	652	1205-1400	652-760
BAg-8	1435	779	1435	779	1435-1650	779-899
BAg-8a	1410	766	1410	766	1410-1600	766-871
BAg-9	1240	671	1325	718	1325-1550	718-843
BAg-10	1275	691	1360	738	1360-1550	738-843
BAg-13	1325	718	1575	857	1575-1775	857-968
BAg-13a	1420	771	1640	893	1600-1800	871-982
BAg-18	1115	602	1325	718	1325-1550	718-843
BAg-19	1400	760	1635	891	1610-1800	877-982
BAg-20	1250	677	1410	766	1410-1600	766-871
BAg-21	1275	691	1475	802	1475-1650	802-899
BAg-22	1260	680	1290	699	1290-1525	699-830
BAg-23	1760	960	1780	970	1780-1900	970-1038
BAg-24	1220	660	1305	750	1305-1550	750-843
BAg-26	1305	705	1475	800	1475-1600	800-870
BAg-27	1125	605	1375	745	1375-1575	745-860
BAg-28	1200	649	1310	710	1310-1550	710-843
BAg-33	1125	607	1260	682	1260-1400	681-760
BAg-34	1200	649	1330	721	1330-1550	721-843
BVAg-0	1761	961	1761	961	1761-1900	961-1038
BVAg-6	1435	779	1602	872	1600-1800	871-982
BVAg-8	1435	779	1435	779	1435-1650	779-899
BVAg-8b	1435	779	1463	795	1470-1650	799-899
BVAg-18	1115	602	1325	718	1325-1550	718-843
BVAg-29	1155	624	1305	707	1305-1450	707-788
BVAg-30	1485	807	1490	810	1490-1700	810-927
BVAg-31	1515	824	1565	852	1565-1625	852-885
BVAg-32	1650	900	1740	950	1740-1800	950-982
			GOLD			
BAu-1	1815	991	1860	1016	1860-2000	1016-1093
BAu-2	1635	891	1635	891	1635-1850	891-1010
BAu-3	1785	974	1885	1029	1885-1995	1029-2091
BAu-4	1740	949	1740	949	1740-1840	949-1004
BAu-5	2075	1135	2130	1166	2130-2250	1166-1232
BAu-6	1845	1007	1915	1046	1915-2050	1046-1121
BVAu-2	1635	891	1635	891	1635-1850	891-1010
BVAu-4	1740	949	1740	949	1740-1840	949-1004
BVAu-7	2015	1102	2050	1121	2050-2110	1121-1154
BVAu-8	2190	1200	2265	1240	2265-2325	1240-1274
			PALLADIUM			
BVPd-1	2245	1230	2555	1235	2255-2285	1235-1252

Table 3.1 (Continued)

AWS Classification	Solidus °F	Solidus °C	Liquidus °F	Liquidus °C	Brazing Temperature Range °F	Brazing Temperature Range °C
ALUMINUM						
BAlSi-2	1070	577	1142	617	1110-1150	599-621
BAlSi-3	970	521	1085	585	1060-1120	571-604
BAlSi-4	1070	577	1080	582	1080-1120	582-604
BAlSi-5	1070	577	1110	599	1090-1120	588-604
BAlSi-7	1038	559	1105	596	1090-1120	588-604
BAlSi-9	1044	562	1080	582	1080-1120	582-604
BAlSi-11	1038	559	1105	596	1090-1120	588-604
COPPER						
BCu-1	1981	1083	1981	1083	2000-2100	1093-1149
BCu-1a	1981	1083	1981	1083	2000-2100	1093-1149
BVCu-1X	1981	1083	1981	1083	2000-2100	1093-1149
BCu-2	1981	1083	1981	1083	2000-2100	1093-1149R
BCuZn-A	1630	888	1650	899	1670-1750	910-954R
BCuZn-C	1590	866	1630	888	1670-1750	910-954R
BCuZn-D	1690	921	1715	935	1720-1800	938-982
BCuP-1	1310	710	1695	924	1450-1700	788-927
BCuP-2	1310	710	1460	793	1350-1550	732-843
BCuP-3	1190	643	1495	813	1325-1500	718-816
BCuP-4	1190	643	1325	718	1275-1450	691-788
BCuP-5	1190	643	1475	802	1300-1500	704-816
BCuP-6	1190	643	1450	788	1350-1500	732-816
BCuP-7	1190	643	1420	771	1300-1500	704-816
NICKEL						
BNi-1	1790	977	1900	1038	1950-2200	1066-1204
BNi-1a	1790	977	1970	1077	1970-2200	1077-1204
BNi-2	1780	971	1830	999	1850-2150	1010-1177
BNi-3	1800	982	1900	1038	1850-2150	1010-1177
BNi-4	1800	982	1950	1066	1850-2150	1010-1177
BNi-5	1975	1079	2075	1135	2100-2200	1149-1204
BNi-6	1610	877	1610	877	1700-2000	927-1093
BNi-7	1630	888	1630	888	1700-2000	927-1093
BNi-8	1800	982	1850	1010	1850-2000	1010-1093
BNi-9	1930	1055	1930	1055	1950-2200	1066-1204
BNi-10	1780	970	2020	1105	2100-2200	1149-1204
BNi-11	1780	970	2003	1095	2100-2200	1149-1204
COBALT						
BCo-1	2050	1120	2100	1149	2100-2250	1149-1232
MAGNESIUM						
BMg-1	830	443	1110	599	1120-1160	604-627

* Solidus and liquidus shown are for the nominal composition in each classification.

Brazing Filler Metal — Base Metal Interaction

When a brazing filler metal wets the base metal, diffusion may occur. This interaction, for the braze joint to be sound, should occur without undesirable diffusion into the base metal, dilution with the base metal, base metal erosion, or the formation of brittle compounds. Factors influencing the extent of these interactions are the following:

(1) The mutual solubility between the filler metal and base metal

(2) The amount of filler metal present

(3) The temperature to which the brazement is exposed

(4) The time of the brazing cycle

(5) Joint geometry, e.g., thick components vs. very thin components, long joints vs. short joints

Some filler metals will diffuse excessively, leading to changes in base metal properties. Diffusion can be controlled by filler metal selection, by application of a minimum quantity of filler metal, and by the use of an appropriate brazing cycle.

For proper capillary flow the filler metal must wet the base metal surfaces. If the filler metal must flow into long capillaries, mutual solubility can change the filler metal composition by alloying. This can raise its temperature and cause it to solidify or stop its flow before completely filling the joint. Compositions of brazing filler metals are formulated to allow for these types of factors. Silver-copper alloys for example, which have limited wettability on ferrous base metals, can have this wetting enhanced by the addition of zinc or tin.

FILLER METAL SELECTION

Criteria for Selection

Factors which should be considered when selecting a brazing filler metal are:

(1) Metallurgical compatibility with the base metals being joined

(2) Heating method and heating rate

(3) Service requirements of the brazed assemblies

(4) Brazing temperature required, e.g. heattreating the part during the brazing cycle

(5) Joint design considerations such as fit up, size of brazed components, length of joint

(6) Form of the filler metal

(7) Aesthetic requirements

(8) Safety considerations

Base Metals

The brazing filler metal must interact with the base metal to produce a braze joint suitable for the particular application. This requires that undesirable metallurgical reactions such as brittle intermetallic compound formation do not occur. It requires that the use of platings, either to enhance wetting or to prevent these undesirable reactions, be taken into consideration. It also requires that consideration be given to any heat treating requirements of the base metals. The brazing cycle design and selection of the filler metal need to either create the required heat treat properties or prevent those that already exist from being degraded. For detailed information, refer to the chapters on the base materials being used. Table 3.2 outlines some suitable filler metal-base metal combinations.

Heating Method

Filler metals with narrow melting ranges, generally less than 50°F (10°C) between solidus and liquidus temperatures, can be used with any heating method. They can also be used in any form and can be preplaced in the joint area. Filler metals that tend to liquate should be used with heating methods that bring the joint to brazing temperature quickly. Furnaces with slow heating rates or assemblies of large mass create the most common heating rate problems with filler metals that tend to liquate. These filler metals may still be able to be used in these situations if the alloy can be added to the joint after it has reached brazing temperature. An example of this would be the face feeding of wire.

Service Requirements

Braze joints may be exposed to various conditions in service. Service condition considerations may include:

(1) Service temperature

(2) Life expectancy of the assembly

(3) Thermal cycling

(4) Stress conditions

(5) Corrosive conditions

(6) Radiation stability

The most frequent service requirements are corrosion or oxidation resistance and adequate strength at service temperatures. Chapter 5 discusses corrosion of brazed joints. Tests under actual service conditions are recommended to confirm performance

Table 3.2
Base Metal-Filler Metal Combinations

	Al & Al Alloys	Mg & Mg Alloys	Cu & Cu Alloys	Carbon & Low Alloy Steels	Cast Iron	Stainless Steel	Ni & Ni Alloys	Ti & Ti Alloys	Be, Zr, V, & Alloys Reactive Metals	W, Mo, Ta, Cb, & Alloys Refractory Metals	Tool Steels
Al & Al Alloys	BAl-Si										
Mg & Mg Alloys	X	BMg									
Cu & Cu Alloys	X	X	BAg, BAu, BCuP, BNi, RBCuZn								
Carbon & Low Alloy Steels	X	X	BAg, BAu, RBCuZn, BNi	BAg, BAu, BCu, RBCuZn, BNi							
Cast Iron	X	X	BAg, BAu, BNi, RBCuZn	BAg, BNi, RBCuZn	BAg, BNi, RBCuZn						
Stainless Steel	BAl-Si	X	BAg, BNi, BAu	BAg, BAu, BCu, BNi, RBCuZn	BAg, BAu, BCu, BNi, RBCuZn	BAg, BAu, BCu, BNi					
Ni & Ni Alloys	BAl-Si	X	BAg, BAu, RBCuZn	BAg, BAu, BCu, BNi, RBCuZn	BAg, BCu, BNi, RBCuZn	BAg, BAu, BCu, BNi	BAg, BAu, BCu, BNi				

(Continued)

Table 3.2 — (Continued)

	Al & Al Alloys	Mg & Mg Alloys	Cu & Cu Alloys	Carbon & Low Alloy Steels	Cast Iron	Stainless Steel	Ni & Ni Alloys	Ti & Ti Alloys	Be, Zr, V, & Alloys Reactive Metals	W, Mo, Ta, Cb, & Alloys Refractory Metals	Tool Steels
Ti & Ti Alloys	BAl-Si	X	BAg*	BAg*	BAg*	BAg*	BAg*	BAg, BAl-Si*			
Be, Zr, V, & Alloys (Reactive Metals)	Y	X	BAg	BAg	BAg	BAg*	BAg*	Y	Y		
W, Mo, Ta, Cb, & Alloys Refractory Metals	X	X	BAg	BAg, BCu, BNi	BAg, BCu, BNi	BAg, BCu, BNi, BAu	BAg, BCu, BNi, BAu	Y	Y	Y	
Tool Steels	X	X	BAg, BAu, BNi, RBCuZn	BAg, BAu, BAu, BCu, RBCuZn	BAg, BAu, RBCuZn, BNi	BCu, BNi, BAg, BAu	BAg, BAu, BCu, BNi, RBCuZn	X	X	X	BAg, BAu, BCu, BNi, RBCuZn

NOTE: Refer to text for information on the specific compositions within each classification.
X - Not recommended; however, special techniques may be practicable for certain dissimilar metal combinations.
Y - Generalizations on these combinations cannot be made. Refer to appropriate individual chapters for usable filler metals.
* - Special brazing filler metals are available and are used successfully for specific metal combinations. Refer to appropriate individual chapters.

of the selected base-filler metal combination. Brazing is often used in cryogenic applications. The limited available data indicates tensile, shear, and impact properties of brazed joints are not adversely affected by low service temperatures. Brazed assemblies sometimes are used in elevated temperature service. Table 3.3 describes some service temperature limitations of various brazing filler metal compositions. Joint strength at elevated temperature is influenced by factors such as joint clearance, amount of filler metal/base metal diffusion, and length of time at brazing temperature. These variables suggest performance testing be performed at required service temperature to determine suitability.

Brazing Temperature

Generally, a filler metal with the lowest brazing temperature suitable for intended service requirements is preferred. Lower brazing temperatures are advantageous because they:

(1) Minimize heating effects on base metals such as annealing, grain growth, and distortion.

Table 3.3
Maximum Service Temperatures Recommended for Various Brazing Filler Metal Compositions

Filler Metal Classification	*Service Temperature °F	
	Continuous Service	Short Time Service
BAlSi	300	400
BCuP	300	400
BAg	300	400
BAu	800	1000
RBCuZn	400	600
BCu	400	900
BMg	250	300
BNi	1850	2000

* The temperatures listed are necessarily conservative to include all filler metal types under a given general classification. In many cases, higher temperatures than those listed may be permissible. The final temperature limitations in specific cases should be based on life tests simulating expected environmental conditions.

(2) Minimize base metal and filler metal interaction during brazing, particularly to avoid erosion of the base metal.

(3) Increase the life of the tooling.

(4) Economize on the heating energy required in the process.

On the other hand, important considerations for selecting a higher temperature filler metal include, but are not limited to the following:

(1) Higher melting filler metal often is more economical to purchase than low melting filler metals because precious metal content may be lower.

(2) Brazing temperatures are sometimes selected to coincide with heat treatment of the base metals such as stress relieving and annealing.

(3) Higher temperatures promote base metal and filler metal interactions which may increase the remelting temperature and physical properties of the brazed joint.

(4) Higher brazing temperatures may be used to permit later thermal processing at lower temperatures to develop mechanical properties of the base metal, or for sequential brazing (step brazing) with a subsequent lower temperature brazing filler metal.

(5) Brazing in a reducing atmosphere or vacuum will more effectively remove oxides when higher temperatures are employed.

(6) Service temperatures may require a higher melting or more oxidation resistant filler metal.

Joint Design

The design of the braze joint has an impact on the selection of the filler metal. Factors such as the choice of base materials, the joint clearances, the mass of the parts and the length of the joint will all influence whether a fluid or sluggish filler metal is selected.

Joint clearances will determine whether a eutectic type or non-eutectic type filler metal is needed. As discussed earlier in this chapter, clearances of .001 to .003 in. (.026 to .080mm) require a filler metal with high fluidity. These are alloys that are eutectic or that have a narrow melt range. Larger clear-

Figure 3.3 — Examples of Preformed Filler Metal Shapes

ances require more sluggish filler metals. The brazing temperature may also need to be raised to obtain proper flow with the more sluggish filler metals. When using a eutectic filler metal, the joint may need to be heated only 50°F (10°C) above the liquidus, whereas with a wide melt range filler metal, the joint may need to heat as much as 150°F (66°C) above the liquidus to get sufficient flow.

Other factors will have an influence on flow. The mass of the parts and the ability to heat them uniformly will affect the flow of the filler metal. A massive part that heats very slowly or unevenly may require a eutectic type filler metal, even with a loose fit, because of an inability to heat fast enough. Base materials can be dissolved into the filler metal, changing its flow characteristics. This can be a major problem when long braze joints are involved that require the filler metal to flow a great distance.

Form Of Filler Metal

The form of filler metal required for a particular application can have an effect on selection of the filler metal. Some filler metals are only available in a limited number of forms while others are common in all forms.

One way of distinguishing between forms is to separate those that have the filler metal added after the joint has been heated from those that have the filler metal preplaced before heating. The most common form of filler metal added after the joint is heated is rod or wire. Rod is added manually while wire can be face fed manually or automatically through a wire feeder.

Many special shapes and forms, designed for the particular application, are utilized as preformed, preplaced filler metals. Rings, washers, formed wires, shims, and powder or paste are the most common. Examples of these preformed shapes can be found in Figure 3.3 Preplacing of the filler metal assures that there is uniform and proper amount of the filler metal in the correct position on each assembly. Preplaced filler metal may be useful with manual brazing, but usually is associated with mechanized brazing procedures. Where joint areas are large, brazing filler metal foil may be located between the faying surfaces. Brazing rings are sometimes inserted into grooves machined into

the work. Enclosing rings in such grooves may be necessary for long sleeve joints or for salt bath dip brazing where it is desirable to avoid melting the ring before the work is heated to the brazing temperature. An example of an application where preforms are commonly used is found in Figure 3.4.

Aesthetic Requirements

Many times a filler metal is selected based on the aesthetic requirements of the joint. A common case is an application in which stainless steel is used as the base metal. Some joints, such as found in food handling equipment, need a good color match to the stainless steel. Jewelry is another application where color match of the filler metal to the base metal is very important.

Some applications require the external surfaces of the joint to be smooth. The main

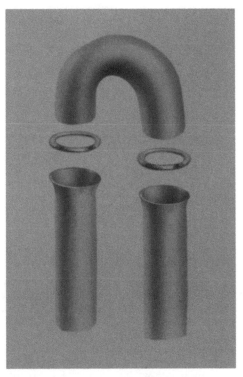

Figure 3.4 — Example of the Use of Preform Rings in the Brazing of Copper Return Bends in the Heating and Air Conditioning Industry

reason is to avoid costly chemical or abrasive cleaning operations to remove excess filler metal from the exterior of the joint. This is done because many consumer products require smooth fillets, particularly when painting or plating is done after brazing. Choosing a filler metal that will not form a large fillet is normally done to achieve this.

Safety Considerations

Hazardous or irritating ingredients may be present in filler metals, fluxes, and gases used in brazing. Generally, fumes and gases produced during heating and melting of filler metals and fluxes are the chief concern, and adequate ventilation is a necessity. A number of AWS BAg classifications contain cadmium. Cadmium oxide produced during brazing is toxic and its fumes must be avoided. As an alternative, cadmium free alloys should be selected whenever possible. Manufacturers' materials safety data sheets (MSDS) should be consulted for detailed information about constituents, fumes and gases generated, exposure limits, and precautions during use. Refer to Chapter 6 for additional safety and health information.

FILLER METAL CLASSIFICATIONS

American Welding Society specification ANSI/A5.8 divides filler metals into various classifications based on chemical composition. The classifications and chemical composition requirements are shown in Table 3.4 through 3.9.

Aluminum-Silicon Filler Metals

The aluminum-silicon filler metals (AWS BAlSi classifications) are used to join brazeable aluminum base metals. Silicon and copper lower the melting point of aluminum and these elements are added to pure aluminum to produce suitable brazing filler metals. These alloys can be used because they melt below the solidus of brazeable base metals. Magnesium is added to some filler metals to facilitate oxide dispersal in vacuum brazing.

Aluminum base metals frequently brazed include many wrought heat-treatable, non-heat-treatable, and some cast alloys.

When selecting a filler metal, choose one with a liquidus below the base metal solidus. A temperature difference of at least 75-100°F (24-38°C) is preferable, although closer ranges are acceptable if care is taken to control the temperature during brazing.

In most aluminum brazing operations a flux is necessary to remove and prevent surface oxides and allow the filler metal to wet the base metal. AWS flux classifications FB1A, FB1B and FB1C are used for torch, furnace and dip brazing respectively. See Chapter 4 on "Fluxes and Atmospheres". Inert gas or vacuum brazing operations may not require use of flux. Aluminum filler metals are supplied in wire or rod for manual applications. Filler metal also may be furnished in strip, preformed rings or shapes, paste, and powder, permitting the alloy to be preplaced for furnace or dip brazing operations. A convenient method of application is aluminum brazing sheet. A layer of brazing alloy is applied to one or both sides of the parent metal. The 5 to 10 percent cladding thickness assists design of brazed assemblies and eliminates preplacing filler metal as a separate operation.

Chapter 25 of this handbook covers base metals, brazing technique, and related aluminum brazing information.

BAlSi-2 brazing filler metal is available as sheet and as a standard cladding on one or both sides of a brazing sheet having a core of either 3003 or 6951 aluminum alloy. See Figure 3.5. It is used for furnace and dip brazing only.

BAlSi-3 is a general-purpose brazing filler metal. It is used with all brazing processes, with some casting alloys, and where limited flow is desired.

BAlSi-4 is a general-purpose brazing filler metal. It is used with all brazing processes requiring a free-flowing filler metal and good corrosion resistance.

BAlSi-5 brazing filler metal is available as sheet and as a standard cladding on one side or both sides of a brazing sheet having a core of 6951 aluminum alloy. BAlSi-5 is used for furnace and dip brazing at a lower temperature than BAlSi-2. The core alloy employed in brazing sheet with this filler

Table 3.4
Chemical Composition Requirements for Aluminum and Magnesium Filler Metals

AWS Classification	UNS Number[b]	Chemical Composition, Weight Percent[a]												Other Elements[c]	
		Si	Cu	Mg	Bi	Fe	Zn	Mn	Cr	Ni	Ti	Be	Al	Each	Total
BAlSi-2	A94343	6.8-8.2	0.25	—	—	0.8	0.20	0.10	—	—	—	—	Remainder	0.05	0.15
BAlSi-3	A94145	9.3-10.7	3.3-4.7	0.15	—	0.8	0.20	0.15	0.15	—	—	—	Remainder	0.05	0.15
BAlSi-4	A94047	11.0-13.0	0.30	0.10	—	0.8	0.20	0.15	—	—	—	—	Remainder	0.05	0.15
BAlSi-5	A94045	9.0-11.0	0.30	0.05	—	0.8	0.20	0.05	—	—	0.20	—	Remainder	0.05	0.15
BAlSi-7	A94004	9.0-10.5	0.25	1.0-2.0	—	0.8	0.20	0.10	—	—	—	—	Remainder	0.05	0.15
BAlSi-9	A94147	11.0-13.0	0.25	0.1-0.5	—	0.8	0.20	0.10	—	—	—	—	Remainder	0.05	0.15
BAlSi-11	A94104	9.0-10.5	0.25	1.0-2.0	0.02-0.20	0.8	0.20	0.10	—	—	—	—	Remainder	0.05	0.15
BMg-1	M19001	0.05	0.05	Remainder	—	0.005	1.7-2.3	0.15-1.5	—	0.005	—	0.0002-0.0008	8.3-9.7	—	0.30

a. Single values are maximum, unless otherwise noted.

b. SAE/ASTM Unified Numbering System for Metals and Alloys.

c. The filler metal shall be analyzed for those specific elements for which values are shown in this Table. If the presence of other elements is indicated in the course of this work, the amount of those elements shall be determined to ensure that they, or their total, do not exceed the limits specified.

Table 3.5
Chemical Composition Requirements for Copper, Copper-Zinc, and Copper-Phosphorus Filler Metals

AWS Classification	UNS Number[b]	Cu	Ag	Zn	Sn	Fe	Mn	Ni	P	Pb	Al	Si	Other Elements Total[c]
					Composition, weight percent[a]								
BCu-1	C14180	99.90 min	—	—	—	—	—	—	0.075	0.02	0.01*	—	0.10
BCu-1a	—	99.90 min[d]	—	—	—	—	—	—	—	—	—	—	0.30[d]
BCu-2[e]	—	86.50 min	—	—	—	—	—	—	—	—	—	—	0.50
RBCuZn-A[f]	C47000	57.0-61.0	—	Remainder	0.25-1.00	*	*	—	—	0.05*	0.01*	*	0.50[f]
RBCuZn-C[f]	C68100	56.0-60.0	—	Remainder	0.80-1.10	0.25-1.20	0.01-0.50	—	—	0.05*	0.01*	0.04-0.15	0.50[f]
RBCuZn-D[f]	C77300	46.0-50.0	—	Remainder	—	—	—	9.0-11.0	0.25	0.05*	0.01*	0.04-0.25	0.50[f]
BCuP-1	C55180	Remainder	—	—	—	—	—	—	4.8-5.2	—	—	—	0.15
BCuP-2	C55181	Remainder	—	—	—	—	—	—	7.0-7.5	—	—	—	0.15
BCuP-3	C55281	Remainder	4.8-5.2	—	—	—	—	—	5.8-6.2	—	—	—	0.15
BCuP-4	C55283	Remainder	5.8-6.2	—	—	—	—	—	7.0-7.5	—	—	—	0.15
BCuP-5	C55284	Remainder	14.5-15.5	—	—	—	—	—	4.8-5.2	—	—	—	0.15
BCuP-6	C55280	Remainder	1.8-2.2	—	—	—	—	—	6.8-7.2	—	—	—	0.15
BCuP-7	C55282	Remainder	4.8-5.2	—	—	—	—	—	6.5-7.0	—	—	—	0.15

a. Single values are maximum, unless noted.

b. SAE/ASTM Unified Numbering System for Metals and Alloys.

c. The filler metal shall be analyzed for those specific elements for those values or asterisks* shown in this Table. If the presence of other elements is indicated in the course of this work, the amount of those elements shall be determined to ensure that their total does not exceed the limit specified.

d. The balance is oxygen, present as cuprous oxide.

e. These chemical composition requirements pertain only to the cuprous oxide powder and do not include requirements for the organic vehicle in which the cuprous oxide is suspended, when applied in paste form.

f. The total of all other elements including those for which a maximum value or asterisk* are shown, shall not exceed the value specified in "Other Elements, Total".

Table 3.6
Chemical Composition Requirements for Silver Filler Metals

AWS Classification	UNS Number[a]	Composition, Weight Percent								
		Ag	Cu	Zn	Cd	Ni	Sn	Li	Mn	Other Elements, Total[b]
BAg-1	P07450	44.0-46.0	14.0-16.0	14.0-18.0	23.0-25.0	—	—	—	—	0.15
BAg-1a	P07500	49.0-51.0	14.5-16.5	14.5-18.5	17.0-19.0	—	—	—	—	0.15
BAg-2	P07350	34.0-36.0	25.0-27.0	19.0-23.0	17.0-19.0	—	—	—	—	0.15
BAg-2a	P07300	29.0-31.0	26.0-28.0	21.0-25.0	19.0-21.0	—	—	—	—	0.15
BAg-3	P07501	49.0-51.0	14.5-16.5	13.5-17.5	15.0-17.0	2.5-3.5	—	—	—	0.15
BAg-4	P07400	39.0-41.0	29.0-31.0	26.0-30.0	—	1.5-2.5	—	—	—	0.15
BAg-5	P07453	44.0-46.0	29.0-31.0	23.0-27.0	—	—	—	—	—	0.15
BAg-6	P07503	49.0-51.0	33.0-35.0	14.0-18.0	—	—	—	—	—	0.15
BAg-7	P07563	55.0-57.0	21.0-23.0	15.0-19.0	—	—	4.5-5.5	—	—	0.15
BAg-8	P07720	71.0-73.0	Remainder	—	—	—	—	—	—	0.15
BAg-8a	P07723	71.0-73.0	Remainder	—	—	—	—	0.25-0.50	—	0.15
BAg-9	P07650	64.0-66.0	19.0-21.0	13.0-17.0	—	—	—	—	0.15	0.15
BAg-10	P07700	69.0-71.0	19.0-21.0	8.0-12.0	—	—	—	—	—	0.15
BAg-13	P07540	53.0-55.0	Remainder	4.0-6.0	—	0.5-1.5	—	—	—	0.15
BAg-13a	P07560	55.0-57.0	Remainder	—	—	1.5-2.5	—	—	—	0.15
BAg-18	P07600	59.0-61.0	Remainder	—	—	—	9.5-10.5	—	—	0.15
BAg-19	P07925	92.0-93.0	Remainder	—	—	—	—	0.15-0.30	—	0.15
BAg-20	P07301	29.0-31.0	37.0-39.0	30.0-34.0	—	—	—	—	—	0.15
BAg-21	P07630	62.0-64.0	27.5-29.5	—	—	2.0-3.0	5.0-7.0	—	—	0.15
BAg-22	P07490	48.0-50.0	15.0-17.0	21.0-25.0	—	4.0-5.0	—	—	7.0-8.0	0.15
BAg-23	P07850	84.0-86.0	—	—	—	—	—	—	Remainder	0.15
BAg-24	P07505	49.0-51.0	19.0-21.0	26.0-30.0	—	1.5-2.5	—	—	—	0.15
BAg-26	P07250	24.0-26.0	37.0-39.0	31.0-35.0	—	1.5-2.5	—	—	1.5-2.5	0.15
BAg-27	P07251	24.0-26.0	34.0-36.0	24.5-28.5	12.5-14.5	—	—	—	—	0.15
BAg-28	P07401	39.0-41.0	29.0-31.0	26.0-30.0	—	—	1.5-2.5	—	—	0.15
BAg-33	P07252	24.0-26.0	29.0-31.0	26.5-28.5	16.5-18.5	—	1.5-2.5	—	—	0.15
BAg-34	P07380	37.0-39.0	31.0-33.0	26.0-30.0	—	—	1.5-2.5	—	—	0.15

a. SAE/ASTM Unified Numbering System for Metals and Alloys.

b. The brazing filler metal shall be analyzed for those specific elements for which values are shown in this table. If the presence of other elements is indicated in the course of this work, the amount of those elements shall be determined to ensure that their total does not exceed the limit specified.

Table 3.7
Chemical Composition Requirements for Gold Filler Metals

AWS Classification	UNS Number[a]	Composition, Weight Percent				
		Au	Cu	Pd	Ni	Other Elements, Total[b]
BAu-1	P00375	37.0-38.0	Remainder	—	—	0.15
BAu-2	P00800	79.5-80.5	Remainder	—	—	0.15
BAu-3	P00350	34.5-35.5	Remainder	—	2.5-3.5	0.15
BAu-4	P00820	81.5-82.5	—		Remainder	0.15
BAu-5	P00300	29.5-30.5	—	33.5-34.5	35.5-36.5	0.15
BAu-6	P00700	69.5-70.5	—	7.5-8.5	21.5-22.5	0.15

a. SAE/ASTM Unified Numbering System for Metals and Alloys.

b. The brazing filler metal shall be analyzed for those specific elements for which values are shown in this table. If the presence of other elements is indicated in the course of this work, the amount of those elements shall be determined to ensure that their total does not exceed the limit specified.

metal cladding can be solution heat treated and aged.

BAlSi-7 is a vacuum brazing filler metal available as a standard cladding on one or both sides of a brazing sheet having a core of 3003 or 6951 aluminum alloy.

BAlSi-9 is a vacuum brazing filler metal available as a standard cladding on one side or both sides of a brazing sheet having a core of 3003 aluminum alloy and is typically used in heat exchanger applications to join fins made from 5000 or 6000 series aluminum alloys.

BAlSi-11 is a brazing sheet clad on one or two sides of alloy 3105 to form a composite vacuum brazing sheet designed for operation in a multizone furnace where the vacuum level is interrupted one or more times during a brazing cycle. The composite can be used in batch-type vacuum brazing furnaces; however, vacuum brazing sheet with a 3003 core is more resistant to erosion. The maximum brazing temperature for the BAlSi-11/3105 composite is 1110°F (595°C).

Magnesium Filler Metals

The BMg-1 classification (ASTM AZ92A) commonly is used for joining AZ10A, K1A, and M1A magnesium base metals. The BMg-1 filler metal is used with torch, dip, or furnace brazing applications and a suitable flux is required. Heating must be closely controlled to avoid melting of the parent metals. Chapter 26 discusses magnesium base metals, brazing techniques, and related topics.

Copper Filler Metals

AWS A5.8 classifies three copper filler metals, BCu-1, BCu1a, and BCu2. These alloys are generally used for brazing carbon and alloy steels, stainless steel, nickel, and copper nickel. These alloys are primarily used in furnace brazing with protective reducing atmospheres without flux. On metals that have constituents with difficult to reduce oxides (chromium, manganese, silicon, titanium, vanadium, and aluminum), a flux may be required. However, pure dry hydrogen, argon, dissociated ammonia, and vacuum atmospheres will adequately handle chromium, manganese, and silicon containing base metals.

BCu1 is available in strip, wire or preforms. BCu1a is available as a powder in various sieve analyses, or as a paste.

BCu-2 is supplied as a paste of copper-oxide suspension in an organic base vehicle.

Copper-Zinc Filler Metals

These filler metals are the AWS RBCuZn-A, RBCuZN-C, and RBCuZn-D classifications. Brazing can be accomplished using torch, furnace, or induction methods. It is important to avoid overheating which can vaporize the zinc, leaving voids in the joint. Flux is required, and a borax/boric acid flux of AWS Type FB3D, or FB3J is commonly used. Flux may be in the form of a powder,

Table 3.8

Chemical Composition Requirements for Nickel-and-Cobalt Filler Metals

AWS Classification	UNS Number[b]	Ni	Cr	B	Si	Fe	C	P	S	Al	Ti	Mn	Cu	Zr	W	Co	Se	Other Elements, Total[c]
							Composition, Weight Percent[a]											
BNi-1	N99600	Remainder	13.0-15.0	2.75-3.50	4.0-5.0	4.0-5.0	0.60-0.90	0.02	0.02	0.05	0.05	—	—	0.05	—	0.10	0.005	0.05
BNi-1a	N99610	Remainder	13.0-15.0	2.75-3.50	4.0-5.0	4.0-5.0	0.06	0.02	0.02	0.05	0.05	—	—	0.05	—	0.10	0.005	0.50
BNi-2	N99620	Remainder	6.0-8.0	2.75-3.50	4.0-5.0	2.5-3.5	0.06	0.02	0.02	0.05	0.05	—	—	0.05	—	0.10	0.005	0.50
BNi-3	N99630	Remainder	—	2.75-3.50	4.0-5.0	0.5	0.06	0.02	0.02	0.05	0.05	—	—	0.05	—	0.10	0.005	0.50
BNi-4	N99640	Remainder	—	1.50-2.20	3.0-4.0	1.5	0.06	0.02	0.02	0.05	0.05	—	—	0.05	—	0.10	0.005	0.50
BNi-5	N99650	Remainder	18.5-19.5	0.03	9.75-10.50	—	0.06	0.02	0.02	0.05	0.05	—	—	0.05	—	0.10	0.005	0.50
BNi-6	N99700	Remainder	—	—	—	—	0.06	10.0-12.0	0.02	0.05	0.05	—	—	0.05	—	0.10	0.005	0.50
BNi-7	N99710	Remainder	13.0-15.0	0.01	0.10	0.2	0.06	9.7-10.5	0.02	0.05	0.05	0.04	—	0.05	—	0.10	0.005	0.50
BNi-8	N99800	Remainder	—	—	6.0-8.0	—	0.06	0.02	0.02	0.05	0.05	21.5-24.5	4.0-5.0	0.05	—	0.10	0.005	0.50
BNi-9	N99612	Remainder	13.5-16.5	3.25-4.0	—	1.5	0.06	0.02	0.02	0.05	0.05	—	—	0.05	—	0.10	0.005	0.50
BNi-10	N99622	Remainder	10.0-13.0	2.0-3.0	3.0-4.0	2.5-4.5	0.40-0.55	0.02	0.02	0.05	0.05	—	—	0.05	15.0-17.0	0.10	0.005	0.50
BNi-11	N99624	Remainder	9.0-11.75	2.2-3.1	3.35-4.25	2.5-4.0	0.30-0.50	0.02	0.02	0.05	0.05	—	—	0.05	11.5-12.75	0.10	0.005	0.50
BCo-1	R39001	16.0-18.0	18.0-20.0	0.70-0.90	7.5-8.5	1.0	0.35-0.45	0.02	0.02	0.05	0.05	—	—	0.05	3.5-4.5	Remainder	0.005	0.50

a. Single values are maximum.

b. SAE/ASTM Unified Numbering System for Metals and Alloys.

c. The filler metal shall be analyzed for those specific elements for which values are shown in this table. If the presence of other elements is indicated in the course of this work, the amount of those elements shall be determined to ensure that their total does not exceed the limit specified.

Table 3.9
Chemical Composition Requirements for Filler Metals for Vacuum Service[a,b,c]

AWS Classification[f]	UNS Number	Ag	Au	Cu	Ni	Co	Sn	Pd	In	Zn	Cd	Pb	P	C
Grade 1														
BVAg-0	P07017	99.95 min	—	0.05	—	—	—	—	—	0.001	0.001	0.002	0.002	0.005
BVAg-6b	P07507	49.5-51.0	—	Remainder	—	—	—	—	—	0.001	0.001	0.002	0.002	0.005
BVAg-8	P07727	71.0-73.0	—	Remainder	—	—	—	—	—	0.001	0.001	0.002	0.002	0.005
BVAg-8b	P07728	70.5-72.5	—	Remainder	0.3-0.7	—	—	—	—	0.001	0.001	0.002	0.002	0.005
BVAg-18	P07607	59.0-61.0	—	Remainder	—	—	9.5-10.5	—	—	0.001	0.001	0.002	0.002	0.005
BVAg-29	P07627	60.5-62.5	—	Remainder	—	—	—	—	14.0-15.0	0.001	0.001	0.002	0.002	0.005
BVAg-30	P07687	67.0-69.0	—	31.0-33.0	—	—	—	4.5-5.5	—	0.001	0.001	0.002	0.002	0.005
BVAg-31	P07587	57.0-59.0	—	20.0-22.0	—	—	—	Remainder	—	0.001	0.001	0.002	0.002	0.005
BVAg-32	P07547	53.0-55.0	—	Remainder	—	—	—	Remainder	—	0.001	0.001	0.002	0.002	0.005
BVAu-2	P00807	—	79.5-80.5	—	Remainder	—	—	—	—	0.001	0.001	0.002	0.002	0.005
BVAu-4	P00827	—	81.5-82.5	—	—	—	—	Remainder	—	0.001	0.001	0.002	0.002	0.005
BVAu-7	P00507	—	49.5-50.5	—	24.5-25.5	0.06	—	—	—	0.001	0.001	0.002	0.002	0.005
BVAu-8	P00927	—	91.0-93.0	—	—	—	—	Remainder	—	0.001	0.001	0.002	0.002	0.005
BVPd-1	P03657	—	—	—	0.06	Remainder	—	64.0-66.0	—	0.001	0.001	0.002	0.002	0.005
Grade 2														
BVAg-0	P07017	99.95 min	—	0.05	—	—	—	—	—	0.002	0.002	0.002	0.002	0.005
BVAg-6	P07507	49.0-51.0	—	Remainder	—	—	—	—	—	0.002	0.002	0.002	0.02	0.005
BVAg-8	P07727	71.0-73.0	—	Remainder	—	—	—	—	—	0.002	0.002	0.002	0.02	0.005
BVAg-8b	P07728	70.5-72.5	—	Remainder	0.3-0.7	—	—	—	—	0.002	0.002	0.002	0.02	0.005
BVAg-18	P07607	59.0-61.0	—	Remainder	—	—	9.5-10.5	—	—	0.002	0.002	0.002	0.02	0.005
BVAg-29	P07627	60.5-62.5	—	Remainder	—	—	—	—	14.0-15.0	0.002	0.002	0.002	0.02	0.005
BVAg-30	P07687	67.0-69.0	—	31.0-33.0	—	—	—	4.5-5.5	—	0.002	0.002	0.002	0.02	0.005
BVAg-31	P07587	57.0-59.0	—	20.0-22.0	—	—	—	Remainder	—	0.002	0.002	0.002	0.002	0.005
BVAg-32	P07547	53.0-55.0	—	Remainder	—	—	—	Remainder	—	0.002	0.002	0.002	0.002	0.005
BVAu-2	P00807	—	79.5-80.5	—	Remainder	—	—	—	—	0.002	0.002	0.002	0.002	0.005
BVAu-4	P00827	—	81.5-82.5	—	—	—	—	Remainder	—	0.002	0.002	0.002	0.002	0.005
BVAu-7	P99507	—	49.5-50.5	—	24.5-25.5	0.06	—	—	—	0.002	0.002	0.002	0.002	0.005
BVAu-8	P00927	—	91.0-93.0	—	—	—	—	Remainder	—	0.002	0.002	0.002	0.002	0.005
BVPd-1	P03657	—	—	—	0.06	Remainder	—	64.0-66.0	—	0.002	0.002	0.002	0.002	0.005
BVCu-1x	C14181	—	—	99.99 min	—	—	—	—	—	0.002	0.002	0.002	0.002	0.005

a. The filler metal shall be analyzed for those specific elements for which values are shown in this table. If the presence of other elements is indicated in the course of this work, the amount of those elements shall be determined.

b. All other elements in addition to those listed in the table above, with a vapor pressure higher than 10^{-7} torr at 932°F (500°C) (such as Mg, Sb, K, Na, Li, Tl, S, Cs, Rb, Se, Te, Sr, and Ca) are limited to 0.001 percent for Grade 1 filler metals and 0.002 percent, each for Grade 2 filler metals. The total of all these high vapor pressure elements (including zinc, cadmium, and lead) is limited to 0.010 percent. The total of other elements not designated as high vapor pressure elements is limited to 0.05 percent, except for BVCu-1x, for which the total shall be 0.015 percent, max.

c. Single values are maximum, unless noted.

Al-Si
BRAZING
ALLOY →

3003 Al

Al-Si
BRAZING
ALLOY →

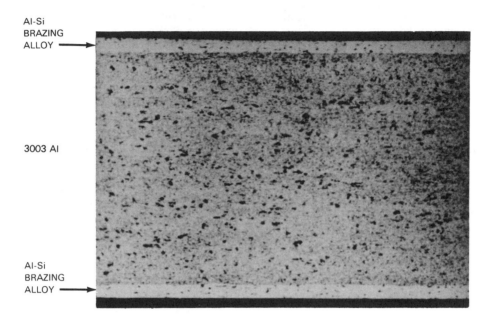

Figure 3.5 — Aluminum Brazing Sheet

paste, or as a coating applied to the rod. A vapor type flux introduced through the torch flame is a popular supplement to powder or paste flux in many production applications.

RBCuZn-A and RBCuZn-C brazing filler metals are used on steels, copper, copper alloys, nickel, nickel-base alloys, and stainless steel where corrosion resistance is not a major requirement. It is used with torch, furnace and induction processes. Joint clearances from 0.002 to 0.005 in (0.05 to 0.13 mm) are suitable.

RBCuZn-D brazing filler metal (called nickel silver) is often used for brazing tungsten carbide. It is also used with steel, nickel, and nickel-base alloys. It can be used with all brazing processes. This filler metal is unsuitable for protective atmosphere furnace brazing.

Copper-Phosphorus Filler Metals

These filler metals (AWS BCuP classification) are primarily used on copper and copper alloy base metals. The phosphorus content deoxidizes copper and these filler metals may be used without a flux on clean unalloyed copper. When brazing brass or bronze, a suitable flux of AWS type FB3A, FB3C, FB3E, FB3F, FB3H or FB4A is required. These alloys should not be used on ferrous or nickel base alloys, or on copper-nickels with more than 10% nickel, to avoid formation of brittle, intermetallic phosphide compounds. Corrosion resistance of copper-phosphorus joints generally is equivalent to copper, except where the joint is exposed to sulfurous gases, compounds, or solutions, e.g. sulfur bearing oil. The copper-phosphorus alloy group includes a variety of compositions with a range of solidus/liquidus temperatures. Some compositions have a wide melting range which make them useful for brazing assemblies with wide or nonuniform clearances. Other compositions have a narrow melting range and the resultant fluidity is helpful when joint clearance is small. Filler metals with wide melting ranges have a tendency to liquate if slowly heated. This should be considered when selecting an alloy and heating method.

The copper-phosphorus alloys are available in wire, rod, preform rings and shapes, strip, powder, and paste. They can be utilized with torch, furnace, resistance and induction heating methods.

BCuP-1 brazing filler metal is particularly suited for resistance brazing applications. This filler metal is somewhat more ductile and less fluid at brazing temperature than other BCuP filler metals containing more phosphorus.

BCuP-2 and -4 brazing filler metals are extremely fluid at brazing temperatures, and will penetrate joints with small clearances. Best results are obtained with clearances of 0.001 to 0.003 in. (0.03 to 0.08 mm).

BCuP-3 and -5 brazing filler metals may be used where very close fits cannot be held. Joint clearances of 0.002 to 0.005 in. (0.06 to 0.13 mm) are recommended.

BCuP-6 brazing filler metal combines some of the properties of BCuP-2 and BCuP-3. It has the ability to fill gaps at the lower end of its brazing range and is very fluid at the high end. Joint clearances of 0.002 to 0.005 in. (0.06 to 0.13 mm) are recommended.

BCuP-7 brazing filler metal is more fluid than BCuP-3 or -5 and has a lower liquidus temperature. It is used extensively in the form of preplaced rings in heat exchanger and tubing joints. Joint clearances of 0.002 to 0.005 in. (0.06 to 0.13 mm) are recommended.

Silver Filler Metals

These filler metals (AWS BAg classification) are used for joining most ferrous and nonferrous metals, except aluminum and magnesium. This classification includes a range of silver based filler metal compositions which may have various additions such as copper, zinc, cadmium, tin, manganese, nickel, and lithium.

Silver, alloyed with copper in a proportion of 72% silver, 28% copper, forms a eutectic with a melting point of 1435°F. This filler metal (BAg-8) can be used to furnace braze non-ferrous base metals in a protective atmosphere. This alloy, however, does not easily wet ferrous metals. The addition of zinc lowers the melting temperature of the silver copper binary alloys and helps wet iron, cobalt, and nickel. Cadmium is also effective in lowering the brazing temperature of these alloys and assists in wetting a variety of base metals. Cadmium oxide present

in brazing fume is poisonous, and cadmium-free filler metals should be utilized wherever possible. Tin can effectively reduce the brazing temperature, and is used to replace zinc or cadmium in filler metals. Nickel is added to assist in wetting tungsten carbides and provides greater corrosion resistance. Brazing alloys containing nickel are especially recommended for joining stainless steels because they reduce susceptibility to interfacial corrosion. Manganese is sometimes added to improve wetting on stainless steel, other nickel-chromium alloys, and cemented carbides. Lithium is effective in reducing oxides of refractory metals to promote filler metal wetting, and improve flow on stainless steels furnace brazed in protective atmospheres.

Flux is required when torch brazing with these filler metals in an oxidizing environment. Mineral fluxes conforming to AWS FB3A, or other classifications as listed in Chapter 4, in powder, paste, or slurry are generally used. Vapor flux introduced through a torch flame also is suitable although filler metal capillary action may be limited with this type application. Vapor (gas) flux is normally used as a supplement to mineral flux types, to improve protection, wetting, and flow.

Silver brazing filler metals are available in numerous forms including: wire, rod, preformed shapes, paste, powder, and strip. Several filler metals are available as a clad or "sandwich" strip with filler metal bonded to both sides of a copper core. This clad strip is popular in brazing carbide tool tips. The copper core absorbs stresses set up by differences in thermal expansion between the carbide and base metal, thus helping to prevent cracking.

BAg-1 brazing filler metal has the lowest brazing temperature range of the BAg filler metals. Because of this it flows freely into tight capillary joints. Its narrow melting range is suitable for rapid or slow methods of heating. This filler metal also contains cadmium and toxic fumes may be formed when it is heated. Refer to the safety information in Chapter 6 for detailed information. BAg-1a brazing filler metal has properties similar to BAg-1. Either composition may be used where low-temperature, free-flowing filler metals are desired. This filler metal also contains cadmium; refer to the safety information in Chapter 6 for detailed information about fume hazards.

BAg-2 brazing filler metal, like BAg-1, is free-flowing and suited for general purpose work. Its broader melting range is helpful where clearances are wide or not uniform. Unless heating is rapid, care must be taken that the lower melting constituents do not separate by liquation. This filler metal contains cadmium and fumes are toxic. Refer to Chapter 6 for safety information on use of the product.

BAg-2a brazing filler metal is similar to BAg-2, but is more economical than BAg-2 since it contains five percent less silver. This filler metal contains cadmium, and fumes formed on heating are toxic. Refer to Chapter 6 for more information.

BAg-3 brazing filler metal is a modification of BAg-1a, i.e. nickel is added. It has good corrosion resistance in marine environments and caustic media, and when used on stainless steel will inhibit crevice (interface) corrosion. Because its nickel content improves wettability on tungsten carbide tool tips, the largest use is to braze carbide tool assemblies. Melting range and low fluidity make BAg-3 suitable for forming larger fillets or filling wide clearances. This filler metal contains cadmium, and toxic fumes are formed when it is heated. Consult Chapter 6 for safety information.

BAg-4 brazing filler metal, like BAg-3, is used extensively for carbide tip brazing, but flows less freely than BAg-3. This filler metal does not contain cadmium.

BAg-5 and -6 brazing filler metals are frequently used for brazing in the electrical industry. They are also used, along with BAg-7 and -24, in the dairy and food industries where the use of cadmium-containing filler metals is prohibited. BAg-5 is an excellent filler metal for brazing brass parts (such as in ships piping, band instruments, lamps, etc.). Since BAg-6 has a broad melting range and is not as free flowing as BAg-1 and -2, it is a better filler metal for filling wide joint clearances or forming large fillets.

BAg-7 brazing filler metal, a cadmium-free substitute for BAg-1, is low-melting with good flow and wetting properties. Typical applications include:

(1) Food equipment where cadmium must be avoided

(2) Minimizing stress corrosion cracking of nickel or nickel-base alloys at low brazing temperatures

(3) Improving color match where the white color will blend with the base metal

BAg-8 brazing filler metal is suitable for furnace brazing in a protective atmosphere without the use of a flux, as well as for brazing procedures requiring a flux. It is usually used on copper or copper alloys. When molten, BAg-8 is very fluid and may flow out over the workpiece surfaces during some furnace brazing applications. It can also be used on stainless steel, nickel-base alloys and carbon steel, although its wetting action on these metals is slow. Higher brazing temperatures will improve flow and wetting.

BAg-8a brazing filler metal is used for brazing in a protective atmosphere and is advantageous when brazing precipitation-hardening and other stainless steels in the 1400 to 1600°F (760 to 870°C) range. The lithium content serves to promote wetting and to increase the flow of the filler metal on difficult-to-braze metals and alloys. Lithium is particularly helpful on base metals containing minor amounts of titanium or aluminum.

BAg-9 and -10 filler metals are used particularly for joining sterling silver. These filler metals have different brazing temperatures and so can be used for step-brazing of successive joints. The color, after brazing, approximates the color of sterling silver.

BAg-13 brazing filler metal is used for service temperatures up to 700°F (370°C). Its low zinc content makes it suitable for furnace brazing.

BAg-13a brazing filler metal is similar to BAg-13, except that it contains no zinc, which is advantageous where volatilization is objectionable in furnace brazing.

BAg-18 brazing filler metal is similar to BAg-8 in its applications. Its tin content helps promote wetting on stainless steel, nickel-base alloys, and carbon steel. BAg-18 has a lower liquidus than BAg-8 and is used in step brazing applications and where fluxless brazing is important.

BAg-19 brazing filler metal is used for the same applications as BAg-8a. BAg-19 is often used in higher brazing temperature applications where the precipitation-hardening heat treatment and brazing are combined.

BAg-20 brazing filler metal possesses good wetting and flow characteristics and has a brazing temperature range higher than the popular Ag-Cu-Zn-Cd compositions. Due to its good brazing properties, freedom from cadmium, and a more economical silver content, new uses for this filler metal are being developed.

BAg-21 brazing filler metal is used in brazing AISI/300 and 400 series stainless steels, as well as the precipitation-hardening nickel and steel alloys. BAg-21 is particularly suited to protective atmosphere furnace brazing because of the absence of zinc and cadmium. It does not require a flux for proper brazing unless the temperatures are low. It requires a rather high brazing temperature, and it flows in a sluggish manner. The nickel content makes it immune to crevice corrosion, particularly on the 400 series stainless steels, by imparting a nickel-rich layer along the fillet edge. It has been used for brazing stainless steel vanes of gas turbine aircraft engines.

BAg-22 is a low-temperature, cadmium-free filler metal with improved strength characteristics over BAg-3, particularly in brazing tungsten carbide tools.

BAg-23 is a high temperature, free flowing filler metal usable both for torch and protective atmosphere furnace brazing. This filler metal is mainly used in brazing stainless steel, nickel-base and cobalt-base alloys for high temperature applications. If this filler metal is used in a high-vacuum atmosphere, a loss of manganese will occur due to its high vapor pressure. Thus a partial pressure vacuum is desirable.

BAg-24 brazing filler metal is low-melting, free-flowing, cadmium-free, and suitable for use in joining low carbon 300 series stainless steels (particularly food handling equipment and hospital utensils), and small tungsten carbide inserts for cutting tools.

BAg-26 brazing filler metal is a low-silver cadmium-free material suitable for carbide and stainless steel brazing. The low brazing temperature and good flow characteristics make it well suited for moderate strength applications.

BAg-27 brazing filler metal is similar to BAg-2 but has lower silver and is somewhat more subject to liquation due to a wider melt range. This filler metal contains cadmium. Toxic fumes are formed on heating. Refer to Chapter 6 for safety information.

BAg-28 brazing filler metal has a lower brazing temperature with a relatively narrower melting range than other cadmium-free classifications with similar silver content. BAg-28 also has free-flowing characteristics.

BAg-33 brazing filler metal was developed to minimize brazing temperature for a filler metal containing 25 percent silver. It has a lower liquidus and, therefore, a narrower melting range than BAg-27. Its higher total zinc plus cadmium content may require more care during brazing. Refer to Chapter 6 for information about safety requirements when brazing with cadmium-bearing alloys.

BAg-34 brazing filler metal is a cadmium-free material with free-flowing characteristics. The brazing temperature range is similar to that of BAg-2 and BAg-2a making it an ideal substitute for these filler metals.

Gold Filler Metals

The AWS BAu series are used for joining parts in electron tube assemblies and for brazing iron, nickel, and cobalt base metals where a greater resistance to oxidation or corrosion is required and where ductility plays an important role in the service of the part. These filler metals are usually used with induction, furnace, or resistance heating in a protective atmosphere (including vacuum).

BAu-1, -2, and -3 brazing filler metals, when used for different joints in the same assembly, permit variation in brazing temperature so that step-brazing can be used.

BAu-4 brazing filler metal is used to braze a wide range of high temperature iron and nickel-base alloys.

BAu-5 brazing filler metal is primarily used for joining heat and corrosion resistant base metals where corrosion-resistant joints with good strength at high temperatures are required. This filler metal is well suited for furnace brazing under protective atmospheres (including vacuum).

BAu-6 brazing filler metal is primarily used for joining of iron and nickel-base superalloys for elevated temperature service conditions. This filler metal is well suited for furnace brazing under protective atmospheres (including vacuum).

Nickel Filler Metals

Brazing filler metals of the BNi classification generally are used for their corrosion resistance and high temperature service properties. These filler materials also are used in subzero applications down to the temperature of liquid oxygen, helium or nitrogen. Atmospheres reducing to both the base and filler metal are recommended. When brazing with an oxyfuel torch, air-atmosphere furnace, or induction coils, a suitable flux of AWS type FB3D, FB1 or FB3J must be used. The BNi series have low vapor pressure, making these alloys ideally suited to vacuum systems and vacuum tube applications.

BNi was the first of the nickel filler metals to be developed. Its composition makes it suitable for brazing nickel, chromium, or iron base metals.

BNi-1a is a low carbon grade (usually held to 0.03 percent or lower) of BNi1. This grade was developed to provide greater corrosion resistance on stainless steel. Tests have indicated, however, corrosion resistance generally is comparable to BNi-1. The BNi-1a does produce stronger joints than BNi-1, but is less fluid at brazing temperature.

BNi-2 is the most widely used of the nickel filler metals. Its properties and uses are similar to BNi-1, yet it can be brazed at lower temperatures.

BNi-3 is used in similar applications as BNi-1 but flows better in marginal atmospheres.

BNi-4 is similar to BNi-3 but exhibits a wider melting range helpful in filling wider clearances or forming fillets.

BNi-5 is used in applications similar to BNi-1, yet is boron-free making it suitable for certain nuclear applications.

BNi-6 is free flowing and used in marginally protective atmospheres.

BNi-7 is often used for brazing honeycomb structures, thin-walled tube assemblies, and for nuclear applications where boron cannot be used.

BNi-8 is used in honeycomb brazements and other stainless and corrosion-resistant materials. The manganese content dictates special brazing procedures. Since manganese oxidizes more readily than chromium, hydrogen, argon, and helium brazing atmospheres must be pure and very dry, with a dew point of -70°F (-57°C) or below. The vacuum atmosphere must have a low pressure and a low leak rate to ensure a very low partial pressure of oxygen.

BNi-9 is a eutectic nickel-chromium-boron filler metal particularly well suited to diffusion brazing applications. Depending upon diffusion time and temperature, the joint remelt temperature can be above 2500°F (1371°C) and, depending on the base metal, the hardness can be as low as 70R_B. With further diffusion time, the grains will grow across the joint and it will appear as all base metal. Since the filler metal is a eutectic, liquation is eliminated. Heavy sections requiring lower heating can be brazed successfully.

BNi-10 is a high-strength filler metal suitable for high temperature applications. The tungsten is a matrix strengthener making the alloy useful for brazing base metals containing cobalt, molybdenum,and tungsten. The filler metal's wide melting range also is useful in brazing cracks. The filler metal layer across the joint acts as a doubler while the lower melting constituent is fluid enough to penetrate the crack to produce a suitable brazement.

BNi-11 is designed for high-temperature brazement applications. The tungsten matrix hardener makes it suitable for brazing base metals containing cobalt, molybdenum, and tungsten. Its wide melting range makes it suitable for slightly larger brazing clearances.

Cobalt Filler Metal

This classification consists of one primary filler metal, BCo-1. This filler metal is used for its high temperature properties and compatibility with cobalt base metals.

Filler Metals for Vacuum Service

This group of filler metals is classified with a BV prefix, i.e., BVAg, BVAu. These filler metals are free of high vapor pressure elements and are manufactured to hold impurities at low levels. They are used for brazing in vacuum atmospheres, and to fabricate vacuum tubes for electronic devices where service life and operating characteristics are of prime importance. Refer to Table

3.9 for the compositions of vacuum service filler metals. In many applications it's important that the filler metal not spatter on areas near the joint. For this reason, the AWS A5.8 specification includes spatter test requirements for vacuum grade classification.

SELECTED READING

Aluminum Association. *Aluminum brazing handbook*, 3rd Edition. Washington, D.C.: Aluminum Association.

Aluminum Company of America. *Brazing alcoa aluminum*. Pittsburgh, PA: Aluminum Company of America.

American Welding Society. *Specification for filler metals for brazing*, ANSI/AWS A5.8-89. Miami, Florida: American Welding Society, 1989.

Brooker, H. R. and Beatson, E. V. *Industrial brazing*, 2nd Edition. London: Newnes-Butterworths, 1975.

Mizuhara, H. and E. Huebel. "Joining ceramic with ductile active filler metal." *Welding Journal* 43(10): 43-51; October 1986.

Mizuhara, H. and Mally, K. "Ceramic-to-metal joining with active brazing filler metals." *Welding Journal* 64(10): 27-32; October 1985.

Swaney, O. W., et. al. "Brazing aluminum automotive heat exchangers in vacuum: process and materials." *Welding Journal* 65 (5): 49; May 1986.

Weymueller, Carl R. "Brazing ceramics to themselves and to metals." *Welding Design and Fabrication* 60(8): 45-48; August 1987.

Winterbottom, W. L. "Process control criteria for brazing aluminum under vacuum." *Welding Journal* 63(10): 33; October 1984.

Keywords — brazing flux, fluxes, atmospheres, controlled atmospheres, surface oxides, oxidation, mineral fluxes, flux removal, vacuum atmosphere, furnace atmosphere

Chapter 4

FLUXES AND ATMOSPHERES

The purpose of a brazing flux[1] is to promote the formation of a brazed joint by protecting the base metal and filler metal from oxidation. The brazing flux may also serve to remove surface oxides and therefore reduce surface tension to promote freer flow of filler metal. In this sense a gas, or a vacuum, are fluxes that may be used to surround the work and provide an active or inert protective atmosphere. For this reason, both fluxes and atmospheres are discussed in this chapter. However, to comply with the usual terminology and because the methods of application and characteristics are quite different, fluxes and atmospheres are discussed and tabulated separately.

When metals are exposed to air, chemical reactions occur. The rate of these reactions generally is accelerated as the temperature increases. The most prevalent reaction leads to formation of oxides, though nitrides and even carbides are formed in some instances. The rate of oxide formation varies with each metal composition and the nature of the oxide. Oxide tenacity, structure, thickness, and resistance to removal or further oxidation are all factors that need consideration. Oxide formation in air on some metals such as aluminum occurs instantaneously, even at, or below, room temperature. In almost all conditions, such surface oxides or other compounds create barriers to the formation of brazed joints.

FLUXES

Fluxes are not designed or intended for the primary removal of oxides, coatings, oil, grease, dirt, or other foreign materials from the parts to be brazed. All parts prior to brazing must be subjected to appropriate cleaning operations as dictated by the particular metals (Chapter 7).

In making a braze, the flux is often needed to combine with, remove, or dissolve those unwanted residual compounds or products of the brazing operation which would otherwise impair filler metal flow.

Oxides are the principal source of surface contamination. The dissolution and removal of oxides during brazing are the most common functions of a flux. So as not to impair metal flow, the flux must also be fluid and displaced readily by the molten brazing filler metal. Wetting ability and viscosity of a flux at brazing temperature, therefore, are important properties.

Under some circumstances, flux also may suppress volatilization of high vapor pressure constituents and the formation of oxide fume condensate in a filler metal.

Some filler metals, such as the lithium-bearing silver or the copper-phosphorus filler metals, are self-fluxing on certain alloys. For example, copper-phosphorus filler metals act as a flux on copper. In most cases,

---

1.. ANSI/AWS A3.0, *Standard Welding Terms and Definitions* defines flux as "Material used to prevent, dissolve, or facilitate removal of oxides or other undesirable surface substances."

use of self-fluxing filler metals alone will result in a joint with a quality equivalent to that of a joint made using flux. Copper-phosphorus (BCuP) filler metals are self-fluxing only in the molten state and will themselves oxidize during the heating and cooling cycles. The use of additional flux is advisable when large sections are to be brazed or where prolonged heating times are contemplated. Lithium- bearing filler metals assist the wetting of stainless and other specialty steels. They generally are used with protective atmospheres.

Constituents of Mineral Fluxes

Many chemical compounds are useful in formulating fluxes. Fluxes, which are generally proprietary, are complex mixtures formulated for specific purposes. When fluxes are heated, reactions may take place between the various ingredients. New compounds may be formed at brazing temperatures that are quite different chemically and physically from the original constituents at the initial room temperature.

For instance, if a fluoborate is a flux ingredient, fluorides may be formed as the ingredients react. The chemistry is especially transient during brazing. Reaction rates of the flux, with oxygen, base metals, filler metals, and any foreign materials present, increase with temperature and quantity of flux. Ingredients of the flux, therefore, must be carefully tailored to suit all the requirements of the braze, including the time at temperature. Attack of the flux on the base metal must be limited while reacting promptly with metal oxides or other tarnish to enable the joint to be satisfactorily formed. Active halide compounds such as chlorides and fluorides, are, for instance, necessary in a flux for aluminum or other highly electropositive metal-bearing alloys (Table 4.1). Some particularly active specialty fluxes must be mixed just before use to avoid reactions between ingredients during storage.

The following list includes the most common ingredients of mineral fluxes. Some comments about their functions follow:

Borates (sodium, potassium, lithium, etc.)
Fused borax
Elemental Boron
Fluoborates (potassium, sodium, etc.)
Fluorides (sodium, potassium, lithium, etc.)
Chlorides (sodium, potassium, lithium)
Acids (boric, calcined boric)
Alkalies (potassium hydroxide, sodium hydroxide)
Wetting agents
Water (as water of hydration, as an addition for paste, slurry and liquid fluxes)

Borates. Borates (e.g., potassium tetraborite, $K_2B_4O_7 \cdot 4H_2O$) are useful in formulating the higher melting fluxes. They have good oxide absorption and provide protection against oxidation for long periods. Most of the borates melt and are effective at temperatures around 1400°F (760°C), or higher. They have a relatively high viscosity. They are usually mixed with other salts to reduce the viscosity of the flux. Higher temperature fluxes usually contain increased borate contents.

Elemental Boron. Fine boron powder may be added to extend the temperature and life of the flux.

Fused Borax. Fused borax ($Na_2O \cdot 2B_2O_3$) is a high melting temperature material that is active at high brazing temperatures.

Fluoborates. Fluoborates are used with other borates or with alkaline compounds, such as carbonates (e.g. potassium carbonate K_2CO_3). Fluoborates react similarly to other fluoride-borate mixtures; however, while they do not provide protection from oxidation to the same extent as other borates, they flow better in the molten state and have greater oxide-dissolving properties. A related class of compounds is fluosilica-borates, which have somewhat higher melting points than the fluoborates, provide good coverage, and improved surface adherence. Their high melting point limits usage. (For safety recommendations for fluorine-bearing compounds, see Chapter 6).

Fluorides. The fluorides (e.g., potassium fluoride, KF) react with most metallic oxides at elevated temperatures and, therefore, often constitute the cleaning component of the brazing flux. They are particularly important in applications where refractory oxides, such as those of chro-

Table 4.1
Brazing Fluxes

Classification	Form	Filler Metal Type	Application	Typical Ingredients	Activity Temperature Range °F	Activity Temperature Range °C	Recommended Base Metals
FB1-A	Powder	BAlSi	For torch or furnace brazing.	Fluorides Chlorides	1080-1140	560-615	All brazeable aluminum alloys.
FB1-B	Powder	BAlSi	For furnace brazing.	Fluorides Chlorides	1040-1140	560-615	All brazeable aluminum alloys.
FB1-C	Powder	BAlSi	For dip brazing with BAlSi	Fluorides Chlorides	1000-1140	540-615	All brazeable aluminum alloys.
FB2-A	Powder	BMg	Because of very limited use of brazing to join magnesium, a detailed classification of brazing fluxes for magnesium is not included.	Fluorides Chlorides	900-1150	480-620	Magnesium alloys whose designators start with AZ.
FB3-A	Paste	BAg and BCuP	General purpose flux for most ferrous and non-ferrous alloys. (Noteable exception Al bronze, etc. See Flux 4A.)	Borates Fluorides	1050-1600	565-870	All brazeable ferrous and non-ferrous metal except those with aluminum or magnesium as a constituent. Also used to braze carbides.
FB3-C	Paste	BAg and BCuP	Similar to 3A but with capability for extended heating times or temperature through use of a deoxidizing additive.	Borates Fluorides Boron	1050-1700	565-925	All brazeable ferrous and non-ferrous metal except those with aluminum or magnesium as a constituent. Also used to braze carbides.
FB3-D	Paste	BAg, BCu, BNi, BAu and RBCuZn	Similar to 3C with a higher active temperature range.	Borates Fluorides	1400-2200	760-1205	All brazeable ferrous and non-ferrous metal except those with aluminum or magnesium as a constituent. Also used to braze carbides.

(Continued)

Table 4.1 (Continued)

Classification	Form	Filler Metal Type	Typical Ingredients	Application	Activity Temperature Range °F	°C	Recommended Base Metals
FB3-E	Liquid	BAg and BCuP	Borates Fluorides	Low activity liquid flux used in brazing jewelry or to augment furnace brazing atmospheres.	1050-1600	565-870	All brazeable ferrous and non-ferrous metal except those with aluminum or magnesium as a constituent. Also used to braze carbides.
FB3-F	Powder	BAg and BCuP	Borates Fluorides	Similar to 3A in a powder form.	1200-1600	650-870	All brazeable ferrous and non-ferrous metal except those with aluminum or magnesium as a constituent. Also used to braze carbides.
FB3-G	Slurry	BAg and BCuP	Borates Fluorides	Similar to 3A in a slurry form.	1050-1600	565-870	All brazeable ferrous and non-ferrous metal except those with aluminum or magnesium as a constituent. Also used to braze carbides.
FB3-H	Slurry	BAg and BCuP	Borates Fluorides Boron	Similar to 3C in a slurry form.	1050-1700	565-925	All brazeable ferrous and non-ferrous metal except those with aluminum or magnesium as a constituent. Also used to braze carbides.
FB3-I	Slurry	BAg, BCu, BNi, BAu and RBCuZn	Borates Fluorides	Similar to 3D in a slurry form.	1400-2200	760-1205	All brazeable ferrous and non-ferrous metal except those with aluminum or magnesium as a constituent. Also used to braze carbides.

(Continued)

Table 4.1 (Continued)

Classification	Form	Filler Metal Type	Typical Ingredients	Application	Activity Temperature Range °F	°C	Recommended Base Metals
FB3-J	Powder	BAg, BCu, BNi, BAu and RBCuZn	Borates Fluorides	Similar to 3D in a slurry form.	1400-2200	760-1205	All brazeable ferrous and non-ferrous metal except those with aluminum or magnesium as a constituent. Also used to braze carbides.
FB3-K	Liquid	BAg, BCuP, and RBCuZn	Borates	Exclusively used in torch brazing by passing fuel gas through a container of flux. Flux applied by the flame.	1400-2200	760-1205	All brazeable ferrous and non-ferrous metal except those with aluminum or magnesium as a constituent. Also used to braze carbides.
FB4-A	Paste	BAg and BCuP	Chlorides Fluorides Borates	General purpose flux for many alloys containing metals that form refractory oxides.	1100-1600	595-870	Brazeable base metals containing up to 9% aluminum (aluminum brass, aluminum bronze, Monel K500). May also have application when minor amounts of Ti, or other metals are present, which form retractory oxides.

Note: The selection of a flux designation for a specific type of work may be based on the filler metal type and the description above, but the information here is generally not adequate for flux selection.

mium and aluminum, are encountered. Fluorides are often added to increase the fluidity of molten borates, thereby facilitating their displacement and improving the capillary flow of the molten brazing filler metal.

Chlorides. Chlorides (eg. potassium chloride, KC1) function in a manner similar to that of the fluorides but have a lower effective temperature range. Chlorides must be used judiciously since, at elevated temperatures, they tend to oxidize the work. Chlorides also melt at lower temperatures, and may depress the melting temperatures of fluoride base fluxes.

Boric Acid. Boric acid (HBO_2) is a principal constituent used in brazing fluxes. It is often used in the hydrated form (H_3BO_3). In the calcined form, the fluxing action is similar to that of boric acid, but it has a somewhat higher melting point. Boric acid has the property of facilitating the removal of the glass-like flux residue after brazing. The melting point of boric acid is best described as being below that of the borates but higher than that of the fluorides.

Alkalies. Potassium and sodium hydroxide (KOH and NaOH) are used sparingly, if at all, because of their deliquescent properties. Even small amounts incorporated in a flux can create difficulty in humid weather and may seriously limit storage life of the flux. Alkalies elevate the useful working temperature of the flux. They may be beneficial additions to fluxes that are used on molybdenum-containing tool steels.

Wetting Agents. Wetting agents are employed in paste, slurry, and liquid fluxes to facilitate the flow or spreading of the flux onto the workpiece prior to brazing. Such wetting agents, when used, are present in quantities which do not adversely affect the normal functions of the brazing flux.

Water. Water is present in brazing fluxes either as water of hydration in the chemicals used in the formulation or as a separate addition for the purpose of making a paste or liquid. Water used to form a paste must be evaluated for adverse impurities and hard waters should be avoided. If the available water is unsuitable, deionized or distilled water should be used.

Other Solvents

Other solvents such as alcohols are sometimes added to commercial fluxes by end users. This is done when a coating of dried flux is desired prior to brazing. Other solvents would enable faster drying than a water based flux.

Selection of a Brazing Flux

The fluxes shown in Table 4.1 are classified by type according to the recommended base metal and filler metals, and by the class, in accordance with the recommended temperature range. Within each type and class, there are numerous commercial fluxes available of proprietary composition. Selection of an appropriate flux within any class and type must be based on careful analysis of the properties or features required for a particular application. For example, within each type and class, longer brazing cycles may require fluxes which emphasize protection. On the other hand, a short brazing cycle may dictate the use of a flux which will promote quick filler metal flow at the minimum temperature. The method of application may also be a factor in selecting the ideal flux.

Some other factors in flux selection are

(1) For dip brazing, water (including water of hydration) must be removed (usually by preheating) prior to immersion in the molten bath.

(2) For resistance brazing, the flux mixture must allow the passing of current. This generally requires a dilute, wet flux.

(3) The effective temperature range of the flux must include the brazing temperature for the specific brazing filler metal involved.

(4) Controlled atmospheres typically may require less active fluxes or less quantity of flux.

(5) Ease of flux residue removal and the consequences of incomplete removal should be considered.

(6) Corrosive action on the base metal or filler metal should be minimized.

(7) The ability of flux to withstand prolonged times (sufficient time to make difficult brazements) at brazing temperature without breaking down.

Form and Methods of Application

Fluxes for brazing generally are available in the form of powder, paste, slurry, or liquid.

The form selected depends upon individual work requirements, the brazing process, and the brazing procedure.

Frequently, powdered flux is mixed with water or alcohol to make a paste. The powder may also be used in the dry form and sprinkled on the joint, although adherence in this form isn't good when the joint is cold. Dry powdered flux may be applied to the heated end of a filler metal rod simply by dipping the rod into the flux container and applying it to preheated surfaces prior to brazing. Dry powdered flux is also used to make the bath for chemical bath dip brazing.

Paste is the most commonly used form for applying brazing flux. Paste is typically applied by brushing on the base metals. It can be applied with good adherence to a joint and filler metal before brazing. A low viscosity slurry or diluted paste flux is used when the flux is to be sprayed on a joint. Flux is also applied by dipping the base metal surface into the paste or slurry flux before assembly. The extent of dilution controls the amount of mineral flux used. Certain flux ingredients will completely dissolve in water to produce a liquid solution called "liquid flux" (Type FB3E).

The particle size of dry flux or paste flux is important because better fluxing action will result when all constituent particles of a flux are small and thoroughly mixed. Stirring, ball milling, or grinding of a flux mixture is helpful if the flux has become lumpy. Preheating of the paste or liquid flux may facilitate application.

Liquid flux of Type FB3-K is used almost exclusively in torch brazing. Fuel gas is passed through the liquid flux container, thereby entraining the flux in the fuel gas. Application of the flame and flux is made where needed. Usually, a small amount of additional preplaced flux is used for the joint surroundings.

Sometimes powdered filler metal is mixed into paste fluxes and they are successfully preplaced together.

Flux Removal

Flux residue generally is removed to avoid corrosion from remaining active chemicals. The residue obtained from a flux, particularly when considerable oxide removal has occurred, is a form of glass. Less formation of glass makes for easier flux residue removal.

Therefore, the work should be clean before applying flux, sufficient flux should be used, and overheating should be avoided.

Removing flux from properly cleaned, brazed parts can usually be accomplished by washing in hot water accompanied by light brushing. Preferably, this rinse should be done immediately after the brazing operation. Following the rinse thorough drying is recommended.

Flux residue removal may be accelerated by immersing the brazed joint in water before it has fully cooled from the brazing temperature. That process creates a thermal shock that cracks off the residue. However, quenching should not be used where the thermal shock involved will impair the strength of the brazed joint.

For residues which are difficult to remove or surfaces which are subject to staining, special practices are required such as using proprietary compounds available for this purpose. (See the specific chapters on base metals for details about pickling solutions and chemical cleaning.)

Additional special practices, such as mechanical methods, can be effective for removing flux residues. Mechanical means used include a fiber brush, wire brush, blast cleaning, or steam jet. Caution should be exercised that the cleaning method is compatible with the base metal. Soft metals such as aluminum are susceptible to embedment of flux residue particles.

After mechanical cleaning it may be desirable to clean again with fluids. Different metal groups are commonly given special treatment following flux cleaning to enhance a desired feature. For example, aluminum assemblies are given a chemical immersion treatment to passivate the surface in order to improve corrosion resistance.

CONTROLLED ATMOSPHERES

Controlled atmospheres are used during brazing to prevent the formation of oxides or other undesirable compounds. In most cases, controlled atmospheres are also used to reduce the oxides that permit the brazing filler metal to wet and flow on clean base metal.

Controlled atmosphere brazing is widely used for the production of high-quality joints. Large tonnages of assemblies of a wide variety of base materials are mass produced by this process. This chapter discusses controlled atmospheres as they pertain to brazing metals. For atmosphere discussion on non-metallic base materials such as ceramics, refer to chapters about specific materials.

Like fluxes, controlled atmospheres are not intended to perform the primary cleaning job for removing oxides, coatings, grease, oil, dirt, or other foreign materials. All parts for brazing should be subjected to appropriate prebraze cleaning operations as dictated by the particular metals (Chapter 7). Controlled atmospheres generally are employed in furnace brazing (Chapter 10); however, they may also be used with induction, resistance, electron beam, or infrared brazing. Even when flux is needed, a controlled atmosphere still may be desirable to minimize the amount of flux required. The use of controlled atmosphere also extends useful life of the fluxes and thereby minimizes postbraze cleaning. In controlled atmosphere applications, postbraze cleaning is generally not necessary. In fact, some types of equipment e.g., metallic muffle furnaces, mesh belts and vacuum systems, may be damaged or contaminated by the use of flux.

Controlled atmospheres avoid formation of oxides and scale over the whole part and often allow finish machining to be done before brazing.

Supplying sufficiently pure reducing atmosphere to a brazing furnace which is in good condition does not guarantee that a proper environment for a braze is attainable. Three examples follow:

(1) A purified hydrogen atmosphere of sufficiently low dew point [for example -60°F (-51°C)] used to braze stainless steel at a temperature of 2000°F (1093°C), still contains enough oxygen to potentially oxidize the stainless steel as it is heated through the lower temperature range. This oxidation may produce a thin film of spongy metal that aids in capillary flow, but if heating is slow a large amount of oxide may form. Sufficient time at an elevated temperature must subsequently be allowed for reduction of the surface oxide.

(2) Oxygen-bearing copper- or silver-based filler metals can contaminate the hydrogen atmosphere surrounding nearby stainless steel surfaces to the extent that a heavy, green oxide will form. Further, the copper or silver filler metal would probably be embrittled and porous after the brazing attempt. This would happen as a result of a reaction between the oxygen contained in the filler metal and the hydrogen atmosphere. The dew point of the hydrogen is increased by reaction of the hydrogen with the small amounts of oxygen in the copper or silver, even though no oxide might appear on their bright, metallic surfaces.

(3) A carbon steel assembly had been copper brazed at 2050°F (1120°C) using an exothermic atmosphere (AWS Type 1). When this same part was brazed in nitrogen-based atmosphere containing 2 percent hydrogen, the copper flowed completely out of the joint. Dew point of the exo gas was +50°F (+10°C), and dew point of the nitrogen-based atmosphere was -10°F (-23°C). Proper braze flow was achieved in a nitrogen-based atmosphere by humidifying the atmosphere to a slightly higher dew point which increases surface tension as the result of oxygen absorption.

Mixtures of some atmospheres with air are explosively combustible. Before heating a furnace chamber containing any such atmosphere, all air must be purged. For example, mixtures of hydrogen with air ranging from 4 to 75 percent by volume hydrogen are explosive. Waste atmosphere must be either burned or rapidly diluted. The safe implementation and use of a protective atmosphere is thoroughly discussed in Code 86-C of the National Fire Protection Association (NFPA).[2.]

The precautions dictated in Chapter 6 should be followed when metals or fluxes containing toxic elements are used. Some atmospheres contain carbon monoxide (CO), which requires special attention as discussed in NFPA Code 86-C. Burning of waste gases or adequate local ventilation is required.

Composition of Atmospheres for Brazing

Table 4.2 lists a number of recommended atmosphere types. Approximate compositions are specified, where applicable.

2.. Code 86-C is available from National Fire Protection Association, Batterymarch Part, Quincy, Massachusetts 02269.

Table 4.2
Atmospheres for Brazing

AWS Brazing Atmosphere Type Number	Source	Maximum Dew Point Incoming Gas	Composition of Atmosphere, Percent				Filler Metals	Base Metals	Remarks
			H_2	N_2	CO	CO_2			
1	Combusted fuel gas (low hydrogen)	Room temperature	5-1	87	5-1	11-12	BAg,[1] BCuP, RBCuZn[1]	Copper, brass[1]	Referred to commonly as exothermic generated atmospheres.
2	Combusted fuel gas (decarburizing)	Room temperature	14-15	70-71	9-10	5-6	BCu, BAg,[1] RBCuZn,[1] BCuP	Copper,[2] brass,[1] low nickel, monel, medium carbon steel[3]	Decarburizes Referred to commonly as endothermic generated atmospheres.
3	Combusted fuel gas, dried	-40°F (-40°C)	15-16	73-75	10-11	—	Same as 2	Same as 2 plus medium and high-carbon steels, monel, nickel alloys	Referred to commonly as endothermic generated atmospheres.
4	Combusted fuel gas, dried (decarburizing)	-40°F (-40°C)	38-40	41-45	17-19	—	Same as 2	Same as 2 plus medium and high-carbon steels	Carburizes
5	Dissociated ammonia	-65°F (-54°C)	75	25	—	—	BAg,[1] BCuP, RBCuZn,[1] BCu, BNi	Same as for 1, 2, 3, 4 plus alloys containing chromium	
6A	Cryogenic or purified $N_2 + H_2$	-90°F (-68°C)	1-30	70-99	—	—	Same as 5	Same as 3	
6B	Cryogenic or purified $N_2 + H_2 + CO$	-20°F (-29°C)	2-20	70-99	1-10	—	Same as 5	Same as 4	
6C	Cryogenic or purified N_2	-90°F (-68°C)	—	100	—	—	Same as 5	Same as 3	
7	Deoxygenated and dried hydrogen	-75°F (-59°C)	100	—	—	—	Same as 5	Same as 5 plus cobalt, chromium, tungsten alloys and carbides[4]	

(Continued)

Table 4.2 (Continued)

AWS Brazing Atmosphere Type Number	Source	Maximum Dew Point Incoming Gas	Composition of Atmosphere, Percent				Filler Metals	Base Metals	Remarks
			H_2	N_2	CO	CO_2			
8	Heated volatile materials	Inorganic vapors (i.e., zinc, cadmium, lithium, volatile fluorides)	—	—	—	—	BAg	Brasses	Special purpose. May be used in conjunction with 1 thru 5 to avoid use of flux.
9	Purified inert gas	Inert gas (e.g., helium, argon, etc.)	—	—	—	—	Same as 5	Same as 5 plus titanium, zirconium, hafnium	Special purpose. Parts must be very clean and atmosphere must be pure.
9A	Purified inert gas + H_2	Inert gas (e.g., helium, argon, etc.)	1-10	—	—	—			

AWS Brazing Atmosphere Type Number	Source	Pressure	H_2	N_2	CO	CO_2	Filler Metals	Base Metals	Remarks
10	Vacuum	Vacuum above 2 torr	—	—	—	—	BCuP, BAg	Cu	
10A	Vacuum	0.5 to 2 torr	—	—	—	—	BCu, BAg	Low carbon steel, Cu	
10B	Vacuum	0.001 to 0.5 torr	—	—	—	—	BCu, BAg	Carbon and low alloy steels, Cu	
10C	Vacuum	1×10^{-3} torr and lower	—	—	—	—	BNi, BAu, BAlSi, Ti alloys	Ht. and corr. resisting steels, Al, Ti, Zr, refractory metals	

Note: AWS Types 6, 7, and 9 include reduced pressures down to 2 torr.

1 Flux required in addition to atmosphere when alloys containing volatile components are used.

2 Copper should be fully deoxidized or oxygen-free. See Chapters 2 and 14.

3 Heating time should be minimized to avoid objectionable decarburization.

4 Flux must be used in addition to the atmosphere if appreciable quantities of aluminum, titanium, silicon, or beryllium are present.

The first four brazing atmospheres listed are generated by passing metered mixtures of hydrocarbon fuel gas and air into a retort for reaction. With natural gas, the ratio of air to gas may range between 9.5 to 1 and 2.5 to 1 for the production of useful atmospheres. For ratios between 5 to 1 and 9.5 to 1, the reaction is exothermic, and the heat liberated is sufficient to support the reaction. Mixtures richer in natural gas than about 5 (air) to 1 (gas) require the addition of heat and the presence of a catalyst in the retort to promote combustion. This type of reaction is termed endothermic. Mixtures leaner than 9.5 to 1 result in a gas product that contains too much oxygen to be suitable for a controlled brazing atmosphere. Following the combustion of such air-gas mixtures, water may be removed either by cooling alone or by the additional use of an absorption-type dryer to reduce the dew point to an acceptable value. Carbon dioxide may be scrubbed out if required.

The AWS No. 5 brazing atmosphere is obtained by the dissociation of dry ammonia (NH_3) into hydrogen and nitrogen in a catalyst-filled heated retort. Drying of the dissociated ammonia may be necessary. The AWS No. 6 brazing atmosphere is a nitrogen-based atmosphere, with its major constituent being cryogenically produced or purified nitrogen. This category of atmosphere is subdivided according to enrichment constituents: (a) hydrogen (b) hydrogen and carbon monoxide (c) none (pure nitrogen). The AWS-6A brazing atmosphere consists of nitrogen plus a small amount of hydrogen. The hydrogen can be supplied in pure form or from dissociated ammonia. AWS-6A atmospheres are often used in lieu of AWS Nos. 1, 2, and 5. The AWS-6B brazing atmosphere consists of nitrogen (N_2) plus additions of hydrogen (H_2) and carbon monoxide (CO). Economical sources of H_2 and CO are endothermic gas and dissociated methanol (CH_3OH). AWS-6B atmospheres are often used in lieu of AWS Nos. 3 and 4, where carbon control is required.

The recommended applications for atmospheres are shown in Table 4.2. Recommendations from suppliers of gases and gas-producing equipment should be consulted for specific jobs. The atmosphere for brazing must be compatible with both the base metals and brazing filler metal for satisfactory results. The equipment used for handling the atmosphere must be such that the atmosphere is supplied at brazing temperature without objectionable contamination.

The following discussion deals with the action of some atmospheres and the atmosphere components:

Hydrogen (H_2). Hydrogen is an active agent for the reduction of most metal oxides at elevated temperatures. Hydrogen can damage some base metals by hydrogen embrittlement, as described in Chapter 2 and in the various chapters on specific base metals.

In materials intended for critical vacuum service, the possibility of hydrogen retention and subsequent outgassing should be investigated.

Carbon Monoxide (CO). Carbon monoxide is an active agent for the reduction of some metal oxides, e.g., those of iron, nickel, cobalt, and copper, at elevated temperatures. Carbon monoxide can serve as a desirable source of carbon, as in some applications on carbon steels, or as an undesirable source of carbon and oxygen in other applications. Carbon monoxide can be generated from oil on the parts at brazing temperatures. Carbon monoxide is toxic, and adequate ventilation must be provided unless waste gas is trapped and burned.

Carbon Dioxide (CO_2) Carbon dioxide is widely present in manufactured brazing atmospheres. A proper ratio of carbon monoxide to carbon dioxide is required to inhibit decarburization and maintain atmosphere stability in brazing steels. At high temperatures carbon monoxide is more stable; at low temperatures it is preferable that carbon dioxide forms. The presence of carbon dioxide, when decomposed may be undesirable as a source of oxygen, carbon, and carbon monoxide. In CO-CO_2 atmospheres the carbon dioxide content of a furnace atmosphere can be undesirably increased by air leakage.

Nitrogen (N_2). Nitrogen is used in a controlled atmosphere to displace air from the furnace and to act as a carrier gas for the other constituents. The typical high purity of nitrogen permits the use of low levels of reducing gases. Nitrogen is inert to most metals, but high levels of nitrogen should be used cautiously when working with metals that are susceptible to nitriding such as chromium, molybdenum, titanium, and zirconium. Nitrogen is non-combustible and non-explosive, therefore desirable from a safety standpoint.

Water Vapor (H_2O). The amount of water vapor in the atmosphere is specified by the dew point (temperature at which moisture in the gas will condense). The relationship between dew point temperature and moisture content of gases by volume is shown in Table 4.3.

Water vapor may be added intentionally by controlled humidification or unintentionally. In the latter case water vapor is added to the atmosphere from air leakage, air carried into the furnace with the work, reduction of metal oxides, leakage from water jackets, contaminated gas lines, diffusion of oxygen through inadequate flame curtains, and other less obvious sources.

Water vapor is generally undesirable during brazing. An excess of water vapor in some instances will cause decarbonization or oxidation. However, a carefully controlled amount of water vapor will aid in the cleaning of carbonaceous material from brazed surfaces and in binder removal of filler metal pastes. In addition, water vapor can be used to inhibit braze flow for filler metal containment. This is particularly beneficial when brazing wide-gapped joints. The amount of water vapor required in the latter instance depends on the amount of reducing gases (primarily hydrogen) present in the atmosphere and base metal surface conditions.

Oxygen (O_2). In addition to sources already mentioned, oxygen may come from gases absorbed on surfaces in the heating chamber. The presence of free oxygen in the brazing atmosphere always is undesirable.

Methane (CH_4). Methane may come from the atmosphere gas as generated or organic materials left on the parts by inadequate cleaning. Methane may also be added deliberately to the gas as a source of carbon and hydrogen.

Sulfur (S). Sulfur or sulfur compounds may be another unintentional contaminant in the atmosphere, and can react with the metal involved in a braze to inhibit wetting. Sulfur compounds can be introduced from fuel gas used in the generation of the atmosphere, air burned with fuel gas, lubricants left on the surface of the parts as a result of inadequate cleaning, brickwork, and other thermal insulation used in the structure of the furnace.

Inorganic Vapors. In equipment designed for their use, vapors such as those of zinc, magnesium, lithium, and fluorine compounds can serve to (1) reduce metal oxide, (2) scavenge the atmosphere of oxygen, or (3) suppress the loss of some of the constituents of alloys that have evolved during brazing. These vapors may be toxic (Chapter 6).

Inert Gases. The inert gases helium (He) and argon (Ar) form no compounds and are therefore suitable for brazing of metals and ceramics. In equipment designed for their use, helium and argon inhibit evaporation of volatile components during brazing and permit the use of weaker retorts than are required for a vacuum.

Table 4.3
Relationship Between Dew Point and Moisture Content

Dew Point Temperature		Moisture Content % by Volume	Moisture Content ppm
°F	°C		
0	18	0.150	1500
-30	-34	0.0329	329
-60	-51	0.0055	55
-80	-62	0.0014	14
-100	-73	0.0002	2

Caution should be exercised when using argon, particularly in pit furnaces, as it is heavier than air. A person climbing down into a pit furnace could suffocate.

Vacuum. An increasing amount of brazing is done in vacuum. Some common units of vacuum measurements are as follows:

1 mm Hg = 1 torr (135 Pa)
1 micron Hg = 1 x 10^{-3} mm Hg (0.135 Pa)
1 micron Hg = 1 x 10^{-3} torr (0.135 Pa)

By removal of gases to a suitably low pressure, including gases which are evolved during heating to brazing temperature, very clean surfaces are obtainable.

Vacuum brazing is particularly useful where base metals are used that adversely react with other atmospheres, or where entrapped fluxes or gases are intolerable. The maximum permissible pressure for successful brazing depends on a number of factors. These include the composition of the base metals, the brazing filler metal, the amount and composition of gas that remains in the evacuated chamber, and the condition of the furnace.

Vacuum brazing is economical for fluxless brazing of many similar and dissimilar base metal combinations. Vacuum is especially suited for brazing very large, continuous areas or complex dense assemblies where (1) solid or liquid fluxes cannot be removed adequately from the interfaces during brazing, (2) gaseous atmospheres are not completely efficient because of their inability to purge occluded gases evolved at close-fitting brazing interfaces.

Vacuum is also suitable for brazing ceramics and reactive and refractory base metals such as titanium, zirconium, columbium, molybdenum, and tantalum. The characteristics of these metals are such that even very small quantities of atmospheric gases may result in embrittlement and sometimes disintegration at brazing temperatures. These metals and their alloys may also be brazed in inert-gas atmospheres if the gases are of sufficiently high purity to avoid contamination and the resultant loss in properties of the metals.

Vacuum brazing has the following advantages and disadvantages compared with other high-purity brazing atmospheres:

(1) Vacuum removes essentially all gases from the brazing area, thereby eliminating the necessity of purifying the supplied atmosphere. Commercial vacuum brazing generally is accomplished at pressure varying from 10^{-6} to .5 torr (10^{-4} to 67.5 Pa) and above, depending upon the materials brazed, the filler metals being used, the area of the brazing interfaces, and the degree to which gases are expelled by the base metals during the brazing cycle.

(2) Certain oxides of the base metal will dissociate in vacuum at brazing temperatures. Vacuum is used widely to braze stainless steel, superalloys, aluminum alloys, and, with special techniques, refractory materials and ceramics.

(3) Difficulties sometimes experienced with contamination of brazing interfaces due to base metal expulsion of gases are minimized in vacuum brazing. Occluded gases are removed from the interfaces immediately upon evolution from the metals.

(4) The low pressure existing around the base and filler metals at elevated temperatures removes volatile impurities and gases from the metals. Frequently, the properties of the base metals themselves are improved. Nevertheless, this characteristic is a disadvantage where elements of the filler metal or base metals volatilize at brazing temperatures because of the low surrounding pressures. This tendency may be corrected by employing partial pressure inert gas according to proper vacuum brazing techniques. Many vacuum furnaces have the ability to operate under a partial pressure of inert gas.

Vacuum Atmosphere Quality

Vacuum pressure is one of many variables that affect atmosphere quality. Two of the major variables are virtual leakage and actual leakage. Actual leakage is the leak(s) that can occur in the vacuum chamber wall, i.e., seal leaks, weld leaks, corrosion penetration leaks, etc. Virtual leakage is present because the entire internals and load are a sorption system.

When evacuating and heating, the air is first removed, then the occluded gases on the surfaces and oils. Following that, the higher vapor-pressure materials are removed that usually condense on the colder areas of the furnace, such as the heat shielding, furnace cage, furnace walls, etc. Finally the included

gases, high vapor pressure materials, and oxides within the heated materials are removed. When pumping down a vacuum furnace that has been open to the atmosphere, the occluded moisture on all materials surfaces is the last to be pumped off, thus the major residual gas that remains is H_2O.

With the many atmosphere variables, it is possible to have a poor vacuum atmosphere even with the very high pumping capacities of many vacuum furnaces. It is possible to have good vacuum pressure (10^{-5} torr) and have a poor quality atmosphere. Likewise with a clean furnace in very good condition, a vacuum of 10^{-3} torr can be satisfactory for brazing titanium or zirconium.

Metal-Metal Oxide Equilibria in Hydrogen

Figure 4.1 is a graphical presentation of metal-metal oxide equilibria. It is based on the thermodynamics of the metal-oxide formation.

Furnace brazing, bright annealing, sintering, and all other metallurgical processes where metallic oxides are reduced in hydrogen atmospheres require a fundamental knowledge of the factors involved in the oxide reducing process. The ability of pure hydrogen to reduce oxides on the surfaces of metals and alloys is determined by three basic factors: (1) temperature, (2) purity of the hydrogen (measured as dew point), and (3) pressure of the gas.

Most applications are at a pressure of one atmosphere. Therefore, only two variable factors remain: temperature and dew point. The diagram presented in Figure 4.1 is a plot of the dew point at which the oxide and the metal are in equilibrium at various temperatures. The 20 curves shown in this diagram define the equilibrium conditions for 20 pure metal-metal oxide systems. The position of 13 additional elements whose curves fall outside the chart is also indicated. The oxide chosen for the calculation

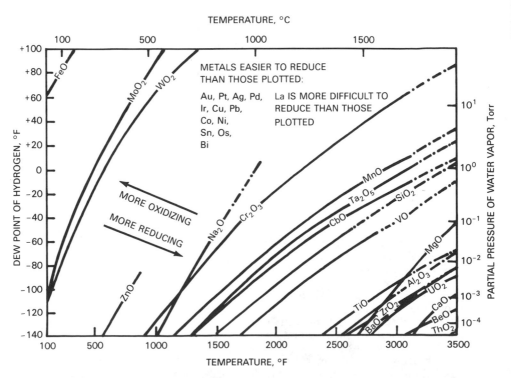

Figure 4.1 — Metal-Metal Oxide Equilibria in Pure Hydrogen Atmospheres

of this diagram represents the most difficult to reduce oxide of each metal.

The metal-metal oxide equilibrium curves (Figure 4.1) slope upward and to the right for each metal. The region above and to the left of each curve represents conditions that are oxidizing for that metal. All points below and to the right of each curve cover the conditions required for reducing the oxides. The diagram, therefore, illustrates that the higher the processing temperature, the higher the dew point (or oxygen content) that can be tolerated for any particular metal. In other words, a given purity of the hydrogen becomes progressively more reducing at progressively higher temperatures.

Use of the Diagram

Use of Figure 4.1 for practical purposes requires first, that the correct curve be selected. For processing an alloy, the element having the most stable oxide (farthest to the right) is the governing curve. Example: for processing chromium stainless steels, the Cr_2O_3 curve applies, since chromium oxides are more stable than those of iron or nickel. Generally, it has been found that, when the most difficult-to-reduce constituent of an alloy is present in more than about 1 atomic percent, a continuous film of its oxide is formed and its curve, therefore, is applicable. Alloys having a concentration progressively lower than 1 atomic percent of the most stable oxide former appear to lie progressively closer to the curve of the next most stable oxide former.

In practice it is necessary to use hydrogen that has a somewhat lower dew point than that indicated by the curve for any given metal. This is partly because the surface becomes oxidized during heating until the temperature is reached that corresponds to the equilibrium temperature for that dew point. The reduction of these oxides formed during heating requires that sufficient time be allowed at conditions sufficiently below, or to the right, of the equilibrium curve. It is also necessary in practice to provide a continuous flow of hydrogen into the work zone during the processing. The purpose of this is to sweep out the out-gassed contaminants and thus maintain the necessary atmosphere purity at the metal surface.

Practice has shown that certain metal oxides can be partially reduced in a high pu-rity inert gas environment. This is possible because of the low partial pressure of oxygen in the atmosphere. A continuously pumped vacuum roughly equivalent to the partial pressure of water vapor as presented on the right-hand ordinate of Figure 4.1 also gives similar results in practice. Since the equilibrium diagram is presented only for H_2/H_2O atmospheres, all the oxygen (O_2) in the hydrogen atmospheres must be converted to H_2O before the dew point is determined.

Special caution is indicated when dealing with metals having a high affinity for H_2. Pure metals or alloys composed principally of titanium, zirconium, hafnium, tantalum, columbium (niobium), and certain others will react immediately with H_2 at sufficiently high temperatures, forming hydrides, or dissolving hydrogen in the base metal.

Such curves as shown in Figure 4.1, and others which have been published (Reference 2), aid in understanding the actions of hydrogen and water vapor in respectively reducing and oxidizing metal oxides and metals. However, they do not portray the complete story involved in the use of controlled atmospheres. They do not indicate the rate at which reduction will occur nor the physical form of the oxide. The curves show whether the removal of these oxides for the assumed chemical reaction is possible under the conditions of gas composition and temperature indicated. In each case, the oxide is *assumed* to be reduced to the metal in one reaction, which may not be the case.

Complex metal oxides may not behave in a manner similar to simple oxides. The oxides of aluminum, magnesium, beryllium, calcium, and titanium are only reducible at very high temperatures and very low dew points. Usually, the equilibrium temperature is far above a useful temperature for brazing. If a base metal contains less than 0.5% of these elements, a satisfactory braze in hydrogen can usually be obtained. If greater amounts of these elements are present in the base metal, a flux or a reactive filler metal (such as those containing titanium) must be used, or a surface plating used, to mask the unreduceable oxides.

The predictions of Figure 4.1 are in rough agreement with actual observations. A notable deviation is that copper oxidizes between

150° and 350°F (66°-177°C) even in a hydrogen atmosphere having a dew point below 0°F (-18°C). Remember that brazing in a controlled atmosphere is subject to considerable error unless extreme precautions are taken to avoid contamination of the atmosphere. Typical contaminations were pointed out earlier in this chapter. Other difficulties include that of analyzing of the atmosphere gas at brazing temperature in the immediate vicinity of the brazing operation.

A rigorous theoretical treatment of metal oxidation would cover not only the reaction of the atmosphere with the metal oxide, but also all other possible reactions. This would include such reactions as metal with oxide and metal with atmosphere, as well as effects due to unstable oxides. The influences of alloying elements further complicate the problem on commercially available metals.

When solutions are discovered to these problems, it will be possible to predict the controlled atmosphere requirement for each brazing application. Meanwhile, suppliers of materials and equipment should be asked for recommendations when difficulties are encountered or maximization of production efficiency is justified.

Many combinations of base metals and brazing filler metals can be used for the efficient production of brazed joints. Many suitable procedures consist of standard controlled atmospheres in lieu of flux and without such precise control of the atmosphere as may seem to be implied during a litany of potential problems.

Quantifying a Furnace Atmosphere

Figure 4.2 shows the effect of varying atmosphere quality on the degree of oxidation and dissociation (reduction) of chromium/chromium oxide in a chromium-containing base metal. As a chromium-containing base metal (not containing elements that more readily react with oxygen such as aluminum or titanium) is heated up in a hydrogen/nitrogen vacuum, or similar atmosphere, the chromium first starts to oxidize. The degree of oxidation is directly proportional to the partial pressure of oxygen in the atmosphere. Note the increasing degree of oxidation in the lower graph as the pressure in the vacuum furnace increases or the dew point of hydrogen increases (gets wetter). This effect can be seen by (1) placing a

metallographically polished piece of AISI 304 stainless steel or other similar chromium-nickel-iron base metal in a furnace and taking the furnace up to 1400°F (760°C), (2) holding for 10 minutes, and then (3) removing the specimen. The degree of discoloration (oxidation) represents the partial pressure of oxygen in the atmosphere and the activity of the atmosphere at that temperature.

In general, oxidation on such a metallographic specimen will start to show up at approximately 1000°F (538°C) and will be dissociated (reduced) at a temperature of 1600° to 1700°F (871° to 927°C) depending on the partial pressure of oxygen around the sample or part.

At the right side of the lower graph, at the point indicated by D, it can be noted that all atmospheres would produce a specimen that is bright and clean.

However, the curve that shows the greatest degree of oxidation (curve in the lower chart) has the smallest pressure to dissociate oxides (the distance D to E from the horizontal line down to the first curve). The larger distance down to the curve 4 (D to F) represents an increase in the pressure to dissociate oxides. This increased pressure to dissociate oxides increases the surface activity of the metal and thus produces better flow of filler metal on the surface of the base metal.

To increase the activity of the upper curve to obtain a similar pressure to dissociate oxides of the lower curve at point F, the temperature would have to be raised to point G.

The area under the curves on the left side of the lower graph can be called the "oxidation area". It is important because it shows how substantial the layer of oxide is that forms on the base metal and filler metal. When using a filler metal with boron, silicon, or phosphorus as the melting point depressant, these elements can be partially or nearly totally removed from the filler metal since they readily combine with the oxygen present and are lost. Upon their removal, the character of the filler metal is changed. The result is that a higher melting residue remains where the filler metal was applied.

Also, the oxidation that takes place at the left side of the lower graph and the subsequent dissociation of the oxide as temperature increases causes an etching of a

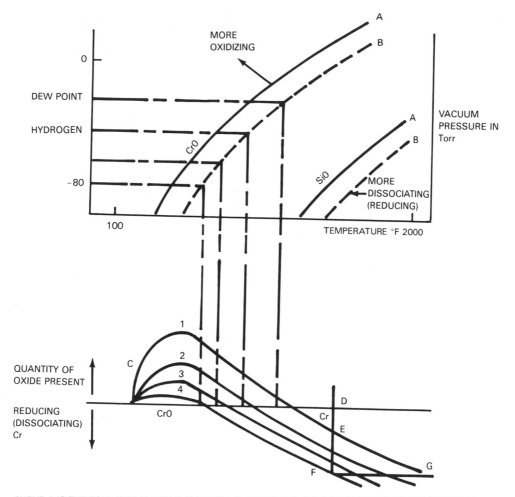

CURVE A IS THE EQUILIBRIUM CURVE MO ⇌ M + O. CURVE B IS A REPRESENTATION OF THE EMPIRICAL CURVE WHERE THE REACTION ALWAYS GOES IN ONE DIRECTION MO → M + O AND THE METAL IS BRIGHT AND CLEAN. C SHOWS A FAMILY OF CURVES SHOWING INCREASING OXIDATION AND HIGHER DISSOCIATING TEMPERATURES AS THE ATMOSPHERE QUALITY IS REDUCED.

Figure 4.2 — Effect of Varying Atmosphere Quality on the Degree of Oxidation and Dissociation (Reduction) of Chromium — Chromium Oxide in a Chromium Containing Base Metal

polished surface. The greater the oxidation at the left of the lower curve, the greater the degree of etching. This phenomenon has been used to macroetch metallographic specimens where the acid required to etch a heat and corrosion resistant stainless alloy would excessively attack a low-alloy steel welded to it.

If parts need to come out of the furnace as bright and shiny as they went in, it is important to have a good atmosphere quality at the lower temperature range as well as the higher temperature range.

Basis for Choice of Atmosphere

On the basis of direct operating costs, the atmosphere listed nearest the top of Table 4.2 for recommended base and filler metals is most cost effective per cubic foot of gas. However, numbers 4, 5, and 6 can be similar in cost depending on the price of hydrocarbon feedstock, electricity, ammonia and nitrogen. Other factors that outweigh the direct costs of the system also should be considered, such as cost of capital, downtime, maintenance, reliability of supply, atmosphere quality and consistency, flexibility, and safety. The footnotes associated with the table are important, and these points are elaborated upon in the text. When base metals susceptible to carburization or decarburization (Chapter 20 and 24) are to be heated in the atmosphere to temperatures at which these reactions take place, a compatible atmosphere must be chosen. In applications involving metals that may be embrittled by hydrogen, or that may combine chemically with hydrogen or nitrogen (e.g., tantalum, chromium, titanium, etc.), the amount of hydrogen or nitrogen present in the atmosphere must be kept below a suitable maximum value.

Probably the most widely used (on a tonnage basis) controlled atmosphere brazing is the use of BCu on steel with AWS Nos. 1, 2, 3, 5, or 6 atmospheres, depending upon the composition and requirements imposed upon the steel.

References

Bannos, T. S. "The effect of atmosphere composition on braze flow." *Heat Treating*, 26-31 April 1984.

Bredzs, N. and Tennenhouse, C. C. "Metal-metal oxide-hydrogen atmosphere chart for brazing or bright metal processing." *Welding Journal* 49(5), Research Supplement: 189s-193s; May 1970.

Kay, W. D. and Peaslee, R. L. "Furnace brazing: how to boost quality and productivity." *Heat Treating*, April 1981.

Sibley, A. T., Ahuja, R. K., and Buck, D. M. "Choosing a nitrogen-based atmosphere." *Metal Powder Report* 36(1): 1981.

Keywords — brazed joints, corrosion resistance, corrosion types, general corrosion attack, localized attack, selective attack, galvanic corrosion, interfacial corrosion, cracking phenomena, corrosion fatigue, stress corrosion cracking, corrosion-erosion, corrosion control

Chapter 5

CORROSION RESISTANCE OF BRAZED JOINTS

INTRODUCTION

Corrosion is usually seen as "rusting" or deterioration of metal as it is exposed to air or water. However, corrosion reactions are much broader in scope. Whereas rusting is corrosion of iron or steel, other materials also suffer deterioration when they react with their environment. In a wider sense, corrosion is any reaction between a material (metal, plastic, or ceramic) and its environment that results in a reduction in the material's capabilities to perform the intended service.

Since corrosion can cause degradation, structural failure, and equipment or plant downtime for replacement or repair, much time and money is spent to protect components. Corrosion occurs on practically all metals in a wide variety of environments. Reactions occur in liquids such as aqueous solutions, molten salts or molten metals and in air, oxygen or other gaseous compounds also at ambient or elevated temperatures. The degradation can be obvious or subtle; the structure may not actually be affected by corrosion, but change in appearance may render the material unusable or cause contamination (as in food, drugs, or paint).

Although compatibility of brazing filler metal itself with the environment is important, when examining the corrosion of brazed joints, the effects of diffusion and alloying with the base metal during brazing (and during service) also influence compatibility. For this reason each brazing filler metal class must be considered in conjunction with the various base materials commonly brazed with it. For specific applications, surveillance testing in actual systems or simulated systems should be performed in order to realistically model actual service conditions.

The deterioration of materials, of course, is not limited to metals. As more brazed ceramics are used in varying environments, more will be written about their corrosion behavior. Currently, the brazing of composites is limited.

Brazed joint design can also greatly influence corrosion response — often through mechanical effects such as complicating stress factors. Brazed joint performance is improved when the joint is designed so that the braze, when stressed, is placed in shear or compression, rather than tension. The stress and corrosion resistance of the brazed joint improves with increasing braze length or depth. Unless specified in this chapter, the joint design of the braze being discussed typically shows the effect of corrosion only; the joint is not load bearing. Examples include lap joints and T-joints.

TYPES OF CORROSION

The types of corrosion affecting brazed joints are often similar to those acting upon the base metals and their alloys. Brazed joints experience attack caused by environmental and geometric factors, selective attack due to the microstructure or composition of the brazed joint, galvanic corrosion, high temperature corrosion, and occasional cracking complicated by corrosion. The materials, stress condition, and the service conditions determine which corrosion mechanisms will be involved. Certain types of corrosion are seen more frequently because of conditions that are inherent to brazing. For example, galvanic corrosion is often of concern because the brazed joint consists of a bond between dissimilar base and filler metals. If the braze is immersed in an electrolyte, galvanic corrosion may occur. The electrolyte may be as simple as water. Various authors classify corrosion in their own way, but the National Association of Corrosion Engineers has published a guide with a consensus classification of nine forms of corrosion, Figure 5.1.

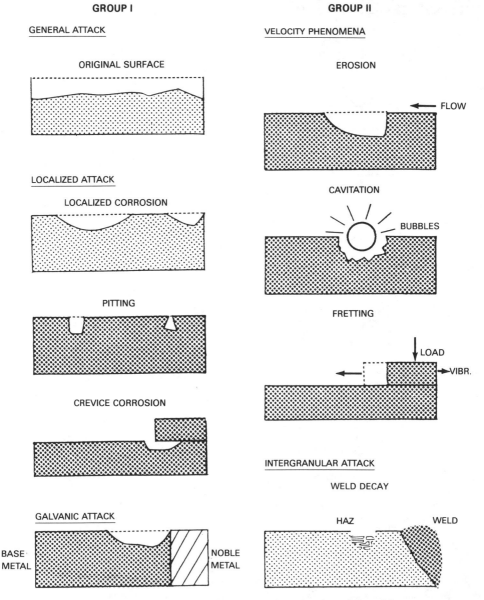

Figure 5.1 — Forms of Corrosion (Mechanics of Attack)

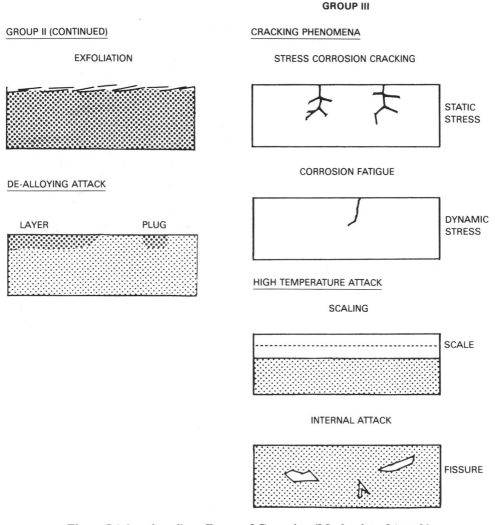

Figure 5.1 (continued) — Forms of Corrosion (Mechanics of Attack)

General Attack

General attack refers to predominantly uniform metal removal over the metal surface. The electrochemical reactions are insensitive to slightly varying surface conditions resulting in anodic and cathodic reactions occurring homogeneously over the surface. The distances between the anodic and cathodic sites are on an atomic scale and protective films are unable to form, Figure 5.1. An example is pickling or acid cleaning in which metal removal can be uniform and rapid, leaving the surface clean. Another ex-

ample is oxidation. The surface is usually covered with scale that is nonprotective, and the attack is frequently uniform in nature. A low rate of uniform attack on the base metal is usually inconsequential. However, if that same rate of attack is only on the braze material, the consequences could quickly become detrimental.

Localized Attack

Attack can be localized because of differences in the microstructure or microcomposition of the metal, or it can be caused by such environmental or geometric nonuniformities as pits, crevices, discontinuous deposits, flow effects, or other conditions surrounding the surface of the metal or braze.

Pitting is also localized but is not necessarily dependent upon the microstructure or composition of one location relative to another. General pitting is usually statistical in nature and may be considered general corrosion. A certain size of exposed area is needed to support and propagate a pit. In other words, without enough area to support the cathodic reaction, the pit will not grow. Thus the corroding system determines the frequency or location of pits. Exceptions do occur. Sometimes pitting will occur around a particular composition at the surface. For example, pitting tends to occur around the MnS inclusions in iron, being fueled by an oxide within the inclusion. A pit may start at imperfections at the surface, such as dirt, scratches, nicks, oxide deposits, grain boundaries, and inclusions, as well as areas of localized segregation due to different phases, differing alloy concentrations in the braze composition, or coring from solidification. In some cases pitting may be caused by a concentration cell.

A concentration cell is an electrolytic cell resulting from an environmental difference between anodic and cathodic areas. The major types of concentration cells are metal-ion, (salt concentration) and oxygen (differential-aeration) cells.

Whenever two pieces of the same metal are immersed in two electrolytes they will develop different potentials. If these two come in contact, current will flow. For example, current will flow if copper immersed in a concentrated copper sulfate solution, comes in contact with another piece of copper, immersed in a dilute copper sulfate solution. Copper in the dilute solution becomes the anode and is dissolved while the copper in the concentrated solution becomes the cathode and copper is plated out.

Differing metal ion concentrations may occur in the same solution, but at differing points, particularly with turbulence at one place and a stagnant condition at another. Metal ions can build up in the stagnant condition, whereas, the turbulence may cause dispersion of metal ions. In the area of turbulence, the concentration of metal ions would be less than in the area of stagnant conditions; thus the area of turbulence would become anodic and corrode. This same situation is possible with a crevice.

However, crevice corrosion occurs most often by the oxygen concentration (differential — aeration) mechanism. The oxygen reaction is a cathodic reaction that supports metallic corrosion at the anode. With differing oxygen concentrations, areas exposed to high levels of oxygen can support corrosion at areas of lower oxygen concentrations. The area inside the crevice is sheltered from a supply of oxygen, so it becomes anodic to the mouth of the crevice. The mouth has a relatively easy access to oxygen resulting in corrosion within the crevice where the oxygen is depleted. Due to electrical resistance effects, however, the corrosion may not extend deeply into the crevice.

A discontinuous deposit or mill scale on a surface makes conditions favorable to development of concentration cells. The metal oxide deposit becomes more noble than the metal and the cathodic reaction occurring at the oxide promotes the rapid corrosion of the exposed metal areas. The greater the area of the metal oxide, the more rapidly the smaller area of exposed metal corrodes.

Selective (Dealloying) Attack

Dealloying attack is selective and is defined as corrosion influenced by the internal structure or differences in composition in the metal or, in this case, the braze.

Many brazed joints consist of two or more phases. Sometimes one or more of the phases will be subjected to selective attack. Occasionally an element is selectively leached from the braze. An example is the leaching (removal) of phosphorus from nickel-phosphorus brazed joints during liquid metal service. Other areas selectively attacked due to structure or composition differences are grain or phase boundaries, precipitates, or the compositionally depleted material immediately surrounding them.

A classical example of selective attack is the dezincification of brass (or Cu-Zn filler metals). In this attack the phase containing higher zinc is removed, leaving an associated residue of one or more of the less active constituents of the alloy. Attack can be quite localized or it can occur in layers. The alloy may be thoroughly corroded, with copper being redeposited along the surface, or the zinc rich phase may be leached from the alloy *in situ*. Its occurrence is promoted by hot, polluted, unaerated water under stagnant conditions, and thick, permeable layers of corrosion products, deposits, or crevices. The same mechanism is involved with the removal of aluminum from copper-aluminum, cobalt from cobalt alloys, nickel from copper-nickel, or the separation of the silver-gold alloys. In an example with a silver braze (Ag-15.6Cu-19.9Zn-18.5Cd) both Zn and Cd were removed. After corrosion, the brittle spongy residue contained 66.3% Ag, 16.5% Zn and 13.8% Cd. Selective attack is less likely under uniform high velocity flow conditions because of less opportunity for concentration buildups.

Galvanic Corrosion

Two dissimilar metals immersed in the same electrolyte will differ in potential. When the two are connected, current will flow between them. A return current will flow in the electrolyte from the less noble metal to the more noble one resulting in corrosion at the anode. Coupling causes a change in potential from that of each metal in the uncoupled condition. Kinetic parameters of the cathodic reaction, relative areas of the anode and cathode, and electrical resistance of the system all affect the rate of corrosion.

The driving force for the current flow, and thus the corrosion, is the potential developed between the two metals. The lowest potential (most electronegative, least corrosion resistant) metal becomes the anode and its corrosion increases. The higher potential (least electronegative, most corrosion resistant) material generally becomes the cathode (unless oxygen controlled) and its corrosion decreases. The behavior of the individual metals when coupled is also affected by the magnitude of the currents on each when coupled, (i.e., their relative areas).

If the open circuit potential of the more noble material is a higher value than the cathodic hydrogen reaction potential (and the oxygen reaction is not involved), it will not corrode. It will provide a surface to support the cathodic reaction that is corroding the less noble metal. If the oxygen reaction is involved, then the open circuit potential of the more noble material must be above the cathodic oxygen reaction potential in order to escape corrosion. Corrosion rate depends on current density, which in turn depends upon the amount of exposed area. In good conducting solutions (such as salt water) a small anode, coupled with a large cathode, will rapidly corrode; whereas a large anode, coupled with a small cathode, will suffer only light, uniform corrosion. However, if the conductivity of the solution is poor, the anodic area nearest the cathode will be preferentially attacked. In atmospheric corrosion involving moisture films, all galvanic effects occur near the junction due to high electrical resistance in air, and for practical purposes the area ratios may be considered one to one.

Since brazed joints consist of dissimilar metals in intimate contact, the possibility of galvanic corrosion should always be considered. In the Electromotive Force (EMF) Series, gold is the most noble and potassium is least. The EMF Series, however, only deals with a portion of the complete corrosion picture and gives only the standard oxidation-reduction potentials that apply to solutions at a concentration of 1 gram-ion per 1000 grams water. The relative position of an element may change in various solutions. Much also depends on the resulting corrosion product and its solubility in the corrosive medium. Soluble corrosion products increase corrosion and insoluble ones may protect or pacify. The rate of electrochemical corrosion (galvanic corrosion included) is determined by the flow of current, and the flow of current is estimated with the help of polarization diagrams (plot of potential versus current).

Different metals and alloys immersed in a specific solution produce their own steady-state potentials. These potentials can be used to set up a galvanic series, such as metals and alloys immersed in sea water.

A galvanic series is more useful for predicting galvanic effects than the EMF series, and a separate galvanic series is required for each environmental condition. Corrosion between materials in the same group is not likely to be appreciable under most conditions. With any other combination, the intensity of the galvanic effects will vary with the distance apart in the series, with the relative areas of the materials forming the couple, and with the polarization effect of the cathodic reaction on the materials. The importance of area ratios should be stressed, because it is current density and not the total current which determines the corrosion rate. It is unwise to combine a small area (i.e., braze) of the anodic member of a couple with a large area (i.e., base metal) of the cathodic member. The reverse relationship is usually not hazardous, unless corrosion is selective toward a phase in the braze or toward the interface.

There are many conditions which may change a corroding system and thereby establish local anodes and cathodes. Metals, alloys and brazes are not normally homogeneous; they contain various phases and precipitates. Potential differences will exist between them. The potential differences vary with the materials and exposure conditions. For example, in an aluminum-magnesium alloy, precipitated particles of a magnesium-aluminum compound are anodic to the matrix in sodium chloride solution but cathodic in a similar solution to which sodium peroxide is added. In cast iron the graphite and cementite are generally cathodic to ferrite. Thus in "graphitic corrosion" the ferrite is dissolved, leaving a porous skeleton of graphite.

In solid solution alloys, there may be potential differences due to concentration gradients of alloying components from point to point. This may be pronounced depending upon solidification patterns. Intergranular attack may be caused or accelerated by potential differences between the grains and grain boundaries. This is one of the principles of etching in metallographic analysis. Precipitates may be anodic and corrode, or a precipitate may deplete its immediate surroundings, making the surroundings anodic, and thus susceptible to corrosion. In austenitic stainless steels the precipitation of chromium carbide at the grain boundaries causes chromium depletion of the adjacent material. Decreasing the chromium makes the material more anodic, causing corrosion of austenitic stainless steels. Similar conditions occur in heat-affected zones of welds and brazed joints whenever the heating cycle causes the carbides to precipitate, depleting surrounding areas of chromium (see "weld decay", Figure 5.1).

Interfacial Corrosion

A brazed joint may be susceptible to interfacial corrosion depending on a number of factors. In addition to the type of base metal and type of filler metal, the brazing technique (i.e., flux or vacuum) and the service environment influence the likelihood of attack. Several of these mechanisms interact and depending upon the particular environmental circumstances, to result in interfacial corrosion.

It is unfortunate that joints in stainless steels joined with silver-containing filler metals employing a flux, the most common technique of all, constitute the circumstances under which joints have failed frequently. Stainless steels of all types may suffer interfacial corrosion attack. Low nickel ferritic and martensitic stainless steels are more susceptible and have a more rapid rate of corrosion than the austenitic grades. With austenitic stainless steels such as types 302, 304, and 316, the use of nickel-containing filler metals reduces the tendency toward interfacial corrosion. These filler metals also slow down or resist interfacial corrosion in type 430 stainless steel. The presence of nickel in the filler alloy is beneficial because it reduces the occurrence of a nickel depleted region at the filler braze to base metal interface.

Fluxless brazing seems to reduce the tendency toward interfacial corrosion in stainless steels.

High Temperature Attack

The types of corrosion at high temperatures are much the same as at low temperatures. At high temperatures, however, diffusion is much more rapid, the temperature differential driving forces are greater, and the resulting corrosion is usually more severe and more rapid.

Oxidation is a common type of high temperature corrosion. Oxidation and tarnish are electrochemical phenomena. Since oxidation is defined as the loss of electrons by a constituent in a chemical reaction, the term oxidation also includes other forms of attack besides combination with oxygen. High temperature attack by carbon and its oxides, sulfur compounds, nitrogen, hydrogen, or halogens is included. Considerable testing has been done on brazes reacting in high temperature air; however, little has been published concerning the effects of the other compounds on brazed joints. The brazing filler metals most resistant to oxidation are those with a high percentage of noble elements or those which form tenacious protective films. Examples include gold and nickel base filler metals respectively.

At high temperatures the protectiveness of external layers of corrosion products may be destroyed by contact with materials with fluxing action such as silicates, fluorides, and lead compounds. In the process of brazing, these fluxes are helpful, but in a service environment, they are not. A molten phase or a volatile reaction is most often damaging. Thus, careful flux removal after brazing is necessary.

Cracking Phenomena

Cracking influenced by corrosion includes situations in which neither stress nor corrosion alone could produce the damage observed within the measured time period. The stress can be either cyclic as in corrosion fatigue or static as in stress corrosion cracking.

Corrosion Fatigue. Corrosion fatigue is defined as the reduction of fatigue resistance due to the presence of a corrosive medium. The appearance of corrosion fatigue normally includes a large area of the fracture covered with corrosion products and a smaller roughened area resulting from the final brittle fracture. The failure is usually transgranular with no branching. Corrosion fatigue is markedly affected by the stress cycle frequency and is most pronounced at low stress frequencies where the exposure time to the corrosive media is greater. It occurs most often in media that promote pitting attack. Possibly the pits act as stress risers to initiate cracks and the corrosion is probably most intensive at the crack tip, giving no stable pit radius for stress reduction. The continued effects increase corrosion and tend to promote the formation of deep notches.

Stress Corrosion Cracking. Stress corrosion cracking is usually defined as a cracking process requiring the simultaneous action of a corrosive agent and sustained tensile stress. This definition does not include corrosion reduced sections that fail by

fast fracture, or corrosion that disintegrates the metal without applied stress. It also does not specify the time involved. Thus, cracking, during brazing which involves action of the liquid brazing filler metal on a highly stressed base metal, could possibly be classed as stress corrosion cracking (although other terms are preferred, i.e. liquid metal embrittlement).

In cracking involving electrochemical mechanisms, stored strain energy tends to increase the activity of metals, making them less noble. In intergranular or interphase corrosion cracking, corrosion occurs at the grain or phase boundary in the absence of stress. However, with stress the attack is usually more localized, more intense, and much more rapid. Plastic deformation, resulting from stress, exposes active metal to corrosion; the rapidly yielding metal at the tip of a crack becomes highly susceptible to rapid corrosion and cracking may proceed intergranularly or intragranularly. A well known example of cracking is the stress corrosion cracking of austenitic stainless steel in slightly impure water.

Several cracking phenomena which can occur during brazing, but before service are mistakenly called stress corrosion cracking. This type of cracking should be kept separate from service-related stress corrosion cracking. Because the occurrences involve cracking, stress, and a brazing environment, they are mentioned here.

Some base metals, such as brasses, nickel-silvers, and phosphor and silicon bronzes, will crack when heated rapidly during brazing. The cracking is caused by stresses in the base metal and may be triggered by contact with the molten brazing filler metal. The cracking can be prevented by stress relieving prior to brazing or heating slowly for stress relief during the brazing cycle and before the brazing filler metal becomes molten. Of course, stresses due to improper fixturing or joint design must be prevented.

Stress cracking or liquid metal embrittlement (LME) is a phenomenon that occurs during brazing with high strength materials (such as stainless steels, nickel alloys and age hardenable alloys) in the highly stressed condition and in contact with molten filler metal. Apparently, the wetting of the surface of the base metal weakens the surface layer and disrupts the equilibrium of the stresses present, then cracks develop rapidly along the grain boundaries. Cracking occurs instantly and is readily visible because the brazing filler metal flows into the cracks. The cracking can be prevented by (1) using annealed rather than hard tempered base metals, (2) removing the stresses by stress relieving or redesigning stress promoting joint fixtures, (3) heating at a slower rate, (4) brazing with a higher melting filler metal to allow stress relieving of the part before the brazing filler metal becomes molten, or (5) selecting a brazing filler metal less likely to induce this type of cracking. Molten flux or a molten corrosion product can also produce similar cracking.

Embrittlement by hydrogen occurs in copper, silver, and palladium if they contain oxygen and are brazed in a hydrogen atmosphere. Hydrogen embrittlement of electrolytic tough pitch copper occurs upon heating when hydrogen diffuses into the base metal, reducing the cuprous oxide at the grain boundaries to metallic copper and steam. Intergranular voids and brittleness are the result. Often the metal is ripped apart due to the force of the reaction. Hydrogen also embrittles titanium, zirconium, tungsten, niobium, and tantalum. However, ductility can usually be restored in tungsten, niobium and tantalum with proper heat treatment after brazing.

Embrittlement during brazing can also be caused by alloying between the filler and base metals (or within the filler metal alone) to form brittle intermetallic compounds or phases. Depending upon the level of stress and other factors, cracking may occur within the brazed joint. An example of this is the phosphorus-rich phase formed in Ni-10P filler metals. If the thermal and mechanical stresses are high during cooling of the braze, cracking will result. Another example is the brittle titanium-aluminum layer formed at the interface of a titanium-to-aluminum braze. Care must be taken to minimize the formation of this brittle layer and to control stresses that can cause cracking.

Corrosion-Erosion

During corrosion-erosion (corrosion involving flow velocity), erosion removes the metal protective coating exposing the metal surface for corrosion. Some examples are impingement, cavitation, and fretting, Fig-

ure 5.1. Attack tends to be localized, rapid and severe. These types of corrosion are aggravated by mechanical effects and are best avoided with proper design and/or system operation.

Impingement (a type of erosion) involves high velocity turbulence of a fluid, usually including solids in suspension, which is directed on a localized area. Surface impingement pits may take a teardrop shape in the direction of impingement, undercutting the metal on the downstream side.

Cavitation is caused by liquid, in contact with a metal surface, being subjected to rapidly alternating pressure. Locally reduced pressure induces boiling which produces small cavities of vapor. When the pressure returns to normal, an implosion of the cavity causes high speed liquid impact. If the implosion contacts the metal, the surface work hardens, roughens, and cracks by fatigue, resulting in rapid, deep spongy pitting.

Fretting involves two tightly fitting surfaces which slip minutely against one another. The rubbing action removes the protective oxide films exposing the active surface to corrosion.

Corrosion-erosion can be equally detrimental to braze and base metal alike since much of the effect is mechanical in nature. However, it will be more severe in the less corrosion resistant material.

THE ROLE OF PROPER BRAZING PROCEDURES IN MINIMIZING CORROSION

The corrosion behavior of brazed joints is a complicated phenomenon. Interacting factors include the brazing procedure, the materials, and the environment. Efforts to minimize the occurrence of corrosion must be made at each step of the brazing process. These steps include proper joint design, selection of materials, brazing process and flux, and final cleaning.

Joint Design

In designing a joint that is to be brazed, attention should be given to geometries that minimize corrosion. For example, if an assembly is to be pickled, the internal geometry of the whole assembly should allow complete exchange of pickling and cleaning solutions. If a lap joint is to be fabricated, the possibility of the geometry leading to the buildup of stagnant solution should be considered. In a butt joint, the efficiency of the joint is maximized by making sure the joint has no defects (i.e., entrapped flux, voids, incomplete joining, or porosity). These defects, in addition to impairing the structural integrity of the joint, may enhance the susceptibility of the joint to corrosion. Entrapped flux may lead to high temperature corrosion; incomplete joining may lead to crevice corrosion.

The concept of distributing the stress in the joint as evenly as possible is also important in joint design. If the stress is concentrated, the likelihood of stress-corrosion cracking exists.

Material Selection

Selection of the materials for a braze joint must involve consideration of the potential for corrosion, in addition to the factors routinely considered. In particular, the appropriate galvanic series for the particular service environment should be consulted if available. Also, attention should be paid to the susceptibility of certain materials to other types of corrosion, such as selective dealloying (or leaching) and general attack. The service environment and selected materials dictate the potential for corrosion.

Brazing Process Selection

Brazing processes are generally classified according to the sources or methods of heating, such as torch, furnace, induction, dip, or infrared. The intensity of the heat source can play a role in the distribution of stress and the formation of alloy phases. If the material is known to be susceptible to stress corrosion cracking, thermal stresses should be minimized. If an alloy system is susceptible to brittle phase formation, the temperature of the brazing process should be selected so that the formation of these phases are suppressed; i.e., times at formation temperature should be avoided or reduced. Susceptibility of filler and base metal systems to brittle phase formation may be determined by consulting the literature or examining phase, TTT, or CCT diagrams.

Torch brazing is a method of brazing which may involve the use of a large low temperature air-gas flame to minimize the possibility of localized overheating. Localized overheating may overstress a filler metal as it solidifies if the metal is sufficiently restrained. Cracking may then result from the overstressing. Furnace or induction heating may be used to keep a uniform temperature distribution around the joint to prevent stress concentration gradients from thermal expansion and contraction to avoid tendencies for stress cracking. Vacuum brazing has the advantage of thoroughly degassing the material and protecting the materials from oxidation and other forms of high temperature corrosion. The cost of vacuum brazing, however, may preclude its use.

Brazing Procedure

The parts must be properly cleaned prior to brazing not only for good wetting and flow but to remove material that may cause corrosion in service. The parts must be protected by fluxing or use of a controlled atmosphere during the heating process to prevent excess oxidation or other forms of high temperature corrosion. The heating process must be selected that will provide the proper brazing temperature and proper heat distribution. After brazing, the braze joint must be cleaned to remove flux residue and any oxide scale formed during the brazing process. Flux removal is particularly important because flux residues are often chemically corrosive and, if not removed, could weaken or cause failure due to corrosion.

CORROSION RESISTANCE OF PARTICULAR BRAZE ALLOY SYSTEMS

The corrosion resistance of braze joints involves the corrosion resistance of the base metals and filler metals individually and collectively.

Silver-Based Filler Metals

Silver-based filler metals can be used to braze most types of base metals. Examples of alloy systems brazed with silver-based filler metals include titanium to titanium, zirconium to zirconium, aluminum to titanium, copper to copper, and stainless steel to stainless steel. In general, the corrosion resistance of silver-based filler metals is similar to that of other nonferrous metals used for their low temperature corrosion resistant properties. These brazing filler metals are unsuitable for use with strong mineral acids, although the attack is slow enough to allow pickling in sulfuric or hydrochloric acids for cleaning.

It should be noted that silver-based braze joints are generally discolored by sulfur-containing substances. Some steel members brazed with silver-based braze alloys, however, have been successfully used in handling sulfite pulps in paper mills.

Tests have shown that joints brazed with silver-based filler alloys have good long-term oxidation resistance at temperatures up to 800°F (427°C). In fact, joints with silver-based filler alloys are being used successfully in engine components which have service environments up to 800°F (427°C). The potential for interfacial corrosion must be considered when brazing stainless steels with silver based filler metals for service in a water environment.

Copper-Based Filler Metals

As a brazing filler metal, copper is stronger and cheaper than silver. However, copper brazing is performed at higher temperatures 2000°F (1100°C) than silver brazing and thus requires a higher temperature flux or controlled atmosphere to prevent oxidation.

As is the case with silver-based filler alloys, many studies of corrosion resistance of copper-based filler alloys have been performed on joints involving stainless steel. Copper filler alloys containing significant amounts of zinc (i.e., RBCuZn) or phosphorus (i.e., BCuP) are not recommended for joining stainless steels because brittle compounds tend to form at the interface between the braze and base metal. The combination of copper-based brazing filler metals and stainless steel base metals are particularly susceptible to liquid-metal embrittlement, particularly if held at brazing temperature for too much time. When copper-based alloys are used to braze stainless steel, care should be taken to quickly cool

through the carbide sensitization range. Dezincification of Cu-Zn filler metals may occur under certain circumstances.

Nickel-Based Filler Metals

Brazed joints produced with nickel-based filler metals are used at temperatures ranging from -328 to above 2192°F (-200 to above 1200°C). The characteristics of nickel make it a good starting point for developing brazing filler metals to withstand high service temperatures. Nickel is ductile, and with proper alloying additions such as chromium, it has excellent strength and oxidation resistance as well as corrosion resistance at elevated temperatures.

Aluminum-Based Filler Metals

Aluminum-based brazing filler metals are used primarily to braze aluminum. Corrosion in an electrolyte is minimized when using Al-Si filler alloys because the electrolytic potential between the base and braze materials is minimal. Aluminum filler metals containing significant levels of zinc or copper exhibit less corrosion resistance, but their resistance is comparable to that of the alloys they are joining.

When brazing with aluminum-based filler alloys, it is most important to thoroughly remove the flux after brazing, because the highly active and hygroscopic nature of aluminum brazing fluxes may cause corrosion by absorbing moisture from the air. Removal may be accomplished by immersion in boiling water or concentrated acid solutions.

Aluminum-based filler metals are also used to join aluminum to titanium or titanium to titanium. The difference in galvanic potential between aluminum and titanium alloys indicates that galvanic corrosion should occur in an electrolyte. However, electropotential studies have shown that the natural oxide passivation films on the titanium and aluminum surfaces greatly reduce the danger of that couple. No susceptibility to stress corrosion has been reported in titanium alloys brazed with aluminum-based filler alloys.

Manganese-Based Filler Metals

Manganese-based filler metals are generally alloyed with nickel when used with stainless steels and Ni base alloys. Most results published on the corrosion behavior of Mn-Ni filler metals indicate that this system exhibits poor corrosion resistance in high temperature environments.

Precious-Metal-Based Filler Alloys

Gold filler alloys are used in vacuum and high temperature heat resisting applications (electronics and engine components) because they are generally free from volatiles and have exceptional resistance to oxidation and other types of corrosion. BAu-4, when used to braze stainless steel or Inconel, is resistant to oxidation by air and sulfur-bearing gases up to 1500°F (816°C). The BAu-4 filler metal was one of three used in brazing the Inconel X-750 thrust chamber of the F-1 engine powering the first stage Saturn V launch vehicle of the Apollo spacecraft system.

APPLICATIONS

Proper brazing techniques must always be employed to avoid preventable problems. Stainless steel must not be sensitized during brazing if corrosion resistance is desired. Stainless steel brazed with copper based filler metal needs to be stress relieved (before or during brazing), and the time the filler metal is molten must be minimized to prevent liquid metal embrittlement during brazing. Joints between copper base metals and filler metals must be designed so that cracking does not occur. Residual flux from brazing should always be considered a source of corrosion and should be carefully removed.

Due to the dissimilar metal nature of brazing, the user should evaluate the compatibility of each filler metal and base metal combination. Compatibility with their service environment should also be evaluated. Galvanic corrosion should be considered. The brazed joint should be designed such that the braze and its various phases are slightly more noble than the base metal. A small amount of metal removed over a large area is not nearly as damaging as the same amount removed from a small area. Keeping the relative nobility of the components of the system similar is usually helpful.

The following briefly summarizes the compatibility of specific brazed joints to the most thoroughly explored environments.

Aqueous Solutions

Room Temperature. Aluminum brazed with BAlSi filler metals is usually as corrosion resistant as the aluminum base metal. The brazed joint will normally be relatively corrosion resistant to water, salt water, and certain other aqueous solutions. Copper joints brazed with silver filler metals, particularly BAg-1, are frequently used in plumbing with good results. In aqueous solutions, the electrochemical characteristics of base metal and filler metals must be considered carefully. For example, copper or brass brazed with BAg-4 is corrosion resistant to tap water, yet nickel-silver or Monel brazed with BAg-4 is not.

When brazing stainless steel, BAg-7 and the silver filler metals containing nickel, such as BAg-3, have more resistance to interface corrosion in water and salt water than several of the others. Silver brazed stainless steels are attacked by acids with time but not by dilute solutions of NaOH.

Stainless steel joints brazed with BCu filler metals containing no silver are usually resistant to water, certain acids, and salt solutions. Copper base metals brazed with copper filler metals are often used for their water, acid, and salt solution corrosion resistance. However, corrosion can occur depending on the electrochemical coupling between base and filler metals. Copper-nickel, copper-silicon, and copper-tin brazed with RBCuZn filler metals are subject to corrosion by dezincification of the brazed joint. Corrosion of copper can also be caused by condensate containing dissolved oxygen and carbon dioxide, i.e., attack under scales formed by hard water.

Stainless steel and nickel alloys brazed with nickel brazing filler metals (BNi) and gold or palladium filler metals (BAu) are relatively resistant to water, salt water, and acids. Increasing chromium in nickel brazing filler metals tends to increase their corrosion resistance.

Elevated Temperature. Copper joints brazed with some of the silver or copper filler metals are resistant to steam up to 300°F (150°C). Several nickel or brazing filler metals have shown resistance to 1290°F (700°C) steam. Gold, platinum, and palladium filler metals have been used at 1470°F (800°C) and above.

In 360°F (182°C) deionized, deoxygenized water, stainless steel brazed with silver filler metals shows resistance similar to or slightly less than stainless steel. Stainless steel joints brazed with copper, nickel, or gold filler metals show more resistance than the stainless steel base metal. Inconel 718 is more resistant than stainless steels, but BNi-6 and BNI-7 are not as resistant to 555°F (290°C) water as is Inconel 718. Stainless steels brazed with BAu-2 or BAu-4 do not resist NaOH at 1100°F (593°C).

Gases

Air. At room temperature, unless high humidity causes galvanic (electrochemical) action, copper alloys and stainless steel brazed with several silver and copper filler metals are resistant to the atmosphere. Certain titanium and aluminum joints brazed with aluminum filler metals will also usually survive normal weather. However, titanium alloys brazed with aluminum may not necessarily be resistant to air with 100 percent humidity or with salt spray.

At elevated temperature, stainless steel, nickel, and copper base metals brazed with filler metals from the silver (BAg) and copper (BCu, BCuP, RBCuZn) groups are usually oxidation resistant at room temperatures up to 750°F (400°C). The use of copper filler metals can usually be extended to 930°F (500°C). Stainless steels and nickel alloys brazed with gold, palladium, and nickel filler metals are normally oxidation resistant up to 1470°F (800°C) and sometimes beyond. Minor constituents (Si, P, B) in nickel filler metals are depleted when exposed to air at these high temperatures.

Oxides of Carbon. Stainless steel joints brazed with Ag-5Pd, Ni-19Cr-10Si, Ni-17Mn-8Si, several copper and several gold filler metals are relatively corrosion resistant to CO_2 plus five percent CO at 1650°F (900°C).

Liquid Metals

The precious metals and the more noble filler metals (BAg, BCu, BCuP, BAu) are rapidly attacked by liquid metals through dissolution.

The resistance of nickel brazing filler metals (BNi) to alkali metals depends on time, temperature, thermal gradient, velocity, and oxygen content of the alkali metal. In general, the higher brazing temperature filler metals show better resistance. BNi2 and BNi5 appear to be compatible with sodium and potassium up to 1300°F (700°C) for 7000 hours. In certain circumstances the use of Ni-Cr-Si-B and Ni-Si-B (and possibly others) can be extended to higher temperatures. The liquid metal tends to deplete (dealloy) the minor constituents (Si, P, B) from the brazed joints. The boron also tends to diffuse into the base metal. The depletion without void formation or embrittleness has not been proved necessarily detrimental; however, other problems with B in the base metal may occur.

Nb-1Zr brazed with several high temperature V-Ta brazing filler metals showed no corrosion when exposed to vacuum and potassium at 1700°F (927°C) for 1000 hrs. At 1500°F (816°C) for 1000 hrs. the Ti-28V-4Be brazing filler metals also showed no deterioration. However, several lower melting Ti-Zr and Zr-Nb brazing filler metals were not resistant to either environment.

Alumina metallized with manganese-molybdenum will be preferentially attacked by potassium at 1600° at (870°C). Of metal-to-ceramic joints tested, beryllia (99.8% BeO) brazed to Nb-1Zr with Zr-25V-15Nb was the most compatible with potassium at 1600°F (817°C).

Molten Fluoride Salts

Silver, copper, gold, and nickel brazing filler metals have all shown good compatibility with various molten fluoride salts. The more noble filler metals such as Au-18Ni, Ag-28Cu, and unalloyed copper showed the best corrosion resistance.

Recommended References

American Society for Metals. *Metals handbook, desk edition.* Metals Park, OH: ASM International.

————. *Metals handbook, corrosion.* Metals Work, Vol. 1113. Metals Park, OH: ASM International, 1987.

American Society for Testing and Materials. *Standard definitions of terms relating to corrosion and corrosion testing.* ASTM G-15-85. Philadelphia: American Society for Testing and Materials.

American Welding Society. "Welding workbook: brazing joint designs and filler metal placement." *Welding Journal,* 51, September 1984.

Atkinson, J. T. and Van Drofferlear, H. *Corrosion and its control, an introduction to the subject.* Houston: National Association of Corrosion Engineers.

Bose, B. and Datta, A. and DeCristofaro, N. "Comparison of gold-nickel with nickel-base metallic glass brazing foils." *Welding Journal.* October 1981, p. 29.

Cole, N. C. "Corrosion resistance of brazed joints." *Source Book on Brazing and Brazing Technology.* Metals Park, OH: American Society for Metals.

————. "Corrosion resistance of brazed joints." Welding Research Council Bulletin No. 247. New York: Welding Research Council, April 1979.

Cole, N. C. and Slaughter, G. M. "Development of brazing filler alloys for molybdenum." *Nuc Tech,* 26, June 1979.

Dillon, C. P. (Ed). *Forms of corrosion, recognition and prevention:* National Association of Corrosion Engineers, 1982.

D'Silva, T. L. "Nickel-palladium base brazing filler metal." *Source Book on Brazing and Brazing Technology.* Metals Park, OH: American Society for Metals, 1980.

Fontana, M. G. and Greene, N. D. *Corrosion engineering,* 2nd Edition. N.Y.: McGraw-Hill Back Co., 1978.

Hattori, T. and Sakamoto, A. "Pitting corrosion property of vacuum brazed 7072 clad aluminum alloy." *Welding Journal,* 339s, October 1982.

Hosking, F. M. "Sodium compatibility of refractory metal alloy-type 304L stainless steel joints." *Welding Journal,* 181s, July 1985.

Kauczor, E. "Erosion-corrosion in pipe systems due to inappropriate brazing and welding practices." *Praktiker* 32(4): April 1980.

McDonald, M. M. "The metallurgy and metallography of braze joints." *International Society Denver Symposium.* July 1985, forthcoming.

Nielsen, K. "Corrosion of soldered and brazed joints in tap water." *Br Corr J.,* 442s, October 1963.

Olson, D. L. Colorado School of Mines. Golden, Colorado, unpublished work, 1977.

Savage, E. I. and Kane, J. J. "Microstructural characterization of nickel braze joints as a function of thermal exposure." *Welding Journal,* 316s, October 1984.

Schmatz, D. J. "Grain boundary penetration during brazing of aluminum." *Welding Journal,* 267s, October 1983.

Schwartz, M. M. (Ed.). *Source book on brazing and brazing technology.* Metals Park, OH: American Society for Metals, 1980.

Sheward, G. E. and Bell, G. R. "Development and evaluation of a Ni-Cr-P brazing filler alloy." *Source book on brazing and brazing technology,* 213. Metals Park, OH: American Society for Metals, 1980.

Stansbury, E. E. Potentiostatic etching. *Applied Metallography.* Van Nostran, 1986.

Takemoto, T. and Okamoto, I. "Effect of composition on the corrosion behavior of stainless steels brazed with silver-based filler metals." *Welding Journal,* 300s, October 1984.

Uhlig, H. H. and Reifie, R. W. *Corrosion and corrosion control, an introduction to corrosion science and engineering.* 3rd Edition. N.Y.: John Wiley and Sons, 1984.

Watson, H. H. "Fluid-tight joints for exacting applications." *Source book on brazing and brazing technology,* 249. Metals Park, OH: American Society for Metals, 1980.

Williams, B. R. "The basics of copper brazing." *Source Book on Brazing and Brazing Technology,* 151. Metals Park, OH: American Society for Metals, 1980.

Winkel, John R. and Childs, Everett L. *Galvanic corrosion of metals and coatings when coupled to uranium in severe environments.* RFP-3486. Golden, Colorado: Rockwell International, October 16, 1983.

Keywords — safety, health, brazing safety, safety and health standards, safe welding practices, protective clothing, respiratory protective equipment, ventilation, mechanical ventilation, air quality, precautionary labeling, material safety data sheets, fire prevention and protection, compressed gas handling, pressure relief devices

CHAPTER 6

SAFETY AND HEALTH

HISTORICAL BACKGROUND

Brazing safety is subject to the provisions of the American National Standard Z49.1,[1] *Safety in Welding and Cutting.* Requirements of this standard apply to brazing, braze welding and soldering and other welding and cutting processes.

The huge demands placed on U.S. production facilities during World War II created a tremendous expansion in the use of welding processes. In mid-1943 it was recognized that some type of code or standard was needed to encourage and increase safety for the welding processes. Under the auspices of the American Standards Association, American War Standard Z49.1, *Safety in Electric and Gas Welding and Cutting Operations* was drafted and published in 1944. That standard was revised several times between 1950 and 1988. Each revision updated the standard in accordance with changing technology and welding practices. The revisions up to and including 1973 were largely evolutionary and closely preserved the format of the original wartime standard.

In 1983, a major rewrite dealt with the vast changes in welding processes that occurred in the 40 years of the standard's existence and clarified the patchwork presentation that had evolved during early revisions. The 1983 edition focused on safety rules to be practiced by the welder and enforced by supervision and management. Material was deleted from previous editions which concentrated on building construction and piping installation — things over which the welder had little control.

The 1988 revision followed the same philosophy. Since then the American Standards Association has become the American National Standards Association and *War Standard ASA Z49.1-1944* has now become *ANSI/AWS Z49.1-1988.* Excerpts from the 1967 edition of ANSI Z49.1 were adopted and are published by the Department of Labor, Occupational Safety and Health Administration in 29 CFR Part 1910, Chapter ZVII; Subpart Q — *Welding, Cutting and Brazing,* Sections 1910.251 through 1910.254.[2] Subpart Z, Section 1910.1000 and Tables Z1, Z2, and Z3 of 29CFR list *Permissible Exposure Limits for Hazardous Materials* while Section 1910.1200, *Hazard Communication,* of 29CFR describes the methods of communicating information on hazardous chemicals to affected employees. All of these sections of the OSHA regulation govern the safety and health requirements of brazing.

1. Published by the American Welding Society, Miami, FL, 1988.

2. OSHA Safety and Health Standards 29 CFR 1910, Subpart I, Sect. 132-140, *Personal Protective Equipment*; Subpart Q, Section 251-254, *Welding, Cutting, and Brazing*; Subpart Z, Section 1000, *Air Contaminants* and Section 1200, *Hazard Communications Standard.* Washington, D.C. U.S. Government Printing Office, 1987.

GENERAL AREA SAFE PRACTICE

Brazing equipment, machines, cables, and other apparatus shall be placed so as not to present a hazard to personnel in work areas, in passageways, on ladders or on stairways. Good housekeeping shall be maintained.

Precautionary signs conforming to the requirements of ANSI Z535.2[3] shall be posted designating the potential applicable hazard(s) and safety requirements.

PROTECTION OF PERSONNEL

Eye And Face Protection

Eye and face protection shall comply with ANSI Standard Z87.1.[4] Goggles or spectacles with filter lenses of shade number 3 or 4 should be worn by operators and helpers for torch brazing. Operators of resistance, induction, or salt bath dip brazing equipment and their helpers should use face shields, spectacles or goggles as appropriate to protect their faces or eyes.

Protective Clothing

Appropriate protective clothing for brazing should provide sufficient coverage and be made of suitable materials to minimize skin burns caused by spatter or radiation. Heavier materials such as woolen or heavy cotton clothing are preferable to lighter materials because they are more difficult to ignite. Cotton clothing should be chemically treated, as required, to reduce combustibility. All clothing shall be free from oil, grease, and combustible solvents. Brazers should wear protective heat resistant gloves made of leather or other suitable materials.

Respiratory Protective Equipment

When controls such as ventilation fail to reduce air contaminants to allowable levels and where the implementation of such controls is not feasible, respiratory protective equipment shall be used to protect personnel from hazardous concentrations of airborne contaminants. Only approved respiratory protection equipment shall be used. Approvals of respiratory equipment are issued by the National Institute of Occupational Safety and Health (NIOSH) or the Mine Safety and Health Administration (MSHA). Selection of the proper equipment shall be in accordance with ANSI Publications Z88.2[5] and Z88.6.[6] As a minimum, compressed air for air supplied respirators shall meet the Grade D requirements of the Compressed Gas Association Publication ANSI/CGA G-7.1.[7]

Ventilation

Adequate ventilation (natural or mechanical) shall be provided for all brazing operations. Adequate ventilation means that hazardous concentrations of airborne contaminants are maintained below the allowable levels for personnel exposure specified by The Occupational Safety and Health Administration (OSHA), The American Conference of Governmental Industrial Hygienists (ACGIH), or other applicable authorities. In cases where the values for allowable exposure limits vary among recognized authorities, the lowest values should be used to achieve the maximum personnel protection.

Adequate ventilation depends upon the following factors:

(1) Volume and configuration of the space in which operations occur.

(2) Number and type of operations generating contaminants.

(3) Allowable levels of specific toxic, asphyxiant, or combustible contaminants being generated.

(4) Natural air flow (rate and direction) and general atmospheric conditions where work is being done.

(5) Location of the brazers' and other persons' breathing zones in relation to the contaminants or their sources.

Fumes and gases from brazing cannot be classified simply. The composition and

---

3. ANSI Z535.2, *Environment and Facility Safety Signs*, New York: American National Standards Institute, 1988.
4. ANSI Z87.1, *Practices for Occupational and Educational Eye and Face Protection*, New York: American National Standards Institute, 1979.

5. ANSI Z88.2, *Practices for Respiratory Protection*, New York: American National Standards Institute, 1980.
6. ANSI Z88.6, *Physical Qualifications for Respiratory Use*, New York: American National Standards Institute, 1984.
7. ANSI/CGA G-7.1, *Commodity Specification for Air*, Arlington, VA: Compressed Gas Association, 1973.

quantity of fumes and gases depend upon such factors as (1) the metals being joined, (2) the brazing process (filler metal, flux, atmosphere, heating method) being used, (3) coatings or surface contaminants on the work, and (4) atmospheric contaminants.

In brazing, the composition of the fumes and gases usually is different from the composition of the consumables. Fume and gas by-products of normal brazing operations usually include those originating from (1) volatilization, (2) reaction or oxidation of consumables, (3) base metal coatings, and (4) atmospheric contaminants. The recommended way to determine adequate ventilation is to sample the composition and quantity of fumes and gases to which personnel are exposed. Sampling procedures shall be in accordance with ANSI/AWS Standard F1.1.[8]

Brazers and other workers shall take precautions to avoid breathing the fume plume directly. This can be done by positioning of the work or the head. Ventilation should direct the plume away from the face. Tests have shown that fume removal is more effective when the air flow is directed from the side of the brazer rather than from the back to front or from the front to back.

Natural Ventilation. Natural ventilation is acceptable for brazing where the necessary precautions are taken to keep the brazer's breathing zone away from the plume and where the sampling of the atmosphere shows that the concentrations of contaminants are below the levels specified by OSHA or ACGIH.

Natural ventilation can meet the specified levels when the necessary precautions are taken to keep the brazer's breathing zone away from the plume and when all of the following conditions are met:

(1) Space of more than 10000 ft³ (284m³) per brazer is provided.

(2) Ceiling height is more than 16 ft (5m).

(3) Brazing is not done in a confined space.

(4) Brazing space does not contain partitions, balconies, or other structural barriers that significantly obstruct cross ventilation.

(5) Low Permissible Exposure Limit (PEL) materials (discussed below) are not present as deliberate constituents, and are not present in excess of allowable limits as determined from air samples per ANSI/AWS F1.1.

If natural ventilation is not sufficient to maintain contaminants below specified levels, mechanical ventilation or respirators shall be provided.

Mechanical Ventilation. Mechanical ventilation includes local exhaust, ventilation, local forced ventilation, and general area mechanical air movement. Local exhaust ventilation is preferred. It includes fixed or movable exhaust hoods placed as near as practicable to the work. The hoods must maintain a capture velocity sufficient to keep airborne contaminants below the limits specified by OSHA or ACGIH. Precautions shall be taken to ensure that excessive levels of contaminants are not dispersed to other work areas.

Local forced ventilation consists of a local air moving system (such as a fan) placed so that it moves the air horizontally across the brazer's face. Such ventilation should produce an approximate velocity of 100 ft per min. (30 m per min.), and be maintained for a distance of approximately 2 ft (0.6m) directly above the work area. Precautions shall be taken to ensure that excessive levels of contaminants are not dispersed to other work areas.

General mechanical ventilation is usually not satisfactory for health hazard control alone. However, it is often beneficial (in addition to local exhaust or local forced ventilation) for (1) maintaining the general background level of airborne contaminants below the specified levels, or (2) preventing the accumulation of combustible or asphyxiant gas mixtures. Examples of general mechanical ventilation are roof exhaust fans, wall exhaust fans and similar local area air movers.

Where permissible and verified by air sampling, air cleaners that have high efficiencies in the collection of submicron particles may be used to recirculate a portion of air that would otherwise be exhausted. Since some of these filters do not remove gases, adequate monitoring must be done to ensure compliance.

---

8. ANSI/AWS F1.1, *Methods for Sampling Airborne Particulates Generated by Welding and Allied Processes*, Miami, Florida: American Welding Society, 1985.

Low Permissible Exposure Limit Materials

Certain materials contained in the fumes generated during brazing operations may have very low OSHA Permissible Exposure Levels (PEL) or ACGIH Threshold Limit Values (TLV). The materials encountered most commonly during brazing are

Antimony	Cobalt	Nickel
Arsenic	Copper	Oxides of Nitrogen
Barium	Fluorides	Selenium
Beryllium	Lead	Silver
Cadmium	Manganese	Vanadium
Chromium	Mercury	

Refer to the Material Safety Data Sheets (MSDSs) provided by the manufacturer to identify the presence of any of the materials listed above. Whenever these materials are encountered in brazing operations, the following special ventilation precautions shall be taken (unless atmospheric sampling and testing under the most adverse conditions has established that the level of the hazardous constituents is within specified limits).

Confined Spaces. Whenever these materials are encountered in confined space brazing operations, local exhaust, and respiratory protection shall be used.

Indoors. Whenever these materials, except beryllium, are encountered in indoor brazing operations, local exhaust mechanical ventilation shall be used. Whenever beryllium is encountered, confined space ventilation requirements shall apply.

Outdoors. Whenever these materials are encountered in outdoor brazing operations, there is a need for respiratory protection equipment.

Adjacent Persons. All persons in the immediate vicinity of brazing operations involving low PEL materials shall be similarly protected as necessary by adequate ventilation or approved respirators, or both.

Other Hazardous Brazing Materials and Conditions

Fluorine Compounds. Fumes and gases from fluorine compounds in brazing fluxes can be dangerous to health, and can burn eyes and skin on contact. (See below for special labeling for fluorine-containing fluxes.) When brazing operations involve fluorine-bearing fluxes in confined spaces, local exhaust mechanical ventilation or respiratory protection shall be provided. When brazing with fluorine-bearing fluxes in open spaces, the need for local exhaust ventilation or respiratory protection will depend on the individual circumstances and the results of air sample tests. Experience has shown that such protection is desirable for brazing in a fixed location.

Oxides Of Nitrogen. Nitrogen dioxide and nitrogen gases are products of combustion from oxyfuel gas torch and furnace brazing systems. Whenever these gases are encountered in a confined space, local exhaust mechanical ventilation and respiratory protection shall be used. When the gases are encountered indoors, local exhaust mechanical ventilation shall be used. During outdoor brazing with a torch, respiratory protection shall be used.

Zinc. Fumes containing zinc compounds may produce symptoms of nausea, dizziness, or fever, sometimes called metal fume fever. Brazing operations involving filler metals, base metals, or coatings containing zinc shall be performed as described above for fluorine compounds.

Cleaning Compounds. Cleaning materials, because of their possible toxic or flammable properties, often require special ventilation precautions. Manufacturers' instructions and Material Safety Data Sheets shall be followed. Degreasing or cleaning operations involving chlorinated hydrocarbons require special attention. They should be so located that vapors from these operations will not reach or be drawn into the atmosphere that surrounds the molten filler metal or the heat source. When such vapors enter the atmosphere of brazing operations, a reaction product having a characteristic objectionable, irritating odor can be produced, including highly toxic phosgene gas. Low levels of exposure can cause feelings of nausea, dizziness, or malaise. Heavy exposure can produce serious health impairments.

Asbestos. Brazing shall not be performed on surfaces where asbestos is present because asbestos is recognized by ACGIH as a human carcinogen.

Confined Spaces. Work in confined spaces (such as pits, tanks, or furnace interiors) requires special precautions. In addition to the requirements for adequate ventilation to keep airborne contaminants in breathing atmospheres below allowable levels, ventilation in confined spaces shall also be sufficient to (1) assure adequate oxygen for life support, (2) prevent the accumulation of flammable mixtures, and (3) prevent oxygen-enriched atmospheres. Asphyxiation causes unconsciousness and death without warning. Oxygen-enriched atmospheres greatly intensify combustion that may cause severe and often fatal burns.

Confined spaces shall not be entered unless (1) they are well ventilated or (2) the brazer is wearing an approved air-supplied breathing apparatus and a similarly equipped attendant is present outside the confined space to ensure the safety of those working within. Adequate ventilation in confined spaces shall be provided not only to protect brazers but also to protect all other personnel in the area.

Before entering confined spaces, tests shall be made for toxic or flammable gases and vapors, and for adequate or excess oxygen. Approved instruments shall be used for the tests. Refer to ANSI Publication Z117.1[9] for further precautions. A continuous monitoring system with audible alarms should be used for confined space work.

Heavier-than-air gases such as argon, carbon dioxide, and nitrogen dioxide may accumulate in a pit, tank bottom, furnace bottom, or low areas near floors. Lighter-than-air gases such as helium, hydrogen, and carbon monoxide, may accumulate in high areas such as tank tops, furnace tops and near ceilings. The same precautions shall apply to these areas as to confined spaces.

Air Quality and Quantity

Only clean respirable air shall be used for ventilation. The quality and quantity of air for ventilation shall maintain personnel exposures to hazardous contaminants below the limits specified by OSHA or ACGIH. As a minimum, compressed air for air-supplied respirators or breathing equipment shall meet the Grade D requirements of ANSI/CGA Publication G-7.1. The compressed air line for respirators shall be a single purpose line that can not be connected to any other supply line. Oxygen or any other gas or mixture of gases except air, shall not be used for ventilation. Air may be natural air or synthesized air for breathing purposes.

When brazing processes are performed in areas immediately threatening to life and health, National Institute for Occupational Safety and Health (NIOSH) or Mine Safety and Health Administration (MSHA) approved positive pressure, self-contained breathing apparatus or airline respirators shall be used. "Areas immediately threatening to life or health" are those areas which persons intend to occupy and the atmospheres are not suitable (or potentially not suitable) to safely sustain life or health. The respirator or breathing apparatus shall have an emergency self-contained air supply of at least five minutes duration in the event that the main air source fails.

Gas cylinders and brazing power sources shall be located outside confined spaces to prevent contamination of the atmosphere within the confined space by possible leaks from the cylinders or fumes from the power source. Portable equipment shall be blocked to prevent accidental movement before brazing in a confined space. Ventilation ducts used to provide local exhaust for brazing operations shall be constructed of noncombustible materials. These ducts shall be inspected as necessary to ensure their proper function and that the internal surfaces are free of excessive or combustible residuals.

When brazing is to be performed near any confined space, personnel must be made aware of the respiratory hazards in the confined space, and they must not enter such spaces without observing the necessary precautions as defined in ANSI Publication Z117.1.

When a person must enter a confined space through a restricted opening, means shall be provided for signaling an outside attendant for help. When body belt-harness systems are used for emergency rescue purposes, they shall be attached to the person's body so that they do not become obstructed

---

9. ANSI Z117.1, *Safety Requirements for Working in Tanks and Other Confined Spaces*, New York: American National Standards Institute, 1977.

in passing through a small exit path while executing a preplanned rescue procedure.

When operations are carried on in confined spaces where atmospheres immediately hazardous to life or health may be present or may develop, an attendant shall be stationed on the outside of the confined space to ensure the safety of those working within. The attendant shall have a preplanned rescue procedure worked out with the worker inside for quickly removing or protecting the worker in case of emergency. The attendant shall observe, or be in constant communication with, the worker inside, and shall be capable of putting rescue operations into effect as required. An air-supplied breathing apparatus for rescue operations shall be available.

Brazing Furnace Atmospheres.

Brazing furnaces are, in many respects, a type of confined space. They employ a variety of atmospheres to exclude oxygen during the brazing process. Such atmospheres may include inert gas, flammable gas, flammable gas combustion products, or vacuum. If brazing furnaces require personnel entry into the furnace or adjacent areas, the provisions of confined spaces above shall be observed. For a more detailed discussion of furnace brazing safety practices, see the section *Brazing Processes and Equipment Safety* below.

PRECAUTIONARY LABELING AND MATERIAL SAFETY DATA SHEETS

Brazing operations pose potential hazards from fumes, gases, electric shock, heat, and radiation. Personnel shall be warned against these hazards, where applicable, by use of adequate precautionary labeling as defined in OSHA 29CFR 1910.1200. Examples of minimum labeling are shown in Figures 6.1 to 6.4. The Material Safety Data Sheet (MSDS) must be available to all personnel.

Resistance and Induction Brazing Processes

As a minimum, the information shown in Figure 6.1, or its equivalent, shall be placed on stock containers of consumable materials and on major equipment, such as power supplies, wire feeders, and controls, used in electrical resistance or induction brazing processes. The information shall be readily visible to the worker and may be on a label, tag, or other printed form as defined in ANSI Publications Z535.2 and Z535.4.

Additional precautionary measures can be added as appropriate for special requirements or when new knowledge is available. First aid information is optional (recommended for major health hazards) and,

---

10. ANSI Z535.4, *Product Safety Signs and Labels*, New York: American National Standards Institute, 1988.

WARNING: PROTECT yourself and others. Read and understand this label.

FUMES AND GASES can be dangerous to your health.
ARC RAYS can injure eyes and burn skin.
ELECTRIC SHOCK can KILL.

- Before use, read and understand the manufacturer's instructions, Material Safety Data Sheets (MSDSs), and your employer's safety practices.
- Keep your head out of the fumes.
- Use enough ventilation, exhaust at the arc, or both, to keep fumes and gases from your breathing zone and the general area.
- Wear correct eye, ear, and body protection.
- Do not touch live electrical parts.
- See American National Standard Z49.1, *Safety in Welding and Cutting*, published by the American Welding Society, 550 N.W. LeJeune Rd., P.O. Box 351040, Miami, Florida 33135; OSHA Safety and Health Standards, 29 CFR 1910, available from U.S. Government Printing Office, Washington, DC 20402.

DO NOT REMOVE THIS LABEL.

Figure 6.1 — Warning Label for Arc Welding Processes and Equipment

WARNING: PROTECT yourself and others. Read and understand this label.

FUMES AND GASES can be dangerous to your health. HEAT RAYS (INFRARED RADIATION from flame or hot metal) can injure eyes.

- Before use, read and understand the manufacturer's instructions, Material Safety Data Sheets (MSDSs), and your employer's safety practices.
- Keep your head out of the fumes.
- Use enough ventilation, exhaust at the arc, or both, to keep fumes and gases from your breathing zone and the general area.
- Wear correct eye, ear, and body protection.
- See American National Standard Z49.1, *Safety in Welding and Cutting*, published by the American Welding Society, 550 N.W. LeJeune Rd., P.O. BOX 351040, Miami, Florida 33135; OSHA Safety and Health Standards, 29 CFR 1910, available from U.S. Government Printing Office, Washington, DC 20402.

DO NOT REMOVE THIS LABEL.

Figure 6.2 — Warning Label for Oxyfuel Gas Processes

when present, shall follow the last precautionary measure. The company name and address and a label identification number shall appear on the label.

Oxyfuel Gas, Furnace, and Dip Brazing Processes

As a minimum, the information shown in Figure 6.2, or its equivalent, shall be placed on stock containers of consumable materials and on major equipment used in oxyfuel gas, furnace, and dip brazing processes. The information shall be readily visible to the worker and may be on a label, tag, or other printed form as defined in ANSI Publications Z535.2 and Z535.4.

Additional precautionary measures can be added as appropriate for special requirements or when new knowledge is available. First aid information is optional (recommended for major health hazards) and, when present, shall follow the last precautionary measure. The company name, address, and label identification number shall appear on the label.

DANGER: CONTAINS CADMIUM. Protect yourself and others. Read and understand this label.

FUMES ARE POISONOUS AND CAN KILL.

- Before use, read, understand and follow the manufacturer's instructions, Material Safety Data Sheets (MSDSs), and your employer's safety practices.
- Do not breathe fumes. Even brief exposure to high concentrations should be avoided.
- Use only enough ventilation, exhaust at the work, or both, to keep fumes from your breathing zone and the general area. If this cannot be done, use air supplied respirators.
- Keep children away when using.
- See American National Standard Z49.1, *Safety in Welding and Cutting*, published by the American Welding Society, 550 N.W. LeJeune Rd., P.O. Box 351040, Miami, Florida 33135; OSHA Safety and Health Standards, 29 CFR 1910, available from U.S. Government Printing Office, Washington, DC 20402.

If chest pain, shortness of breath, cough, or fever develop after use, obtain medical help immediately.

DO NOT REMOVE THIS LABEL.

Figure 6.3 — Warning Label for Brazing Filler Metals Containing Cadmium

WARNING: CONTAINS FLUORIDES. Protect yourself and others. Read and understand this label.

FUMES AND GASES CAN BE DANGEROUS TO YOUR HEALTH. BURNS EYES AND SKIN ON CONTACT. CAN BE FATAL IF SWALLOWED.

- Before use, read, understand and follow the manufacturer's instructions, Material Safety Data Sheets (MSDSs), and your employer's safety practices.
- Keep your head out of the fume.
- Use enough ventilation, exhaust at the work, or both, to keep fumes and gases from your breathing zone and the general area.
- Avoid contact of flux with eyes and skin.
- Do not take internally.
- Keep out of reach of children.
- See American National Standard Z49.1, *Safety in Welding and Cutting*, published by the American Welding Society, 550 N.W. LeJeune Rd., P.O. Box 351040, Miami, Florida 33135; OSHA Safety and Health Standards, 29 CFR 1910, available from U.S. Government Printing Office, Washington, DC 20402.

First Aid: If contact in eyes, flush immediately with clean water for at least 15 minutes. If swallowed, induce vomiting. Never give anything by mouth to an unconscious person. Call a physician.

DO NOT REMOVE THIS LABEL.

Figure 6.4 — Warning Label for Brazing and Gas Welding Fluxes Containing Fluorides

Filler Metals Containing Cadmium

As a minimum, brazing filler metals containing cadmium as a constituent greater than 0.1 percent by weight shall carry the information shown in Figure 6.3, or its equivalent on tags, boxes, or other containers, and on any coils or wire or strip not supplied to the user in a labeled container. Label requirements will also conform to ANSI Publication Z535.4.

Brazing Fluxes Containing Fluorides

As a minimum, brazing fluxes containing fluorine compounds shall have precautionary information as shown in Figure 6.4, or its equivalent, on tags, boxes, jars, or other containers. Labels for other fluxes shall conform to the requirements of ANSI Publication Z129.1.[11]

Material Safety Data Sheets (MSDS)

The suppliers of brazing materials shall provide Material Safety Data Sheets, or equivalent, which identify those hazardous materials present in their products. The MSDS shall be prepared and distributed to users in accordance with OSHA 29CFR 1910.1200.

A number of potentially hazardous materials may be present in fluxes, filler metals, coatings, and atmospheres used in brazing processes. When the fumes or gases form a product containing a component whose PEL or TLV will be exceeded before the general brazing fume TLV of 5 mg/m^3 is reached, the component shall be identified on the MSDS. These include, but are not limited to, the low PEL materials listed earlier.

FIRE PREVENTION AND PROTECTION

For detailed information on fire prevention and protection in brazing processes, NFPA Publication 51B[12] should be consulted.

Preferably brazing should be done in specially designated areas which have been designed and constructed to minimize fire

---

11. ANSI Z129.1, *Precautionary Labeling for Hazardous Industrial Chemical*, New York: American National Standards Institute, 1988.

12. ANSI/NFPA 51B, *Standard for Fire Prevention in Use of Cutting and Welding Processes*, National Fire Protection Association, 1977.

risk. No brazing shall be done unless the atmosphere is either nonflammable or unless combustion products (e.g., furnace atmospheres) are confined and protected to prevent fire hazards. (See Brazing Furnaces and Atmospheres above).

Sufficient fire extinguishing equipment shall be ready for use where brazing work is being done. The fire extinguishing equipment may be pails of water, buckets of sand, hose, portable extinguishers, or an automatic sprinkler system, depending upon the nature and quantity of the combustion material exposed. Before brazing is begun in a location not specifically designated for such purposes, inspection and authorization by a responsible person shall be required.

When brazing containers that have held flammable materials, there is the possibility of explosions, fires, and the release of toxic vapors. Do not begin work until fully familiar with American Welding Society Publication F4.1.[13]

HANDLING OF COMPRESSED GASES

Gas Cylinders And Containers

Gases used in the brazing industry are packaged in containers, usually called cylinders or tanks.[14] All portable cylinders used for storage and shipment of compressed gases shall be designed, constructed, and maintained in accordance with regulations of the U.S. Department of Transportation (DOT) Regulation 49 CFR.[15] Compliance will be indicated by markings on the cylinder, usually on the top shoulder, listing the applicable DOT specification number.

No one except the owner of the gas cylinder or persons authorized by the owner shall fill a cylinder. No person other than the authorized gas supplier shall mix gases in a cylinder or transfer gases from one cylinder to another.

Cylinders shall be stored, secured, and protected from falling, bumps, falling objects, combustibles, and the weather. Cylinders must never become part of an electrical circuit by being a ground connection. Oxygen cylinders should be stored away from fuel gas cylinders. Cylinders shall be kept in areas where temperatures do not exceed 130°F (54°C). Damaged or corroded cylinders shall not be used. (See ANSI/NFPA Publication 51[16] for detailed storage regulations).

Acetylene and liquefied gas cylinders shall be stored and used in the upright position. It is preferable that other gas cylinders also be stored upright.

Before using gas from a cylinder, the contents should be identified by the label thereon, legibly marked with either the chemical or trade name of the gas in accordance with ANSI/CGA Publication C-4.[17] Contents must never be identified by any other means such as cylinder color, banding, or shape, because these may vary among manufacturers, geographical area, or product line. If a label is missing from a cylinder, the contents should not be used. A valve protection cap is provided on all cylinders with a water weight capacity over 30 lb. (13.6Kg). The cap shall be completely threaded in place except when the cylinder is in use.

Withdrawal of Gas

Many gases in high pressure cylinders are filled to pressures of 2000 psig (13 800 kPa) or more. Unless the equipment is to be used with the gas is designed to withstand full cylinder pressure, an approved pressure-reducing regulator shall be used to withdraw gas from a cylinder or manifold. Simple needle valves never should be used. A pressure-relief or safety valve, rated to function at less than the maximum allowable pres-

---

13. AWS F4.1, *Recommended Safe Practices for the Preparation for Welding and Cutting of Containers That Have Held Hazardous Substances*, Miami: American Welding Society, 1988.

14. For additional information on compressed gases, refer to the Compressed Gas Association, Inc., *Handbook of Compressed Gases*, 2nd Ed., Arlington, VA: Van Nostrand Reinhold Co., 1981.

15. D.O.T. 49CFR, *Transportation*, Part 173, Washington, D.C.: U.S. Government Printing Office, 1987.

16. ANSI/NFPA 51, *Oxygen-Fuel Gas Systems for Welding, Cutting and Allied Processes*, Quincy, Mass: National Fire Protection Association, 1983.

17. ANSI Z48.1/CGA C-4 *Method of Marking Portable Compressed Gas Containers to Identify the Material Contained*, 3rd Ed., Arlington, VA: Compressed Gas Association, 1978.

sure of the brazing equipment, also should be employed. The function of the relief valve is to prevent failure of the equipment at pressures in excess of working limits if the regulator should fail in service.

Valves on cylinders containing high pressure gas, particularly oxygen, always should be opened slowly to avoid the high temperature of adiabatic decompression, which can occur if the valves are opened rapidly. In this case of oxygen, the intense heat can ignite the valve seat which, in turn, may cause the metal to melt or burn. The cylinder valve outlet should be pointed away from the operator and other persons when opening the valve to avoid injury should a fire occur.

Prior to connecting a gas cylinder to a pressure regulator or a manifold, the valve outlet should be cleaned of dirt, moisture, and other foreign matter by first wiping it with a clean, oil-free cloth. Then the valve should be opened momentarily and closed immediately. This is known as cracking the cylinder valve. Fuel gas cylinders must never be cracked near sources of ignition (i.e., spark and flames) while smoking or in confined spaces.

A regulator should be drained of gas pressure prior to connecting it to a cylinder and also after closing the cylinder valve upon shutdown of operation.[18] The outlet threads on cylinder valves are standardized for specific gases so that only regulators or manifolds with similar threads can be attached. Standard valve thread connections for gases normally used for welding, brazing, and allied processes are given in Table 6.1.

It is preferable not to open valves on low-pressure fuel gas cylinders more than one turn. This usually provides adequate flow and permits rapid closure of the cylinder valves in an emergency. High pressure cylinder valves, on the other hand, usually must be opened fully to backseat the packing and prevent leakage during use.

The cylinder valve should be closed after each use and when an empty cylinder is to be returned to the supplier. This prevents loss of product through undetected leaks

that might develop while the cylinder is unattached. It also avoids hazards that might be caused by leaks and prevents backflow of contaminants into the cylinder. It is advisable to return cylinders with at least 25 psig (40 psia) of the contents remaining. This prevents possible atmospheric contamination during shipment.

Pressure Relief Devices

Pressure relief devices on cylinders should not be tampered with. These are intended to provide protection in the event the cylinder is subjected to a hostile environment, usually fire or other source of heat. Such environments may raise the pressure within the cylinder. The safety devices are designed to prevent cylinder pressures from exceeding safe limits.

Regulators

A pressure-reducing regulator should always be used when withdrawing gas from a gas cylinder. A regulator shall be used only for the gas and pressure given on the label.

Regulators should not be used with other gases or at other pressures even though the valve outlet threads of cylinders may be the same. The threaded connections to the regulator must not be forced. Mismatch of threads between a gas cylinder and regulator, or between the regulator and hose connection indicates an improper regulator or hose connection for the gas.

Use of adaptors to change the cylinder connection thread is not recommended because of the danger of using an incorrect regulator or contaminating the regulator. For example, gases that are oil-contaminated can deposit an oily film on the internal parts of the regulator. This film can contaminate an oil-free gas or result in fire or explosion in the case of oxygen.

The threads and connection glands of regulators should be inspected for dirt or damage before use. If a hose or cylinder connection leaks, it should not be forced with excessive torque. Damaged regulators and components should be repaired by properly trained mechanics or returned to the manufacturer for repair.

A suitable valve or flowmeter should be used to control gas flow from a regulator.

18. Gas regulators should meet the requirements of the CGA E-4, *Standard for Gas Regulators for Welding and Cutting*, 2nd Ed., Arlington, VA: Compressed Gas Association Inc., 1987.

Table 6.1
Compressed Gas Association
Standard Connectors for Compressed Gas Cylinders for Brazing[a]

Gas	Cylinder Limitation	Standard Connection[b]	Limited Standard Connection[c]	Alternator Standard Connection[d]
Acetylene	Over 50 cu ft (1.39 m³)	510		
	Between 35 (970L) and 75 cu ft (2.08 m³)		300	
			520	
	Approx. 10 cu ft (2802)		200	
Air (R729)	Up to 3000 psig (20 680 KPa) Threaded	346	590	
	" " Yoke	950		
	3001-5500 psig (20 690-37 900 kPa)	347		
	5501-7500 psig (38 000-51 700 kPa)	702		
Argon	Up to 3000 psig (20 680 kPa)	580		
	3001-5500 psig (20 690-37 900 kPa)	680		
	5501-7500 psig (38 0000-51 700 kPa)	677		677
Butane (R600)		510		
Carbon Dioxide (R744)	Threaded	320		
	Yoke	940		
Helium	Up to 3000 psig (20 680 kPa) Threaded	580		
	" " Yoke	930		
	3001-5500 psig (20 690-37 900 kPa)	680		
	5501-75000 psig (38 000-51 700 kPa)	677		
Hydrogen	Up to 3000 psig (20 680 kPa)	350		
	3001-5500 psig (20 690-37 900 kPa)	695		677
	5501-7500 psig (38 000-51 700 kPa)	703		677
Methylacetylene-Propadiene (MPS)		510		

(Continued)

Table 6.1 (Continued)

Gas	Cylinder Limitation	Standard Connection[b]	Limited Standard Connection[c]	Alternate Standard Connection[d]
Natural Gas	Up to 3000 psig (20 680 kPa)	350		
	3001-5500 psig (20 690-37 900 kPa)	695		677
	5001-7500 psig (38 000-51 700 kPa)	703		677
Nitrogen	Up to 3000 psig (20 680 kPa) Threaded	580		
	" " Yoke	960		
	3001-5500 psig (20 690-37 900 kPa)	680		677
	5501-7500 psig (38 000-51 700 kPa)	677		
Oxygen	Up to 3000 psig (20 680 kPa) Threaded	540		
	" " Yoke	870		
	3001-4000 psig (20 690-27 580 kPa)	577		
	4001-5500 psig (27 590-37 900 kPa)	701		
Propane (R290)		510	600	
Propylene (R1270)		510	600	

a. ANSI/CGA V-1 "Standard for Compressed Gas Cylinder Valve Outlet and Inlet Connections", Arlington, VA: Compressed Gas Association, 1987.

b. A standard connection is the connection recommended for a particular gas.

c. A limited standard connection is one used at present, is considered safe, and is not expected to be discontinued.

d. An alternate standard connection is one that is scheduled to become obsolete on 1/1/92.

Manifolds

A manifold may be used when a gas is needed without interruption or at a delivery rate higher than that available from a single cylinder. A manifold must be suitable for the specific gas and operating pressure, and be leak-tight. The components should be approved for such purpose, and used only for the gas and pressure for which they are approved. Oxygen and fuel gas manifolds must meet specific design and safety requirements.[19]

Piping and fittings for acetylene and methylacetylene-propadiene (MPS) manifolds must not be pure copper. These fuel gases react with copper under certain conditions to form unstable copper acetylide. This compound may detonate under shock or heat. However, copper alloys containing less than 70 percent copper may be used in systems for these gases. Piping systems must contain an appropriate over-pressure relief valve. Each fuel gas cylinder lead should incorporate a backflow check valve and flash arrester. Backflow check valves must also be installed in each station outlet where both fuel gas and oxygen are provided for a welding, cutting, or preheating torch.

Piping System Safety Relief Valves

Unless a piping system is specifically designed and constructed to withstand the full pressure of the compressed gas source supplying it, the piping system must always be protected with sufficient safety pressure relief devices to prevent excessive pressure in the system that may rupture a component. Such pressure relief devices may be relief valves or bursting discs. A pressure reducing regulator must never be solely relied upon to prevent over-pressurization of the system. A pressure relief device must be placed in every part of the system which could be exposed to the full source supply pressure while isolated (blocked) from other protective relief devices (such as by a closed valve).

Some pressure regulators incorporate safety relief valves. These valves are designed for the protection of the regulator only and seldom are adequate to protect a downstream system.

In cryogenic piping systems, relief devices should be located in every section of the system where liquified gas may become trapped. Upon warming, such liquids can vaporize to gas at very high pressure. Pressure relief devices protecting fuel gas piping systems, or systems containing other hazardous gases, should be vented to safe locations.

Oxygen

Oxygen itself is nonflammable but it supports the combustion of flammable materials. It can initiate combustion and vigorously accelerate it. Oxygen cylinders and liquid containers should, therefore, not be stored in the vicinity of combustibles or with cylinders of fuel gas. Oxygen should never be used as a substitute for compressed air. Oxygen should always be referred to by name, never as air. Oil, grease, and combustible dusts may ignite spontaneously on contact with oxygen. Hence, all systems and apparatus for oxygen service must be kept free of these materials. Valves, piping, or system components that have not been expressly manufactured for oxygen service must be cleaned and approved for this service before use.[20]

Apparatus that has been manufactured and labeled expressly for oxygen service must be kept in the clean condition as originally received. Oxygen valves, regulators, and apparatus must never be lubricated with oil. If lubrication is required, is must be done only with lubricants that are compatible with oxygen.

Oxygen must never be used to power compressed air tools because these are almost always oil lubricated. Similarly, oxygen must not be used to blow dirt from work and clothing because they are often contaminated with oil or grease.

Only clean clothing should be worn when working with oxygen systems. Similarly, oxygen must not be used to ventilate confined spaces. Severe burns can result from ignition of clothing or the hair in an oxygen-rich atmosphere.

19. ANSI/NFPA 51, *Oxygen-Fuel Gas Systems for Welding, Cutting and Allied Processes*, Quincy, Mass.: National Fire Protection Association, 1983.

20. ANSI/NFPA 70, *National Electric Code*, Quincy, Mass: National Fire Protection Association, 1987.

Fuel Gases

Fuel gases commonly used in brazing are acetylene, methylacetylene-propadiene (MPS), natural gas, propane, and propylene. Hydrogen is used in a few applications. These gases should always be referred to by name.

Acetylene in cylinders is dissolved in a solvent to permit safe storage under pressure. In the free state, acetylene never should be used at pressures over 15 psig (30 psia). At pressures above 15 psig (30 psia), acetylene may dissociate with explosive violence. Acetylene and methylacetylene-propadiene (MPS) gases never should be used in contact with unalloyed silver or mercury, or, as stated previously, with unalloyed copper. These gases react with the former two metals in the same manner as with copper to form unstable compounds that may detonate under shock or heat. As mentioned before, the valves on fuel gas cylinders should never be cracked to clean the outlet near to possible sources of ignition or in confined spaces.

When used for a brazing furnace atmosphere, a fuel gas must be burned or vented to a safe location prior to filling the furnace or retort with the fuel gas. The equipment first shall be purged with a nonflammable gas, such as nitrogen or argon, to prevent formation of an explosive air fuel mixture.

Special attention shall be given when using hydrogen. Flames of hydrogen may be difficult to see and could pose a hazard to parts of the body, clothes, or other combustibles.

Fuel Gas Fires

The best procedure for avoiding fire from a gas or liquid fuel, is to keep it contained within the system, that is, prevent leaks. All fuel systems should be checked carefully for leaks upon assembly and at frequent intervals thereafter. Fuel gas cylinders should be examined for leaks, especially at fuse plugs, safety devices, and valve packings.

In the event of a fuel fire, one of the most effective means for controlling the fire is to shut off the fuel valve, if accessible. Fuel gas valves should not be opened beyond the point necessary to provide adequate flow while permitting rapid shutoff in an emergency. In most cases, this is less than one turn. If the immediate valve controlling the burning gas is inaccessible, another upstream valve often will be effective.

Most fuel gases in cylinders are in liquid form or dissolved in liquids. Therefore, the cylinders should always be used in the upright position to prevent liquid surges into the system. A fuel gas cylinder can develop a leak and sometimes result in a fire.

In case of fire, the fire alarm should be sounded and trained fire personnel summoned immediately. A small fire in the vicinity of a cylinder valve or a safety device should be extinguished, if possible, by closing the valve or by the use of water, wet cloths, or fire extinguishers. If the leak cannot be stopped, the cylinder should be removed by trained fire personnel to a safe outdoor location, and then the supplier should be notified. A warning sign should be posted, and no smoking or other ignition sources should be permitted in the area.

In the case of a large fire at a fuel gas cylinder, the fire alarm should be actuated, and all personnel evacuated from the area. The cylinder should be kept wet by fire personnel with a heavy stream of water to keep it cool. It usually is better to allow the fire to continue to burn and consume the issuing gas rather than attempt to extinguish the flame. If the fire is extinguished, there is danger that the escaping gas may reignite with explosive violence.

ELECTRICAL SAFETY

Electric Shock

Electric shock can cause sudden death. Injuries and fatalities from electric shock in brazing operations can occur if proper precautionary measures are not followed. Many brazing operations employ some type of electrical equipment. For example, automatic oxyfuel gas brazing machines use electric motor drives, controls, and systems. With proper training and equipment, electrical accidents can be avoided.

Shock Mechanism

Electric shock occurs when an electric current of sufficient magnitude to create an adverse effect passes through the body. The severity of the shock depends mainly on the amount of current, the duration of flow, the path of flow, and the state of health of the person. The amount of current depends upon the applied voltage and the resistance of the body path. The frequency of alternating current may also be a factor.

Shock currents greater than about 6 milliamperes (mA) are considered primary because they are capable of causing direct physiological harm. Steady state currents between 0.5 and 6 mA are considered secondary shock currents. Secondary shock currents are defined as those capable of causing involuntary muscular reactions without normally causing direct physiological harm. The 0.5 mA level is called the perception threshold because it is the point at which most people just begin to feel the tingle from the current. The level of current sensation varies with the weight of the individual and, to some extent, between men and women.

Shock Sources

Most electrical equipment can be a shock hazard if improperly used or maintained. Shock can occur from lightning induced voltage surges in power distribution systems. Even earth grounds can attain high potential relative to true ground during severe transient phenomenon. Such circumstances, however, are rare.

Most electric shock in the brazing industry occurs as the result of accidental contact with bare or poorly insulated conductors operating at 80 to 575 volts. Operators, therefore, must take precautions against contacting bare elements in the circuits.

Electrical resistance usually is reduced in the presence of water or moisture. Electrical hazards are more severe under such circumstances. Brazers must take special precautions to avoid shock when working under damp or wet conditions, including perspiration.

Wearers of Pacemakers

The technology of heart pacemakers and the extent to which they are influenced by other electrical devices constantly is changing. It is impossible to make a general statement concerning the possible effects of brazing operations on such devices. Wearers of pacemakers or other electronic equipment vital to life should check with the device manufacturer or their clinician to determine whether a hazard exists.

Equipment Selection

Electric shock hazards are minimized by (1) proper equipment installation and maintenance, (2) good operator practices, (3) proper operation clothing and body protection, and (4) the use of equipment designed for the job and situation. Equipment should meet applicable NEMA or ANSI standards, such as ANSI/UL551, *Safety Standard for Transformer Type Arc Welding Machines*, American National Standards Institute, New York, latest edition. If a significant amount of brazing work is to be done under electrically hazardous conditions, automatic machine controls that reduce the no-load (open circuit) voltage to a safe level are recommended.

Personnel Training

A good safety training program is essential. Operators must be fully instructed in electrical safety by a competent person before being allowed to commence operations. As a minimum, this training should include the points covered in ANSI/AWS Publication Z49.1. Persons should not operate electrical equipment until they have been trained properly.

Installation

Equipment should be installed in a clean, dry area. When this is not possible, it should be guarded adequately from dirt and moisture. Installation shall be made in accordance with the requirements of ANSI/NFPA Publication 70 and local codes. This includes necessary disconnects, fusing, and type of incoming power lines.

Terminals for leads, electrodes, and power cables must be shielded from accidental contact by personnel or by metal objects, such as vehicles and crane hooks. Connections between leads and power supplies may be guarded using (1) dead front construction and receptacles for plug connections, (2) terminals located in a recessed opening or under a non-removable hinged cover, (3) insulating sleeves, or (4) other equivalent mechanical means.

Grounding

The frame or chassis of all electrically powered machines must be connected to a good electrical ground. Grounding can be done by locating the machine on a grounded metal floor or platen, or by connecting it to a properly grounded building frame or other satisfactory ground. Chains, wire ropes, cranes, hoists, and elevators shall not be used as grounding connectors or to carry current.

Connectors for portable control devices, such as push buttons to be carried by the operator, must not be connected to circuits with operating voltages above about 120V. Exposed metal parts of portable control devices operating on circuits above 50V shall be grounded by a grounding conductor in the control cable. Controls using intrinsically safe voltages below 30V are recommended.

Cables and Connections

Electrical connections shall be tight, and checked periodically for tightness. Magnetic work clamps shall be free of adherent metal particles and spatter on contact surfaces. Coiled leads should be spread out before use to avoid overheating and damage to the insulation. Jobs alternately requiring long and short leads should be equipped with insulated cable connectors to permit idle lengths to be disconnected when not needed.

Equipment, cables, fuses, plugs, and receptacles shall be used within their current carrying and duty cycle capacities. Operation of apparatus above its current rating or duty cycle leads to overheating and rapid deterioration of insulation and other parts. Leads should be the flexible type of cable designed especially for the rigors of industrial service. Insulation on cables used with high voltages or high frequency oscillators shall provide adequate protection. Cable insulation shall be kept in good condition, and cables repaired or replaced promptly, when necessary.

Operations

Brazers should not allow the metal parts of electrodes, electrode holders, or torches to touch their bare skin or any wet covering of the body. Dry gloves in good condition shall be worn. The insulation on electrode holders shall be kept in good repair. Electrode holders shall not be cooled by immersion in water. If water cooled-resistance brazing guns or holders are used, they should be free of water leaks and condensation that would adversely affect the brazer's safety.

When brazing is to be performed under damp or wet conditions, including heavy perspiration, the worker must wear dry gloves and clothing in good condition to prevent electric shock. As a minimum, the brazer should be protected from electrically conductive surfaces, including the earth, by rubber soled shoes. Preferably an insulating layer such as a rubber mat or wooden board should be used. Rings and jewelry should be removed before brazing to decrease the possibility of electric shock.

Modification and Maintenance

Electrical equipment should be inspected frequently and kept dust and dirt free to ensure efficient safe operation. Electrical coil ventilating ducts should be similarly inspected. It is good practice to occasionally blow out the equipment with clean, dry compressed air at low pressure using adequate safety precautions. Air filters in the ventilating systems of electrical components are not recommended unless provided by the manufacturer or the equipment. The reduction of air flow resulting from the use of an air filter on equipment not so designed can subject internal components to an overheating condition and subsequent failure. Machines that have become wet should be dried thoroughly and properly retested before being used.

Prevention of Fires

Fires resulting from electrical brazing equipment generally are caused by overheating of electrical components, flying sparks or spatter from the brazing. Most precautions against electrical shock also are applicable to the prevention of fires caused by overheating of equipment.

BRAZING PROCESSES AND EQUIPMENT SAFETY

All brazing equipment shall meet product specifications and approval criteria of the jurisdictional authority responsible for such equipment. The equipment shall be kept in good working condition, inspected as necessary to be sure it is in good working condition. When equipment is found to be defective (incapable of reliable, safe operation), it shall be promptly withdrawn from service, or repaired by qualified personnel.

All equipment shall be operated in accordance with the manufacturer's recommendation and instructions, provided they are consistent with ANSI/AWS Publication Z49.1.

OXYFUEL GAS BRAZING SAFETY

Hazards associated with oxyfuel gas brazing include: (1) injury from improper selection and use of equipment and consumable materials, (2) inhalation of toxic fumes and gases emitted during the brazing process, (3) exposure to heat rays from the flame or hot metal, and (4) fire from contact of the flame or metal spatter with combustible materials. Electric shock is a potential hazard from equipment used in automated oxyfuel gas brazing. Items 2, 3, 4, and electric shock have been discussed in detail earlier in this chapter. The major equipment items involved in oxyfuel gas brazing are torches, hoses, and hose connections, flashback arresters, and gas cylinder equipment (discussed earlier).

Torches

All welding and cutting torches should be of an approved type[21]. They should be kept in good working order, and serviced at regular intervals by the manufacturer or qualified technicians. A torch must be used only with the fuel gas for which it is designated. The fuel gas and oxygen pressures should be those recommended by the torch manufacturer.

The procedure recommended by the manufacturer should be followed when lighting and extinguishing the torch. The torch should be lighted only with a friction lighter, pilot light, or similar ignition source. Matches or cigarette lighters should not be used.

Hoses

Hoses used should be only those specified for oxyfuel gas brazing systems. Generally,

these hoses are manufactured in accordance with RMA/CGA Publication IP-7[22] The fuel gas hose is usually red with left-hand threaded fittings. A green hose with right-hand threaded fittings is generally used for oxygen.[23] Hoses should be free of oil and grease, and in good condition. When parallel lengths are taped together for convenience, no more than 4 in. (101.6 mm) of any 12 in. (304.8 mm) section of hose should be covered.

Only proper ferrules and clamps should be used to secure hose to fittings. Long runs of hose should be avoided. Excess hose should be coiled to prevent kinks and tangles, but it should not be wrapped around cylinders or cylinder carts while in use.

Backfire and Flashback

A backfire during brazing is a momentary retrogression of the flame back into the tip. It usually results in a momentary flame-out followed by reignition of the normal tip flame, and is accompanied by a pop or bang, depending upon the size of tip. In severe cases, however, hot products of combustion within the tip may be forced back into the torch and even the hoses. Occasionally, such backfires ignite the inner liner of hoses (especially oxygen), resulting in the hoses burning through. Such backfires can cause injury.

A flashback is an occurrence initiated by a backfire where the flame continues to burn inside the equipment instead of being re-established at the tip. Flashbacks result in very rapid internal heating of the equipment, and can destroy it quickly. A flashback usually is recognized by a whistling or squealing sound. The equipment will heat up rapidly and sparks may issue from the tip. The flashback should be extinguished by turning off the torch valves as quickly as possible. Different manufacturers may recommend shutting off either the fuel or oxygen first, but the most important concern is to get both valves closed quickly.

21. CGA E-5, *Torch Standard for Welding and Cutting*, 1st Ed., Arlington, VA: Compressed Gas Association, 1977.

22. RMA/CGA IP7, *Specification for Rubber Welding Hose*, Washington, DC: Compressed Gas Association, 1982.

23. CGA E-1, *Standard Connections for Regulator Outlets, Torches and Fitted Hose for Welding and Cutting Equipment*, 2nd Ed.; Arlington, VA: Compressed Gas Association, 1980.

Backfires and flashbacks ordinarily are not a concern if the apparatus is operated in accordance with the manufacturer's instructions. Generally, they occur from allowing a tip to become overheated, such as by flame backwash, forcing the tip into the work, or providing insufficient gas flow for the size of the tip in use. If frequent backfiring or flashbacks are experienced, the cause should be investigated. There is probably something wrong with the equipment or operation.

Hose lines should be purged before lighting oxyfuel gas equipment to prevent backfires or flashbacks. Purging flushes out any combustible mixtures of fuel and oxygen or fuel and air in the hoses. It is done by first opening either the fuel or oxygen valve on the torch and allowing that gas to flow for several seconds to clear that hose of any possible gas mixtures. That valve is then closed, and the other valve is opened to allow the other gas to flow for a similar period of time. Purging should always be done before any brazing tip is lighted. The purge stream should not be directed towards any flame or source of ignition. Torches should not be purged in confined spaces because of possible explosion of accumulated gases.

Hose Line Safety Devices

Reverse flow check valves and flashback arresters for hose line service are available. These devices can prevent backflow of gases and flashbacks into hoses provided they are operating properly. They should be used strictly in accordance with the manufacturer's instructions, and maintained regularly in accordance with the manufacturer's recommendations.

Shutdown Procedures

When oxyfuel gas operations are completed the equipment should always be completely shut down and the gas pressures drained from the system. The fuel gas line should be closed first to avoid flashback. Cylinder or supply valves should then be closed, fuel gas first. Finally, drain the fuel gas line, then the oxygen line by opening, draining, and closing the torch and regulator valves. The equipment should not be left unattended until shutdown has been completed.

Oxyfuel gas cylinders or equipment connected to cylinders shall always be stored in well ventilated spaces; never in confined areas or unventilated cabinets. Even small gas leaks in confined spaces can result in explosive mixtures that might be set off with disastrous results. For the same reason, oxyfuel gas cylinders should never be transported in enclosed vehicles, particularly not in closed vans or the trunks of automobiles.

Consumables

Consumables such as filler metals, fluxes, fuel gas and oxygen should be selected and used in a way to provide safe working conditions. Filler metals containing cadmium shall be used only where proper ventilation or respirator protection is provided as defined in the ventilation section above. Likewise, proper ventilation or respirators must be provided to prevent overexposure to zinc oxide fumes (from filler metals), oxides of nitrogen and carbon monoxide (from combusted gases) and fluoride gases (from fluxes). Some combinations of filler metals or of fluxes and gases are incompatible. For instance, filler metals containing manganese often will spatter when exposed to an oxyacetylene flame. Fluxes containing high sodium content will produce a yellow-orange glare, and fluxes with high magnesium impurity will cause spatter when exposed to the flame.

INDUCTION BRAZING SAFETY

The potential hazards involved with the induction brazing process are (1) fumes and gases, (2) electric shock, (3) burns from infrared radiation and spatter, and (4) fire. These hazards have been discussed previously. Precautions to avoid injury or death from such hazards should be followed. Precautionary label information, such as in Figures 6.1 through 6.4 and instructions for the safe operation of induction brazing equipment follow.

Warning

Voltages within this equipment are high enough to endanger human life. Safety interlock switches are present to disconnect the plate power when the generator panel is removed and to safeguard operator and maintenance personnel. *Bypassing or tam-*

pering with the generator cover interlock switch is prohibited. Disconnect all power before changing tubes or making any internal adjustments.

Safety Instructions

Read the manufacturer's instructions before unpacking the generator. This generator has been equipped with adequate features for your protection. However, due to the presence of dangerously high voltages within the cabinet, adjustment and repairs should only be performed by a competent and experienced technician. The following precautions should be rigidly adhered to:

• Danger — Do not operate generator without cover in place - high voltage inside.
• Do not touch the load coil or load coil leads when the power is on. The high frequency voltages can produce arcs causing severe burns. Guards must be in place to prevent contact. Note: Where long load coil lead lengths are required, coaxial lines are available to meet both safety and electrical requirements.
• Do not touch the workpiece during heating or until sufficiently cooled after heating as severe burns can result.
• Do not block sides and rear of generator cabinet. Air intake and exhaust openings are located at sides and rear of generator cabinet.
• Do not operate generator if blower is not functioning.
• Do not operate generator with a work coil having a no-load current reading on the plate meter (located on front panel) exceeding 1.0 amperes.
• Do not alter time setting while timer is operating.
• Do not place overload breaker in the "on" position without a work coil secured in place.
• Do not apply power to new tubes until after proper "break-in" period.
• Do not operate generator without air filters.
• Do not operate generator if wearing a heart pacemaker. Proof is *not* available that the pacemaker will *not* be affected.
• Do not service generator unless the safety interlock switch is engaged and the plate power is off. Due to the hazards associated with AC circuits, it is recommended that all AC power service be disconnected from the equipment during servicing.
• Do not service generator until capacitors have been discharged, as residual charges will remain in the capacitors for some time after shut-down of the equipment.
• Do not steam clean or wash the generator equipment. Follow the equipment manufacturer's cleaning instructions.

RESISTANCE BRAZING SAFETY

The main hazards of the resistance brazing process and equipment are

(1) Electric shock from contact with high voltage terminals of components
(2) Ejection of small particles of molten metal from the braze
(3) Crushing of some part of the body between the electrodes or other moving components of the machine
(4) Fumes and gases from the consumables or base metal surface coatings

Mechanical Considerations

Initiating devices on resistance brazing equipment, such as push buttons and switches, must be positioned or guarded to prevent the operator from inadvertently activating them. With some machines, the operator's hands can be expected to pass under the point of operation during loading and unloading. These machines must be effectively guarded by proximity-sensing devices, latches, blocks, barriers, dual hand controls, or similar accessories that prevent (1) the hands from passing under the point of operation or (2) the ram from moving while the hands are under the point of operation.

The support system of suspended portable brazing equipment, with the exception of the gun assembly, must be capable of withstanding the total mechanical shock load in the event of failure of any component of the system. Devices, such as cables, chains, or clamps, are considered satisfactory.

Guarding should be provided around the mounting and actuating mechanism of the movable arm of a gun if it can cause injury to the operator's hands. If suitable guarding cannot be achieved, the gun should have two handles and operating switches that must be actuated to energize the machine.

One or more emergency stop buttons should be provided on all brazing machines, with a minimum of one at each operator position. If the brazer controls the flow of current by some type of switch (e.g., a hand grip, foot pedal, push button, etc.) it should be spring loaded to return to the OFF position when released. In other words, the equipment must be configured to require the brazer to take positive action to initiate current flow; current flow must cease when the brazer ceases the positive action. In addition, there should be a master switch for shutting off all power to the transformer.

Similarly, the electrodes that contact the braze joint components shall be spring loaded to separate from the components if the brazer interrupts the brazing cycle.

Eye protection against expelled metal particles shall be provided by a guard of suitable fire-resistant material or by the use of approved personal protective eye wear.

Electrical Considerations

All external braze initiating control circuits should operate at a maximum of about 120VAC for stationary equipment and about 36VAC for portable equipment.

Resistance brazing equipment containing high voltage capacitors shall have adequate electrical insulation and be enclosed completely. All enclosure doors shall be provided with suitable interlocks that are wired into the circuit. The interlocks shall effectively interrupt power and discharge all high voltage capacitors when the door or panel is opened. In addition, a manually operated switch or suitable positive device should be provided to assure complete discharge of all high voltage capacitors. The doors or panels should be kept locked except during maintenance.

If cables are used to carry alternating current (AC) from the transformer to the brazing electrodes, these cables shall be positioned (permanently, if possible) as close to each other as practical. Close positioning of AC cables minimizes circuit impedance, thus insuring that the circuit's protective device will trip as required if there is an electrical fault. In addition, close positioning cancels the alternating magnetic field around the cables. This alternating magnetic field would heat any metal object in the vicinity; if hot metal touched the cable, it would damage the cable insulation.

The power transformer must be grounded and fused properly. Grounding prevents dangerous electrical shocks to personnel in the event of an electrical short. If there is a surge of current for any reason, an adequate fuse or circuit breaker will guarantee rapid shutdown of the equipment.

The brazing transformer secondary circuit should be grounded by one of the following methods:

(1) Permanent grounding
(2) Connection of a grounding reactor across the secondary winding with a reactor tap to ground

As an alternative on stationary machines, an isolation contactor may be used to open all of the primary lines.

Installation and Maintenance

All equipment should be installed in conformance with the latest edition of ANSI/NFPA Publication 70, *National Electric Code*. The equipment should be installed and inspected by qualified personnel. At frequent intervals all current-carrying parts of the brazing equipment should be inspected visually for wear, damage, or any other improper condition. For example, if the equipment is used during three shifts, it should be inspected at the beginning of each shift.

FURNACE BRAZING SAFETY

Potential hazards in the operation of brazing furnaces are

(1) Asphyxiation of personnel entering or working in adjacent areas where there is insufficient oxygen in the atmosphere to support life
(2) Development of explosive mixtures of flammable gas and air within the furnace during purging or venting of atmosphere
(3) Accumulation of hazardous fumes or gases in the work area due to the brazing process

A large number of brazing furnaces of many different designs and types are in use that employ protective brazing atmospheres. There also are a variety of protective atmospheres that can be used for brazing. Some of these atmospheres are flammable, and none will support life. Adequate precautions are required. For instances, where a brazing furnace atmosphere is a flammable gas or the combustion products of a flammable gas, procedures shall be followed to assure that an explosive mixture of the atmosphere and air is not produced inside, or outside the furnace. Atmosphere from within brazing furnaces shall be exhausted to a location where it will not expose personnel to hazard.

Furnaces

The furnace manufacturer designs a furnace with a specific atmosphere in mind, and thus develops a specific procedure to purge air from the furnace and to remove the atmosphere. It is essential, therefore, to contact the equipment manufacturer for the recommended purging procedure. The various types of brazing furnaces generally use gas or vacuum atmospheres.

Gas atmosphere furnaces are generally of two types, batch or continuous. Both types essentially can use any of the gas atmospheres, whether they be inert or combustible. For a more detailed discussion of brazing furnaces and atmospheres and the safety procedures and equipment required refer to ANSI/NFPA Publications 86C[24] and 86D.[25]

Purging

Oxygen and air must be removed from protective brazing atmospheres to insure a safe effective furnace brazing operation.

Gas Atmosphere Furnaces. The furnace or retort that will be filled with a flammable brazing atmosphere, such as AWS brazing atmospheres Numbers two through seven, shall be purged to remove oxygen and air by one of the following methods:

---

24. ANSI/NFPA 86C, *Industrial Furnaces Using A Special Processing Atmosphere*, Quincy, Mass: National Fire Protection Association, 1985.
25. ANSI/NFPA 86D, *Industrial Furnaces Using Vacuum as an Atmosphere*, Quincy, Mass: National Fire Protection Association, 1985.

(1) Below 1400°F (760°C), purging shall reduce the oxygen below the explosive level by one of the following methods:

(a) Evacuate to a pressure below 5mm of mercury absolete (660Pa) and backfill to working pressure with the flammable atmosphere.

(b) Purge with a neutral or inert atmosphere for a minimum of four volumes. Then test to ensure a safe oxygen level, and subsequently purge with a minimum of four volumes of the flammable atmosphere while at room temperature.

Purging is usually done at a temperature of under 200°F (93°C). Argon, being heavier than air, nitrogen, or nitrogen containing atmospheres, should be piped into the container at the bottom and exhausted at the top. Similarly, hydrogen and hydrogen-containing atmospheres, being lighter than air, should be piped into the container at the top with the exhausted at the bottom.

(2) Above 1400°F (760°C), heated furnaces or retorts may be purged with flammable gas atmospheres that enter the 1400°F (760°C) chamber area. Under these conditions, the air will be progressively consumed, thus eliminating an explosive condition.

In all cases, adequate area ventilation shall be provided to exhaust and discharge all explosive or toxic gases which may emanate from such furnace purging and brazing operations to a safe location. When a flammable atmosphere is burned completely in, or at, the furnace during the heating cycle, the exhaust requirement may diminish.

Vacuum Atmosphere Furnaces. On occasion, combustible gases such as hydrogen may be introduced into a vacuum brazing furnace. When using combustible gases, it is essential that the gas ballast valves on mechanical vacuum pumps be plugged or connected to a source of inert gas. The operating procedure must be reviewed with individual pump and furnace manufacturer.

Furnace Maintenance or Shut Down

Prior to the start-up of a cycle, or at the end of a cycle, it is often necessary to adjust parts in the load or put thermocouples on parts. This can be particularly hazardous in a pit furnace if the argon or nitrogen has not been adequately removed at the end of the brazing cycle. A leaking or improperly closed gas valve also could present a hazard of asphyxiation. Before anyone leans into, or enters, a furnace for any reason, all residual protective atmosphere shall be removed. This can be accomplished as follows:

(1) For a pit furnace with a heavier than air atmosphere, insert a flexible tube from the suction side of a blower into the bottom of the pit.

(2) For a pit furnace with a lighter than air atmosphere, repeat procedure A except insert the tube at the top of the pit furnace.

(3) For batch horizontal or continuous furnaces, open all doors and put a large fan at one end.

In all cases, it is essential to assure a continuing source of fresh air during the time that personnel are in the furnace. The air flow should be kept running at all times when someone is in the furnace during shutdown to assure that there is no seepage of atmosphere into the furnace. All atmosphere valves should be closed and padlocked in the closed position.

Care should be exercised to use nonsparking tools, and to avoid ignition conditions during interior maintenance or cleaning of vacuum atmosphere furnaces that are involved in brazing with aluminum filler metals containing magnesium. Magnesium surface deposits, present as condensate from the brazing process, are extremely flammable. For detailed information on the processing and handling of magnesium refer to ANSI/NFPA Publication 480-1987.[26]

Specialized Equipment and Newer Equipment And Processing Procedures

Recent advancements in furnaces and atmospheres may require deviations from the requirements noted above. Such deviations shall be permissible when recommended by the furnace manufacturer or atmosphere source, provided such techniques are demonstrated to be safe operations and to comply with recognized industry regulations,

such as ANSI/NFPA 86C and ANSI/NFPA 86D.

DIP BRAZING SAFETY

The principal hazards encountered in salt bath or molten filler metal dip brazing processes include (1) fumes and gases from brazing consumables, (2) burns or fire from spatter or contact with molten salt or filler metal, (3) gases from vaporized salt, and (4) electric shock. These hazards are discussed in the sections on protection of personnel and electrical safety.

In dip brazing, the parts to be immersed in the bath must be completely dry to prevent spatter of the salt or molten filler metal and possible burns or fire. The presence of moisture on the parts, e.g., from an insufficiently dried flux or paste, will cause an instantaneous generation of steam within the bath that may expel the contents of the bath with explosive force and create a serious burn hazard. Predrying of the parts, commonly done by preheating in an air furnace, will prevent this problem. The filler metal or salt supplier should be contacted for recommended drying procedures. If supplementary flux is necessary, it must be adequately dried to remove not only attached moisture but also water of hydration to avoid explosion hazards.

REFERENCE AND SUPPLEMENTARY READING

(1) American Conference of Governmental Industrial Hygienists (ACV4GIH). *Threshold limit values for chemical substances and physical agents in the working environment.* Cincinnati: American Conference of Government Industrial Hygienists.

(2) American National Standards Institute (ANSI). ANSI/AWS Z49.1, *Safety in Welding and Cutting* (available from the American Welding Society). Miami: American Welding Society, 1988.

(3) ———. ANSI Z535.2, *Environment and facility safety signs.* New York, NY: American National Standards Institute, 1988.

26. ANSI/NFPA 480-1987, *Storage and Processing of Magnesium*, Quincy, MA: National Fire Protection Association, 1987.

(4) ———. ANSI Z87.1, *Practice for occupational and educational eye and face protection.* New York, NY: American National Standards Institute, 1979.

(5) ———. ANSI Z88.2, *Practices for respiratory protection.* New York, NY: American National Standards Institute, 1980.

(6) ———. ANSI Z88.6, *Physical qualifications for respiratory use.* New York, NY: American National Standards Institute, 1984.

(7) ———. ANSI Z117.1, *Safety requirements for working in tanks and other confined spaces.* New York, NY: American National Standards Institute, 1977.

(8) ———. ANSI Z535.4, *Product safety signs and labels.* New York, NY: American National Standards Institute, 1988.

(9) ———. ANSI Z129.1, *Precautionary labeling for hazardous industrial chemical.* New York, NY: American National Standards Institute, 1987.

(10) American Society for Testing and Materials (ASTM). ASTM D22.04, *Sampling and analysis of industrial atmosphere.* Philadelphia, Pennsylvania: American Society for Testing and Materials, 1988.

(11) ASM International. *Metals handbook*, Vol. 4, 389-416. Materials Park, Ohio: ASM, 1981.

(12) American Welding Society (AWS). F4.1, *Recommended safe practices for the preparation for welding and cutting of containers that have held hazardous substances.* Miami: American Welding Society.

(13) ———. *Effects of welding on health I, II, III, IV.* Miami: American Welding Society, (1979, 1981, 1983)

(14) ———. *Fumes and gases in the welding environment.* Miami: American Welding Society, 1979.

(15) ———. *The welding environment.* Miami: American Welding Society, 1973.

(16) ———. ANSI/AWS F1.3, *Evaluating contaminants in the welding environment; a sampling strategy guide.* Miami: American Welding Society, 1985.

(17) ———. ANSI/AWS F1.1, *Method for sampling airborne particles generated in welding and allied procedures.* Miami: American Welding Society, 1985.

(18) ———. ANSI/AWS F1.2, *Laboratory method for measuring fume generation rates and total fume emission of welding and allied processes.* Miami: American Welding Society, 1985.

(19) ———. ANSI/AWS F1.4-87, *Methods for analysis of airborne particulates generated by welding and allied processes.* Miami: American Welding Society, 1986.

(20) ———. ANSI/AWS F1.5-87, *Methods for sampling and analyzing gases from welding and allied processes.* Miami: American Welding Society, 1987.

(21) Balchin, N. C., *Health and safety in welding and allied processes*, 3rd Edition. England: The Welding Institute, 1983.

(22) Barthold, L. O., *Electrostatic effects of overhead transmission lines, Part I — Hazards and effects.* IEEE Transactions, Power Apparatus and Systems, Vol. PAS-91, pp. 422-444. 1972.

(23) Cadmium Council, Inc. *Using cadmium safely.* New York, NY: Cadmium Council, Inc., 1986.

(24) Compressed Gas Association, Inc. *Handbook of compressed gases*, 2nd Edition. Arlington, Virginia: Van Nostrand Reinhold Co., 1981.

(25) ———. CGA E-4, *Standard for gas regulators for welding and cutting.* Arlington, Virginia: Van Nostrand Reinhold Co., 1980.

(26) ———. ANSI/CGA mG-7.1, *Commodity specification for air.* Arlington, Virginia: Van Nostrand Reinhold Co., 1973.

(27) ———. ANSI Z48.1/CGA C-4, *Method of working portable compressed gas containers to identify the material contained*, 34d Edition. Arlington, Virginia: Van Nostrand Reinhold Co., 1978.

(28) ———. ANSI/CGA V-1, *Standard for compressed gas cylinder valve outlet and inlet connections.* Arlington, Virginia: Van Nostrand Reinhold Co., 1980.

(29) ———. CGA G-4.1, *Cleaning equipment for oxygen service*, 3rd Edition. Arlington, Virginia: Van Nostrand Reinhold Co., 1985.

(30) ———. CGA E-5, *Torch standard for welding and cutting*, 1st Edition. Arlington, Virginia: Van Nostrand Reinhold Co., 1980.

(31) ———. CGA E-1, *Standard connections for regulator outlets, torches, and fitted hose for welding and cutting equipment*, 2nd Edition. Arlington, Virginia: Van Nostrand Reinhold Co., 1980.

(32) ———. CGA — SB-4, *Handling acetylene cylinders in fire situations*, Arlington, Virginia: Van Nostrand Reinhold Co., 1972 (Reaffirmed 1980).

(33) ———. CGA P-1, *Safe handling of compressed gases in containers*, 7th Edition. Arlington, Virginia: Van Nostrand Reinhold Co., 1984.

(34) Dalziel, Charles F., Effects of electric current on man. *ASEE Journal*, 18-23, June 1973.

(35) Department of Transportation (D.O.T). 49 CFR, *Transportation, Part 173*. Washington, DCJ: U.S. Government Printing Office, 1987.

(36) Fireman's Fund Insurance Companies. *Welding fume control with mechanical ventilation*, 2nd Edition. San Francisco, California: Fireman's Fund Insurance Companies, 1981.

(37) National Institute for Occupational Safety and Health (NIOSH). *Criteria for a recommended standard . . . .occupational exposure to industrial fluorides*. Washington, DC: National Institute for Occupational Safety and Health, 1976.

(38) ———. *Criteria for a recommended standard . . . occupational exposure to cadmium*, 1976.

(39) *Criteria for a recommended standard . . . occupational exposure to oxides of nitrogen*, 1976.

(40) ———. *Criteria document-working in confined spaces*. Cincinnati, Ohio: NIOSH Publications Dissemination, 1979.

(41) ———. *Workers guide to confined spaces*, 1979.

(42) ———. ANSI/NFPA 51B, *Cutting and welding processes*. Quincy, MA: National Fire Protection Association, 1977.

(43) ———. ANSI/NFPA 51, *Oxygen-fuel gas systems for welding, cutting, and allied processes*. Quincy, MA: National Fire Protection Association, 1983.

(44) ———. ANSI/NFPA 70, *National electric code*. Quincy, MA: National Fire Protection Association, 1987.

(45) ———. ANSI/NFPA 86C, *Industrial furnaces using a special processing atmosphere*. Quincy, MA: National Fire Protection Association, 1984.

(46) ———. ANSI/NFPA 86D, *industrial furnaces using vacuum as an atmosphere*. Quincy, MA: National Fire Protection Association, 1985.

(47) ———. OSHA *Safety and Health Standards*, 29 CFR 1910, Subpart I, Sect. 132-140, Personal protective equipment; Subpart Q, Sect. 251-254, Welding, Cutting and Brazing; Subpart Z, Sect. 1000, Air contaminants; and Sect. 12200, Hazard Communications Standard. Washington, DC: U.S. Government Printing Office, 1987.

(48) Welding Institute. *The facts about fume*. England: The Welding Institute, 1976.

Keywords — precleaning, surface preparation, chemical cleaning, mechanical cleaning, thermal treatments, braze flow inhibitors, stop-off materials

Chapter 7

PRECLEANING AND SURFACE PREPARATION

Clean surfaces are essential to the formation of sound brazed joints of uniform quality. Capillary attraction is enhanced when the various components and filler metal are free of grease, oil, dirt, and scale prior to brazing. While fluxes and atmospheres can provide some cleaning, it is not their primary purpose. They should not be used as a substitute for precleaning operations.

Fluxes and brazing atmospheres are used to:

(1) Remove, penetrate, or prevent the formation of oxides at the joint during preheating and brazing

(2) Reduce the surface tension of the filler metal

(3) Form a protective covering over solidifying brazed joints

Their effectiveness in removing existing oxides in addition to their primary functions may be inadequate (see Chapter 4).

The length of time that cleaning remains effective depends on the metals involved, the atmospheric conditions, the amount and method of handling the parts, storage conditions, and similar factors. Cleaning techniques for specific materials are specified in the chapters on those particular base materials. It is recommended that brazing be accomplished as soon as possible after the material has been cleaned. Maintaining part flow (first in — first out) is essential, particularly for atmosphere or vacuum brazing processes. Protection from contamination during storage is equally important. Maximum storage time and conditions must be determined for each product.

The choice of a cleaning process depends on:

(1) Nature of the contaminant

(2) Specific base metal to be cleaned

(3) Degree of cleanliness required for brazing

(4) Part configuration

In general, cleaning processes are employed to remove substances in the following order:

(1) Pigmented drawing and forming compounds

(2) Oils, greases and waxes

(3) Scales, oxides and smut

(4) Abrasive and grinding residues

Pigmented compounds and some cutting and machining oils are difficult to remove. They require multiple cleaning methods such as hot emulsion sprays, hot alkaline soaks and hot rinses. Elapsed time between the application of these compounds and the cleaning process should be kept as short as possible to reduce the difficulties in cleaning.

CHEMICAL CLEANING

Chemical surface cleaning methods range from simple manual immersions to complex multi-stage automatic operations. Nine chemical cleaning methods and materials commonly used are:

(1) Solvent cleaning in a soak or spray operation with petroleum or chlorinated solvents for removing oil and grease

(2) Vapor degreasing operations with chlorinated and trichlorotrifluoroethane solvents that clean by soaking and condensation of the hot vapor on the work, for removing oil and grease

(3) Emulsion cleaning with mixtures of water, hydrocarbons, fatty acids and wetting agents, for general soil removal

(4) Acid cleaning with phosphate type acid cleaners

(5) Detergent cleaning with buffering salts, wetting agents and/or soaps, for general soil removal

(6) Alkaline cleaning with commercial mixtures of silicates, phosphates, carbonates, wetting agents and, in some cases, hydroxides, for light oxide removal from materials such as stainless steel, nickel base, and cobalt base alloys

(7) Electrolytic cleaning (anodic, cathodic and periodic reversal) for cleaning such as removal of tarnish, or for metal coating removal when base material is endangered by acid dipping

(8) Acid dipping and pickling with mixtures of mineral acids and acid salts for oxide removal and light metal removal

(9) Molten salt bath pickling, both electrolytic and nonelectrolytic, for heavy scale removal

Mechanical action is an extremely important factor in removing soil and increases the speed and efficiency of the cleaning process. It can be accomplished by stirring or agitating the cleaning solution with air, mechanical stirring, circulation, ultrasonic agitation, or spray. The selection of the chemical cleaning agent depends upon the nature of the contaminant, the base metal, the surface condition, and the joint design. Many commercial products are available for use with the various cleaning methods. Regardless of the cleaning agent or the method used, it is very important that all residue or surface films be removed from the cleaned parts by adequate rinsing to prevent the formation of other equally undesirable films on the faying surfaces.

MECHANICAL CLEANING

Mechanical cleaning methods, such as grinding, filing, machining, blasting, and wire brushing, are also used to remove objectionable surface conditions and roughen faying surfaces in preparation for brazing. If a power-driven wire wheel is used, care is needed to prevent burnishing. Burnishing can result in surface oxide embedment which interferes with the proper wetting of the base metal by the filler metal. Furthermore, bristle selection must be compatible with the base metal.

When rolling, fine grinding, or lapping has produced a base metal surface that is too smooth, the filler metal may not effectively wet the faying surfaces. In this case, the parts can be roughened slightly by mechanical abrasion for improved wetting.

Blasting techniques are used on the faying surfaces of parts to be brazed in order to remove any surface oxides. Blasting also roughens the mating surfaces to increase capillary attraction of the brazing filler metal. The blasting material must be clean and should not leave a deposit on the surfaces to be joined that will impair brazing or restrict filler metal flow. The materials should be fragmented rather than spherical, thus lightly roughening rather than peening the blasted part. The operation should be performed in such a way that delicate parts are not distorted or otherwise harmed. In addition the size of the fragmented grit and the intensity of the carrier media should be given due consideration as fatigue life may be impaired.

Types of blasting material are

(1) Chilled cast iron fragmented shot or hardened steel fragmented shot

(2) Stainless steel grits or powders

(3) Modified nickel-base braze filler metal grit

(4) Glass beads

(5) Silicon carbide

(6) Alumina, zirconia, sand

The use of these materials should be seriously considered, especially as they relate to the base materials and application of the brazed assemblies. Caution must be used with these materials due to the fact they may embed in base metals and create difficulty in wetting the surface with braze filler alloys. Wet blasting is acceptable using grits from the above listing, but proper cleaning and drying of parts is required to prevent surface contamination.

THERMAL TREATMENTS

Thermal treatments can be utilized to clean surfaces that are difficult to clean by other methods. Parts can be thermally cycled to temperatures near or above the final braze temperature to reduce oxides and remove contaminates by outgassing. Chapter 4 provides specific information on oxide reduction.

PRECOATINGS AND FINISHES

For some applications, parts to be brazed are precoated by electrodeposition, hot dip coating, flame spraying, and cladding methods. Precoatings and finishes frequently are used to insure wetting and flow on base metals containing constituents, such as aluminum, titanium, or other additions, that form stable oxides during heating and are therefore difficult to wet with most filler metals. Precoatings also protect clean surfaces and prevent the formation of oxides on base metals in storage and during the heating process. When brazing dissimilar metals, precoatings can reduce the tendency of a filler metal to wet and flow preferentially on one of the base metals. Occasionally, precoatings are used on refractory metals to prevent the rapid diffusion of filler metal constituents into the base metal and the subsequent formation of brittle intermetallic compounds.

The selection and thickness of the precoating depends on the base metal, filler metal, and brazing technique. Electroplatings of copper on steel and low-stress nickel on stainless steels are coatings that are often used. Brush plating is the technique used to apply electroplates locally at the joint areas. To prevent blistering of plating during heating to braze temperature, proper procedures must be followed with all types of plating processes to activate the surface prior to applying the electrodeposit. It is common technique to heat-cycle the plated parts prior to brazing to determine plating quality. Electroless deposition (autocatalytic) of nickel should not be employed to precoat materials that are to be brazed at temperatures above 1600°F (870°C). The phosphorus content of the nickel plating provides a eutectic at 1616°F (880°C) and melting of the coating may interfere with wetting and flow.

BRAZE FLOW INHIBITORS

Stop-off materials are used to restrict filler metal to joint areas and prevent wetting or flow at other areas of the base metal, such as in holes and threads, and between mating surfaces that are not to be brazed. Commercial preparations generally consist of oxides of aluminum, magnesium, titanium and the rare earth minerals that are stable in vacuum or reducing atmospheres. They are provided in water slurries or organic binder mixtures.

Stop-off materials are applied by brushing or with hypodermic needles on areas where flow and wetting of brazing filler is to be prevented. Small quantities are very effective in preventing wetting, and excessive applications may result in costly removal operations.

Atmosphere brazing accounts for the widest use of stop-off materials. They are occasionally used in torch and induction brazing operations but their effectiveness is reduced if fluxes are employed.

MAINTAINING CLEANLINESS

Cleanliness is of great importance when brazing is performed in a protective atmosphere without a flux. Although chemical fluxes can handle more residual oxides than are generated during the brazing cycle, atmospheres often cannot handle these same conditions. Therefore, precleaning must be more thorough and the components, once clean, must be protected and preserved. A common technique is to use a "clean room" for final handling and assembling of the cleaned parts. Clean rooms are areas physically separated from the rest of the shop and provided with a means for controlling the atmosphere. Sometimes a plastic "bubble" or tent with low pressure circulating air is installed to reduce or eliminate airborne contamination. Clean rooms are widely used as final assembly areas for brazed honeycomb assemblies, stainless steel heat exchangers used with jet engine fuel systems, and electronic components prior to brazing. Workers in clean rooms may be required to wear special lint-free clothing and gloves to prevent the contamination of parts. White, lint-free, cotton gloves are more satisfactory than thin rubber gloves that can be slit while working with thin sheet metals. On occasion the cotton gloves are reinforced by covering them with nylon gloves.

Not all atmosphere or vacuum brazing processes require clean rooms for successful brazing. Using suitable closed containers to protect parts and establishing firm guidelines about storage time between process steps may be sufficient to insure clean parts. Maintaining inventory control to use parts on a first in-first out basis can greatly reduce degradation of cleaned surfaces.

Keywords — assembly, fixturing, self-fixturing methods, pressurized bellows fixture, pressure differential fixturing, differential thermal expansion fixturing, creep forming, fixture design

Chapter 8

ASSEMBLY AND FIXTURING

Components to be joined by brazing must be assembled and held in fixed relation to each other, and maintained in position throughout the brazing cycle. The faying surfaces must be spaced with the proper joint clearance. This clearance must be maintained during the brazing cycle so that the brazing filler metal can flow into and fill the joint at brazing temperature. The method of assembly and fixturing for brazing is therefore very important.

ASSEMBLY

Assembly of parts for brazing depends on the brazing process to be used, the materials being joined, and the configuration of individual parts in a brazement. Components are often assembled and held together by intermittent resistance spot welds, capacitor-discharge welds, tack welds made by gas tungsten arc, EB or laser welding, rivets, or straps of thin foil. Assembly by welding requires the use of a flux or an inert atmosphere at weld locations to prevent surface oxidation that would inhibit the flow of brazing filler metal. Aluminum sheet metal components to be brazed are sometimes held together with interlocking tabs and slots. Complex sheet metal assemblies are commonly built-up from individual subassemblies. The subassemblies are held together either with tie rods or with strips that are welded directly to the appropriate subassembly or attached to fixture plates. The connecting rods or strips are removed after

brazing. Figure 8.1 shows a brazed plate oil-cooler assembly made from various subassemblies that are held together for brazing in the fixture shown.

Cylindrical parts, tubing, and solid members are assembled for brazing using a variety of methods, such as staking, expanding, flaring, spinning, swaging, knurling, and dimpling. In all cases, the fixture should be as simple as possible to ease part removal and minimize costs. The assembly method should provide uniform joint clearance that can be maintained throughout the brazing operation so that proper joint fill will occur. When joining dissimilar metals, it may be necessary to control room temperature clearance to provide extremely tight or overlapping fitups. This will then allow for proper braze clearance at the brazing temperature. To accomplish this fit-up, it would require heating one component and cooling the other to allow assembly. The force of gravity and the part weight should be used wherever possible to assist in holding the parts together, to control the flow of the brazing filler metal during brazing, and to reduce the amount of fixturing required.

Complex structures usually require elaborate assembly methods. One requirement is that an acceptable braze joint clearance must be held at the proper brazing temperature for the brazing filler metal, the base metal, and the brazing process. Recommended joint clearances are given in Table 2.1. Minimum joint clearance may be maintained using shims of wire, ribbon, or screen material that are compatible with the base

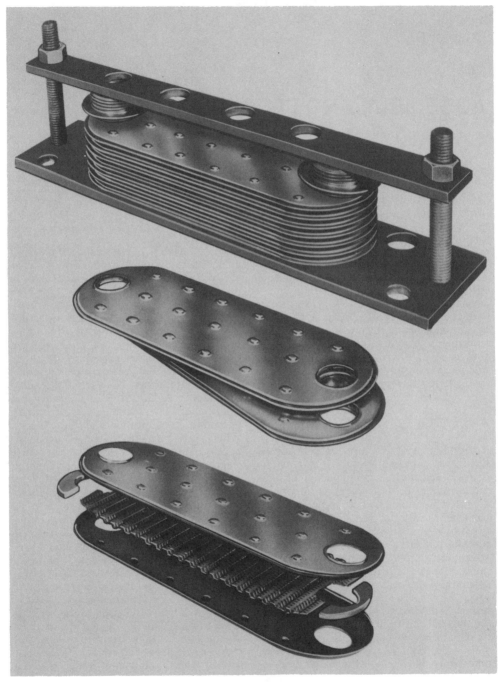

Figure 8.1 — Plate-Type Oil Cooler Assembly Fixtured With Tie Rods

metal and the brazing filler metal. Dimpling, pinning, gluing, and crimping may also be used to locate or fit parts within a component. Assembly of the components for brazing and maintenance of joint clearances should be considered during the design of the brazement to minimize the need for external fixturing.

In many applications, the brazement design may require that air passages, cooling holes, sealing surfaces, machined contours, and other areas must be free of brazing filler metal upon completion of the brazing operation. A variety of braze stop-off methods and filler metal placement techniques are available to prevent the flow of the brazing filler metal into such areas of the assembly.

Honeycomb structures, both sandwich type and open face, may be assembled by selective resistance welding of the honeycomb to the brazing substrate with or without preplaced brazing filler metal at the joint. Expansion rings are sometimes used for fixturing of open-faced honeycomb.

Examples of methods used to assemble parts for brazing without the use of auxiliary fixtures are shown in Figure 8.2. In these cases, brazing is accomplished by placing the assembled part on a plate, rack, or tray and subjecting the part to the prescribed brazing cycle. Self-fixturing assemblies, in some cases, may require the use of fixtures because of the complexity of the parts, location of the brazed joints, or distribution of the mass of material. The fixturing may consist of clamps, springs, support blocks, channels, rings, or other devices to keep the parts in proper position for brazing. Mild steel, stainless steel, machinable ceramics, and graphite are often used for primary or auxiliary fixturing. Care must be taken to prevent carburizing or contamination of the parts from the fixturing, in which case protective coatings or interlayers will be required.

FIXTURING

Because applications requiring braze fixtures vary widely, it is difficult to be specific about fixture designs. However, some basic design rules are given below.

(1) Keep fixture design simple and with minimum contact to the parts. Use thin sections consistent with rigidity and durability requirements. Both of these will minimize any heat sink effects from the fixture.

(2) Avoid use of bolts or screws in heated fixtures in high temperature applications because they tend to relax during heating and often pressure weld in place. One way to reduce bonding is to stop-off the threads and slot the nuts to facilitate their removal.

(3) Design springs or clamps to withstand the brazing temperature. Otherwise, relaxation will take place during the brazing operation. A spring can be preloaded to allow for relaxation at brazing temperature.

(4) Fabricate furnace brazing fixtures using components of uniform thickness for even heating. Use open structure designs without thick sections so that uniform heating and cooling can occur.

(5) Avoid the use of dissimilar materials where differences in thermal expansion could affect final assembly dimensions. An exception is where thermal expansion differences are used to obtain the correct joint gap, intentionally form the parts, or both.

(6) To ensure stability by relieving stresses, subject a fixture to the brazing environment and temperature before it is used for brazing of hardware.

(7) Fixtures should be designed for easy dimensional inspection. Because fixtures will change dimensions on each cycle, periodic inspection is required. The degree of change depends on the type of base metal, the thermal cycle, and the number of cycles.

(8) Where practical, use nickel alloys containing aluminum and titanium for fixture components. They develop protective and stable oxide films upon heating in air which inhibit wetting by molten brazing filler metals.

(9) Do not specify fixture materials that might react at elevated temperatures with the assembly being brazed while they are in intimate contact. For example, nickel alloys are unsatisfactory fixture materials for furnace brazing of titanium, or vice-versa, because a nickel-titanium eutectic forms at about 1730°F (945°C). This temperature is lower than the brazing temperatures normally used for titanium or nickel brazements.

(10) Design fixturing to apply localized pressure to braze components using flexible refractory blanket-type pads, steel balls, pins, bellows, or weighted levers.

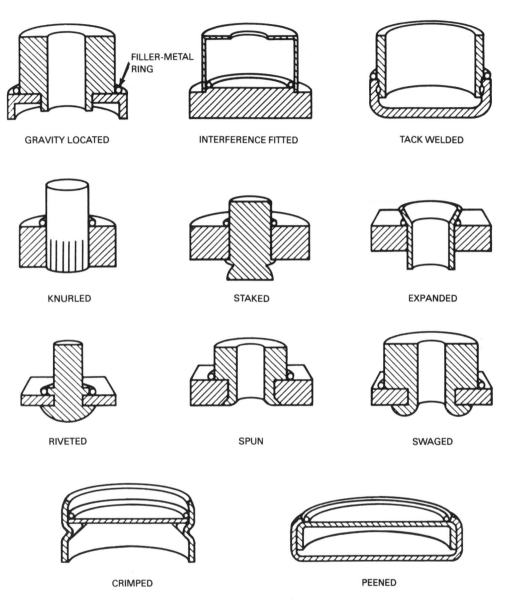

FILLER-METAL RING

GRAVITY LOCATED

INTERFERENCE FITTED

TACK WELDED

KNURLED

STAKED

EXPANDED

RIVETED

SPUN

SWAGED

CRIMPED

PEENED

Figure 8.2 — Typical Self-Fixturing Methods for Brazed Assemblies

(11) Use materials that have sufficiently high strength at temperature to prevent dimensional change due to frequent thermal cycles.

The brazing of complex assemblies often requires fixtures to position the parts, to maintain surface contours or, in special cases, to creep form the brazement to dimensional requirements during the brazing cycle. Creep forming consists of applying external loading to the fixture during brazing to force the components to conform to the contour of the fixture. This can be accomplished by weighted fixtures, hot pressing, or differential thermal expansion of fixture components.

Determining the fixture design best suited for a particular brazing application requires careful selection of fixture material. The choice must be based on (1) the brazing process to be used, (2) the base materials in the assembly, and (3) the required brazing temperature and atmosphere. When brazing in a vacuum, the only limitations on fixture materials are that they be stable at brazing temperature, possess compatible thermal expansion properties, and do not outgas and contaminate the brazing atmosphere.

Whenever metal fixtures are used for brazing, fixture surfaces that contact the part being brazed should be adequately protected to prevent brazing the assembly to the fixture. Suitable methods include the use of stop-off materials, refractory fibers, oxides, and plasma coatings.

Brazing fixture materials must not interact with the brazing atmosphere to produce undesirable products that are detrimental to the flow of the brazing filler metal or to the mechanical properties of the base metal.

Graphite is used for fixturing because it has good thermal conductivity and dimensional stability but low abrasion resistance. At elevated temperatures, graphite diffuses into some metals (carburization), reacts with water vapor, or forms methane in a hydrogen-containing atmosphere to produce gas-phase carburization and contaminants. Silicon carbide is also used for braze fixturing, but it has a lower thermal conductivity than carbon and has the same detrimental characteristics.

Induction brazing often requires that the fixturing be located near the inductor. In such cases, the choice of fixture material is limited to those materials that remain relatively unaffected by a magnetic field. Nonmetals, such as graphite, fiberglass, and ceramics are commonly used for fixturing. With a few exceptions, metals that are readily heated by induced currents should be avoided unless a portion of the fixture is intentionally heated to ensure uniform heating of the joint.

In molten chemical-bath dip brazing, the fixtures must be free of moisture to avoid explosion by steam. Therefore, the fixture materials must not absorb moisture when exposed to the atmosphere. They also must not react with the molten bath. The buoyancy effect of the molten salt requires that fixtures hold the components in position during brazing. The mass effect of the components and fixture on the cooling of the chemical bath during brazing should be

considered, and the components and fixture should be preheated if necessary.

Heat and corrosion resistant alloys are best suited for fixtures that are to be used repeatedly. Periodic cleaning of the fixtures is also necessary. Furthermore, the fixture design should facilitate draining as it is removed from the molten bath.

The high thermal coefficient of expansion of aluminum and magnesium require that fixtures for brazements of these metals be designed with two considerations in mind: (1) the difference in thermal expansion, and (2) the differential heating and cooling between the fixture and base metal during the brazing cycle. High temperature nickel alloy springs are preferred because they have suitable spring characteristics at brazing temperature and good corrosion resistance. Thermal reconditioning of the springs may be required in critical applications.

Fixtures for torch brazing must not interfere with the torch flame, the operator's view, or the application of the brazing filler metal.

Brazing of assemblies such as contours, which require maintenance of close tolerances, need very stable fixturing. Metal fixtures normally lack the stability required for repeated use and will require constant maintenance. Therefore, ceramic fixturing is often used for holding close tolerances. Castable ceramics, such as fused silica, are excellent. They possess superior dimensional stability and can be cast to close tolerances without machining. Their primary drawbacks are poor thermal conductivity and porosity that can hold moisture. Usually a bake-out cycle to remove moisture is required before each use. Alumina is also used for braze fixturing, but it is difficult to fabricate to close tolerances, is hard to machine, and has poor thermal shock properties.

Fixturing honeycomb can also be done by self-jigging using a stored energy (capacitor discharge) spot welder to hold the honeycomb to the backing strip. This method is frequently used for segmented as well as circular structure.

Another method of fixturing for brazing is based on the use of pressure differentials from external pressurization. One such fixturing method involves the use of a welded thin metal retort bag that is evacuated and then placed between the assembly and a

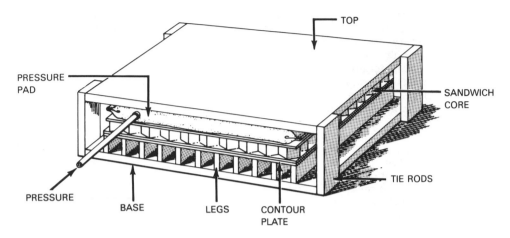

Figure 8.3 — Pressurized Bellows Fixture

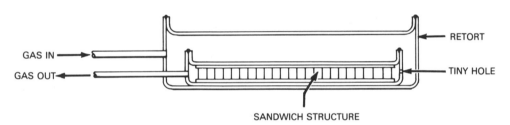

Figure 8.4 — Pressure Differential Fixturing

rigid backup structure. During the brazing cycle, the pressure pad is inflated to a low pressure. Figure 8.3 shows this fixturing method for producing honeycomb sandwich structures.

Another method is to use two retort chambers to create an external pressure on the part to be brazed. Figure 8.4 illustrates one possible system for utilizing this method.

The use of differential thermal expansion can also be used to bring surfaces into contact for brazing. Figure 8.5 illustrates this method for making honeycomb seals. In brazing the 360 degree honeycomb seal to the inner part, a metal band of a higher coefficient of thermal expansion can be used to apply pressure, thus bringing the honeycomb into intimate contact with the inner part at the brazing temperature.

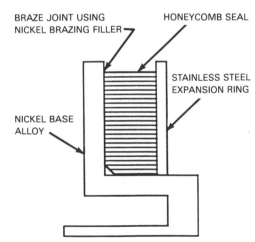

Figure 8.5 — Using Differential Thermal Expansion Fixturing to Braze a 360° Honeycomb Seal

Keywords — torch brazing, manual torch brazing, manual torch brazing equipment, torch brazing fixtures, mechanized torch brazing equipment, torch tips, rotary turntables, torch brazing safe practices

Chapter 9

TORCH BRAZING

PROCESS DESCRIPTION

Torch brazing is a brazing process where the heat required to melt and flow the filler metal is furnished by a fuel gas flame. This includes any operation where a fuel gas, such as acetylene, hydrogen, or propane, is combusted with oxygen or air to form a flame which is used to directly heat the assembly. For ease of communication, the term *oxyfuel gas* will be used to describe operations of this nature. The process involves the broad heating of the assembly by one or more flames, with the use of torch or workpiece movement, or both, to achieve proper heating and both surfaces to be joined are brought to brazing temperature uniformly.

Torch brazing is performed in air without the need for protective gas, normally requiring the use of flux. An exception to this is the use of phosphorus bearing filler metals on pure copper base metal where the phosphorus combines with the oxides present to create an oxide-free, wettable joint surface. The finished brazement must then undergo post-braze cleaning operations to remove flux residue and heat scale. The oxidizing environment limits the base metals being joined to those which are not highly reactive and do not require special atmosphere environments during the braze cycle.

Torch brazing is a widely used process which requires relatively low initial investment for equipment, particularly in the case of manual brazing, and operator skill is readily acquired. The equipment is portable and can be used in other operations that require flame heating. The same equipment used in oxyfuel gas welding can be used in brazing. The process lends itself to the use of low melting temperature filler metals which, because of their excellent flow characteristics, simplify brazer qualification.

Torch brazing is a flexible process where any degree of automation can be achieved. For small quantities of joints, a single operator with a hand-held torch is sufficient. When the labor cost in manual torch brazing becomes prohibitive in larger production quantities, semi- or fully-automated systems are available.

Filler metal selection ranges from the low temperature silver base filler metals to the high temperature nickel and copper base filler metal systems. Filler metal in the forms of wire, preformed shapes, or paste are all acceptable choices in torch brazing, facilitating the selection of the optimum material for each application.

APPLICATIONS

Torch brazing is found in virtually all industries where brazing is performed and is not limited by production quantity. One joint to many millions of joints can be economically made with torch brazing.

Manual torch brazing is used on low volume production applications and on large assemblies where multiple joints at various locations need to be brazed with localized heating. It is used in field installation and repair situations. A brazer using manual torch brazing is shown in Figure 9.1.

Semi- or fully-automated torch brazing is used when production volume renders manual brazing uneconomical. In these cases, some level of automation is required. Moderate volume production can use simple shuttle systems with manually placed components, filler metal, and flux. Larger volumes can be accommodated on rotary turntables or in-line conveyors with the capability of heating at a very high rate and with the potential of automating all opera-

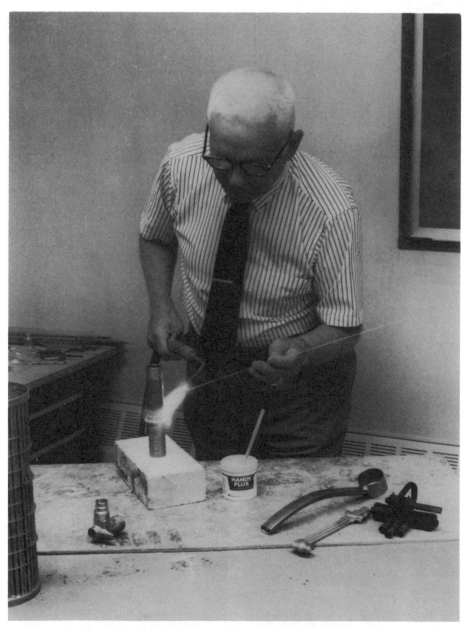

Figure 9.1 — Brazer Using Manual Torch Brazing

tions relating to component and braze material loading and unloading.

The heating, air conditioning and refrigeration industries utilize torch brazing to join a large number of tubular connections found in these products. Manual torch brazing is used to make the final connections on air conditioning units as they move along an assembly line. Automated torch brazing is used to braze the multiple return-bend joints on slab heat exchanger units.

Torch brazing is found in the tube-and-shell heat exchanger industries. Tube to endplate joints and endplate to shell joints are made by this method. In the compressor industry, tubes are joined to compressor shells either manually or on semi-automated indexing tables using torch brazing equipment. In the plumbing ware industry, the majority of brass and copper faucet assemblies are joined by automated torch brazing. Millions of joints are made annually by torch heating of such products as carbide tools, automotive components, furniture assemblies, houseware products, and valving. Examples of products that contain braze joints are shown in Figure 9.2.

ADVANTAGES AND LIMITATIONS

Advantages

The advantages of torch brazing are as follows:

(1) Any volume of joints can be made. Small quantities can be brazed manually and large quantities can be brazed automatically. Automating a torch brazing operation is readily accomplished.

(2) Manual torch brazing equipment is portable, facilitating field installation applications.

(3) Capital investment is small for oxyfuel heating equipment used in manual torch brazing.

(4) Oxyfuel gas equipment used for manual torch brazing can be used for other purposes (e.g. welding and cutting).

(5) The fuel gas flame can be adjusted to be carburizing, reducing, neutral, or oxidizing in nature, depending on the application. Caution should be employed when using hydrogen as the flame is virtually invisible.

(6) Torch heating is most practical for large assemblies, such as air conditioning units, where multiple tubular joints are made at various locations and the heating of those joints needs to be localized.

(7) Brazing filler metal can be preplaced or face fed.

(8) A wide range of fuel gases can be used depending on cost, availability, and the amount of heat required.

(9) All base materials which are not degraded in an oxidizing environment, and for which fluxes are available, can be torch brazed.

(10) Odd joint configurations which do not lend themselves to fixed heat patterns can be torch brazed.

(11) Joints with dissimilar cross sections or dissimilar materials can be brazed by proper location and amount of heating,

Figure 9.2 — Examples of Brazed Assemblies

controlled by the movement of one or more torches.

(12) On large components with small or multiple braze joints, torch heating provides a localized heat source.

(13) On small components without large heat sinks, torch heating is rapid, minimizing liquation of the filler metal.

(14) Through proper torch movement, heating the joint can be controlled to flow the filler metal from the cool side to the hot side of the joint by capillary attraction.

(15) Joints with poor fit-up can be brazed by controlling filler metal flow with skilled manipulation of a manual torch.

(16) A wide selection of filler metals from low-temperature silver base filler metals to high-temperature gold and nickel base filler metals can be used. All filler metal forms are available.

(17) Torch brazing skill is readily acquired by brazers.

Limitations

The limitations of torch brazing are as follows:

(1) Torch brazing is performed in an oxidizing environment, requiring some degree of post-braze cleaning to remove heat scale and flux residues.

(2) The use of flux results in joints with a higher degree of joint porosity than made in a controlled atmosphere without the use of flux.

(3) Brazements of extremely large mass cannot be made in an economical fashion with the localized heating of a torch because of the excessive heating time needed.

(4) Some base materials such as titanium and zirconium are not brazeable with a torch because of their highly reactive nature.

(5) Manual torch brazing is labor intensive.

(6) In general, the maximum temperature at which torch brazing is practical is 1800°F (982°C).

(7) Torch brazing filler metals which contain cadmium pose a health risk.

(8) The use of flux creates the potential for corrosion from flux residue left on the brazement.

EQUIPMENT

The equipment used in torch brazing, either manual or automated, has several functions to perform. It must be capable of supplying oxygen (or air) and fuel gas to the torch tip at the correct mixture and the correct rate of flow. These variables will affect the flame temperature, rate of heating, and nature of the flame atmosphere. In the case of automation, the equipment must allow the brazement to be fluxed, heated, and cooled in such a manner that will produce sound braze joints.

Manual Torch Brazing Equipment

The equipment used in manual torch brazing is essentially the same as that used in oxyfuel gas welding. The main differences are in the choice of torch tip and the heating method. The major components of this equipment are the gas supplies, the gas delivery hoses or piping to the torch, the torch itself, and other related items. The selection of the specific oxyfuel gas combination determines the type and size of equipment. This equipment must be appropriate to the heat required for the specific application. A typical torch brazing setup is shown in Figure 9.3.

Gas Storage. The fuel gases used in manual torch brazing are typically stored in (1) individual cylinders, (2) manifolded cylinders, or (3) bulk storage.

Individual cylinders are appropriate when usage is not large or where a portable supply is needed. Cylinders are labeled and designed for specific gases. Acetylene cylinders contain acetone, and the rate of fuel gas withdrawn from these cylinders is limited because acetone can be drawn out with the acetylene. The maximum withdrawal rate per hour is generally considered to be one seventh of the cylinder's capacity. For the withdrawal rate of the specific gas being used, consult the cylinder or gas supplier.

Manifolded cylinders are used when moderate volumes of gas are being consumed. These can be either portable or permanent in nature, with the gas piped to brazing stations and with all safety devices being utilized.

Bulk storage is an alternative when large volumes of gas are required. The gas is

Figure 9.3 — Example of Manual Torch Brazing Equipment

piped to the brazing stations utilizing, as with the manifolded system, all required safety devices. The system must be designed to prevent pressure fluctuations throughout the system.

Regulators. Pressure regulators reduce the gas pressure from the source of supply to a usable reduced pressure, even when the source pressure changes. Regulators handle specific ranges of pressure and are available in two different types, single-stage and two-stage.

Single-stage regulators step down the cylinder pressure to deliver pressure in one step. This is a less expensive type and typically requires adjustment by the brazer as the cylinder pressure decreases.

Two-stage regulators step down the cylinder pressure to deliver pressure in two steps. These regulators maintain the original delivery pressure setting as a constant until the useful pressure range of the cylinder has been exhausted.

Gas cylinders are normally used with a regulator which has two gauges. The inlet gauge of the regulator measures the cylinder pressure. The outlet gauge measures the delivery pressure to the torch.

Connections on the supply and delivery sides of the regulator vary with the type of cylinder being used, the gas, and the regulator capacity. The different connections prevent users from connecting inappropriate regulators to cylinders and hoses.

Hoses. Hoses used in oxyfuel equipment are flexible and can withstand relatively high pressures. They give the brazer mobility and allow the gas cylinders to be kept away from the actual brazing area. The hoses are color coded to keep the fuel gas and oxygen from being connected improperly. All fuel gas hoses are colored red and the connections have left handed threads. Oxygen or air hoses are green or black and have right handed threads. In addition, the left hand threaded nuts have a groove cut into the outside for ease of identification.

Torches. Torches designed for manual torch brazing consist of a torch body, a mixer, and a tip. The torch body acts as the handle and has valves to control the flow of oxygen and fuel gas through the torch. A mixer provides mixing of the fuel gas and oxygen for proper combustion. A torch tip provides the desired flame size and profile. It enables the brazer to direct the flame on the brazement effectively.

The torch body provides a comfortable way for the brazer to control flame movement. The control valves adjust gas flow to match the application requirements. It also allows the oxygen-fuel gas ratio to be varied, providing the ability to use a carburizing, reducing, neutral, or oxidizing flame, depending on the application. Torch bodies are produced in a wide range of sizes to accommodate a variety of heating requirements. For example, small torches which allow for acetylene flow to approximately 30 ft.^{3}/hr. (0.85 m.^{3}/hr.), up to large torches with acetylene flow up to 400 ft.^{3}/hr. (11.33 m.^{3}/hr.), are available.

In addition to the proper mixing of fuel-gas and oxygen, the mixer acts as a heat sink, helping prevent flashbacks of flame into the torch. Mixer designs are either of the positive pressure type or the injector type. In the positive pressure type, the fuel gas is delivered at a minimum pressure of 1 psig and the oxygen pressure is generally the same. Mixing is accomplished by flowing the fuel gas into the oxygen stream creating an efficient mixing turbulence. The injector mixer is used where the fuel gas is supplied to the torch at pressures below 1 psig. A venturi style design is used to aspirate the fuel gas into the mixing chamber at a flow rate suitable for proper combustion.

Torch Tips. Tips for torch brazing vary depending on the torch size, size of the workpiece to be heated, and fuel gas selection. They are made typically of copper alloys which possess high thermal conductivities, minimizing the potential for overheating. Tips used with acetylene or hydrogen have a flat face across the orifice while those used with propane or other fuel gases have recessed faces at the orifice. Orifice diameters are specified in a variety of ways, such as drill sizes or a manufacturer's number. A difficulty arises in specifying a tip size for a particular application because torch size and mixer design will affect tip performance. The manufacturer's recommendations should be followed. A variety of tips commonly used in manual torch brazing are shown in Figure 9.4.

Accessory Equipment. Flash back arresters are valves which prevent a torch flame from travelling into the torch or hoses as the result of an equipment problem or operator misuse. Left unchecked, flashback can cause an explosion. The two most common types of arrestors are the mechanical check valve and the hydraulic valve. Check valves are inserted between either the hoses and the torch or the hoses and the regulators. The hydraulic valve type is more reliable because flashbacks have been known to burn through mechanical check valves.

Another necessary item is a cleaner for torch tips which become clogged by soot, flux, or other materials. Constricted tips cause inconsistent heating and can result in overheating of the tip to the point where flashbacks can occur. In addition, safety items such as goggles, gloves, and an apron should be used.

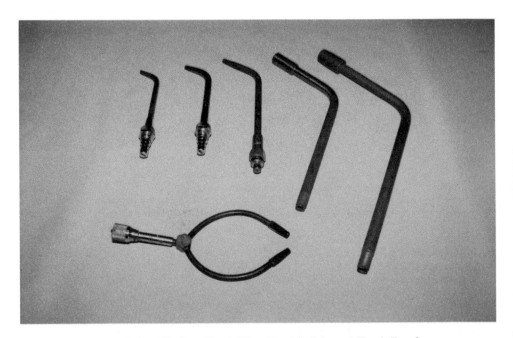

Figure 9.4 — Various Torch Tips Used in Manual Torch Brazing

Fixtures. Fixtures for manual torch brazing should be designed with the same criteria as for all brazing fixtures (see Chapter 8). The fixturing should have a minimum contact with the assembly. It should not be degraded by torch flame and flux, and should not act as a heat sink. The fixtures should be designed to accommodate the thermal expansion of the components being brazed.

A major consideration when designing a fixture for torch brazing is that the fixture not interfere with proper movement of the torch flame and uniform heating of the joint area. The brazer needs to have a good view of the joint.

Mechanized Torch Brazing Equipment

With relatively large production quantities, part or all of the torch brazing operation may be mechanized. The degree to which the process is automated depends on such factors as (1) component size and configuration, (2) joint accessibility, (3) desired production rate, (4) the labor needed to handle the components, and (5) filler metal to be used.

Some of the considerations that must be taken into account regardless of the type of equipment being used are as follows:

(1) Assembly and Loading of Components. Assembly is done either prior to components reaching the machine or as the machine operator loads the individual components into the fixtures. In either case, the assembly needs to be properly positioned in the fixtures. Automatic loading can be accomplished by various methods such as vibratory feeders, pick and place devices, or robots.

(2) Filler Metal Placement. Preform filler metal shapes are placed in their proper joint location during assembly. Some off-line loading of preforms into components prior to reaching the machine can reduce the number of items the operator needs to handle. Paste filler metal can be placed by a pneumatic application system which is either manually operated at the time of component assembly or actuated automatically when the parts index to an application station. Filler metal wire can be automatically fed into the joint, near the end of the heating cycle, after the assembly has been raised to proper brazing temperature.

(3) Flux Placement. Fluxes are applied manually, through an automatic flux dispenser, or as a component of a brazing paste. Manual application of flux is done either by dipping a component into flux before assembly or by brush. Manually applied fluxes are typically a pasty consistency while automatically dispensed fluxes are a slurry designed for spraying.

(4) Fixturing. The fixtures on mechanized torch brazing systems should utilize the same concepts described in the previous section on manual torch brazing fixtures. In addition, the fixtures need to be quickly and easily loaded and unloaded since the operator or device has a fixed, limited time to get the parts into or out of the fixture.

(5) Heat Pattern. Heat pattern refers to the number of heat stations, the number of torches per station, the intensity of flame at each station, and the locations where the flames impinge on the brazement at each station. The amount of heat used depends on the production rate desired and the time required to load all components. Sufficient heat to achieve a three-second dwell time at each station on a rotary table may be impractical if the components cannot be loaded in that three-second period. The number of torches and their locations are determined by the requirement that all components reach brazing temperature uniformly and that each braze joint is heated in a way to promote capillary flow.

(6) Brazement Cooling. Typically, after the brazement leaves the last heating station, it is allowed to cool at a still air station until the filler metal solidifies. Forced air and water quenching are methods used to cool the brazement further before it is removed from the machine. Any number of cooling stations using a combination of still air, forced air, and water quenching can be used.

(7) Unloading. The completed brazements are moved from the machine either manually or automatically. Pick and place devices, robots, or simple ejection mechanisms can be used. Manual removal is possible, depending on the time available to the operator since this person is typically doing the loading also.

The equipment is normally designed to use multiple torches. By increasing the number of torches used to heat the assembly, the rate of heating and the production rate can be increased. The three basic types of equipment utilizing this concept are rotary turntables, in-line conveyors, and shuttles.

Rotary Turntables

The most common design of mechanized torch brazing equipment is the rotary turntable. An example of this equipment is shown in Figure 9.5. A turntable consists of a chassis, drive system, combustion system, tooling plate, fixtures, and torch heat pattern. Drive systems are typically of the indexing type but continuous drives can be used. These machines use progressive heating where the parts index through multiple heat stations, increasing in temperature, until brazing is completed at the last heat station. For example, if a turntable has five heat stations with 12 seconds per station (excluding indexing time for simplicity in this example), the assembly sees a total heating time of one minute. The operator or an automated device unloads a finished part each 12 seconds compared to one minute or more that it would take a manual brazer to accomplish the same braze. The operating sequence for the machine in Figure 9.5 is shown in Figure 9.6.

Rotary turntables can be designed with any number of index stations. The most common designs have a fixed number of indexing station positions. For greater flexibility, continuous drive systems or programmable drives, which can be set for any number of stations, can be used.

The chassis needs to be sturdy and supply stable, vibration free operation. Table tops are typically made of stainless steel to stand up to the corrosivity of the brazing fluxes used. It also acts as a trough for cooling water when used. The drive needs to supply consistent and accurate positioning of the brazements in the heat pattern. It needs smooth acceleration and deceleration through the indexing cycle to prevent misalignment of the fixtures in the heat pattern and to avoid

Figure 9.5 — Semi-Automatic Rotary Indexing Torch Brazing machine

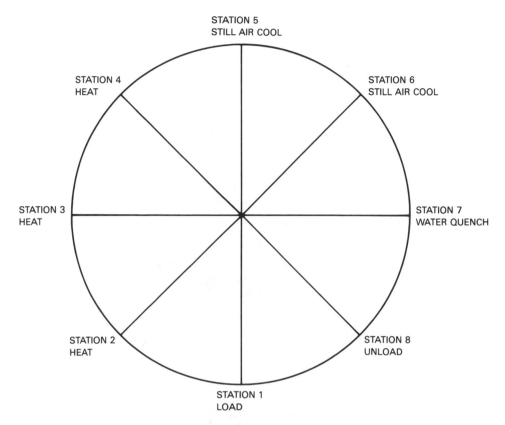

Figure 9.6 — Schematic of 8 Station Rotary Indexing Machine Pictured in Figure 9.5

movement of the filler metal during solidification. Barrel-cam and roller gear drives accomplish these tasks effectively.

The fixtures for rotary turntables should be made from corrosion and heat resistant materials. Stainless steel and other heat resistant materials are recommended. By using simple inserts, families of similar brazements with the same joint size and location can be brazed on the same machine.

On most rotary turntables, the heat patterns are dedicated to a particular assembly. Manifolds are secured in place and stainless steel tubing with fixed torches are used to provide a fixed heat pattern. Components index through the various heat stations while being held firmly by fixtures bolted to the indexing tooling plate. It is possible to incorporate torch movement if the application calls for it. It is also possible to have fixtures that rotate the assemblies in the heat pattern, as in the case of large cylindrical joints where uniform heating with fixed torches is difficult to achieve.

Inexpensive rotary turntables can be designed using continuous drive systems and a linear arrangement of torches. Simple fixturing and ease of operation are additional benefits of this type of machine. Figure 9.7 shows this type of turntable.

In-line Conveyors. In-line conveyor brazing equipment consists of a conveyor, a guide system for the assemblies, and a combustion system. Figure 9.8 shows a typical in-line conveyor system. Components are assembled with filler metal and flux in place, and then the assembly is placed on the conveyor. The parts are moved between the guide system through a manifold-burner arrangement that brings the assembly to brazing temperature. The conveyor system

Figure 9.7 — Low Cost Continuous Speed Brazing Machine

transports the assemblies out of the heat zone, and the parts are then removed after cooling enough to solidify the filler metal and allow handling.

The combustion system is designed to match the desired production rate. The length of the manifold and the number of burners per manifold can be varied. Longer manifolds allow faster belt speeds to be used. There are limits to the size of manifold systems and multiple manifolds can be used for assemblies requiring more heat.

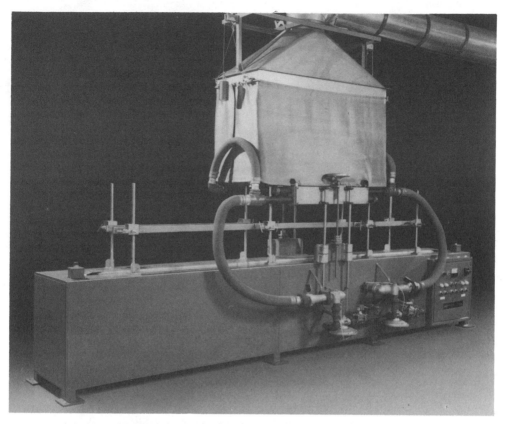

Figure 9.8 — In-Line Brazing Conveyor System

Shuttles. Shuttle systems are inexpensive machines for low volume production requirements. A typical shuttle is pictured in Figure 9.9.

Shuttles provide a fixed heat pattern in a single station, multiple torch arrangement. This allows for faster production rates than a brazer with a single station, single torch arrangement. Shuttles are also faster than a brazer with a single torch but slower than that of a rotary or in-line system. In shuttle systems, either the torches or the fixtured assembly moves. A typical sequence of operation would have the operator loading the components, the filler metal, and the flux into the fixture. The fixtured assembly is then moved into the heat source. After filler metal flow is completed, the assemblies are moved out of the heat source, the filler metal solidifies, and the assembly is then quenched. Upon completion of cooling, the operator removes the brazed assembly and repeats the process.

Paste Dispensers

Braze paste consists of a mixture of powdered filler metal, binder, and possibly a flux. Dispensing of paste eliminates the need for separate operations to preplace preforms, apply flux, or manually apply the paste. Air operated equipment is available to perform this function. A typical dispenser is shown in Figure 9.10.

In automatic dispensing on a rotary turntable, a fixtured assembly actuates a microswitch, that in turn initiates a paste dispensing cycle. This takes place at a station between the component loading station and the first heat station. The dispensing gun moves into position to de-

Figure 9.9 — Low Volume Brazing Shuttle Machine

posit a fixed amount of braze paste at the proper position at the joint. After dispensing and before the table indexes, the gun retracts. The paste can be obtained in containers that are then emptied into a reservoir or in disposable cartridges which load directly into the dispensing unit. Dispensing units can also be hand held, with the size of the paste deposit controlled either by the operator or by a timer.

Slurry Flux Dispensers. In applications that use preforms or wire filler metal, the application of a slurry flux can be accomplished through an automated dispenser. This will eliminate the operator's need to dip or brush the flux onto the brazement. A slurry flux dispenser is shown in Figure 9.11.

Fluxes used in these dispensers are manufactured specifically to be sprayed from these units. The flux needs to be mixed well, typically by agitation for several minutes in a paint shaker, prior to pouring it into the dispenser. The unit constantly recirculates the flux, keeping it mixed. On a rotary turntable, a fixtured assembly activates a microswitch that in turn initiates a flux dispensing cycle. This operation is located at a station between the component loading station and the first heat station. The units are available in hand held, manually operated versions also.

Vapor Flux Dispensers. Another type of flux dispenser is the vapor flux type. This equipment, shown in Figure 9.12, introduces a vapor flux into the fuel gas line. The flux is present in the flame and provides a degree of protection from heat scale and discoloration.

Wire Feeders. Wire feeding units provide a means of applying filler metal automatically. A typical unit is shown in Figure 9.13. On a rotary turntable, a fixtured assembly activates a microswitch that initiates a wire feeding cycle. The wire is fed at the final

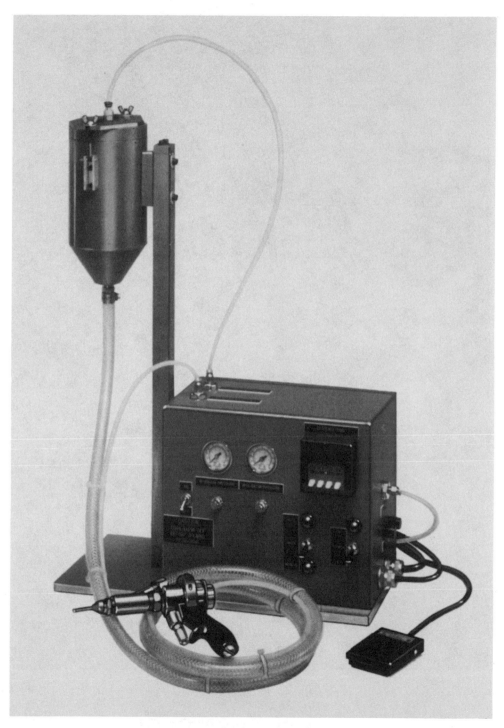

Figure 9.10 — Brazing Paste Dispenser

Figure 9.11 — Brazing Flux Dispenser

ficult to control because of its close proximity to the intense heat at the final heat stations. Also, the joint needs higher heat input than with braze paste or preforms since the wire causes a chilling effect when it comes in contact with the joint.

FUEL GASES

The most commonly used fuel gases for torch brazing are acetylene, natural gas, propane, propylene, and certain proprietary mixtures. These gases are combusted with pure oxygen or air. Maximum heat is obtained with oxygen-fuel gas combustion. Manual torch brazing most commonly uses oxygen and acetylene while mechanized torch brazing most commonly uses natural gas and compressed air. The larger volume of fuel gas consumed in mechanized operations makes lower cost natural gas a logical choice. The progressive heating employed in mechanized torch brazing permits the use of the lower-temperature natural gas-compressed air flame. The fast rate of heating with an oxygen-acetylene torch can reduce labor costs on short run manual torch brazing applications. Oxygen-natural gas heating is used on mechanized equipment when high heating rates are required.

General Characteristics

Table 9.1 lists some important characteristics of the common fuel gases. The information provides a means of comparison. Flame temperature forms the main basis, along with cost, for selecting a fuel gas.

The precautions in the use and safe handling of these fuel gases are given in Chapter 6, Safety and Health. Specifics on each of these gases, including the equations for complete combustion, are included here.

Acetylene. This fuel gas is generally preferred because the high flame reaches 5660°F (3127°C). It is supplied in cylinders containing partly separated cells and is dissolved in acetone, a solvent capable of absorbing 25 times its own volume of acetylene. This method of supply is the best known method of assuring safety of this highly explosive gas. When not absorbed in acetone, acelylene is unstable and may explode, even without the presence of oxygen,

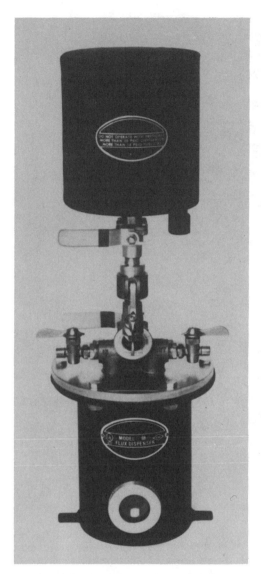

Figure 9.12 — Vapor Flux Dispenser

heat station. Some designs have the wire feed mechanism in a fixed position and others move it in and out of feed position to avoid premature melting of the filler metal and to avoid interference with the fixtures during indexing. The filler metal is fed into the joint, melting on contact. A fixed amount is fed and then the wire is retracted to prevent melting and balling up of the exposed wire end. This equipment can be dif-

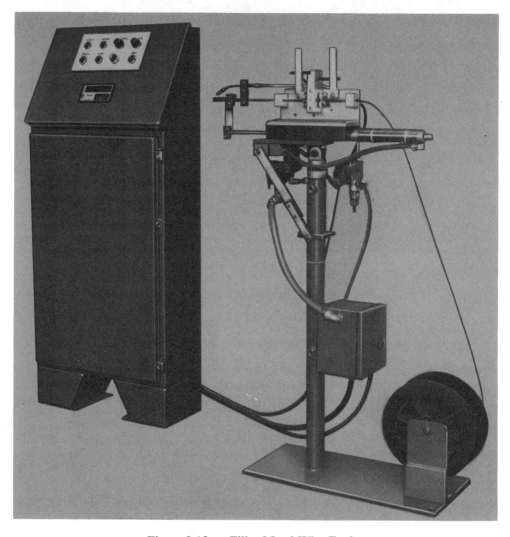

Figure 9.13 — Filler Metal Wire Feeder

at temperatures of 1435°F (780°C) and higher or at pressures above 30 psig (207KPa). The accepted safe practice for acetylene is never to use it at pressures in excess of 15 psig (103KPa).

Acetylene reacts with oxygen for complete combustion according to the equation $C_2H_2 + 2.5 \, O_2 \rightarrow 2CO_2 + H_2O$.

Methylacetylene-Propadiene. These gases are mixtures of various fuel gases. The flame temperatures are lower than that of acetylene and more stable. The equation for combustion is: $C_3H_4 + 4\,O_2 \rightarrow 3CO_2 + 2H_2O$.

Propane. Although not as commonly used as acetylene and natural gas, propane is delivered in bulk quantities. Where reliable sources of natural gas are not available it offers a good alternative. Its combustion equation is: $C_3H_8 + 5 \, O_2 \rightarrow 3CO_2 + 4 \, H_2O$.

Natural Gas. Widely used in mechanized torch brazing, it is readily available and var-

Table 9.1
Characteristics of the Common Fuel Gases

Fuel gas	Formula	Specific gravity 15.6°C (60°F) Air = 1	Volume to weight ratio (15.6°C)		Oxygen-to-fuel gas combination ratio[a]	Flame temperature for oxygen[b]		Primary		Secondary		Total	
			m³/kg	ft³/lb		°C	°F	MJ/m³	Btu/ft³	MJ/m³	Btu/ft³	MJ/m³	Btu/ft³
Acetylene	C_2H_2	0.906	0.91	14.6	2.5	3087	5589	19	507	36	963	55	1470
Propane	C_3H_8	1.52	0.54	8.7	5.0	2526	4579	10	255	94	2243	104	2498
Methylacetylene-propadiene (MPS)[c]	C_3H_4	1.48	0.55	8.9	4.0	2927	5301	21	571	70	1889	91	2460
Propylene	C_3H_6	1.48	0.55	8.9	4.5	2900	5250	16	438	73	1962	89	2400
Natural gas (methane)	CH_4	0.62	1.44	23.6	2.0	2538	4600	0.4	11	37	989	37	1000
Hydrogen	H_2	0.07	11.77	188.7	0.5		4820					12	325

a. The volume units of oxygen required to completely burn a unit volume of fuel gas. A portion of the oxygen is obtained from the atmosphere.
b. The temperature of the neutral flame.
c. May contain significant amounts of saturated hydrocarbons.

ies in composition depending on its source. It is comprised mainly of methane. Its combustion equation is: $CH_4 + 2\,O_2 \rightarrow CO_2 + 2\,H_2O$.

Hydrogen. Combustion of this gas produces a nearly invisible flame and it is not widely used in torch brazing. Its combustion equation is: $2H_2 + O_2 \rightarrow 2H_2O$.

Flame Characteristics

As the ratio of oxygen to fuel gas changes, the nature of the flame and the resulting application for each flame condition changes. The most common flame conditions used in torch brazing are reducing and neutral. Oxidizing flames are not recommended for torch brazing. Figure 9.14 shows the different flame conditions encountered with various oxygen to fuel gas combinations. The best flame for a particular base metal combination is given in the chapter on a specific base metal system.

Fuel Gas Flame. When hydrocarbon fuel gases are burned without oxygen or air added through the torch, they typically produce a yellowish flame. Soot particles are usually present because the oxygen in the air is not sufficient to support complete combustion. This flame is not useful in brazing.

Carburizing Flame. As oxygen is added to the fuel gas flame, the sooting disappears and the flame becomes luminous. As the oxygen content is increased, the luminous part of the flame becomes smaller and is centered near the torch tip. A blue zone, consisting of an excess of fuel gas, forms around the outside edge of the flame. This flame can be used in brazing.

Reducing Flame. As the oxygen content is increased further, the luminous area becomes smaller and consists of an inner cone with a feathery trail extended out toward the flame end. This condition indicates a slight excess of fuel gas and is an excellent flame for brazing.

Neutral Flame. When the oxygen addition reaches the ratio necessary for the fuel gas to be completely combusted, the feather that extended out from the bright inner cone disappears. This flame is used in brazing when

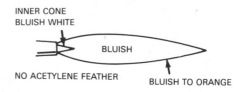

NEUTRAL FLAME

INNER CONE
BLUISH WHITE

BLUISH

NO ACETYLENE FEATHER

BLUISH TO ORANGE

OXIDIZING FLAME

SHARP INNER CONE
BLUISH WHITE

BLUISH TO ORANGE

INNER CONE TWO-TENTHS
SHORTER THAN CONE OF
NEUTRAL FLAME

NEARLY
COLORLESS

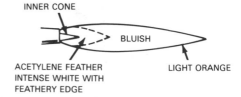

CARBURIZING (EXCESS ACETYLENE) FLAME

INNER CONE

BLUISH

ACETYLENE FEATHER
INTENSE WHITE WITH
FEATHERY EDGE

LIGHT ORANGE

Figure 9.14 — Flame Conditions for Oxy-Acetylene Are Shown. Other Oxyfuel Flames Are Similar. (Permission to reprint granted by J.W. Harris Company, Inc., Cincinnati, Ohio)

arc excess of carbon in the reducing flame is detrimental to the base metals or when maximum flame temperature is required.

Oxidizing Flame. When the oxygen-to-fuel gas ratio exceeds that needed for complete combustion, the flame becomes oxidizing. The flame produces a hissing sound. The inner cone in the flame will appear to be constricted. Oxidizing flames are not recommended in brazing.

JOINT DESIGN

Joint design requirements for torch brazing follow the same principles that apply to all braze joints (See Chapter 2). Additional design considerations in torch brazing are (1) the joint area needs to be accessible to localized torch heating, (2) flux will be used (except for copper to copper joints using phosphorous bearing filler metals), and (3) the constituents of filler metals used must be known.

Torch brazing allows heat to be concentrated at specific locations. This is particularly true of manual torch brazing. Assemblies that have components of differing mass and cross section can have the heat directed toward the larger components to avoid overheating of smaller ones. Joints made from materials with widely differing coefficients of thermal expansion can have the heat concentrated on either the high or low expanding metal, depending on where each is located, to insure that capillary attraction is maintained at brazing temperature. The joint design should capitalize on this heating flexibility.

Filler metals used in torch brazing contain melting point depressants such as zinc, cadmium, phosphorus, or tin. The fluxes are salts with a relatively high molten viscosity at brazing temperatures. These filler metals and fluxes require a joint clearance at brazing temperature of 0.001 to 0.005 inch (0.0254 to 0.1270 mm) for adequate capillary flow. When face feeding is used, it is recommended that the filler metal be applied at the coolest section and allowed to flow toward the hottest section of the joint. Considering these characteristics, the filler metal form, and placement should be designed to promote capillary flow towards the hotter section of the joint.

MATERIALS

Most base metal systems can be brazed with a torch. Reactive metals, such as titanium and zirconium, are not recommended for torch brazing because compatible fluxes are not available. Stainless steels can be torch brazed but it is recommended that the stabilized grades, such as 321 or 347, or low carbon grades, such as 304L and 316L, be used. Use of the low carbon grades minimizes the possibility of carbide sensitization. Further information can be found in chapters covering specific base metal systems.

Filler Metals

The commonly used filler metal systems in torch brazing are the BAg, BCuP, and RBCuZn systems. The BAg filler metals melt at temperatures generally from 1100°F (593°C) to 1500°F (815°C). The RBCuZn filler metals melt at various temperatures between 1500°F (815°C) and 1800°F (982°C). They are available with a variety of flow characteristics which should be considered when designing the assembly and process. The BCuP filler metals are used predominantly on copper and copper alloys in the 1300øF (704°C) to 1700°F (926°C) range.

The BAg filler metals, although higher in cost than the copper-zinc filler metals, can be utilized at higher production rates because of their low brazing temperature range and the rapid heating available with a torch. The high cost of BAg filler metals can be offset by other means such as automation, lower energy costs, higher production rates, and the ease of operator training. These filler metals can be used for torch brazing most ferrous and non-ferrous metals. They have good brazing properties, are suitable for preplacement in the joint or for manual feeding, and are available in all forms. Flux is required on all base metals.

Brazing filler metals of the BCuP classifications are used primarily for joining copper and copper alloys, although some have limited use on silver, tungsten and molybdenum. These filler metals should not be used on ferrous or nickel-base alloys or on copper-nickel alloys containing more than 10 percent nickel because brittle intermetallic compounds may form. BCuP filler metals are self-fluxing on pure copper; however, a flux is recommended when they are used on all other base metals, including alloys of copper. Their corrosion resistance is satisfactory, except when the joint is in contact with sulfur bearing environments. They are all available in wire, powder, preforms, and paste. BCuP-5 filler metal is available in strip, shims, and preform stampings.

Brazing filler metals of the RBCuZn classifications are used for joining various ferrous and non-ferrous metals. They have brazing temperatures considerably higher than the BAg and BCuP filler metals. Care must be taken to avoid overheating these filler metals. Zinc vaporization can occur, raising the melting point of the molten filler metal and creating voids which can weaken the joint or cause leakage.

More information on filler metals and their application can be found in Chapter 3 and in chapters covering specific base metal systems.

Fluxes

The fluxes used in torch brazing are of the fluoride-borate type. The AWS classifications and typical applications are shown in Table 9.2. The FB3A fluxes are for general purpose use on most ferrous and non-ferrous base metals. Stainless steels, carbides and other more refractory base metals often require the FB3C fluxes. Both types are used in paste or sprayable forms. Various fluxes are available for use with the low temperature BAg filler metals, the BCuP filler metals, and the higher temperature RBCuZn filler metals.

A vapor flux can be prepelled to the fuel gas, and propelled in the flame impinging on the assembly. In a limited number of applications, this vapor flux can be used alone. It is used also to help minimize the extent of discoloration and oxidation on the brazed assembly. This facilitates removal of normal flux residue and makes post braze cleaning easier. In the case of hand-fed wire, vapor flux eliminates the need to continually dip the wire into flux for additional fluxing. Additional information on fluxes and their application can be found in Chapter 4 and in chapters covering specific base metal systems.

PROCESS TECHNIQUES

Proper adjustment of the flame is essential for satisfactory results in torch brazing. Generally, a slightly reducing flame is desirable. One exception is the case of oxygen bearing copper, where a neutral or slightly oxidizing flame is employed to prevent rupture of the copper during heating. Reaction of the oxygen in the copper with the gases in a reducing flame can result in small pockets of steam forming inside the metal from reduction of the internal oxides.

Adjustment of the oxyacetylene flame is relatively simple and may be maintained by observing the characteristics of the flame as described earlier. However, the proper flame adjustment with some types of fuel gases is more difficult since no marked visual change in flame characteristics takes place. The adjustment can be made and controlled by the use of flowmeters. The outer envelope of the flame, not the inner cone, should be impinged on the brazement. This technique is different from that used in oxyfuel gas welding where intense heat is concentrated in the specific area being welded. The need to preheat areas adjacent to the joint will depend upon the size, shape, and thermal conductivity of the metals and the type of joint. Improper heating, primarily due to incorrect manipulation of the torch, may cause cracking, excessive filler metal-base metal interaction, or oxidation of the base metal and brazing filler metal.

Proper heating is required to offset the high thermal conductivity of joint members and to avoid stress cracking due to thermal shock. It is important that the joint be brought to a uniform temperature within the brazing temperature range so that the filler metal will flow freely and fill the joint. Figure 9.15 shows the various stages that a low temperature, general purpose brazing flux goes through during the heating cycle.

Overheating can be avoided by using a flux with a melting temperature slightly below the brazing temperature range of the filler metal. The melting of the flux can then serve as an indicator of the approach to the proper brazing temperature. This is of particular importance if the brazing filler metal is to be face fed in rod form. As soon as the flux is completely fluid the filler metal is touched to the joint and is applied in sufficient quantity until it flows completely throughout the joint. Heating is then stopped. This technique allows the molten flux to act as a temperature guide, and the heat from the assembly causes the filler metal to melt and flow.

It is poor practice to apply the flame directly to the filler metal. Overheating the filler metal is likely to result and fumes may be evolved. When heat is not supplied to the base materials in a sufficient amount a "cold joint" can occur. This is characterized by a lumpy joint appearance, perhaps even a balling up of filler metal.

Table 9.2
Standard Specifications for Brazing Fluxes

AWS Brazing Flux Classification	Fed. Spec. O-F-449d 02/06/85	Society of Automotive Engineers ASM	Application
FB3A	Type B	3410G	All purpose, low temperature flux for use in brazing both ferrous and non-ferrous metals and alloys.
FB3C		3411B	For brazing high chromium stainless steels, tungsten and chromium carbides, and molybdenum alloys.
FB3D		3417	Used where brazing temperatures go into the 1600° to 2000°F (870°-1100°C) range or for considerable time above 1450°F.
FB3E			Liquid flux with limited fluxing ability, used for brazing in furnaces with poor atmospheres or joining jewelry parts above 1160°F (625°C).
FB3F			Dry powder flux used in the 1200°F-1500°F (650°C-815°C) range.
FB3G			Dispensable slurry type flux using similar salts as FB3A.
FB3H			Dispensable slurry type flux using similar salts as FB3C.
FB4A	Type A		For brazing aluminum bronze and other alloys containing small amounts of aluminum and/or titanium.

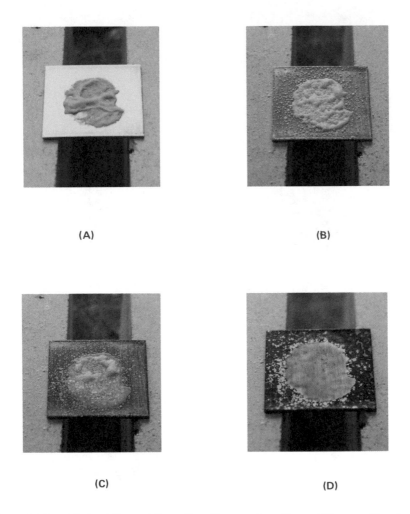

(A) (B)

(C) (D)

Figure 9.15 — Various Stages That a Low Temperature General Purpose Brazing Flux Goes Through During the Heating Cycles. The Photographs Show the Progressive Effects of Heating on a Flux. (a) Shows the Flux as Applied at Room Temperature, (b) Shows Flux at 212°F After the Water Has Been Driven Off, (c) Shows Flux at 600°F Giving it an "Eggs in the Frying Pan" Appearance, and (d) Shows Flux at 1100°F, Clear and Ready to Braze

Maintaining a uniform temperature is desirable. To accomplish this, auxiliary, multiple tip, or multiflame torches may be necessary. In mechanized operations, the parts can be moved, oscillated, or rotated as they pass through the heating zone. Alternatively, the flames can be moved around the parts. In some operations, torches or burners are positioned to heat the parts over the entire joint area. In any case, the purpose is to reach and maintain a uniform temperature at the joint.

Filler metal is commonly face fed in manual torch brazing operations. However, in mechanized operations where higher production is desired, proper brazed joint design will permit the use of preplaced filler metal in the form of preforms or paste. Preplaced filler metal should be positioned to avoid premature melting. Paste should be applied in such a way that the force of the flames does not blow it off the assembly.

Brazing should be performed in an environment with minimal air movement. Drafts should be avoided because they can cause uneven brazing temperatures. Filler metal should be applied so that it will flow toward the hottest section of the joint and, at brazing temperature, the heat should be directed to pull the filler metal in the desired direction through the joint.

When manually face feeding filler metal wire into the joint, the wire is dipped in flux, and then applied to the joint. Heat is then directed to flow the applied filler metal throughout the joint. When brazing small and thin parts with face fed wire, the brazing temperature should be sufficiently above the liquidus temperature of the filler metal to compensate for the chilling of the joint when the filler metal is touched to it.

At the start of the heating cycle, care should be taken to avoid blowing the flux off of the assembly with the flame. When using boron-modified fluxes, it is difficult to see through the murky nature of the molten flux. Operators should be cautioned not to wipe this flux away to improve their view of the joint. This practice can result in oxidized sections of the joint and insufficient filler metal flow at those areas.

When heating a tube-to-socket joint, the flame should be concentrated first on the inner member. This expands it and tightens the fit. The heating then is concentrated on the outer, more massive component and the inner tube is now heated by conduction. If a preform is being used, the heat is directed to draw the filler metal out of the joint. If face feeding of wire is being used, the braze joint should be heated in a manner that will draw the filler metal in to joint.

After completing any brazement, the joint must be allowed to air cool to below the solidus temperature of the brazing filler metal. This still air cooling must be accomplished without disturbing the joint through vibration or movement. Once the joint solidifies it may be quenched by immersing in or spraying with water or by using a forced air cool. The allowable rate of cooling will depend on the coefficients of thermal expansion of the base materials and the thermal shock that the joint can withstand.

Carbide brazed joints, for example, must be still air cooled to prevent cracking. To facilitate post-braze cleaning of the assembly, the quench water can contain commercially available flux and scale removal compounds.

SAFE PRACTICES

The safety hazards in torch brazing arise from (1) the improper use of oxyfuel gas equipment, (2) exposure to toxic fumes and gases emitted during the brazing operation and (3) the close proximity of the brazer to the heat, flame, and hot metal associated with the process. This subject is covered extensively in Chapter 6 and in ANSI Publication Z49.1, *Safety in Welding and Cutting*, published by the American Welding Society. These references, along with manufacturer's instructions on the proper use of all equipment and materials, should be an integral part of the planning and design of any torch brazing operation.

SUPPLEMENTARY READING

American Society for Metals. *Metals handbook: welding, brazing and soldering*, Volume 6, 9th Edition, 950-965.36. Metals Park, Ohio: American Society for Metals, 1983.

American Welding Society. *Welding handbook*, Volume 2, 7th Edition. Miami: American Welding Society, 1978.

—— *C3.4, Specification for torch brazing*, Miami: American Welding Society, 1990.

—— *ANSI/ASC Z49.1, Safety in welding and cutting*. Miami: American Welding Society, 1988.

Handy and Hartman. *The brazing book*, New York: Handy and Hartman, 1978.

Schwartz, M. M. *Brazing*, 189-195. Metals Park, Ohio: ASM International, 1987.

The Aluminum Association. *Aluminum brazing handbook*, 3rd Edition. Washington, D.C.: The Aluminum Association, 1979.

Keywords — furnace brazing, fluxless furnace brazing, brazing furnaces, batch-type furnaces, continuous type furnaces, retort/bell-type furnaces, cold wall vacuum furnace, protective atmosphere furnace brazing, hot wall vacuum furnaces, cleaning for furnace brazing

Chapter 10

FURNACE BRAZING

INTRODUCTION

Furnace brazing offers two prime advantages: (1) protective atmosphere brazing that substitutes high purity gases or vacuum for mineral fluxes and (2) the ability to control and record every stage of the heating and cooling cycles with computerized instrumentation. The latter has facilitated "repeatability" of the brazing cycle for high production of quality parts.

HISTORY

It was the problem of flux entrapment that first motivated engineers to turn to furnace heating when brazing filler metals were of the non-ferrous (silver, aluminum) types having low temperature melting ranges.

Furnace brazing is a process in which assembled components, with filler metal preplaced, are placed in a furnace. The furnace is purged with a gaseous atmosphere or evacuated of air to provide a specified low partial pressure of air. It is then heated to a temperature above the liquidus of the filler metal but less than the melting point of the base metals. The brazements are then cooled or quenched by appropriate methods to minimize distortion and produce the required properties in the filler and base materials. This cycle is designed to produce the required melting and solidification of the filler metal to join the components without melting or injuring the base metals.

In the 1920's the introduction of copper (BCu) as a braze filler metal into batch-type furnaces with exo- and endothermic gaseous atmospheres was a giant step into fluxless furnace brazing. The success of furnace copper brazing of carbon and low alloy steels in batch-type furnaces led to widespread use of continuous belt-type furnaces. Better reducing atmospheres, such as dissociated ammonia, improved the process and facilitated bright annealing and copper brazing of stainless steels. In recent years nitrogen base atmospheres (controlled mixtures of high purity nitrogen and hydrogen gases) have literally rejuvenated the use of continuous belt-driven furnaces.

The combination of the ultra-dry hydrogen atmosphere and furnace equipment resulted in the development of many nickel-base braze filler metals. It also led to a new group of filler metals that sometimes includes large percentages of such precious metals as gold, palladium, and platinum. New equipment served the aircraft engine and aerospace industries, and also many metal working industries involved with brazements for food handling and vacuum equipment, ultrasonics, instrumentation, and cryogenics.

Development of a fluxless method for brazing aluminum centered on versatile vacuum furnace equipment, which allows all types of brazing (aluminum, copper, nickel, silver, precious metal) to be performed on almost all known materials.

FURNACE CLASSIFICATION

Furnaces for brazing can be oxyfuel fired or electrically heated. All furnaces, whether they furnish heat directly or indirectly, must provide a uniform work load temperature. As an example, brazing of high temperature nickel base alloys at 2000 F (1093°C) and higher generally requires a hot zone uniformity of ± 15°F (± 8°C). When brazing aluminum alloys, the temperature uniformity must be ± 5°F (± 3°C). Therefore, it is important that each work load be monitored with a minimum of two thermocouples attached to or imbedded in the parts to be brazed.

Furnaces used for brazing processes are classified as (1) batch-type with either air or controlled atmosphere, (2) continuous-type with controlled atmosphere, (3) retort/bell-type with controlled atmosphere, (4) batch-type vacuum, or (5) semi-continuous and continuous type.

BATCH-TYPE FURNACE

An old-fashioned batch-type furnace (See Figure 10.1) which is refractory lined can be gas or oil fired, or heated by electricity. Initial brazements, placed directly onto the hearth, were not only coated with flux in the braze area but entirely covered with flux to prevent heavy oxidation to the base materials (usually low carbon steels). The atmosphere, as far as it existed, was the result of the by-products of combustion of air, gas, and fuel.

The addition of metallic muffles to refractory lined furnaces eventually shielded the work load from atmospheres created by combustion of air, gas, and fuel. The introduction of controlled atmospheres of generated gas (such as exothermic and endothermic) into the muffles along with a positive pressure flow to flush the brazing zone of contaminants from reduction, helped eliminate base metal oxidation. Dissociated ammonia (75% hydrogen and 25% nitrogen) was used as a high reducing gas to make stainless steel bright and clean as it emerged from the furnace.

CONTINUOUS-TYPE FURNACE

The need to process brazed assemblies faster than was possible with the batch-type furnace led to brazing in semi-continuous and/or continuous furnaces. The most common form is the conveyor-type furnace with mesh belt or roller hearth. With this system, brazements must pass through three zones in which a positive pressure of controlled atmospheric gas is continuously introduced. Both entrance and exit are equipped with flame curtains. Assemblies (with braze filler) proceed through the first zone where parts are preheated to a uniform temperature. Next is the "hot zone", or brazing area. The cooling zone under atmosphere is usually longer in order to provide clean non-oxidized work at the exit end.

Brazements are either loaded individually on the belt or in trays or baskets at the entrance end. Success depends upon the mass of the parts, speed of the conveyor belt, and the set temperature for the desired braze. With this system it is advisable to process a few assemblies with predetermined settings, and then make any needed adjustments to belt speed and temperature settings.

Although many such furnaces were built with refractory linings throughout the entire structure and on a horizontal plane, a much better atmosphere prevailed in a later model called a "humpback". It contained metal muffles and elevated the hot zone or brazing area above the entrance and exit zones. See Figure 10.2.

The introduction of nitrogen base atmospheres has given new life to continuous-type furnaces. Preparation and preconditioning is important — and so is a minimization of shutdowns. If a continuous-type furnace must be shut down from a production standpoint, it is better to idle the furnace. Restarting requires preconditioning at higher temperatures with atmosphere in order to bake out moisture and reduce metal oxides that have re-formed on the elements or muffles.

RETORT/BELL-TYPE FURNACES

Development of the nickel braze process along with the purification of hydrogen re-

Figure 10.1 — Old Batch-Type Furnace, Refractory Lined

sulted in the need for a new type of furnace with an inner container (retort/bell) made of a heat resistant alloy. The retort/bell is "sealed" from outside air and products of combustion in order to avoid contaminating the purified hydrogen atmosphere. Welded seals were originally used on box or tube types of retorts for experimental brazing in batch-type furnaces, with only ingress and egress tubes for the atmosphere. The welded retorts proved to be expensive, and for production quantities the retort/bell was based on a sand sealed design.

Sand sealed retorts/bells (Figure 10.3) consist of a two-piece construction. The top section is a "cover" or "hood" and the bottom section the "base". The cover is lowered over the work load and latched in such a way that when the cover is raised the base and the work load are raised also. Silica sand provides the seal at the gap between the bottom of the cover and the base. Ultra dry hydrogen enters at the top of the cover, and with a continuous and positive pressure flow, retorts/bells are purged of air. After purging, hydrogen exiting through the sand seal is ignited prior to lowering the retort/bell into the preheated pit-type furnace. This positive flow of hydrogen continues through heating, brazing, and

Figure 10.2 — Hump Mesh-Belt Furnace Brazes Light-Weight Stainless Assemblies Continuously with Uniform Time and Temperature

cooling cycles. This type of sand sealed retort/bell is still used to process many brazements. However, the use of water cooled retorts with O-ring seals and clamps provides for a much tighter seal than sand seals and is used for improved quality through better control of dew point. The need to braze production assemblies for the aircraft and aerospace industries required large retorts/bells with an operating system that is somewhat reversed. Figure 10.4 shows a typical multiple retort/bell set-up that provides semi-continuous operation. One retort/bell containing a work load can be heated in the furnace through the brazing cycle, while another retort/bell is cooling to room temperature or is being unloaded. Still another retort/bell is either in the loading stage or undergoing purging.

Brazing sources that used either type of equipment were faced with three disadvantages: (1) danger of explosive mixtures of hydrogen and air at the end of the purge and cooling operation, (2) rising cost of hydrogen and excessive expense of indirect heating and

cooling, and (3) slow cooling of stainless steel work loads was not compatible with physical and metallurgical properties of the base materials (unless they were of the stabilized or low carbon grades). The first two disadvantages are somewhat alleviated by using relatively inexpensive and inert nitrogen gas in lieu of hydrogen during purge and cooling cycles. However, caution must be exercised in switching to ultra-dry hydrogen prior to heating above 500°F (260°C) since there could be a reaction of retained nitrogen with nickel base filler metals containing boron in powder or paste applications. Other disadvantages are not fully overcome with any type of furnace heating that involves indirect heating and cooling of the work load. These problems were only solved with the introduction of cold wall vacuum furnaces with gas quenching.

Still another Retort/Bell-type furnace was called the Hydrogen-Vacuum (H_2V) furnace and is pictured in Figure 10.5. This retort/bell eliminated the sand seal at the bottom and included a water cooled O-ring seal

Figure 10.3 — Sand Sealed Retorts/Bells Consisting of Hood and Base

on top. The water cooled head also acts as the door of the pit furnace. Air is vacuum-pumped out before introducing the hydrogen and lowering the retort into a preheated furnace for cycling to the brazing temperature. During the cooling cycle the hydrogen is evacuated and the work load is cooled in vacuum, or the hydrogen exhaust is closed and a pressure of approximately two psi (14 KPa) is maintained using a regulator. These furnaces are still used by many specialists in hydrogen brazing.

BATCH-TYPE VACUUM FURNACES

The batch-type vacuum furnace came in two types: (1) "Hot Wall" and (2) "Cold Wall".

Hot Wall Vacuum Furnaces

As shown in Figure 10.6, the hot wall vacuum furnace is similar to the H_2V atmo-sphere furnace in which a vacuum pump purges air from the work zone and then introduces the hydrogen atmosphere for the heating and brazing cycle. In the hot wall vacuum furnace, a much better vacuum is created, and heating and brazing takes place in this atmosphere. Back-filling with argon or nitrogen gas speeds the cooling rate of the work load. Maximum operating temperature is usually limited to approximately 2100°F (1149°C). Later models adopted the double wall retort/bell which raised the maximum operating temperature.

Primary disadvantages of "hot wall" vacuum furnaces are indirect heating and the slow, static gas cooling of the work load. Slow cooling rates can prove detrimental to the physical and corrosion resistant properties of stainless steels.

Almost no "hot wall" vacuum furnaces are manufactured today. However, there are still some in operation that continue to provide satisfactory brazements, especially on aluminum assemblies.

Cold Wall Vacuum Furnaces

These are the most popular furnaces of the present high technology era and will probably continue well into the future to be the most versatile furnace for brazing.

Cold wall furnaces are: (1) Horizontal with front loading (Figure 10.7), (2) Vertical with top loading or bottom loading, and (3) "Clam Shell" with front loading (Figure 10.8). Horizontal furnaces with side loading are ideal for the brazing of small assemblies which can be placed in stacked baskets or on tiers of work grids. Unloading work from a horizontal vacuum furnace and immediately loading another batch not only provides semi-continuous operation, but also reduces open-door time and excessive contamination from room atmosphere. This is especially important in vacuum brazing of aluminum in batch-type furnaces. Vertical cold wall furnaces with bottom loading are ideal for large brazements to assure more uniform heating and cooling.

High capital equipment costs are usually the biggest obstacle to investing in a modern, cold wall vacuum furnace. However, versatility, safety, and quality with a well-equipped vacuum furnace are only a few of the many advantages which overshadow the initial investment cost. Following are fac-

Figure 10.4 — Electric Bell-Type Furnace (Right) Equipment Includes Three Bases, Water-Jacketed Cooling Bell (Center) and Steel Retorts with Sand Seals (Left)

tors and some essential pieces of equipment that yield brazements with a high degree of product reliability.

Heating. A cold wall vacuum furnace is a completely insulated furnace in which the work load is heated directly by radiation. The radiation comes from electric heating elements that usually encircle the work load. The heating elements can be metallic, usually molybdenum, tantalum, tungsten, nichrome, or nonmetallic materials such as carbon, graphite, and silicon carbide. Selection of the proper material will depend upon operating temperature range, compatibility with base and braze filler metals, and economic factors. Molybdenum and carbon are the most popular selections for high temperature vacuum furnaces that will operate in the 3000°F (1649°C) range.

Insulation. The cold wall vacuum furnace features double-wall construction and water cooling of the vessel. Since the heating of

the work load is from within, however, the insulation of the hot zone is also important in order to prevent heat loss. Following are two distinctly different insulation systems that are currently in use.

All-Metallic Shield Pack. This consists of multi-layered metal sheets which form a series of reflective shields. The type of shield material used depends on the temperature capacity of the furnace. For high temperature, molybdenum is commonly used for the innermost shields, backed by stainless steel shields with molybdenum hangers.

Advantages of the all-metallic shield pack are that it is highly reflective, features low heat storage which promotes faster cooling, and the surface area is small relative to fibrous insulation (hygroscopic) so that absorption of moisture and contaminating gases is reduced thereby reducing pump down time. Disadvantages are the initial costs of materials and labor and replacement and maintenance costs.

Figure 10.5 — Hydrogen-Vacuum (H₂V) Furnace Eliminated the Sand Seal at the Bottom and Included a Water-Cooled O-Ring Seal on Top. Water-Cooled Head also Acts as the Door of the Pit Furnace

Economical Non-Metallic Insulation Pack. This system employs fibrous insulation material such as alumina and graphite felt capped at the innermost area with laminated graphite. Advantages of the non-metallic insulation pack are low cost, better insulation properties, and minimum susceptibility to contamination.

Figure 10.6 — Hot Wall Vacuum Furnace with Double Wall

VACUUM PUMPING SYSTEM

An essential part of a vacuum furnace is a dependable pumping system. A four-stage system including a mechanical pump, cascade blowers, and holding and diffusion pumps should attain very low pressures in the 10^{-3} to 10^{-6} torr range. Such a four-stage pumping system is shown in Figure 10.9. Cryogenic pumps are sometimes substituted for diffusion pumps with various degrees of success. When a vacuum furnace is purchased for the sole purpose of copper brazing low carbon steels, the diffusion pump is not needed. (However, the port is usually blanked off so that if the furnace is needed later to braze stainless and high temperature base materials, the diffusion pump can be added.)

Pumping down to a suitable vacuum and maintaining a leak-free vacuum system throughout the brazing cycle are essential. Leak tests should be taken on each furnace on a scheduled basis. Specifications will vary on

maximum "leak-up rates" from 5 to 20 microns per hour. For brazing certain base materials (precipitation hardenable with percentages of titanium or aluminum, or both, the leak-up rate should be at 5 microns or less per hour. It is also recommended that the furnace and furnace fixtures be preconditioned at higher temperatures than the temperature planned for the braze cycle. Repetitious copper brazing assemblies of carbon steel are not affected by a higher ($>$20 microns per hour) leak-up rate since they are done with partial pressure gas operation. A leak check is always advisable and if a high rate is recorded the leak should be repaired as soon as possible.

PARTIAL PRESSURE GAS OPERATION

When brazing in vacuum (low pressures) at high temperatures, knowing the vapor pres-

Figure 10.7 — Cold Wall Vacuum Furnaces, Horizontal with Front Load

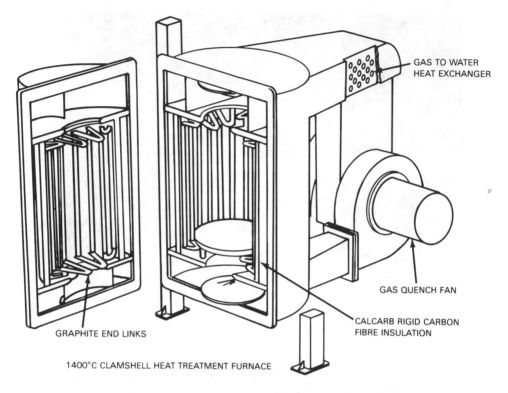

GAS TO WATER
HEAT EXCHANGER

GAS QUENCH FAN

CALCARB RIGID CARBON
FIBRE INSULATION

GRAPHITE END LINKS

1400°C CLAMSHELL HEAT TREATMENT FURNACE

Figure 10.8(A) — 1400C Clamshell Heat Treatment Furnace

sures of all the elements is important, whether they are 100 percent or just a small percentage of the base metals or the braze filler metals that are in the work load. (Vapor pressure is defined as: "That pressure exerted at a given temperature at which the material is in equilibrium with its own vapor".) If either the vacuum is increased (lower pressure) or temperature is raised, the material will vaporize and the vapors will condense on the colder parts of the hot zone. Condensation of these metallic elements can take place on the ceramic insulators of the electrical connections and, in time, can render them conductive. This, of course, can cause short circuits.

The solution is to provide a partial pressure gas operation in which high-purity gases such as argon, nitrogen, or hydrogen are introduced into the work zone at a constant flow to produce a vacuum pressure that is above the vapor pressure of the ma-

terial in question. Partial pressures between 200-500 microns will suppress any vaporization through the higher brazing temperature. The system will hold the partial pressure until the chamber is backfilled with an inert gas (argon or nitrogen) for the cooling or quenching cycle.

Materials and braze filler metals that contain such elements as cadmium, lead, or zinc should not be processed in vacuum because partial pressures are not effective in suppressing vaporization at brazing temperature.

Partial pressure gases should be of high purity, or from cryogenic cylinders or on-site installations. They should also be compatible with the braze filler metal and the base materials. The use of nitrogen is not recommended if base metals contain even the smallest percentage of titanium, zirconium, beryllium, or aluminum.

Figure 10.8(B) — Clamshell High Vacuum Heat Treatment Furnace with Door Open

INSTRUMENTATION

According to AWS C3.6, *Specification on Furnace Brazing*, all brazing furnaces shall have automatic temperature controlling and recording devices in good working order capable of controlling the temperature of the furnace to the requirements of the specification. They shall have adequate heating capacity to accomplish uniform heating of the workload at the rate required to prevent both unacceptable thermal distortion of the assemblies and liquation of the filler metal. They shall also be properly maintained in good working order.

Figure 10.10 shows the front panel of a control cabinet containing vacuum gauges, temperature instrumentation, and temperature controllers.

Programmable microprocessors can control and provide a braze cycle through the heating, brazing, cooling, and quenching phases to specific temperatures and holding times. Such special events as a "partial pressure gas operation" can also be programmed as to where and when the gas must be introduced.

The center of instrumentation and record keeping is a digital process recorder that provides a complete account of furnace temperature, work load temperatures (any quantity), and vacuum in microns or torr. If so programmed, charts can provide heat number, customer, part number, quantity of parts, type of braze process, and date. Furnace charts are an indispensable aid for troubleshooting a load that produced an unusually large number of defective brazements.

The process recorder can eliminate the tedious work of interpolating and hand-recording the temperatures of each thermocouple every five minutes for Temperature Uniformity Surveys, as specified by military and aerospace material specifications.

Figure 10.9 — Four-Stage Vacuum Pumping System

COOLING SYSTEMS

Rapid cooling of a work load of brazements is essential to the metallurgical properties and chemical corrosion resistance of the base materials when they are heat processed at braze temperatures well above their critical temperatures. Faster cooling rates also reduce furnace time and save money.

Backfilling with inert gases promotes convective transfer of heat from the work to the water-cooled shell. Early cold wall vacuum furnaces were equipped with a conglomerate of small fans, bungs, baffles, and copper finned coils to cool and direct the flow of backfill gases through the work load and accelerate heat removal. However, because of fluctuations of gas flow through the work load, most cooling systems did not work well.

To increase the rate of heat removal, the mass flow had to be increased. Mass flow is the product of the gas moved multiplied by the velocity of the gas. The gas flow can

therefore be increased by (1) increasing the pressure of the backfill gas in the system and (2) increasing the velocity of gas circulating through the work load to absorb heat from the work.

While there are many different cooling/quenching systems, furnace manufacturers seem to favor two:

(1) An internal gas quenching system in which the inert backfill gases are introduced into a plenum behind the insulated hot zone. By means of baffles and nozzles, the gas is directed at high velocity in an orbital fashion at the work load. The hot gases are recirculated through an external heat exchanger and then the cooled gases are forced back into the hot zone by means of an external blower.

(2) Vacuum furnaces with advanced turbocharged gas quenching integrated with full-circle heat treating. These introduce inert backfill gas into the hot zone in an orbital pattern. With a large internal turbo fan this gas is

Figure 10.10 — Front Panel of a Control Cabinet Containing Vacuum Gauges, Temperature Instrumentation, and Over-Temperature Controllers

recirculated through the work load. The hot gases are recirculated through an internal heat exchanger and through the fan re-enter the hot zone of the vacuum furnace.

Although furnaces with such cooling systems can achieve extremely rapid quenching rates by positive pressure gas quenching, these rapid rates are usually not needed for most brazements. In fact, the turbulent action of the cooling gases at higher pressures can create distortion. Positive pressure gas quenching requires that the furnace be equipped with positive pressure clamps on the doors, and operators must follow stringent safety procedures. High pressure vacuum furnaces must be frequently tested and often re-coded as to performance at high pressure. These furnaces operate at from 2 bar to 7 bar when cooling. Higher pressures are being investigated.

The following gases can be used for backfilling for cooling/quenching. They should be compatible with the base metals and braze filler metals.

(1) Argon. An inert gas available in high pressure or cryogenic cylinders, or at on-site installations. It is the most preferred gas and is readily available. However, it is expensive and has the slowest cooling rate of the gases listed here.

(2) Nitrogen. A relatively inert gas except to the reactive elements. Nitrogen is available in high pressure or cryogenic cylinders, or at on-site installations. It is gaining popularity because it cools faster than argon and costs less. Nitrogen should not be used if the base materials contain even small percentages of titanium, beryllium, zirconium, or aluminum. However, it can be used for cooling aluminum brazements.

(3) Helium. An inert gas available in high pressure cylinders. It is rarely used because of lack of availability and high cost. Helium has a faster cooling rate than both argon and nitrogen.

(4) Hydrogen. A highly reducing gas available in high pressure cylinders (banks, cascades). It has the fastest cooling rate of the four gases listed. The prime disadvan-

tage is that hydrogen can be explosive when mixed with air and ignited.

SEMI-CONTINUOUS VACUUM FURNACES

For a long time aluminum was vacuum brazed in high temperature cold wall vacuum furnaces despite the fact that it was a very expensive operation unless production was high. A single furnace load of aluminum would require three furnace cycles: (1) bake-out and pre-conditioning of furnace, (2) aluminum braze cycle, and (3) bake-out to rid furnace of contaminants. Since vacuum furnaces for aluminum brazing did not actually require high temperatures, the heating elements in cold wall vacuum furnaces eventually reverted to nickel-chrome with temperature limitations of 1200°F (649°C), thereby constituting a more economical batch-type vacuum furnace.

The introduction of vacuum brazing for aluminum base materials in the mid-1960s proved to be very timely for such lightweight automotive parts as radiators, evaporators, heat exchangers, oil coolers, etc.

The success of a fluxless brazing process in the batch-type cold wall vacuum furnaces led to semi-type vacuum brazing by the late 1970s. Furnaces of this type usually employ three chambers: (1) a loading chamber (2) the hot zone chamber and (3) removal chamber. See Figures 10.11(A), 10.11(B), 10.11(C) and 10.11(D). Such vacuum furnaces have been known to run for over a year without a major overhaul.

FLUXES EMPLOYED IN FURNACE BRAZING

The terms "fluxless brazing" and "protective atmosphere furnace brazing" are used synonymously. However, prior to the introduction of protective atmospheres, mineral fluxes were essential to low temperature brazing with silver and aluminum braze filler metals. Furnace brazing with flux, with or without the aid of a gaseous atmosphere, continues to be called out on outdated specifications and engineering drawings which contribute to retention of flux regardless of the method of heating.

Figure 10.11(A) — Batch Aluminum Brazing Furnace for Brazing Evaporators, Radiators, and Air Charge Coolers

Figure 10.11(B) — Another View of the Batch Aluminum Brazing Furnace

Brasses, bronzes, and other base materials that contain elements such as zinc, cadmium, lead, magnesium (except in vacuum brazing of aluminum), etc. with high vapor pressures are not conducive to vacuum furnace brazing even with partial pressure gas operation. These base materials can be brazed successfully in a gaseous protective atmosphere furnace with a flux coating of the base metal at the braze joint. The mineral flux will somewhat suppress the vaporization of the high vapor pressure elements. Faster heating, such as in torch or induction methods, has proven to be very effective in the brazing of brasses and bronzes, with the silver base braze filler metals having the lowest melting and flow points.

The introduction of precipitation hardening steels, some of which contain aluminum and titanium of more than 1 percent of the reactive elements, presented special problems. Upon heating in an ultra dry hydrogen atmosphere (the best atmosphere at that time), very stable oxides would form and act as a barrier to the wetting and flow of the molten braze filler metal. To accomplish a satisfactory braze on these steels it was necessary to employ a high temperature flux in the braze joint and adjacent areas. This "bandage" technique contributed not only to braze joint failures, but also to the lowering of metallurgical properties such as corrosion resistance of the base materials. Nickel plating of faying and adjacent surfaces of the braze joint to prevent oxide formation eliminated the need for flux in brazing this class of stainless steels.

The vacuum brazing process for aluminum that is employed today is "fluxless" and depends upon the vaporization of magnesium below the brazing temperature to act as a "getter" to residual moisture. In the early years of development of this process, small quantities of magnesium powder were placed

Figure 10.11(C) — Batch Aluminum Brazing Furnace for Radiators, Condensors, Evaporators, and Air Charge Coolers

within the hot zone. Manufacturers of aluminum and aluminum braze clad have increased the magnesium content to suitable percentages so that powder additions of magnesium are not necessary. This has helped considerably in reducing maintenance problems due to condensation of magnesium, a high vapor pressure, metallic element.

PROTECTIVE ATMOSPHERE FURNACE BRAZING

The introduction of protective atmospheres in lieu of fluxes initiated the era of protective atmosphere furnace brazing for base metals of carbon steel, alloy steels, stainless steels, and others. The process produces brazements with a high degree of quality and reliability due to elimination of porosity and voids caused by flux entrapment.

CLEANING

The precleaning of detail parts for furnace brazing is the same as for all the other methods of brazing. All surfaces of the base materials should be free of oil, grease, paint, oxides, dirt, scale, or other foreign substances that can interfere with the brazing process or contaminate the braze joint. Burrs shall be removed as required prior to final cleaning to permit proper assembly and filler metal flow. Special chemical or mechanical cleaning methods for specific base materials are discussed in the relevant chapters for those materials. Additional information is also available in Chapter 7 on Precleaning and Surface Preparation. Gaseous protective and vacuum atmospheres do perform the final cleaning functions during the brazing cycle, but if overburdened with cleaning of dirty parts that should have

Figure 10.11(D) — Continuous Aluminum Brazing Furnace for Evaporators

been cleaned prior to brazing, the braze quality suffers.

DESIGN AND BRAZE JOINT CLEARANCE

Design and braze joint clearance for assemblies that are to be protective atmosphere furnace brazed are discussed in Chapter 2 on Design. However, there have been misconceptions about braze clearance and increasing braze clearance when employing certain braze filler metals where base metal and braze filler metal interaction occurs.

Brazing clearance can be said to have a volume relationship since the larger the clearance between faying surfaces, the larger the amount of braze filler metal that must be applied at the joint to be brazed. Also, the surface condition of the faying sur-

faces is important to both the capillarity and wetting action of the molten brazing filler metal. Pickled surfaces of sheet metal details are ideal. Machined details should not have surfaces that are mirror-like since this condition lowers capillarity and flow of the filler metal. This can be remedied by metallic grit blasting with a triangulated grit (SAE 120) which is compatible with the base material. For example, to prevent rusting of stainless steels, a stainless blasting material should be used. Refractory blasting grits such as sand or aluminum oxide and similar types should not be used since they create a refractory barrier which inhibits the flow and wetting action of the molten braze filler metal and leaves non-metallic inclusions which lower the joint strength of the brazement.

After brazing, a visual examination of the surfaces that were blasted with the refractory materials may show a "superfluous flow" due to the roughened areas. However, tests indicate there is a drastic reduction in braze joint strength due to the imparted refractory layer. This could prove disastrous if the brazement is designed and intended to function in a highly stressed application.

Finally, upon rapid heating to the braze temperature, the flow of molten braze filler metal by capillarity through the joint is instantaneous. Braze filler metals with a large temperature differential between the solidus (melting point) and the liquidus (flow point) when heated slowly may result in "liquation". Liquation results when the low melting constituents of these brazing filler metals melt and flow at a temperature just above the solidus. This advance flow of molten braze metal completely fills and completes the braze, leaving the higher melting constituents at the application area in a liquid-solid stage. When the liquidus temperature is attained, the higher melting constituents may or may not fully melt. However, since the joint may not be fully filled, there may be a need for this additional metal. The unmelted residue will remain at the point of filler metal application. Formation of the unmelted residue can be objectionable. A similar unmelted residue will result with some filler metals when the protective atmosphere has a higher than desirable partial pressure of oxygen.

To successfully braze areas which are larger than a normal braze joint, it is best to consider (1) designing for intimate fit-up, (2) using free flowing (eutectic or eutectic-type) braze filler metals, and (3) processing these assemblies in the best quality protective atmosphere. One of the first aircraft engine parts that was brazed with a nickel base braze filler metal (BNi-2) is pictured in Figure 10.12. Actually the original design specified "resistance spot welding" to join alternate corrugations to straight sections of sheet metal. In actual operation, however, thermal fatigue and stress corrosion contributed to failures at the spot welds. Recommended remedial procedure was to apply the BNi-2 braze paste to over 100 joints, each of which was about 0.5 sq. in. (12.7 sq. mm). The nickel braze procedure eliminated the service problem, but ironically a weldment that specified resistance spot welding then became a brazement — with resistance spot being merely the means

for positioning and holding parts together for furnace brazing.

The benefits of nickel braze on the combustion liners did not end there. The lower or exhaust end of the combustion liner was designed to have a doubler section spot welded on 0.375 in. (9.53 mm) centers for faster heat removal. The purpose was to retard deterioration of the thinner [0.050 in. (1.27 mm)] cobalt base sheet metal by oxidation. The extreme lower end contained a doubler section which measured 3 in. by 9 in. (76.2 mm by 228.6 mm). Placement of nickel braze paste along the 9 in. (228.6 mm) length was sufficient to exhibit complete flow through to the opposite edge. The exceptional flow of braze material provided a homogeneous structure at this transition zone with ideal conditions for thermal transfer.

Brazing large areas with pre-placement of braze filler metals that are not of the free

Figure 10.12 — Combustion Liner, One of the First Aircraft Engine Parts Brazed with a Nickel Base Braze Filler Metal

flowing type may necessitate consulting with suppliers or those experienced with design and application of those materials. When shims of braze filler metals are preplaced, adjustments in the amount of shim and the heating cycles are very important to the success of the brazement.

ASSEMBLY AND FIXTURING

A prime advantage of furnace brazing is uniformity of heating with minimum distortion. This is extremely important to the assembly operation. Detail parts made from similar base materials that are designed for close fitting and self positioning assembly are ideal; however, it is desirable to maintain this relationship during the entire handling process from braze filler application to furnace brazing. Staking, gas tungsten arc tack welding, and spot welding are some of the methods used to maintain the relationship of details and to provide fixturing for the brazing operation. Assembly fixtures are preferred over brazing fixtures, especially when brazing at elevated temperatures. Sometimes furnace fixturing is essential to compensate for a differential of the coefficients of expansion of the base materials or when parts are very complicated in design.

A certain condition can occur when dissimilar metals are in contact with each other and are heated, particularly when they are oxide free and in a protective atmosphere furnace. Whether they are parts being brazed or the furnace fixturing (baskets, screens, work grids, hearth, etc.), these materials will interact and form a new alloy. The new alloy may be eutectic that will melt at a much lower temperature than the melting points of the original materials. Some

eutectic-forming materials, along with their critical melting temperatures, are listed in Figure 10.13. Processing temperatures near or above these points should be carefully considered. Ceramic cloths placed at contact areas will prevent interaction. For additional information on assembly and fixturing, please refer to Chapter 8.

ATMOSPHERES FOR FURNACE BRAZING

The selection of the proper atmosphere to employ for protective atmosphere furnace brazing can become rather complicated. Not only must the type of atmosphere and compatibility with the base materials and the braze filler metal be considered, but also the furnace equipment that is available and its capability for maintaining the quality of the atmosphere throughout the entire braze operation.

Chapter 4 contains a thorough evaluation of furnace atmospheres and components. Some of the most relevant information will be expanded upon in this chapter. The AWS designated atmospheres as they are listed in Chapter 4 can be grouped into three classifications (1) active or reducing, (2) inert or relatively inert, and (3) vacuum, which can react as #1 or #2 depending on the type of materials being processed.

Active or Reducing Atmospheres

Prior to the introduction of vacuum furnaces, practically all protective atmospheres employed in furnace brazing contained active components essential to the braze process. The active components of the initial atmospheres, for instance, were hydrogen and carbon monoxide. Both reacted with metallic oxides to create a

MATERIAL COMBINATIONS	*APPROXIMATE EUTECTIC MELTING POINTS*	
Moly/Nickel	2310°F	(1266°C)
Moly/Titanium	2210°F	(1210°C)
Moly/Carbon	2700°F	(1482°C)
Nickel/Carbon	2130°F	(1166°C)
Nickel/Tantalum	2450°F	(1343°C)
Nickel/Titanium	1730°F	(943°C)

Figure 10.13 — Eutectic Forming Materials in Vacuum Atmospheres

chemical reaction as the parts were heated. The chemical reaction produced gases which, along with other gases containing contaminants, were flushed from the braze area with a positive flow of gaseous atmosphere. This is indicative of the similarity of atmospheres and fluxes and also establishes two advantages of atmosphere brazing: (1) uniform heating and cooling of the entire assembly is essential to uniform filler metal flow throughout the braze joint, and (2) cleaning of entire assembly and removal of residual fluxes (corrosive) is not needed because the parts emerge from the brazing operation bright and clean.

The initial atmospheres (AWS Types 1 through 4) were generated from the combustion of fuel gases, and are very economical to produce. Later, a gaseous atmosphere (AWS Type 5), generated from ammonia and dried, offered a 75 percent hydrogen content which was very effective for the copper brazing and bright annealing of many of the stainless and high temperature alloys. In recent years, the nitrogen base atmospheres (AWS Types 6A and 6B) have actually rejuvenated interest for the copper brazing and bright annealing of carbon, low alloy, tool, and stainless steels. The nitrogen base atmospheres are basically mixtures of high purity nitrogen with smaller percentages (1-30%) of hydrogen, or with even smaller percentages (2-10%) of hydrogen. The flexibility of adjusting the ratios of these nitrogen/hydrogen atmospheres (Type 6A and 6B) to suit the conditions of heating, brazing, and cooling on the wide variety of base metals is a decided advantage. The higher cost of high purity gases is offset by the elimination of generating equipment and costly maintenance.

Cylinder hydrogen (97-100% and AWS Type 6) was available for furnace atmosphere, but its prime usage was for the atomic hydrogen welding process (precedent to the gas tungsten arc welding process). This hydrogen atmosphere proved instrumental in the development of nickel base braze filler metals. The ultra dry hydrogen (AWS Type 7) atmosphere was valuable in introducing the cobalt and precious metal groups of braze filler metals, and it remains as the best reducing gaseous atmospheres for the nickel brazing of stainless steels. For a better understanding of the hydrogen atmosphere, see Chapter 4 on fluxes and atmospheres.

Inert or Relatively Inert Gases

Argon and helium (AWS Type 9) of high purity and low dew point are inert gases that can be used as protective atmospheres for all types of furnace brazing in retort/bell type furnaces. Since neither of these gases are active and therefore will not reduce surface oxides, they are frequently employed in mixtures with ultra dry hydrogen atmosphere (AWS Type 7). When they are used without the addition of hydrogen, it is necessary that the furnace interior, fixturing, and assemblies provide the ultimate in cleanliness prior to the furnace braze operation.

Although nitrogen (AWS Type 9) is not an inert gas, it is considered relatively inert to all materials except those that contain reactive elements such as aluminum, titanium, and tantalum. Pure nitrogen is not usually employed as a brazing atmosphere, but it has gained popularity as a purging and backfill gas for hydrogen brazing operations due to its inert qualities (safety) and low cost (economy). It is, however, used as a brazing atmosphere when brazing copper base metal with silver filler metals.

Argon and nitrogen gases are used extensively with vacuum furnaces for brazing and heat treating: (1) to create partial pressures in order to suppress the volatilization of the elements in either the base material or braze filler metal when temperatures exceed their vapor pressures; and (2) as a backfill gas for more rapid cooling/quenching of the work load. Hydrogen (AWS Type 7) can also be used, but specific safety precautions must be observed. Again the choice of gas to employ will depend upon its compatibility with the base materials and the braze filler metal. This is the main reason that argon is mandatory for certain aircraft and aerospace functions. Nitrogen is popular for certain applications because it is more economical than argon and provides a faster quench rate.

The initial use of such atmosphere gases as hydrogen (AWS Type 7) and argon (AWS Type 9) in high pressure cylinders presented some problems due to variations in purity and moisture content. The need to provide a large quantity of "on site" supply created banks of cylinders with "pigtail" connections. Each connection became a potential source of leakage, not only for gas to escape, but for air to seep into the gas supply. The installation of large cascade containers with a minimum of connections has reduced this source of contaminants entering the furnace.

Delivery of gases with ultra high purity and low moisture content (and certified accordingly) has become standard practice for many critical brazements. However, argon and nitrogen in cryogenic containers or "on site" installations are ideal for vacuum furnace brazing operations.

Vacuum Atmospheres

Vacuum is defined as a space absolutely devoid of matter; therefore a vacuum does not contain any active components which can create a chemical reaction. The reaction that takes place in a vacuum is a physical one in which the low pressures (high vacuum), combined with sufficiently high temperatures, dissociates the metallic oxides and produces atomically clean surfaces. Such surfaces are ideal not only for brazing, but also for coating, plating, etc. Degassing of materials can only be accomplished in vacuum; therefore, vacuum is valuable in removing hydrogen from titanium base materials which are to be employed for critical applications.

Zero absolute pressure or the perfect vacuum has not been attained, and even the vacuum in intergalactic space contains at least a few particles of matter per cubic meter. On earth, a vacuum is defined as any pressure less than atmosphere, and atmospheric pressure is considered zero gage pressure. The pressure in high vacuum chambers refers to pressure above absolute zero and is measured in Torr, usually from 1×10^{-6} Torr up to 1×10^{-3} Torr (1 micron).

Conversion of pressure (vacuum) units that are conducive to vacuum atmospheres is shown in Table 10.1.

Vacuum furnace brazing of some base materials and braze filler metals such as copper and silver alloys require the employment of gases such as argon and nitrogen for partial pressure gas operation and as a quenching medium for rapid cooling. Hydrogen (AWS Type 7) is also used in vacuum for some braze operations. However, it is used more frequently as an active atmosphere for the cleaning of furnace hot zone and fixturing in a bake-out cycle prior to the braze cycling of critical brazements.

TECHNIQUES FOR PROTECTIVE ATMOSPHERE FURNACE BRAZING

Such techniques as direct heating and cooling/gas quenching of work load, partial gas operation, proper employment of work load sensors (TCs), double pump-downs, and high temperature "bake-outs" of hot zone and fixtures have facilitated continued development of the furnace brazing process.

In addition, the introduction of the protective atmosphere furnace method of copper brazing (by substituting a suitable atmosphere for the mineral fluxes and practically eliminating braze clearance) produced superior braze quality and brazements which were virtually free of distortion. Today protective atmosphere furnace brazing with copper braze filler metals continues to be the most economical method of joining carbon steels when large quantities are to be processed.

Development of the furnace nickel braze process with BNi-2 braze filler metal provided a highly reliable method of joining assemblies with not only structural strength, but heat and corrosion resistance. Two major advantages of the furnace nickel braze process are that it:

(1) Provides uniformity of heat and cooling/quenching of brazements, which minimizes distortion and allows many applications requiring finish machined details to be assembled and brazed

(2) Offers opportunity to select a braze filler metal to perform the brazing and heat treating operations either concurrently or conjunctively

(3) Provides a higher quality brazement at an economical cost

Table 10.1
Conversion of Pressure Units

UNIT	Torr	micron (μ)	Atmosphere (technical)	Atmosphere (physical)	Microbar (μ bar)	Millibar (mbar)	Bar (bar)
1 Torr	1	10^3	1.3595×10^{-3}	1.3158×10^{-3}	1333×21	1.33321	1.3332×10^{-3}
1 micron (μ)	10^{-3}	1	1.3595×10^{-6}	1.3158×10^{-6}	1.33321	1.3332×10^3	1.3332×10^{-6}
1 atmosphere (tech.)	735×6	7.356×10^5	1	0.9678	9.807×10^5	0.9807×10^2	0.9807
1 atmosphere (phys.)	760	7.6×10^5	1.033	1	1.013×10^6	1.013×10^3	1.013
1 microbar (μ bar)	7.501×10^{-4}	0.7501	1.02×10^{-6}	9.8698×10^7	1	10^{-3}	10^{-6}
1 millibar (mbar)	0.7501	7.501×10^2	1.02×10^{-3}	9.8698×10^4	10^3	1	10^3
1 bar abs. atmos.	750×1	7.501×10^5	1.02	0.98698	10^6	10^3	1

CLEANING OF DETAIL PARTS FOR FURNACE BRAZING

Cleaning is a vital step towards attaining satisfactory brazements. Blasting materials, chemical solutions, and other recommended procedures have been stipulated in chapter 7. The final cleaning operation occurs while heating the parts when surface oxides are reduced or volatilized, enabling the molten braze filler metal to flow and wet the faying surfaces. While employing fluxes with torch and induction methods of brazing, the final cleaning operation is locally concentrated at the faying and adjacent surfaces of the braze joint. However, with brazing in protective atmospheres the final cleaning takes place over the entire surface area of the total work load, including the furnace interior and fixturing.

Sand Castings. The outer surfaces (skins) of castings are usually permeated with molding compounds during pouring. Long imposed cleaning procedures at most foundries are usually accomplished by shaker hearth (vibration), followed by high pressure blasting with either silica sand or aluminum oxide, two refractory materials which are not recommended for blasting on parts to be furnace brazed. Annealing of castings is usually done in air or disassociated ammonia atmosphere followed by water quenching. Again, blasting to clean the castings is conducted with the undesirable silica sand or aluminum oxide. These surface contaminants must be removed by chemical or mechanical methods before parts are brazed. Absorbed gases such as nitrogen and oxygen can also cause brazing problems. To eliminate these problems, furnace cleaning (out gassing) can be accomplished in a simulated furnace brazing cycle prior to the actual brazing cycle.

Stainless Steel Tubing. Tubing may undergo a number of drawing and process anneal operations. Drawing lubricants may not be completely removed by vapor degreasing and can remain on O.D. and I.D. during the next annealing operation. Final cleaning operations usually consist of blasting with aluminum oxide (fine mesh), followed by polishing with polishing compounds. These operations leave oxides on the tubing surface which may result in poor flow of brazing filler metals.

In the early days of protective atmosphere furnace brazing of stainless steel hypodermic tubing, problems existed due to burrs left from hand de-burring operations. These problems were overcome with the introduction of electrolytic cutting machines which could cut tubing without leaving burrs. This operation involved the submerging of tubing in the electrolyte (salt solution bath), with the likelihood of salt residues remaining on the tubes where brazing was to be performed.

From a practical standpoint, it is best to assume that contamination problems may exist and to therefore provide additional operations prior to assembly such as:

(1) Either machining surfaces of castings which are to be brazed or sanding them with metal carbide disks or cloth.

(2) Bright annealing castings and tubing in a high vacuum (1 micron). This is an excellent choice since a vacuum can produce surfaces that are atomically clean. Other atmospheres are suitable for base materials which can be brazed in them.

(3) Specify with request for certification that tube lengths must be repeatedly flushed in water to remove all salt residues.

A very thorough cleaning of all surfaces of detail parts that will be subjected to the protective atmosphere furnace methods of brazing is important.

SELECTION OF BRAZE FILLER METALS FOR PROTECTIVE ATMOSPHERE FURNACE BRAZING

Throughout the book and in publications distributed by manufacturers there are recommendations as to the many filler metals that are compatible with specific base materials, and those that can be used to furnace braze. This information may have its shortcomings in that there is no distinction as to "protective atmosphere" furnace brazing or "furnace brazing in a gaseous atmosphere" with additions of fluxes. Upon the introduction of vacuum atmospheres, updated information has distinguished those filler metals that do not contain elements such as zinc, lead, cadmium, etc., which have high vapor pressures. Such filler metals with high vapor pressures are not conducive to furnace brazing in a vacuum atmosphere since the volatilization of these elements will reduce service life of the furnace.

The proper selection of filler metal and its form to be used in the protective atmosphere method of furnace brazing should be based upon the following factors:

(1) Compatibility with base materials to assure the metallurgical integrity of the brazement so it will function properly throughout its service life.

(2) Filler metal form (powder for paste and spray applications, wire, sheet, foil, pre-forms, transfer tape, etc.) available to best satisfy the ease of application and size of braze joint.

(3) The cost of the filler metal: although the amount of filler metal is sometimes considered insignificant, most of those assemblies that are furnace brazed contain a multitude of braze joints. This, together with large production quantities, can require substantial amounts of filler metal.

Throughout the remainder of this section the primary classifications of filler metals (BAg, BCu, BNi, and BAu) will be discussed for the protective atmosphere furnace brazing of carbon and low alloy steels, copper and copper alloys, high temperature alloys and corrosion resistant alloys. Special or newly developed filler metals for protective atmosphere furnace brazing of the more exotic base materials are mentioned in respective chapters which bear their name. Aluminum filler metals are discussed in Chapter 25.

BAg (Silver Base) Filler Metal Classification

The early transfer of silver brazing of carbon and low alloy steels from torch brazing to protective atmosphere furnace brazing to take advantage of the uniformity of heat was not considered successful for three reasons: (1) Lack of active atmosphere gases; (2) furnaces were incapable of maintaining atmosphere purity and dew point at brazing temperatures and, (3) existing furnaces provided very slow heat transfer to the work load.

Many publications state that the functions of fluxes and atmospheres are comparable; however, with protective atmosphere brazing not only the entire brazement must be heated to the brazing temperature, but the entire work load, including the furnace fixturing when used, and furnace interior. Therefore, it was necessary to supplement gaseous atmospheres with the addition of fluxes, especially at the braze joint, in order to protect the silver braze filler metals from oxidation and gassing off of zinc and cadmium. With the introduction of more active atmospheres, such as exothermic, dissociated ammonia, ultra dry hydrogen, and vacuum, it appeared that carbon and low alloy steels might be furnace brazed. Experience proved, however, that brazing at low temperatures (1200 to 1600°F [650 to 870°]) caused poor flow and wetting as well as oxidation of stainless steels.

The early desire to join stainless steels led to the use of low melting silver filler metals along with flux at the braze joint. Flux was essential to reduce the highly refractory oxide of chromium and to produce flow and wetting of the silver braze filler metal. Nickel plating of the stainless steels was sometimes employed when using silver filler metals for protective atmosphere furnace brazing. Both of these methods resulted in some success, depending upon the service conditions to which the stainless steel brazement was subjected. The two metallurgical phenomena which threaten service performance of a stainless steel brazement are: (1) carbide precipitation or sensitization of the stainless steel, and (2) interface corrosion. These conditions are attributed to: (1) brazing temperatures within the carbide precipitation range (800 to 1600°F [425 to 870°C]) and (2) the employment of flux, responsible for the depletion of a nickel rich layer at the faying surfaces. Carbide precipitation and interface corrosion are discussed in Chapter 24 on Stainless Steels.

One silver braze filler metal (BAg-13) emerged as a "high temperature" filler metal although its liquidus was only 1575°F (855°C) and the alloy contained five percent zinc. This filler metal was employed quite successfully for protective atmosphere furnace brazing of the AISI 400 series martensitic stainless steels in ultra dry hydrogen atmosphere at temperatures in the range of 1850 to 1900°F (1010 to 1040°C) which enabled brazing and austenitizing (hardening) cycles to be run concurrently. A brazing temperature of 1900°F (1040°C) also coincided with solution annealing temperatures for the AISI 300 series of stainless steels, and most of the precipitation hardening stainless steels. Unfortunately the zinc was vaporized in the pure dry hydrogen atmosphere, creating a health hazard. This also changed the chemistry and melting temperature of the filler metal.

The introduction of the vacuum furnace brought about the development of BAg-13a, very similar to BAg-13, but without the zinc content. Due to vaporization of silver at the 1800-1900°F (980 to 1040°C) brazing temperatures and 10-3 torr vacuum, it is necessary to employ a partial pressure gas operation in which an inert gas such as argon is introduced into the hot zone at a temperature below 1650°F (900°C). Furnace brazing of stainless steels in a vacuum atmosphere is another plus for this method of brazing in that parts can be cooled very rapidly by gas quenching, in many cases preventing chromium-carbide precipitation in the standard grades. Standard grades of stainless steels are frequently used by most commercial and high technology metal working industries on the basis of availability and cost. However, for some applications it is necessary to specify low carbon (L) or stabilized grades.

Although the furnace brazing of stainless steels with silver filler metal, BAg-13a, is successful (especially for the AISI 400 martensitic steels), the nickel base filler metal, BNi-2, is more compatible with the stainless steels for chemical and physical characteristics. Therefore the latter has become the prime choice for the high temperature furnace brazing of stainless steels. However, furnace brazing stainless steels with silver filler metal, BAg-13a, is a viable process and is firmly entrenched in specifications and engineering drawings. This is especially true of brazements for the military. Figure 10.14 depicts the paste application of the silver filler metal, BAg-13a, with automatic air-operated dispensing equipment to a helmet harness which is constructed of AISI 304 S.S. tubing, clips, and threaded fittings. The latter have been bright annealed, positioned in assembly fixtures, and spot welded for intimate fit-up. This is a brazement that is painted before assembly into a pilot's helmet, so that neither the base material nor the braze joint is subjected to even the mildest of corrosive media.

Since silver filler metals were introduced at the conception of the brazing process, they have been employed as probably the

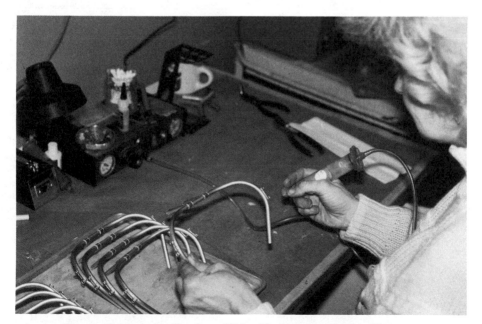

Figure 10.14 — Application of BAg-13a Paste to Helmet Harness

best, and in the early years, as the only filler metals for brazing copper and copper alloy base metals. Early attempts to convert to protective atmosphere furnace brazing were not successful; however successful brazements did emerge with the development of such high purity atmospheres as ultra dry hydrogen, argon, and vacuum, together with pre-placement of filler metals within the braze joint. The most suitable silver filler metals for brazing of copper are those that do not contain the high vapor pressure elements such as zinc and cadmium. Some of these low vapor pressure filler metals can be identified by the presence of tin (Sn) content.

The most important factors to consider when brazing copper base metal are:

(1) Electrolytic tough pitch copper should not be brazed in atmospheres containing hydrogen. If the type of copper is not known, vacuum or dry nitrogen would be the best atmosphere, followed by the inert gases (argon, helium). When brazing in a vacuum atmosphere at a temperature above 1650°F (900°C), a partial pressure of any of these gases should be introduced at the 1650°F (900°C) level or below in order to prevent volatilization of silver. Even at the lower temperatures, there may be some benefit in employing partial pressures of these gases.

(2) It is important to closely control the brazing temperature and time at brazing temperature relative to the liquidus of the silver filler metal. Prolonged periods of heating can cause an alloying effect of the silver with the copper which can lead to erosion of the base material. The erosion of heavy sections of copper may be slight, but erosion of thin copper sheet metal of copper could prove disastrous.

Two of the most popular silver filler metals for the protective atmosphere furnace brazing of copper assemblies are BAg-8 and BAg-8a, the eutectic compositions of copper and silver with solidus and liquidus of 1435°F (779°C) and 1410°F (766°C), respectively. The BAg-8a contains a small percentage (0.3%) of lithium which lowers the eutectic, the binary composition (72 percent Ag and 28 percent Cu) of BAg-8. These eutectic filler metals are ideal for brazing assemblies such as cold plates and heat exchangers which have intimate fit-up and require large areas of braze coverage or contain a multitude of joints to be brazed.

Figure 10.15 shows the partial assembly of a cold plate of oxygen free copper brazed using preplaced BAg-8a foil filler metal in the form of a strip 1 in. (25.4 mm) wide by 0.002 in (0.05 mm) thick. The foil was fastened to the solid face of the cold plate with instant adhesive. These cold plates were silver brazed in ultra dry hydrogen atmospheres and now can also be brazed in a vacuum atmosphere furnace. The foil is available in widths up to 6 in (1.52 mm).

Figure 10.15 — Cold Plate of O.F.H.C. Copper Partial Assembly with BAg-8

Electrolytic tough pitch copper could be substituted for the oxygen free copper if brazing is to be performed in vacuum, nitrogen, argon, or helium. Fixturing is not needed due to gravity weight loading of this sandwich-type brazement. Since the cold plate must be water tight at the outer extremities, it is either pressure tested or vacuum leak tested. With similar assemblies in which the material selection is a five percent chrome-copper, it is advisable to copper or nickel plate the faying surfaces in order to prevent the formation of the highly refractory oxide of chromium on the surface, thereby providing a surface that is readily wet by the filler metal.

Figure 10.16 depicts checkerboard fixturing of heat exchangers of copper which have been silver brazed with filler metal BAg-8a in shim form. Fixturing plays a very important part in maintaining the final overall dimensions of the heat exchanger.

Most brazements with varying degrees of mass are processed in batch-type furnaces, which are capable of putting all details in equilibrium by holding at temperatures slightly below the solidus of the braze filler metal. They are then heated very rapidly to the brazing temperature.

BCu (Copper Base) Filler Metal Classification

Copper was the first filler metal employed for furnace brazing in a protective atmosphere at high temperatures. Furnace brazing of carbon and low alloy steels with BCu braze filler metal has been adapted to furnaces with continuous belt drives for processing high production quantities of brazements. Brazing of stainless steels with copper filler metal became common as protective atmospheres improved with the introduction of disassociated ammonia, nitrogen base gas mixtures, ultra dry hydrogen, and vacuum. Furnace brazing at temperatures in the solution annealing range, followed by rapid cooling, reduces the probability of carbide precipitation and interface erosion in the standard grades of stainless steels. The only question is whether the copper brazement will function satisfactorily in the intended environment.

While the nickel base filler metal, BNi-2, is best suited to braze stainless steels, braze-

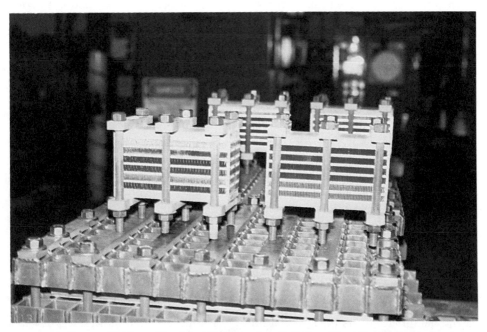

Figure 10.16 — Checkboard Fixturing of Copper Heat Exchangers

ments with copper have become entrenched in specifications and engineering drawings. The brazing of any base materials containing a high percentage of chromium are best processed in ultra dry hydrogen atmosphere or in vacuum. When furnace brazing with BCu filler metal in a vacuum atmosphere, the high temperature of brazing requires a partial pressure of argon or nitrogen to be introduced into the hot zone prior to the work zone temperature attaining 1650°F (900°C). Nitrogen-hydrogen and dissociated ammonia atmospheres are used for high production continuous furnace brazing.

Not only is copper (BCu) the most economical of all the brazing filler metals, but the availability in all forms (wire, sheet, performs, powder, plating, clad, etc.) affords the process engineer great versatility to satisfy any design or size. For example, small (1 in. [25 mm] long x .120 in. [3 mm] diameter parts can be assembled by press fitting, copper plating details, and parts brazing above 2000°F (1093°C). Applying braze filler metals in paste form with automatic, air operated dispensing equipment is the most popular means of applying filler metal to assemblies.

When processed in such highly active atmospheres as ultra dry hydrogen or in vacuum, paste applications of copper filler metal may produce black residues which require an additional cleaning operation to conform to cosmetic standards, or to allow another operation such as plating. Binders, powders, and some copper paste applications can contribute to this problem. Use of such solid metallic forms as wire, sheet, preforms, and plating provide braze fillets of exceptional quality and cleanliness.

While the protective atmosphere furnace brazing of carbon and low alloy steels adapted readily to the continuous type of furnace for high production, not all brazements can be processed with this method of heating. Some are best processed in batch-type furnaces employing ultra dry hydrogen or vacuum atmospheres. Not only should the alloy content of the base materials be of concern, but substantial variance in section size of the components should be considered in making the proper selection. A batch-type furnace offers the advantage of heating to and holding at a temperature slightly below the melting point of the copper filler metal to attain temperature equilibrium of the entire assembly prior to heating rapidly to the brazing temperature.

Mold plates for compressing and forming construction blocks of cinders (cinder blocks) are subjected to severe abrasive wear. The best solution is to copper braze a 0.125 in. (3.2 mm) thick plate of T-15, high speed tool steel to both faces of a carbon steel plate measuring 9 in. x 17 in. x 1.5 in. (23 cm x 43 cm x 3.8 cm). Cladding of both faces is economical, and a handy source for a replacement when the first side becomes worn beyond use. Such copper brazements increase the production of cinder blocks many times over other hard surfacing methods, and at lower cost.

Figure 10.17 depicts a furnace load of oil coolers to be employed in farm tractors which have been copper brazed in vacuum with partial pressure of nitrogen at 2050°F (1121°C) and nitrogen gas quenched. Details were fabricated from copper clad AISI 409 stainless steel. During the early years of vacuum brazing, the continuous furnace was considered the logical and economical choice for this high production part. However, the gaseous atmospheres available for continuous-type furnaces did not perform their function on the AISI 409 base material as well as the vacuum atmosphere. In recent years, the rising cost of copper-clad materials has forced a switch to paste applications. With improvements to copper powder and pastes, and heating cycle adjustments, objectionable black residues at the areas of copper paste application have been minimized and sometimes eliminated.

BNi (Nickel Base) Filler Metal Classification

The introduction of nickel base filler metals has elevated the protective atmosphere furnace method of brazing to a very reliable fabrication process for joining corrosion and heat resistant base materials. Not only have such brazements performed well in high temperature and corrosive environments, but they are also compatible with almost every metal possessing a higher solidus than the liquidus of the nickel filler metal selected.

The nickel brazing process was developed at an aircraft engine manufacturing plant

Figure 10.17 — Furnace Load of Copper Brazed Oil Coolers

for the prime purpose of supplementing electric arc and resistance welding to fabricate jet engine assemblies. In the early stages of the braze process it was determined that the diffusion and interaction of the nickel base filler metals with the base materials containing nickel created a completely new alloy with chemical and physical properties some place between the original filler metal and base material. That the original hardness (Rc 60) was still present in large braze fillets was at first thought to be a problem. However when machining through the small clearances of the brazed joint it was noted that the filler metal was soft. The hard undiffused filler metal in the joint and fillet can reduce the strength of the joint.

In diffusion brazing, the essential variables are time, temperature, and quantity of filler metal. Diffusion increases directly with increasing time and temperature but inversely with quantity of filler metal (clearances). Quantity of filler metal is the most important of the three. When the remaining filler metal in the joint is a single-phase (unetchable) or is no longer present, the remelt temperature can be in the 2500°F (1371°C) range. When there is a two-phase structure in the joint, the center phase

is lower melting and the hardness and remelt temperature approaches that of the original filler metal. Various nickel base filler metals have various hardnesses. The BNi-2 filler metal has a narrow melting range of 50°F (28°C) and thus is a filler metal that is good for smaller clearances of 0.002 in. (0.05 mm) or less, and usually completely diffuses to the single-phase joint structure.

The advantages of BNi-2 filler metal were first realized on microexamination of the braze joints of a combustion liner that had been in operational flight. The braze joint proper had been replaced with a grain structure that was homogenized. The high temperatures of operation were responsible for this improved condition, so project engineers injected a diffusion cycle after the braze cycle, and there were no doubts as to the ability of this brazement to satisfy the strength requirement at elevated temperatures. The BNi-2 filler metal is the best choice for protective atmosphere furnace brazing for a wide variety of brazements constructed of almost every type of base material. The BNi-4 filler metal was introduced as "free flowing" although it does have a melting range [100°F (55°C)] double that of BNi-2, and it is conducive to brazing in marginal hydrogen atmospheres. Those who accept and employ this nickel filler metal produce satisfactory brazements, but the absence of chromium, the prime element which contributes to corrosion resistance, may diminish its use for brazements to be employed in a corrosive environment.

There are three nickel base filler metals that do not contain boron: BNi-5, BNi-6, and BNi-7. These are useful for assemblies inside nuclear reactors, and for applications in which diffusion of base materials must be minimized to prevent erosion. The BNi-5 filler metal has a melting range of 100°F (55°C) and it also possesses one of the highest liquidus temperatures [2075°F (1135 C)] of the nickel base filler metals. The BNi-6 and BNi-7 filler metals contain phosphorus and are eutectic at 1610°F (875°C) and 1630°F (890°C), respectively. Both of these filler metals can be employed to braze iron or high nickel base alloys when adequate diffusion cycles are used. However, difficulty may be encountered where welding is specified in close proximity of the braze joint or when welding over the superfluous flow of filler metal. The BNi-6 and BNi-7 are the two filler metals that do not require the ultra dry hydrogen or vacuum atmosphere. Satisfactory brazements have been accomplished in dry dissociated ammonia and dry nitrogen-base atmospheres. The BNi-6 filler metal can also be used in very high dew point atmospheres such as exothermic atmospheres. The BNi-7 and BNi-2 filler metals have been used to furnace braze copper base metal to the stainless steels without the need to nickel or copper plate the stainless. These dissimilar metals are brazed at 1900 to 1925°F (1040 to 1050°C), but size and design of the brazement are very important. Eight-inch (20.3 cm) diameter plates with one-inch (2.5 cm) thicknesses of these dissimilar materials have been satisfactorily brazed by preplacing metallic foil or transfer tape of BNi-2 filler metal between the copper and stainless steel plates.

Nickel base filler metals such as BNi-1, BNi-1a, BNi-3, BNi-4, and BNi-5 have a wide melting range between the solidus and liquidus which makes them susceptible to liquation, especially during slow heating between their solidus and liquidus temperatures. The melting and separation of the low melting constituents of the applied filler metal by capillarity can quickly fill a shallow joint having a minimal braze joint clearance. The remaining filler metal now has a higher melting temperature and may or may not melt, but since capillarity is not afforded, the molten filler metal will simply add to the fillet size. The large fillet size is undesirable as it is hard and may be susceptible to cracking if the brazement is subjected to stresses. BNi-2 was developed from the low melting phase extracted from the original BNi-1 nickel base filler metal. BNi-2 has the narrowest temperature range [50°F (28°C)] between the solidus and liquidus, thus providing an almost eutectic and very free-flowing filler metal. When brazing filler metals having a brazing range of 1850 to 1950°F (1010 to 1065°C) are used, AISI 400 martinsitic stainless steels and D-2 tool steels (12% Cr) can be brazed and austenitized concurrently. Brazing temperatures can be raised to be compatible with solution annealing temperatures of many of the stainless steels and other high temperature alloys. However, thickness of materials being joined must be considered due to base metal diffusion of this or any other nickel base filler metal that contains boron. Some prime contractors who had originally objected to employment of the nickel base filler metals, and specified the gold-nickel filler (BAu-4), were forced to revert to the nickel base filler metal, BNi-2, due to the escalating cost of gold. At the outset of the nickel braze process at aircraft engine manufacturing plants, project engineers chose from the variety of nickel filler metals and the predominant ones were higher melting with a wide melting range between the solidus and liquidus (BNi-1, BNi-1a, BNi-3, BNi-4, and BNi-5). The objective of the brazement was to satisfy the quality requirements of visual and braze area coverage of the military specification and to perform satisfactorily in the high temperature and corrosive environment of jet engine operation.

BNi-2 became the most widely employed nickel braze filler metal because of three factors: (1) compatibility with almost every base material, (2) availability in every form (except plating) and (3) low cost.

One of the biggest advantages of the nickel braze process is the ability to assemble and join finish machined details which will not require further machining operations after brazing. Pictured in Figure 10.18 is a special vacuum system consisting of a multitude of various components that have been brazed with BNi-2 filler metal in one braze cycle. No other process could fabricate this unit distortion free and vacuum tight with complete penetration through every joint to fill the smallest of crevices at inaccessible areas. The benefits of vacuum processing for surface cleaning and the ability to remove undesirable elements (outgas) dissolved in the stainless steel have created a market for mass spectrometer gas sampling systems designed for the impurity analysis of hydrogen and its isotopes. For these units, all details are electrochemical polished to mirrorlike surfaces. Although highly polished surfaces can retard the flow of the filler metal and act as a stop-off for molten copper filler metal, the nickel base filler metals with the ability to interact and diffuse with the base materials can be used to braze joints of highly polished assemblies. Such an assembled unit is pictured in Figure 10.19.

The seat areas of globe valves offer a challenge for the nickel braze process because some areas are inaccessible for welding and machining or grinding operations. Cracking may result from hard facing with cobalt-base electrodes on small diameter seats. Established design practices center on conventional round seat rings which are either cast or hot pressed metal and positioned in recessed counterbores in the seat area of the body. The different coefficients of expansion may create problems and the joint clearances must be calculated for the brazing temperature. Thus the joint clearance must be calculated for every combination and size. Nickel brazing with BNi-2 filler metal has been very successful, but to reduce costs, a new concept for braze joints in compression was developed. Figure 10.20 depicts two cross sectional views of both the conventional and conical seat rings. The conical seat rings are castings and are assembled without any machining or grinding. Major advantages include: (1) self-positioning and adjusting; (2) uniform wall thickness and less weight (plus lower cost and less dependence upon cobalt supply); (3) premachining that is not critical and can

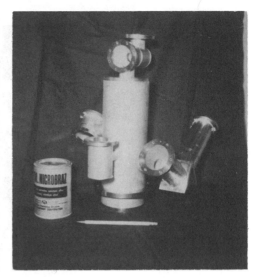

Figure 10.18 — Special Vacuum System Consisting of Various Components Brazed with BNi-2 Filler Metal in One Braze Cycle

be accomplished with a multi-fluted drill; and (4) reduction of grinding costs after brazing. Inspection is visual (top side) and the underside can usually be inspected for complete flow with the aid of small inspection (dental) mirrors. The rework rate is very low. Figure 10.21 shows a load of globe valve bodies which are positioned for brazing in a vacuum furnace.

Figure 10.22 shows an assortment of ferro-magnetic separator screens of AISI 430 stainless steel that have been brazed with BNi-2. The screen is formed by wrapping a straight wire and a corrugated wire together until the desired diameter is attained. One tack weld at the start of the wrap and one tack weld at the end are all that holds the assembled screen together for braze application. Braze powder can be applied by sprinkling onto a screen that has been lightly wetted with a binder of plastic and solvent, or by a spraying operation in which both powder and acrylic binder are combined on the screen, Figure 10.23. Mesh and thickness will vary with size and width of the wires. Screens of 36 in. (91.4 cm) diameter have been brazed, but precautionary procedures for handling must be instituted.

Figure 10.19 — Mass Spectrometer Gas Sampling System Designed for the Impurity Analysis of Hydrogen and its Isotopes

A series of these screens is stacked in a magnetic field so that any magnetic particles in the material being processed through the screens will be separated and accumulate at the brazed intersections.

Figure 10.24 is a tank bottom for an ultrasonic cleaner in which 72 nose pieces (2.25 in. x .25 in. [5.7 cm x .6 cm] thick) of AISI 304 stainless steel are brazed with BNi-2 to AISI 316 stainless steel in vacuum at 1925°F (1052°C) and nitrogen gas quenched to ambient. The nose pieces are positioned with instant adhesive or spot welded with capacitor discharge equipment which does not produce objectionable pressure indentations. BNi-2 in paste form is applied completely around each nose piece. A visual inspection of braze fillet (Figure 10.25) and the full impression of each nose piece on the underside of the sheet metal are excellent indicators of complete braze coverage. (During the early years of brazing with BNi-2 filler metal, microexamination confirmed 99.9 percent braze coverage on these units.) Transducers will be mounted on the nose pieces and tested for ultrasonic transmission of 27 KH_2 and 40 KH_2 prior to gas tungsten arc welding the tank bottom to the water tank. Immersible models are also brazed in the same manner.

The versatility of BNi-2 filler metal is shown by the manufacture of cutting tools from 2.12 in. (5.4 cm) discs of carbon steel. Preforms from transfer tape (BNi-2) are placed on both sides of the disc, and dipped in a container of tungsten carbide particles, one side at a time. After brazing, the discs are coated with tungsten carbide particles imbedded in a matrix of BNi-2 filler metal. Figure 10.26 shows the furnace setup and spacing of these cutting tools.

Gas and liquid chromatographs contain a multitude of connections employing hypodermic tubing, and BNi-2 filler metal is the ideal filler metal. The use of stop-off materials is forbidden and cleanliness is essential for all analytical equipment. The use of electrolytic cutting to eliminate burrs when cutting small diameter tubing was an important step in transferring from torch and induction brazing methods with silver filler metals and flux to protective atmosphere brazing in vacuum employing BNi-2 filler metal.

The single point tool "staking" method has been changed to ring staking, in which

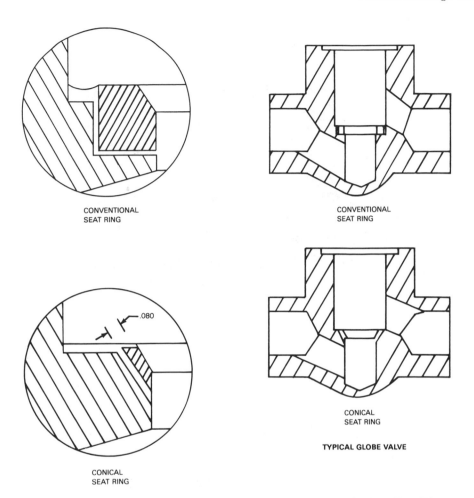

Figure 10.20 — Cross-Sectional Views of Conventional and Conical Seat Rings

the entire joint is peened to a slight interference fit. The I.D. of the tubing on many of the assemblies is critical for gas flow, and must be checked with pin gauges past the vicinity of the ring staking. Figure 10.27 depicts the fixturing and tooling for this type of operation. The prime contractor specifies that a braze fillet be visible at one end of joint (usually the staked end).

It is recommended that base materials that contain more than 0.5 percent of such highly reactive elements as titanium and aluminum be electrolytically nickel plated before brazing. It is advisable to nickel braze with BNi-2 filler metal at 1900°F (1038°C) and not to hold brazements at

the brazing temperature for prolonged periods since the nickel plating may dissipate.

Figure 10.28 shows brazements that have been brazed in both the ultra dry hydrogen and vacuum atmospheres with the BNi-2 filler metal and are employed in equipment used to process potable liquids, dairy products, meats, and food. Brazing fillets must be free of porosity since this condition provides bacteria traps which can cause contamination.

For the metering device for measuring the flow of liquids such as beverages, fruit juices, water, etc., there are three sizes, 1 in., 2 in., and 4 in. (2.54 cm, 5.1 cm, and

Figure 10.21 — Load of Globe Valve Bodies Positioned for Brazing in a Vacuum Furnace

10.0 cm). Base material is AISI 304 stainless steel. There are 64 separate applications of braze paste on each brazement. The final product will have 164 joints to inspect.

For the meat chopper blade, the back plate is a low alloy steel and the chopper bits are D-2 tool steel. Brazing and austenitizing are conducted concurrently and the result-

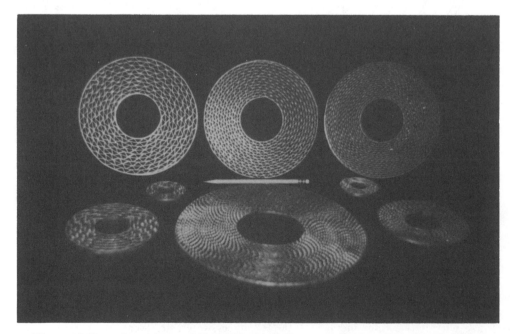

Figure 10.22 — Assortment of Ferro-Magnetic Separator Screens Brazed with BNi-2

Figure 10.25 — Close-up of Braze Fillet of Nose Piece

Figure 10.23 — Spray Application of Ferro-Magnetic Separator Screens

ing hardness of the chopper bits is Rc 64, the same hardness as the perforated plate of D-2 against which the bits operate. The choppers are subsequently tempered to Rc 54 to reduce wear on the perforated plates.

For the milk claw for a milking machine, all details are AISI 304 stainless steel, and this unit contains braze joints that become a part of the vacuum system. All joints must be free of voids and porosity.

A "potty" fire extinguisher is pictured in Figure 10.29. The name is derived from the fact that F.A.A. regulations require that a unit of this type be mounted on the partition above the waste basket in the lavatories of every commercial aircraft. The two spray nozzles which are counterbored at their tips and the one filler tube are AISI 304 stainless steel. The two half stampings which are ma-

chined for a "slip fit" and the mounting bracket are carbon steel. The tubes are staked at appropriate heights for location in a fixture, half stampings are "snapped" together and mounting brackets are positioned and held together by means of two gas tungsten arc tack welds. BNi-2 filler metal is applied in paste form, and brazing is accomplished at 1900 to 1925°F (1038 to 1051°C). Time at temperature is 15 minutes, followed by nitrogen quenching to room temperature. Handling for visual inspection is with cotton or surgical gloves since the first operation performed by the prime contractor on this brazement is to spray a primer coat of paint over the entire unit.

Nickel filler metals are primarily selected for corrosion and heat resistant applications. However, because nickel filler metals are better for filling longer gaps, they are also an excellent choice for substituting for the BCu filler metals on assemblies of carbon and low alloy, high strength steels whenever the fit-up is problematic. It should be noted that the high temperature of brazing the high strength steels with the BNi filler metals, as with the BCu filler metals, will cause grain growth. Brazements should be subjected to normalizing (grain refinement) so that subsequent heat treating can restore proper metallurgical properties for which the brazement was designed.

BAu (Precious Metals) Filler Metal Classification

This group of braze filler metals was introduced shortly after the nickel braze process. The ultra-dry hydrogen atmosphere, and

Figure 10.24 — Tank Bottom for Ultrasonic Cleaners (72 Nose Pieces)

Figure 10.26 — Furnace Set-up of Tungsten Carbide Particles in a Matrix of BNi-2

the sand seal retort/bell furnace (which maintained the purity of the atmosphere) provided successful brazing of stainless and heat and corrosion resistant alloys. Although the precious filler metals have a much lower rate of interaction with nickel and corrosion resistant alloys than nickel base filler metals, BAu-4 (Au 82% and Ni 18%) filler metal produced ideal diffusion and physical properties for sheet metal fabrications when brazing temperatures were maintained between 1850 and 1900°F. Higher temperatures in combination with very thin sections could result in excess interaction and cause erosion of the thin materials at the points of the braze filler metal application. This filler metal, BAu-4, having a eutectic temperature of 1742°F (950°C), was therefore adopted by some aircraft engine manufacturers that were not satisfied with nickel base filler metals for fabrication of jet engine parts.

The BAu-4 filler metal is compatible with most base metals. The only disadvantage is its high cost, which fluctuates daily with the precious metals market. For example, from 1961, when gold was first introduced as a braze filler metal, until 1980, gold escalated

from $32 per troy ounce to $700 per troy ounce. Wide savings in market prices have caused delays in delivery until the market stabilizes.

Although precious metal filler metals, especially BAu-4, served the intended functions, escalating prices prompted substitution of more economical filler metals. Development of the rapid solidification process to produce foil and preforms of nickel base filler metal, BNi-2, helped in this changeover.

A unique application of the BAu-4 filler metal was developed for a repeater assembly (Figure 10.30) composed of carbon steel details and employed for the trans-oceanic cable. Copper (BCu-1) was the logical filler metal to furnace braze this assembly, which is subsequently gold plated to protect the unit from sea water corrosion. Environmental testing of these units indicated that gold plating provided excellent corrosion protection to the carbon steel, but that there was rapid corrosion at the braze fillets. Engineers decided that the roughness of the copper brazed fillets was responsible for the erratic and unreliable gold plating. Fortunately, the BAu-4 braze filler metal had recently been intro-

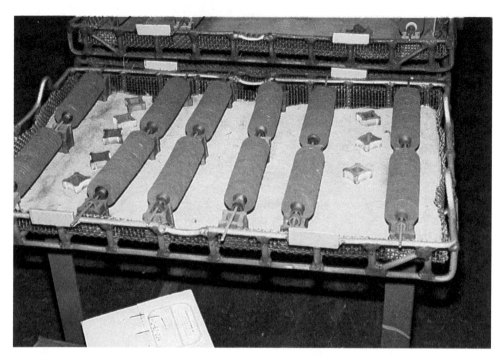

Figure 10.27 — In Ring Staking, the Entire Joint is Peened to a Slight Interference Fit. Fixturing and Tooling for This Type of Operation are Shown.

duced, so this filler metal joined other selected filler metals such as BCu-1, BAg-13 and BNi-2 to braze assemblies for gold plating and testing. Furnace brazing was accomplished in ultra dry hydrogen atmospheres. With an 82 percent composition of gold in the braze fillets, BAu-4 was determined to be the proper filler metal to braze this carbon steel assembly for a salt water environment.

There are many other precious metal filler metals some containing platinum and palladium. Many of them are employed to furnace braze the ultra high melting and exotic alloys.

FURNACES FOR PROTECTIVE ATMOSPHERE BRAZING

The modern day vacuum furnace is the most versatile furnace that can be employed for brazing almost every base metal. Regardless of the type, it is essential that the furnace and associated equipment be main-

tained with a systematic preventative maintenance program. It is also important that furnace operators are fully acquainted with the thermal characteristics of each furnace for proper placement of work load sensors. This section will concentrate on the following activities related to the operation of furnaces and auxiliary equipment:

(1) Care, cleaning, and maintenance of furnaces
(2) Maintenance of high purity gaseous and vacuum systems
(3) Thermal control and establishment of heating, brazing, and cooling cycles

Care, Cleaning, and Maintenance of Furnaces

Gaseous and vacuum atmospheres have a much tougher job than fluxes since they must reduce or vaporize surface oxides from the entire work load, plus those condensed on the interior of the furnace and fixtures. In order to maintain atmosphere quality, furnace in-

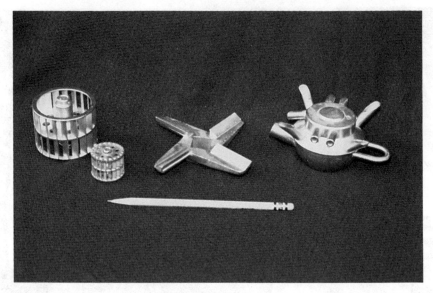

Figure 10.28 — Brazements in Both Ultra-Dry Hydrogen and Vacuum Atmospheres with BNi-2 Filler Metal. Shown are Metering Device, Meat Chopper and Milk Claw

teriors and fixtures must be cleaned just as parts to be brazed are cleaned. Any furnace that has been idled and exposed to air, or a furnace that has been cycled repeatedly below 1800°F (980°C), is susceptible to oxide formation and accumulation of vapor-deposited contaminants.

To remove these oxides and contaminants, it is necessary to subject the furnace to "bake-out" cycles. For batch type furnaces in which ultra dry hydrogen is the prime atmosphere, the "bake out" cycle consists of heating the retort/bell and furnace fixturing to an elevated temperature and holding at that temperature for sufficient time (established from previous experience). In the bake-out of a vacuum furnace, a high flow of hydrogen is sometimes introduced to aid the reduction of oxides. Before attempting this type of cleaning, the age of the furnace must be considered and the furnace manufacturer must be consulted. Excellent references are *NFPA 86D, Industrial Furnaces Using Vacuum as an Atmosphere*,[1] which covers the use of combustible gas atmospheres in vacuum furnaces, and *NFPA 86C, Industrial Furnaces Using a Special Process Atmosphere*. In the case of continuous furnaces, the essential consideration is usually the conditioning of the furnace and belt. For both batch type or continuous type furnaces, scheduling of work loads to take advantage of the condition of the furnace can be effective in eliminating unnecessary bake-out cycles.

"Leaks" in the furnaces employed for brazing may result in loss of expensive atmosphere gases. However, in positive flow or high pressure systems it also indicates the ingress of atmospheric air. For retort/bell type of furnace equipment, a visual inspection during removal from the heating furnace, or during the cooling cycle, will detect leakage areas. The sign of such leakage is escaping hydrogen gas burning on exposure to air. Vacuum furnaces can be checked for leaks by performing a leak-up rate test. Manufacturers of vacuum furnaces specify that new furnaces have a leak-up rate of less than five microns per hour. Specifications issued by the military and prime contractors may specify a leak-up rate "not to exceed 20 microns per hour", but in many specifications this applies to all the brazing processes (silver, copper, nickel, precious metals, aluminum) and the base materials

1. National Fire Protection Association, 1 Batterymarch Park, P.O. Box 9101, Quincy, MA 02269-9101.

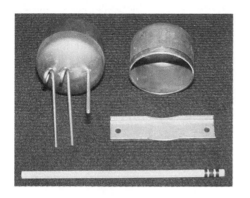

Figure 10.29 — "Potty" Fire Extinguisher Designed for Mounting in the Lavatories of Commercial Aircraft. Mounting Brackets are Held Together by Two Gas Tungsten Arc Tack Welds.

involved. A leak-up rate of 20 microns per hour might be satisfactory for copper brazing a load of carbon steel assemblies; however, if base materials contain fairly high percentages of the refractory or reactive elements, the integrity of the brazements is usually in jeopardy. Brazing operation accustomed to maintaining high quality work will perform leak-up rate checks at the slightest decline in quality or appearance. The same sources will seek out the leaks, repair them, institute a bake-out cycle, and perform a leak-up rate checkup prior to using the furnace for a brazing cycle on assemblies which demand a high standard for quality and appearance. Vacuum furnace manufacturers are an excellent source of information for the best areas to check for leaks.

Maintenance of High Purity Gaseous and Vacuum Systems

Even extremely small leaks in the piping systems that convey furnace atmosphere gases will allow air diffusion back into the piping system and contaminate the hot zones of gaseous and vacuum furnaces. The furnace manufacturer will usually dictate whether soldering or brazing will be performed on permanent connections and the specific type of tubing or piping to be employed. A few connections will remain which are not permanent and may loosen and create a leak. Helium mass spectrometer leak detectors have detected leaks through an elbow (casting), so it is advisable to check all fittings in addition to all connections. In a vacuum furnace, the switch from a neutral gas (nitrogen) that was employed for partial pressure operation and gas quenching in the previous braze cycle, to an inert gas (argon) which is to be employed for the next braze cycle, necessitates a complete purge of the gas supply system with argon before the furnace doors are closed for the start of the pump down cycle.

During the employment of ultra dry hydrogen atmospheres in the Retort/Bell type furnaces, hydrogen was the purge gas and 8 volumes of the retort was required for the purge before subjecting the Retort/Bell to the heating cycles in the gas fired or electric furnace. A proper volumetric flow rate is also required during the brazing cycle to flush out the contaminants outgassed from the load. This could be 10, 15, or 20 volumes per hour. As the size of the brazements or work loads increased, it was inevitable that the size of furnace equipment would also have to increase. The cost of large quantities of hydrogen and safety considerations led engineers to use relatively inert nitrogen gas for purging and replace it with hydrogen at some low temperature as the work load was heated to the brazing temperature. Misjudgment or delay of this replacement resulted in "de-wetting" and "balling-up" of the nickel base filler metals

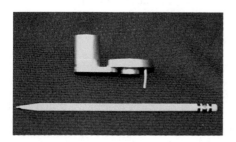

Figure 10.30 — Repeater Assembly Necessitated a Unique Application of BAu-4 Filler Metal

containing boron that had been applied in paste form. These conditions indicated that hydrogen should have been introduced at a much lower temperature. When properly controlled to eliminate nitrogen at the proper time, excellent brazed joints are obtained. Paste or metallic foil applications of the same nickel base filler metals, however, produce excellent brazements when they are processed in a vacuum furnace with a partial pressure gas operation with nitrogen. When the H_2V furnace was developed later, a vacuum pump evacuated the air from the retort, and then the ultra dry hydrogen was introduced as the atmosphere. This system is vacuum tight and thus gas flowing on cooling is not available.

Thermal Control And Establishment of Heating, Brazing, And Cooling/Quenching Cycles

Thermal control and the establishment of sound thermal cycles constitute an important operation which can determine the success or failure of entire work loads. Therefore it is necessary to know the following: (1) pyrometry or temperature measurements, (2) heating characteristics of the furnace, and (3) characteristics of heat transfer to the work load. Surveys for temperature uniformity of the furnace were adapted from MIL-H-6875, *Heat Treatment of Steels*, and became entrenched in brazing specifications instituted by the military technical societies and many prime contractors that do business with the federal government. Since their implementation for thermal processes conducted in batch type furnaces, these surveys have been conducted under the following stipulations:

(1) Without a work load (an "empty" furnace)

(2) Cycled in such a manner as to reflect the normal operating characteristics of the furnace

(3) Before thermal equilibrium is reached, none of the temperature readings should exceed the maximum temperature of the range being surveyed

(4) Once the equilibrium temperature has been reached, readings of all the temperature sensors must be taken every five minutes for a minimum of 30 minutes.

In order to establish reliable thermal cycles for brazing and heat treating, it is essential to know when the entire work load attains uniformity of brazing/heat treat temperatures. This information can only be obtained by conducting surveys with full work loads at temperatures compatible with brazing cycles. In conducting temperature uniformity surveys, as well as production brazing cycles, the selection and proper use of the temperature sensors (thermocouples) become a vital part of all types of brazements. Therefore, it is advisable to understand some basic facts about thermocouples (T/Cs).

First of all, the importance of the measuring junction (formerly called the "hot junction") must be understood, along with its location on the thermocouples. The latest technical information about pyrometry establishes that the measuring junction that can provide error-free temperature measurements is formed by the welding of each wire of the T/C as close together as possible into the surface of the part to be monitored. This, of course, is not a practical or economical method of monitoring production quantities of brazements. The second best measuring junction is thought to be the butt-welded junction of the two wires (contact to be maintained with the surface of the part to be monitored); however, once again, that is not practical or economical. The most practical and economical temperature sensor that gained prominence with the Report/Bell furnace (and served both as the furnace temperature controller and the work load temperature sensor) was the alumel-chromel type with a twisted end, and this tip (at that time) was presumed to be the "hot junction". With the work load T/C operating in a highly reducing gaseous atmosphere and the furnace temperature controller sensor functioning in an oxidizing atmosphere, the inability of the work load to attain the braze temperature as set by the furnace controller indicated that the measuring junction (formerly called "hot junction") was not always at the tip. Instead, it was located at any point closer to the controller where the alumel and chromel wires may have twisted or contacted. For this reason it is highly recommended at least two T/Cs per load be used. Some companies employ three T/Cs; one at the top perimeter, one at the center and one at the bottom perimeter of the load. Using this procedure, the desired temperature of the work load is always monitored and controlled even if one T/C shorts out. When the set temperature of brazing is indicated at all T/Cs, it is safe to presume that the entire work load has attained temperature uniformity at the specified brazing temperature.

The cold wall vacuum furnace introduced the direct heating of the work load, and heat transfer to the work load was again by radiation. With the passage of time and experience, heat transfer by radiation became the most rapid means of heating work loads to the brazing temperature. Although the initial heat is radiated from the heating elements, as heating progresses all shielding (metallic or graphitic), fixtures, and parts radiate this heat. Most of the cold wall vacuum furnaces are constructed with the heating elements in an orbital configuration. However, on some special applications, heating elements have been added to the doors. Bottom loading vacuum furnaces can be equipped with auxiliary heating devices called "spikes". Heating elements added to the spikes provide for controlled heating of the I.D. of a large assembly which may otherwise be subjected to differential expansion of base materials and fixturing.

The conversion to direct heating, heat transfer by radiation, and requirements for higher brazing temperatures in vacuum furnaces focused attention on the need for changes in the type and placement of the furnace and work load temperature sensors (T/Cs). Type "S" T/Cs (platinum-rhodium, butt-welded and encased in high purity alumina) are usually employed as the furnace controller, and the distance of the measuring junction into the working zone and beyond the heating elements is usually dictated by furnace manufacturers. An identical thermocouple is inserted very close to the furnace controller T/C, and serves as a safety valve or "overtemp warning sensor". Furnace operators must take extra care during loading and unloading of large-sized brazements to avoid damage to this type of ceramic encased thermocouple.

Depending upon the proximity of the measuring junction to the heating elements, a thermocouple used to measure temperature, or temperature uniformity, in a vacuum furnace can produce erratic temperature measurements.

There is no better test to illustrate the direct radiation effects on bare, twisted T/C wires than to mount a copper wire and a block (2 in. x 2 in.) of copper in close proximity to a heating element in a furnace load or empty furnace which will be heated to 1900°F (1030°C). Perhaps surprisingly, the copper wire melts and the copper block remains intact. Copper has a melting point of 1981°F (1083°C). There is no better method of shielding the measuring junction of bare, twisted T/Cs than to imbed it on the inside of a brazement. Or if the design of the brazement is not receptive, "heat sinks" should be provided to mount adjacent to brazements at the perimeter of the working zone. Heat sinks should represent the mass or section size of each work load: section of tubing with closed end for sheet metal, and bar stock with a drilled hole for brazements of thicker section size. Temperature uniformity surveys of a work load conducted in this manner truly reflect the normal operating characteristics of the furnace and provide the answer to the locations where: (1) work attains the set brazing temperature the fastest, and (2) work attains the set brazing temperature the slowest. Work load T/Cs imbedded in the brazement or "heat sinks" at these two locations can provide temperature uniformity surveys of the entire work load on every brazing cycle.

The load T/Cs, generally alumel chromel, must be watched carefully as they are life limited, particularly in a vacuum furnace. Some companies limit the use of alumel chromel to two times above 2000°F (1093°C), and five times below 2000°F (1093°C). Preferential attack of the grain boundaries (aluminum vaporization) in the alumel is the reason for the limited use.

Modern instrumentation such as microprocessors and process recorders can provide a furnace chart with a complete picture of the set temperatures, the "soak" times at the set temperatures, the vacuum levels in microns, quenching rates, plus information pertinent to the work load such as: heat number, date, time, part number, quantity, braze process, braze filler metal, etc. This is truly a complete history which can aid problem solving of furnace loads containing braze discrepancies that are normally attributed to conditions that may exist during the brazing cycle. The establishment of a heating, brazing, and cooling cycle after "pump down" for various furnace loads in which the filler metal is BNi-2 is illustrated in Figure 10.31.

Heating rates to either the brazing temperature for very small or light loads or the equilibrium temperature for brazements that contain a variance in section size, are usually set at a rate of 30°F (17°C) per minute. Once the equilibrium temperature (temperature slightly below the solidus of the filler metal) has been attained as indicated by the two work load TCs, heating can proceed at the same or faster rate to the brazing temperature. For furnace loads in which the filler metals are copper (BCu-1) or the "high temperature" silver alloy (BAg-13a), the introduction of argon or nitrogen gases for partial pressure gas operation should take place when the furnace temperature reaches approximately 1650°F (900°C).

Selection of the brazing temperature on corrosion resistant and high nickel alloys is anywhere between 1875°F (1024°C) and 2100°F (1149°C). The higher temperatures and longer soak times at temperature are selected for high diffusion rates on heavier sections, but temperatures at the lower end of this range, along with very short soak times are preferable to prevent erosion on very thin sections at the areas of braze filler (BNi-2) application. Soak times for brazements containing copper (BCu), "high temperature" silver (BAg 13a), and similar nonferrous filler metals are minimal, and longer soak times can only add to the cost.

Controlled cooling/quenching cycles are a great advantage of thermal processing in a cold wall vacuum furnace. Brazements that will possibly contain a large amount of molten braze filler metal at exterior joints at the end of a soak at brazing temperature will be vacuum furnace or static gas cooled to below the solidus (work load T/Cs) and then inert gas (argon, nitrogen) quenched to room temperature. The gas quenching directly from the brazing (and austenitizing) temperature is ideal for base materials such as martinsitic stainless steels and D-2 tool steels in which the brazing range of the filler metal is compatible with the austenitizing (hardening) temperatures of these base materials, thereby processing the brazing and hardening concurrently. There is still an-

other cooling cycle in which the brazements are vacuum furnace cooled to below the solidus of the braze filler metal, reheated to the solidus temperature or slightly below, and soaked at this temperature for controlled diffusion of the base materials to enhance the physical strength of the braze joints. Although diffusion cycles have been, and still are required by some aircraft engine manufacturers, brazing today in vacuum furnaces with close fit-up produces the highest quality braze joints, (free of voids and porosity, and completely diffused) with exceptional strengths, thus reducing the cost of processing.

Although most brazing cycles will stipulate: "Pump down to a hard vacuum", some attention must be paid to the construction of the assemblies in the work load. In some cases a "double or even a triple pump down to a hard vacuum" would be in order. Design for all methods of brazing require that assemblies be vented. As parts are heated, the pressure of trapped air is increased and venting at the braze joint will reject the flow of filler metal from the exterior end of the joint. Protective atmosphere furnace brazing of nonvented assemblies in a hydrogen atmosphere can prove disastrous when heat is applied to an explosive mixture of air and hydrogen.

While explosive mixtures cannot form during heating in vacuum brazing, the pumping system may not be capable of removing all the air in an assembly through the braze joint. It is advisable to employ a "stitch" method of braze paste application and resort to "double or triple pump downs". Vacuum gauges will indicate pressure/vacuum of the working zone, but these reading are not representative of the actual conditions at the nonvented sections of the assemblies. Double or triple pump downs are simply duplication of pumping down the working zone with the mechanical pump, and then the diffusion pump to one micron, backfilling with the inert quenching gas (argon, nitrogen) and then repeating the pump down and backfilling operations before the power for heating is turned on. Repeated pump downs and backfilling will remove or replace air with the inert gas which is better for the braze cycle than if air remained.

Basically, three methods of protective atmosphere furnace brazing have been devel-

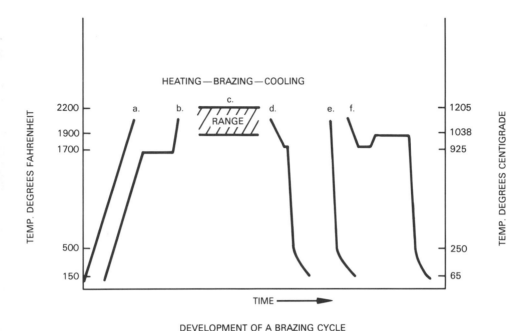

DEVELOPMENT OF A BRAZING CYCLE

Figure 10.31 — A Heating, Brazing, and Cooling Cycle After "Pump Down" is Shown for a Typical Furnace Load in Which the Filler Metal is BNi-2. The Cycle Includes (a) Direct Rise to Brazing Temperature, Usually for Small Thin Parts of Uniform Thickness, (b) Ramp Up to Holding Temperature, Then a Rapid Rise to Brazing Temperature, Usually for Parts With Varying Mass, (c) BNi-2 Brazing Temperature Range, (d) a Slow Cool to a Specific Heat Treat Temperature, Then Quenched to Room Temperature, (e) Rapid Quench to Room Temperature, (f) Diffusion Cycle

oped in continuous type furnaces. They have worked extremely well for brazing large production quantities in belt driven or conveyor type continuous furnaces. They are: (1) copper brazing of carbon and alloy steels, (2) aluminum brazing of aluminum in vacuum, and (3) nickel or copper brazing of stainless steels.

One important requirement prior to establishing braze cycles is pre-conditioning the furnace and attaining a suitable dew point, vacuum or atmosphere quality throughout the pre-heat, work, and cooling zones. For a copper braze cycle, the temperature could be set at 25 to 150°F (14 to 83°C) above the liquidus of the copper filler metal (1981°F [1083°C]), but the belt speed should be adjusted so that the work load and braze filler metal will momentarily attain a temperature slightly above the liquidus of the filler metal

and complete the braze. A sample lot of brazements is made and visual inspection determines whether or not adjustments to the furnace control temperature or belt speed are necessary prior to processing large production quantities.

For aluminum braze cycles in a semi-continuous vacuum furnace, the furnace controller is usually set at 1200°F (650°C). This is not only higher than the liquidus of the aluminum filler metal, but higher than the liquidus of the base material. Therefore, the timing of the speed of the conveyor and the temperature imparted to the brazements are critical. Once the correct settings for temperature and belt/conveyor speed have been attained on initial brazements, interruption of production quantities is discouraged and precise process control records must be maintained.

Keywords — induction brazing, induction brazing equipment, induction coil designs, induction generators, induction brazing in controlled environment, induction joining

Chapter 11

INDUCTION BRAZING

PROCESS DESCRIPTION

In many designs, metal components may be joined by induction brazing to fabricate the final assembly. The components at the joint area, including the part surfaces to be joined and the joining alloy, are heated selectively to the brazing temperature. This is done by induced electrical energy in the joint using an induction coil or inductor.

Heating occurs primarily as result of resistance to the flow of induced current, sometimes called I^2R losses, as shown in Figure 11.1. This happens in each part, if it is an electrical conductor, when it is placed in an electromagnetic field established by rapid alternating current flow in the induction coil.

The induced current is greatest at the surface of the part, decreasing exponentially to the interior as shown in Figure 11.2. Therefore surface heat is possible with high power density (kW/in.^{2} surface) at high frequency.

The actual depth of heating depends on both the frequency and the rate of heating since conduction occurs as soon as the surface begins to heat. Relatively low power densities of 0.5 to 1.5 kW/in.^{2} are common in induction brazing applications to avoid excessive surface heating. The principal objective is to bring the joint area to uniform brazing temperature while localizing the heated area.

ADVANTAGES AND LIMITATIONS

Advantages

Selective Heating. Localized heating makes it possible to join the parts without heating the complete assembly. This often minimizes metallurgical changes, such as loss of strength by tempering or annealing, when joining high strength components.

Such selective heating frequently reduces unwanted part distortion.

Precise Heat Control. Precise repetition of a processing cycle provides uniform joints with smooth fillets. It produces joints

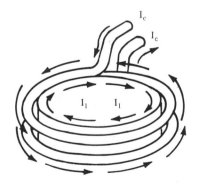

I_c - COIL CURRENT
I_I - INDUCED CURRENT IN WORK PIECE

Figure 11.1 — Induced Current "I_I" in Conducting Workpiece Produced by Electromagnetic Field

201

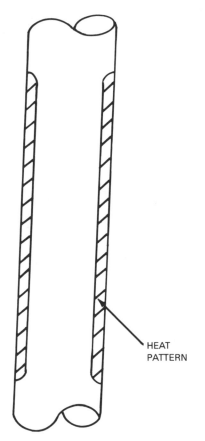

HEAT
PATTERN

Figure 11.2 — Heat Pattern Created by Resistance to Flow of Induced Current

that are identical in appearance with a predetermined minimum consumption of brazing alloy. When several joints in close proximity are required, precise heat control and localized heating permit sequential brazing using filler metals with decreasing brazing temperatures.

Rapid Heating. Normal induction heating cycles generally permit heating in air, while minimizing discoloration and avoiding scale. Cleaning of the brazed joint is also facilitated. When necessary, induction brazing can be conducted in controlled atmosphere or vacuum, as shown in Figure 11.3.

Adaptable To Production Line Processing. A primary advantage of induction brazing

is its adaptability to production line methods. It permits strategic arrangement of equipment in the assembly line. If desirable, heating may be accomplished by remote electrical command and control. Automated and semi-automated production lines with preplaced filler metal preforms also permit the use of less skilled labor.

Fixture Life and Simplification. The use of induction brazing frequently reduces and simplifies holding fixtures. Reduced heating increases fixture life and maintains accuracy in alignment of components to be joined.

Limitations

Complex Assemblies. The design of inductors often can make it possible to successfully heat geometrically-difficult joint areas. On the other hand, complex assemblies involving several brazed joints may not be viable arrangements or are so difficult to fixture that furnace brazing is the preferred process.

Fit Between Parts. Induction brazing requires that the fit (tolerance) between pieces to be joined be reasonably close and free of substantial burrs in order to ensure sound joints. This is true because the supply of brazing alloy is limited to the filler metal preform, rather than being hand-fed as in torch brazing. Alloys with a significant difference between the solidus and liquidus temperatures flow less freely in the brazing temperature range. This characteristic helps fill the spaces between stamped structural components that fit poorly. Excessive variations in clearance may prevent filling the joint and result in incomplete bonding.

Initial Cost of Equipment. Induction brazing is particularly adaptable to production applications in semi- or fully-automatic handling of parts. Systems designed to handle such production may require (a) an induction generator, (b) production handling arrangements, and (c) accessory equipment such as matching network and water cooling systems. All these come at considerable initial cost. Induction brazing may not be economically justified when (1) the number of parts to be

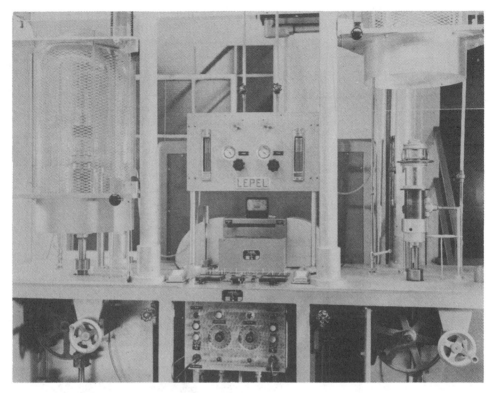

Figure 11.3 — Bell-Jar Arrangement for Induction Brazing Large Electron Tube Assemblies in Reducing Atmosphere Without Flux. (Courtesy of Lepel Corp.)

processed is limited, (2) induction heating equipment is not available, and (3) special considerations such as very localized heating are not necessary.

A small number of parts may be processed economically if an induction generator is available and the joint design requires only a simple inductor wound from copper tubing.

Specialized Knowledge. Achieving optimum system operation in induction brazing applications requires not only proper selection of a generator and appropriate part handling equipment, but also appropriate coil designs for the heat patterns. Appropriate electrical impedence matching between the load (workpiece, coils, leads) and the generator is also required for maximum performance. While technically not more difficult than other brazing methods,

greater specialized knowledge, often supplied by induction heating specialists, may be required.

APPLICATIONS

Induction brazing has been used extensively in the production of consumer and industrial products, structural assemblies, electric and electronic products, mining equipment, machine and hand tools, military and ordnance equipment, and aerospace components.

Figures 11.4 through 11.11, illustrate typical applications. Figure 11.4 shows typical plumbing fixtures composed of brass castings assembled by induction brazing, together with the induction coil and rotary handling fixture used in production. The castings are joined simultaneously in a

4-position semi-automated indexing fixture, using braze filler metal preforms and appropriate flux.

Figure 11.5 illustrates an induction coilwork arrangement for simultaneous brazing of (a) a brass adapter to an austenitic stainless steel shaft, and (b) a brass adapter to a copper-beryllium bellows in a bellows assembly. A copper shield is used to avoid destructive heating in the Cu-Be bellows. A separate induction soldering operation (not shown) joins a brass flange to the bellows to complete the assembly.

Figure 11.6 illustrates use of a pie-wound induction coil to simultaneously join eight copper tubes to a cast brass header. In operation, properly cleaned copper tube ends are assembled in the header; braze alloy rings are preplaced; the joint areas are prefluxed; and then the assembly is placed over the induction coil for heating to produce the joint.

An arrangement for assembly of machined parts and tubes by induction brazing is shown in Figure 11.7. Two operations using double layer induction coils assist production of cutting torches. The first operation involves two joints and the second four joints. Note the position of preplaced braze alloy preforms.

Figure 11.8 illustrates typical joints on structural members produced by induction brazing. Sections of steel tubing are joined to form structural frames for tricycles, wheel chairs, and other products. Split-type solenoid coils provide a uniform heat pattern and produce uniformly strong joints with smooth fillets requiring little or no finishing.

Induction brazing frequently is used for the assembly of electronic and electrical products. Figure 11.9, for example, shows a new design using induction brazing for producing small, compact generator housings for engine-driven sources of dc power in the aircraft and missile industry. The two-position handling fixture, shown with a combination pie-wound and external induction

Figure 11.4 — View of Brass Plumbing Castings Assembled with Preforms and Flux on Holding Fixture at Station 1; Within Induction Coil at Station 3; and Spray-Jet Coaling at Station 4

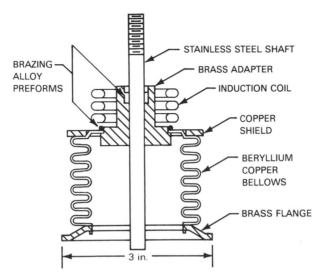

BRAZING ALLOY PREFORMS

STAINLESS STEEL SHAFT

BRASS ADAPTER

INDUCTION COIL

COPPER SHIELD

BERYLLIUM COPPER BELLOWS

BRASS FLANGE

3 in.

Figure 11.5 — Schematic Diagram Showing Induction Coil — Work Arrangement for Simultaneous Brazing of (1) Brass Adapter to Stainless Steel Shaft and (2) Brass Adapter to Bellows (Courtesy of Lepel Corp.)

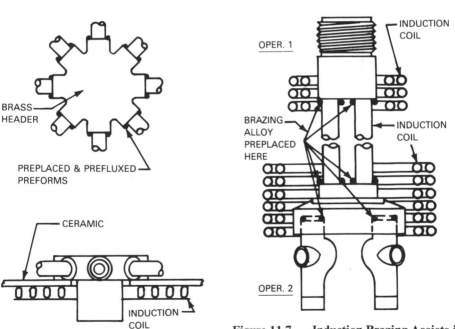

BRASS HEADER

PREPLACED & PREFLUXED PREFORMS

CERAMIC

INDUCTION COIL

Figure 11.6 — Pie-Wound Induction Coil Simultaneously Joins 8 Copper Tubes to a Cast Brass Header. (Courtesy Lepel Corp.)

OPER. 1

INDUCTION COIL

BRAZING ALLOY PREPLACED HERE

INDUCTION COIL

OPER. 2

Figure 11.7 — Induction Brazing Assists in the Production of Cutting Torches in 2 Operations Involving 6 Joints. (Courtesy Lepel Corp.)

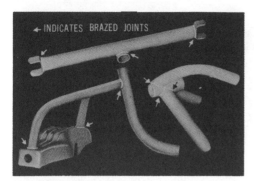

Figure 11.8 — Typical Joints on Structural Members Produced by Induction Brazing Carbon Steel Tube Sections With Silver-Base Filler Alloy. (Courtesy of Lepel Corp.)

coil, combines an Armco iron housing, a pearlitic malleable iron head, and a brazing filler metal preform to form the generator housing assembly.

Figure 11.10 illustrates a multiple-position fixture for joining a tube to a flange in the fabrication of wave guides. The actual operation involved is shown in the inset. Pressure from spring-loaded pins forces the tubes into proper position in the flange as the preplaced braze filler metal preforms melt. Note the internal induction coil. Tube-flange assemblies may be brazed or soldered. Ceramic arbors, that are not heated inductively, aid in accurate assembly.

EQUIPMENT

Successful induction brazing requires a suitable distribution of heat in all elements of the joint. This is primarily dependent on the coil design, the heating rate, the electrical and thermal conductivity of the materials being joined, and the masses of the parts involved.

In addition, a successful braze depends on the selection of a proper brazing filler metal and a flux, if used.

Coil Design

In many respects, the design of induction coils to provide an appropriate heat pattern

for a specific brazed joint is a matter of experience or experiment, or both. This is particularly true when a complex joint is involved. Basic coil designs and some guiding principles are also helpful.

Figure 11.11 shows several basic work coil designs. Coils are made from copper tubing, generally 3/16 to 3/8 in. in diameter, round or flattened. Coils also may have square or rectangular cross sections, depending on the configuration of the parts being brazed, current flowing in the coil, and water-cooling requirements.

Coils vary in design from a very efficient external solenoid coil (A) to a less efficient internal coil (F). The electromagnetic field outside the solenoid coil in Figure 11.11 (A) is much lower than in the part. A plate concentrator (transformer-type) coil (B) focuses a high-density electromagnetic field in a small area of the workpiece. A split-solenoid type coil (D) provides uniform heating for perpendicular tubular sections to be joined. A conveyor-type coil (C) permits continuous transport of the parts to be brazed. There are, of course, many variations of size, shape, contour, number of turns, turn spacings, and other features related to each of these basic designs to adapt it to various part geometries. In practice, coils are designed for each specific application.

In some instances, solid single-turn or multiple-station inductors may be used, as shown in Figure 11.12. Such inductors are machined from solid copper bar into which holes have been drilled and water cooling arrangements designed.

Combinations of the basic designs may be developed to adjust the heat pattern, i.e., external and internal or pie-wound and external, to create uniform heating at the joint area when small to large cross section parts must be brazed.

Heat patterns depend ultimately upon the strength of the electromagnetic field that is created by circulating currents in the induction coil. The number of coil turns per unit length, the spacing between turns, and the coupling distance between the coil and the part dictate the strength of the electromagnetic field. Thus, minor adjustments in the heat pattern can be made by variations in the spacing between turns and the coupling. The effects of such adjustments on the heat pat-

Figure 11.9 — Arrangement for Induction Brazing Generator Housing Assembly Showing (1) 450 kHz Induction Generator for Heating; (2) 2-Station Production Machine for Handling (3) Fixture for Aligning Housing and head in Raised Position in Induction Coil. Inset Shows Combination Pie-Wound and External Coil-Housing Assembly Relationship. (Courtesy of Lepel Corp.)

tern are illustrated in Figure 1 on page 968 of the *ASM Metals Handbook, Vol. 6, 9th Ed. Welding, Brazing and Soldering.*

Induction Generator — Type and Size

Temperature of the joint area may be influenced significantly by the rate of heating. The rate depends on size of the generator and the ability to control the alternating current in the coil. Lower power generally decreases the heating rate and provides time for thermal conduction to equalize temperatures in the brazing area. Power densities of 0.5 to 1.5 kW per in.^{2} are common. Decreased heating rates may be used, if necessary, and usually will result in decreased production. The size of generator required depends upon the size and mass of the components to be joined.

Three basic types of generators are in use: motor generator, solid state, and tube oscillator units. Solid-state units, operating at frequencies up to 10 kHz, generally have replaced motor generator units. Solid-state units with power up to several hundred kilowatts are available with output frequencies up to 50 kHz. Units that can produce frequencies of 100-200 kHz recently have been introduced.

Tube oscillator generators used in brazing applications operate at frequencies of 150 to 450 kHz and are available at power levels up to 200 kW and higher for special applications. Periodically, tube oscillators operating in the megacycle range (2 to 8 MHz) are useful in brazing parts of very small cross section.

The role of operating frequency in brazing applications is considerably less com-

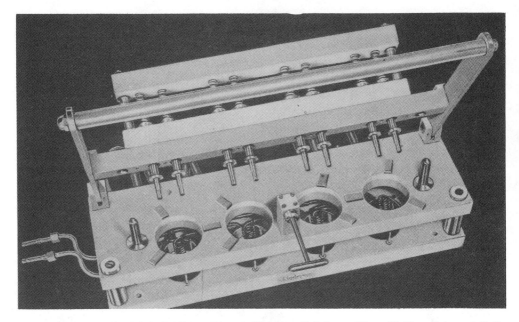

Figure 11.10 — Multiple Position Fixture Used to Fabricate Wave Guides. Internal Coil Provides Beating for Tube-to-Flange Joint. (Courtesy Lepel Corp.)

pared to surface hardening applications. This is true because the objective in brazing applications is to heat the joint area to a uniform temperature rather than to heat the surface. The exception is the brazing of thin parts where frequencies of 450 kHz are more efficient than lower frequencies. Tube generators are used extensively in brazing metal tubes for structural parts. In contrast, solid state generators operating at 10 kHz or lower are used for heavy sections.

Generator Size

Since the rate of heating can be measured by the available power of the generator, determination of the generator capacity needed to achieve a projected production schedule is important. Ultimately, the minimum size generator required for a given application is best determined by actual trial. A preliminary estimate may be made, however, by considering: (a) the power absorbed by the pieces to be joined, (b) radiation from the heated piece, and (c) the production rate required, assuming good coil design and electrical matching between the generator and the work load.

The kilowatts of power absorbed by the work is readily calculated from the relationship

$$P = \frac{W \cdot T \cdot C}{0.95 \, t}$$

where W = pounds of material heated (both by the induction coil and by thermal conduction away from the heated joint); T = temperature rise in °F; C = mean specific heat (see Table 11.1); and t = heating time in seconds required to meet production requirements.

Actually, it is difficult to estimate the weight of material heated in most brazing applications because heat is conducted away from the joint area, and a temperature gradient is created for some distance be-

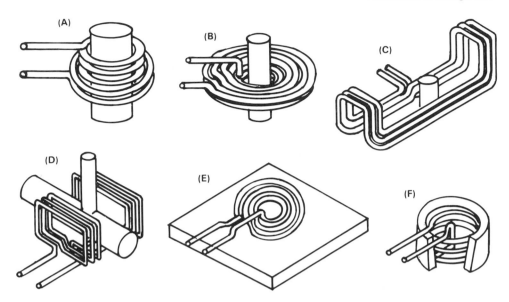

Figure 11.11 — Basic Induction Coil Designs Fabricated From Copper Tubing: (A) External Solenoid; (B) Plate-Type; (C) Conveyor; (D) Split-Solenoid; (E) Pie-Wound and (F) Internal Solenoid. (Courtesy of Lepel Corp.)

Table 11.1
Approximate Mean Specific Heat From 70°F to Higher Temperatures

Material	1400°F or Less	1100°F	1600°F	2000°F
Aluminum (1100)	0.23	0.24		
Beryllium Copper	0.10 (212F)			
Brass (70-30)	0.98	0.10	0.10	
Phosphor Bronze, 5%	0.09			
Copper (ETP)	0.092	0.092	0.092	
Gold	0.031	0.032	0.032	
Graphite	0.25	0.31	0.35	0.37
Iron	0.11	0.13	0.16	0.16
Magnesium	0.24			
Molybdenum	0.066			
Monel	0.13			
Nichrome	0.11			
Nickel	0.13			
Platinum	0.032			
Silver	0.056			
Stainless Steel (300 Series)	0.12			
Stainless Steel (400 Series)	0.11			
Steel (Carbon or low alloy)	Use data for Iron			
Titanum	0.13	0.14	0.16	0.17
Tungsten	0.033			
Zirconium	0.067			

(A) SINGLE-TURN INDUCTOR

(B) TWO-STATION INDUCTOR

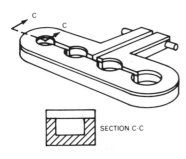

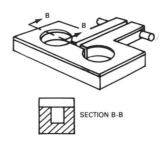

(C) FOUR-STATION INDUCTOR

(D) FOUR-STATION INDUCTOR W/TUBING

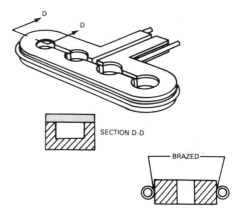

Figure 11.12 — Inductors Made From Solid Copper Bar Indicating Fabrication and Water-Cooling Arrangements: (A) Single Turn; (B) Two-Station; (C) Four-Station Internal Cooling; and (D) Four-Station With External Cooling

yond the induction heated zone. Furthermore, the above relationship does not account for the energy required to compensate for radiation losses. Since the temperatures most often used in brazing are relatively low [1150-1500°F (621.-815°C)] radiation losses are relatively small and generally can be neglected in the preliminary estimate.

Improper selection of power level is a frequent source of difficulty in induction brazing. Surprisingly, the difficulty is often a result of too-rapid heating of the assembly. Very fast heating exaggerates the effects of

electrical and thermal conduction when dissimilar materials, such as steel-to-copper, are involved in the joint. Recognizing the basic objective that the temperatures of the surfaces of the parts to be joined and of the braze filler metal should be as uniform as possible, slower heating rates may be desirable. Another problem that arises with high power densities is that electromagnetic forces may cause relative movement of the parts to be joined. This too can result in an unacceptable assembly.

In the same way, slower heating rates help compensate for any significant differences in masses of the components involved. In brazing a thin tube to a heavy section, for example, slow heating provides time for thermal conduction to help achieve more uniform temperature. To correct the potential problem of large mass differences, generators should have stepless, full range power control which is variable from essentially zero to full output.

In many cases, it is sufficient to know the material and the production rate (pounds per hour) to determine absorbed power in kW from Figure 11.13. Here, the absorbed energy per pound of material is plotted as a function of temperature. The starting point is the nominal room temperature of 68°F (20°C) and the variation of heat capacity with temperature is taken into account. From the figure, a value in kWh/lb is read, and that value multiplied by the production rate in lbs/h to obtain absorbed power in kilowatts.

> *Example*: Copper is heated from 68° to 1000°F at the rate of 200 pounds/hour. What is the absorbed power?
>
> *Answer*: The curve for copper intersects the vertical line representing 1000°F at a level of 0.028 kWh/lb. Multiplying 200 lbs. by 0.028 kWh/lb. results in 5.6 kW absorbed power.

Matching Devices

The actual energy absorbed by the load depends upon optimum transfer of energy from the power supply to the load (coil, workpiece, and coil leads). Optimum transfer is achieved when the output impedance of the power supply equals the input impendance of the load coil. Basically, the circuit that constitutes the load should be balanced, or as close as possible to electrical resonance with the circuit of the power supply for each application.

Most generators provide for easy internal adjustments of inductance or capacitance to assist in matching. In addition, external devices such as matching transformers or variable capacitance may be employed.

Fixtures and Handling

Most applications involving induction brazing require fixtures to hold the parts to be joined in proper alignment and permit movement into and out of the induction coil. Mechanical design of such holding fixtures is based on good engineering principles, including clamping arrangements, and locating pins.

Special attention is required, however, for the materials used in all fixtures and handling equipment near the induction coil. As shown in Figure 11.1, the electromagnetic field associated with the induction coil is not contained within the inductor itself but extends outside the coil, diminishing with distance. There is always the possibility of a fixture or auxiliary equipment heating inductively if it is an electrical conductor and located too close to the coil.

Nonconducting, heat-resisting materials, such as manufactured and natural occurring ceramics, quartz, and glass-filled polymer composites, are not affected by the electromagnetic field and may be used close to the coil. Many of these materials have the additional advantage that they can be machined and fired, or cast to shape. When metallic materials must be used for fixtures or handling equipment, nonmagnetic materials, such as austenitic stainless steels and aluminum, should be used. Closed metal loops in a fixture, coupled inductively to the induction coil or leads, are a common error in fixture design and should be avoided. It should also be kept in mind that for safety, it is required that induction coils be cast into a ceramic or wrapped with insulation. Bare coils are dangerous.

Figure 11.14 illustrates a conveyor-type induction coil used for continuous movement of parts through the coil.

FILLER METALS AND FLUXES

Filler Metals

Basically, brazing involves fusion of a brazing filler metal between surfaces of parts to be joined. If the surfaces are clean, intimate contact is established and the liquid brazing filler metal alloys with each surface forming a joint upon solidification.

Many commercial filler metals are available to join various materials and to provide for special requirements in specific applications. In addition, many experimental alloys have been reported in the literature.

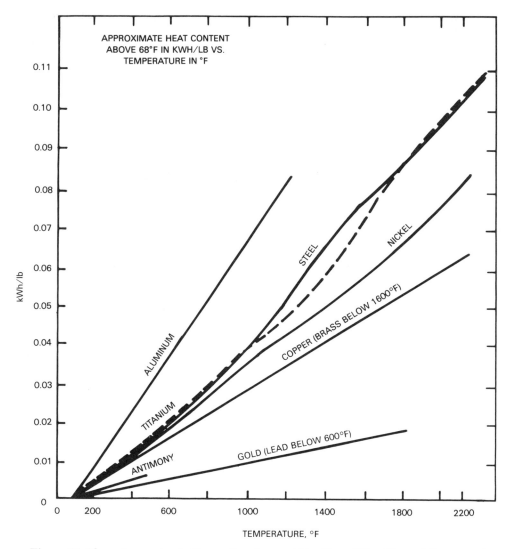

Figure 11.13 — Approximate Power Requirement Per Pound/Per Hour of Brazement

Basic requirements for a suitable filler metal include:

(1) Ability to wet and alloy with the surfaces to be joined

(2) Fusion below the melting point of the parts to be joined

(3) Appropriate fluidity for the filler metal to flow through the joint by capillary action

(4) Suitable strength, electrical conductivity and corrosion resistance in the joint to satisfy the mechanical, electrical, and chemical properties of the application.

Many different materials can be joined by induction brazing. These include carbon and alloy steels, stainless steel, cast iron, and cemented carbides, as well as copper and copper alloys; nickel, cobalt, and heat-resistant alloys; titanium, and zirconium alloys; molybdenum alloys; and more recently, ceramics and even graphite. The need for the brazing alloy to fuse below the melting point of the metals and to alloy with the surfaces to be joined demands many different

Figure 11.14 — Turntable With Conveyor-Type Coil for Continuous Movement of Parts Through Induction Coil for Heating. If Necessary, Parts are Rotated for More Uniform Heating. (Courtesy of Lepel Corp.)

filler metals. Brazing filler metals are discussed in detail in Chapter 3.

Silver-base filler alloys are used very extensively to braze both ferrous and nonferrous alloys by induction brazing. Several of the most common brazing filler metals are shown in Table 11.2

As indicated, induction brazing temperatures generally vary from 1150° to 1550°F (621 to 843°C). There is considerable variation in the difference between solidus and liquidus temperatures of the filler metals which may have significant influence on fluidity or flow. Other silver and copper-base filler metals not shown, but occasionally used, in induction brazing vary in brazing temperatures up to the melting point of pure copper, 1981°F (1083°C). The low brazing temperature indicated for BAg-1 filler metal permits joining at a relatively low temperature, thus minimizing energy consumption, limiting metallurgical changes in the parts being brazed, results in less oxidation of the workpiece, and the for-

Table 11.2
Silver-Base Filler Metals Frequently Used in Induction Brazing

AWS Classification	Composition, %					Temperature, F		
	Ag	Cu	Zn	Cd	Ni	Solidus	Liquidus	Brazing
BAg-1....	44-46	14-16	14-18	23-25	...	1125	1145	1145-1400
BAg-2....	34-36	25-27	19-23	17-19	...	1125	1295	1295-1550
BAg-3....	49-51	14.5-16.5	13.5-17.5	15-17	2.5-3.5	1170	1270	1270-150
BAg-5....	44-46	29-11	24-26	...	...	1225	1370	1370-155

mation of less difficult-to-remove flux residues. The narrow melting range provides good flow. Filler metal BAg-2 exhibits a wider melting range of 1125 to 1295°F (542 to 702°C), is more sluggish with respect to flow, and is useful for filling up or filleting poorly fitted joints. The addition of nickel in BAg-3 filler metal helps promote wetting and improves bonding when joining cemented carbides to steel and also stainless steel to itself or to carbon steel. A ternary alloy containing approximately 15% silver-80% copper and 5% phosphorus (BCuP-5) frequently is used for joining copper and copper alloys since it is self-fluxing with copper. BCuP-5 brazing filler metal should not be used with ferrous metals because it forms a brittle iron phosphide.

The eutectic filler metal, 72% silver-28% copper, melting at approximately 1435°F (780°C), is used frequently when induction brazing in a controlled atmosphere because it does not contain volatile components such as cadmium or zinc. Filler metals with high silver content of 65 to 72% provide a good color match when joining silver parts in the jewelry industry. For metals requiring high joining temperatures, Cu-Au or Cu filler metal may provide advantages in atmosphere and sequential brazing.

Brazing Fluxes

Precleaning is essential in making uniformly sound brazed joints of high integrity. Surfaces to be brazed should be cleaned chemically prior to heating to remove such things as heat treatment scale, corrosion products, and grease. The precleaned joint area should be fluxed as soon as possible to avoid contamination from handling or exposure and oxidation during the process of heating in air. Brazed joints in cast iron parts are more dependably gas-tight, liquid-tight and stronger if the parts are first electrolytically treated to remove graphitic carbon from the surface before fluxing and induction brazing.

Fluxes containing fluorides and alkali salts, preferably potassium, are generally used for induction brazing, particularly with silver brazing filler metals. These fluxes normally are used in paste form and applied by brush or spray in automated handling equipment. They become fluid and active below 1100°F (593°C), dissolving residual oxides and protecting the metal surfaces to be joined. Such fluxes also promote wetting and better flow of the brazing alloy upon melting. Many of the commercial fluxes available for brazing are proprietary formulations. The brazing alloy manufacturer should be consulted regarding a satisfactory flux for a specific alloy.

PROCESS CONSIDERATIONS

Joint Design

Joints to be brazed by induction heating require special attention to (1) the heating pattern, (2) method of preplacing the joining alloy, (3) clearances between mating parts, and (4) thermal conductivity and expansion characteristics of the materials to be joined.

Figure 11.15 shows several coil and joint designs used successfully in induction brazing applications. Figures 11.15(A) and (K) illustrate recommended positions for preplacement of the brazing filler metal. For best results, preforms of the brazing filler metal should not form closed loops when subject to inductive coupling from the work coil. In addition, preplaced filler metal should be electromagnetically shielded to avoid melting it before the joint surfaces are at brazing temperatures. Preforms can be shielded by placing them inside the assembly, as shown in Figures 11.15(A), (E), (F), (G), and (I), or by recessing a component as in Figures (F) and (H). Chambers or grooves frequently can be incorporated on one of the components to hold the preforms in place. Parts to be brazed may be pre-assembled mechanically by spot welding, tab assembly, or knurled press-fits to facilitate handling.

Stress concentration and high residual stresses in the joint may be serious when the joint members are stronger than the brazing filler metal, or when differential contraction takes place with dissimilar materials. The marked difference in the thermal expansion between tungsten carbide (3.3 x 10^{-6}/°F) and carbon steel (6.7 x 10^{-6}/°F) may be particularly troublesome when the carbide is placed in tension. This frequently results in cracked carbide. In such instances, a

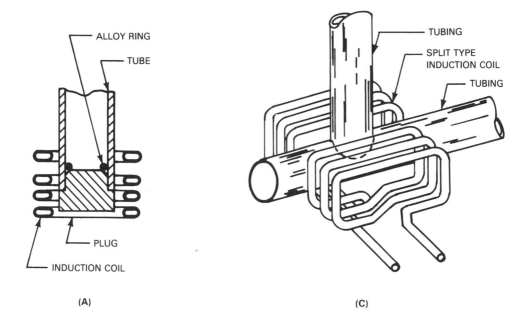

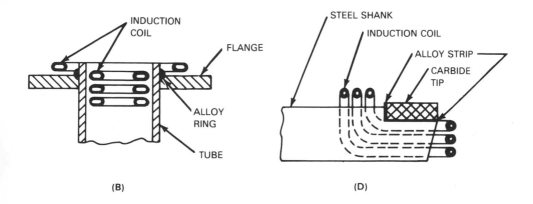

Figure 11.15 — Typical Joint and Coil Designs for Induction Joining Showing Suggested Positions for Pre-Placement of the Filler Metal Preforms. (Courtesy of Lepel Corp.)

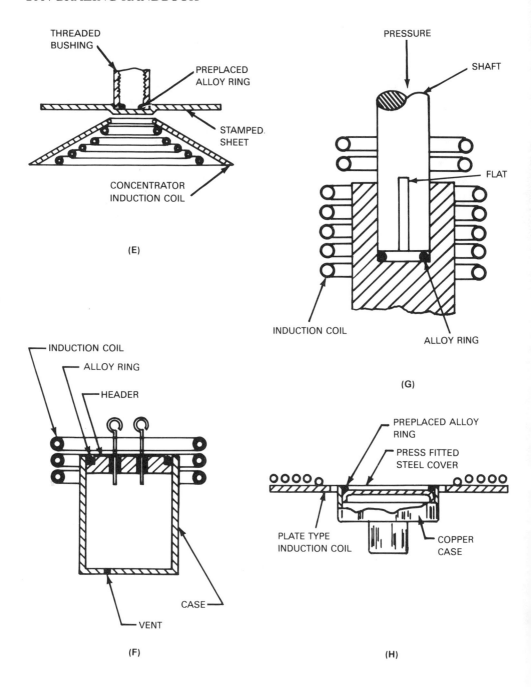

Figure 11.15 (Continued) — Typical Joint and Coil Designs for Induction Joining Showing Suggested Positions for Pre-Placement of the Filler Metal Preforms. (Courtesy of Lepel Corp.)

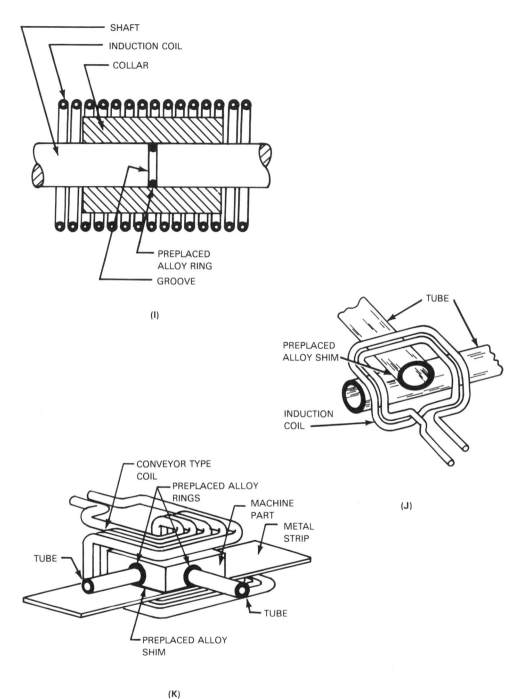

SHAFT

INDUCTION COIL

COLLAR

PREPLACED
ALLOY RING

GROOVE

(I)

TUBE

PREPLACED
ALLOY SHIM

INDUCTION
COIL

(J)

CONVEYOR TYPE
COIL

PREPLACED ALLOY
RINGS

MACHINE
PART

METAL
STRIP

TUBE

TUBE

PREPLACED ALLOY
SHIM

(K)

Figure 11.15 (Continued) — Typical Joint and Coil Designs for Induction Joining Showing Suggested Positions for Pre-Placement of the Filler Metal Preforms. (Courtesy of Lepel Corp.)

"sandwich" braze is helpful, utilizing a clad brazing strip, i.e., copper clad with silver brazing filler metal on both sides. In such a braze, stresses are minimized as a result of plastic deformation in the low yield strength copper layer.

Joint Clearance Between Parts

A joint clearance of 0.0015 to 0.002 in. (0.038 to 0.051 mm) is commonly used between parts to be induction brazed with silver-base filler metals if the parts being brazed are of the same material, i.e., steel. Since metal parts to be brazed may expand at different rates if differentially heated or if made of dissimilar materials, allowance should be made for differential expansion to assure the proper joint clearance when heated to brazing temperature.

A well-designed brazement should allow for the effect of joint thickness on the strength of the *total* brazed joint. In other words, the combined strength of the components being joined, the filler metal, and the alloyed interfaces together should be measured, rather than just the strength of the filler metal alone. The effect of joint thickness upon joint strength is illustrated in Figure 11.16. That illustration shows the change in tensile strength produced by a variation in joint thickness using a silver-base brazing alloy with tensile strength of 65 000 psi to braze pieces of stainless steel of 160 000 psi tensile strength. Tensile strengths of low carbon steel parts joined with silver-base brazing alloys may range from 50 000 to 60 000 psi.

Filler Metal Preforms

Filler metal preforms are used almost universally to accommodate the rapid and localized heating associated with induction brazing. The brazing filler metals in Table 11.2, as well as others, are available in various forms including wire, strip, and many shapes of blanked and formed preforms.

Some brazing filler metals also are available as a cladding on a base metal, i.e., as copper, clad on both sides with thin layers (0.0025 to 0.010 in.) of filler metal. Such clad materials can be utilized where the joint is subjected to considerable stress while in use or from contraction during cooling.

Preforms permit pre-assembly for automated operations and control the amount of filler metal used. They conserve material and produce uniform parts of good appearance.

BRAZING IN CONTROLLED ATMOSPHERE

The rapid advance of technology has created the need to join by brazing many materials such as titanium, zirconium, molybdenum, ceramics, and graphite. Such materials often must be processed in a controlled environment, i.e., in a reducing atmosphere or vacuum to avoid oxidation and volatilization. Brazing in a reducing atmosphere may avoid the use of flux in the assembly of some critical electronic and aerospace parts, and thus eliminate flux removal problems.

Induction brazing in a controlled environment enjoys the same advantages of induction heating, including highly localized and rapid heating with precise control for repetitive results. Figure 11.17 shows a workable arrangement for vacuum brazing a molybdenum assembly by induction. The 4-station fixture allows evacuation of three Vycor tubes enclosing fixtured molybdenum parts, while a single-turn induction coil is moved over the 4th assembly for localized heating to form the braze. A nickel-molybdenum alloy powder is used at 2550°F (1400°C) at a reduced pressure of 4 x 10⁻⁶ mm of Hg to prevent oxidation of the molybdenum and volatilization of the oxide. In this application, the induction coil is outside the nonconductive Vycor tube which does not interfere with the electromagnetic field established by the energized coil.

Figure 11.3 illustrates a bell-jar arrangement for induction brazing large electron tubes in a reducing atmosphere without flux. A two-position fixture is shown served by a 450 kHz induction generator. A transfer switch initiates heating at one station while the other station is unloaded and re-

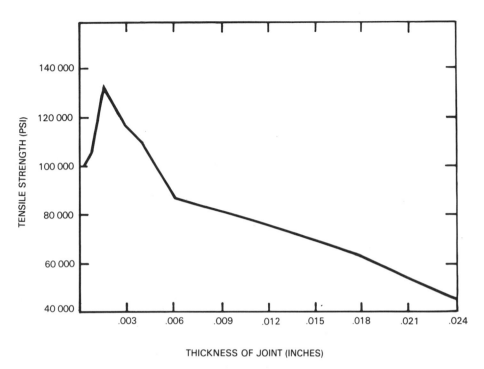

Figure 11.16 — Influence of Joint Thickness on Tensile Strength. Silver Brazing Alloy Used (65 000 psi) to Join Stainless Steel Sheet of 160 000 psi Tensile Strength. (Courtesy of Lepel Corp.)

loaded. In this arrangement a purified dry reducing gas under pressure, such as hydrogen, enters the bell-jar over a diffuser plate at the top. It then displaces the air, and surrounds the part to be heated. Upon heating the joint area by an induction coil within the bell-jar, oxide films are removed from the metal surfaces to be joined and on the brazing filler metal. Bell-jars are counterweighted to permit movement upward for access to the parts for fixturing at one station. Meanwhile, the second bell-jar, containing the atmosphere, is brought into position over the assembly for heating. Proper ventilation exhausts excess gas.

Sequential Induction Brazing

Figure 11.18 shows a number of critical brazed joints in the assembly of large electron tubes without flux, using the equipment in Figure 11.3. This example illustrates the unique potential of highly localized heating in sequential induction

brazing where filler metals with decreasing melting temperatures are used to achieve successive joints in close proximity.

For example, in one type of electron tube, induction brazing often starts with copper brazing at 2000°F (1093°C). The next joint may be gold brazed at 1925°F (1052°C). Silver alloy brazes then follow with filler metals having decreasing brazing temperature at 1650°F, 1475°F, 1380°F, and 1295°F (900°C, 802°C, 749°C, and 702°C, respectively). Succeeding joints are brazed at progressively lower temperatures to prevent damage to joints previously completed.

Simultaneous Brazing and Hardening

For special applications, brazing and hardening may be done in a single induction heating operation. For example, a carbide tip may be brazed to a steel shank and the shank hardened at the same time.

Figure 11.17A — Multiple Position Arrangement for Vacuum Brazing Molybedenum Elements in Cathode Assembly. Inset Shows Cathode Assembly in Vycor Envelope for Induction Heating. (Courtesy of Lepel Corp.)

Figure 11.17B — Enlargement of Figure 11.17A Inset Photo of Cathode Assembly in Vycor Envelope

Proper selection of the brazing filler metal, generally one with a brazing temperature range of 1500° to 1650°F (815° to 900°C), is essential to provide for austenitizing the steel shank without excessive grain growth. Processing then involves: (1) fixturing the carbide, steel shank and preplaced brazing filler metal (fluxed); (2) heating to the brazing temperature to flow the filler metal; (3) cooling to the transformation temperature for the steel shank [depends on type of steel, generally 1200°F (650°C) or less] to allow the brazing joint to solidify; (4) quenching in a suitable quench medium to harden; and (5) tempering to the hardness desired, if necessary. Best results are obtained if the joint is designed in such a

way that the carbide tip is in compression as a result of differential contraction. Most carbide tools for the mining industry are copper brazed at 1900°F (1038°C), cooled to the 1600-1650°F (870-900°C) range, and then quenched either in oil or an adequate water-polymer quench solution.

Figure 11.19 shows highly automated equipment used to produce mining tools by simultaneous brazing and hardening. Such equipment: (1) loads component parts; (2) heats the whole tool for brazing and austenitizing the steel in the shank; (3) provides a time delay for the braze to solidify; (4) supplies quenching for hardening the shank; and (5) discharges the assembly. The inset shows a tool withdrawn from the in-

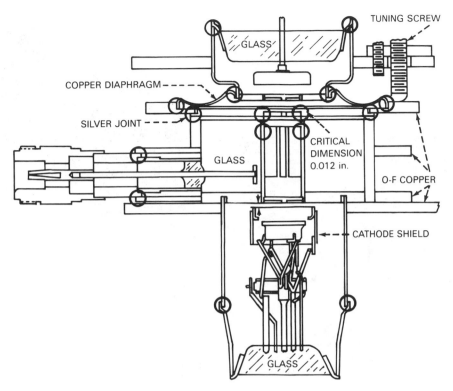

Figure 11.18 — Schematic Diagram Illustrating Induction Brazed Joints Produced Without Flux in the Assembly of Electron Tubes in a Controlled Atmosphere Using Sequential Brazing. Highly Localized Heating Plus Use of Various Alloys With Progressively Decreasing Melting Temperatures Produce Sound Joints Without Destroying Previously Brazed Parts. (Courtesy Lepel Corp.)

duction coil for transfer on the rotary table to the quench.

Safety

During the past 50 years induction brazing has been used increasingly for industrial processing. Safety improvements have been concerned with (1) electrical equipment; (2) the content of corrosive fluxes; (3) hot materials; (4) potentially explosive atmospheres for controlled atmosphere brazing; and (5) chemicals used in cleaning and processing.

Induction heating equipment follows NEMA standards. Extensive precautions have been taken to minimize exposure to potentially dangerous situations for operating and maintenance personnel. Great advances have been realized by (1) providing safety switches for access to generators, (2) coating, encapsulating or mechanically covering induction coils, and (3) isolating leads and other electrical accessory equipment.

Toxic fumes and gases from the fluxes and some filler materials, as well as some materials to be joined, can endanger workers in the absence of proper ventilation. Most brazing fluxes give off corrosive fumes upon heating; some brazing filler materials contain volatile cadmium, consid-

Figure 11.19 — Automated Arrangement for Induction Brazing and Hardening Mining Tools in a Single Heating Operation. Inset Shows Induction Coil in Raised Position with Heated Tool Prior to Time-Delay Quench. (Courtesy of Lepel Corp.)

ered a health hazard; and occasionally materials to be brazed may provide toxic fumes.

Handling hot materials and cleaning chemicals requires proper instruction and equipment. In cases where induction brazing requires an explosive atmosphere, as with hydrogen, proper precautions should be taken for ventilation and use of protective equipment.

Appropriate standards and local ordinances should be consulted for specific information.

REFERENCES

American Society for Metals. *Metals Handbook,* Vol. 6, 9th Ed. Metals Park, OH: American Society for Metals, 1983.

American Welding Society. *Specification for filler metals for brazing,* A5.8-89. Miami: American Welding Society, 1989.

———. *Welding Handbook,* Vol. 2, 8th Ed. Miami: American Welding Society, 1991.

Keywords — dip brazing, molten metal dip brazing, molten flux dip brazing, salt bath furnace, dip brazing and heat treating, salt bath maintenance, post braze cleaning of dip brazed assemblies

Chapter 12

DIP BRAZING

INTRODUCTION

The process of dip brazing is divided into two basic techniques, namely dipping or immersing the parts to be brazed into a molten filler metal, or dipping the part to be brazed into a molten salt. In both cases the temperature of the molten bath is below the solidus of the parent metal but above the liquidus of the filler metal.

Whereas dip brazing in molten metal is limited in application, dip brazing in molten salt is widely used for brazing aluminum, ferrous, and copper alloys.

MOLTEN METAL DIP BRAZING

There are several variations to this technique. One way to apply molten filler metal from a molten bath is to coat the assembled parts, wires, etc. with flux, by either dipping or spraying, then thoroughly drying prior to placing the assembly into the molten filler metal. A second method of molten metal dip brazing is to dip the assemblies into a molten flux bath and then into the molten filler metal. In both of these techniques, the molten filler metal flows into the capillary gaps of the assembled joints to produce the braze. Molten metal dip brazing is mostly used in the electronics industry for brazing wire and small components.

MOLTEN FLUX DIP BRAZING

The types of work which can be dip brazed in molten salt baths may be grouped as follows:

(1) Brazing of aluminum alloys at temperatures of 1000°F to 1140°F (540°C to 615°C). Chloride base salts containing fluorides act as both heating mediums and fluxing agents for these applications.

(2) Brazing of ferrous and non-ferrous assemblies with BAg filler metals at temperatures between 1250°F and 1850°F (677°C and 1010°C). Chloride base salts with suitable working ranges can be used with separate fluxing required.

(3) Brazing of copper with self-fluxing BCuP filler metals at temperatures between 1500°F and 1600°F (815°C and 870°C). A chloride base salt is used.

(4) Brazing of ferrous assemblies with RBCuZn filler metals in self-fluxing, cyanide or chloride (plus borax) base salt baths operating at 1675°F to 1750°F (913°C to 954°C). No dezincification takes place.

(5) Brazing of ferrous assemblies with BNi filler metals at 1800°F to 2200°F (982°C to 1204°C). Chloride base salt mixtures are used for this application.

(6) Brazing of ferrous assemblies with BCu filler metals at 2050°F (1120°C). Chloride base salt mixtures are used for this application.

GENERAL CONSIDERATIONS FOR MOLTEN FLUX DIP BRAZING

Advantages

(1) Strong, smooth, continuous joints are formed quickly.

(2) Selective brazing is practical.

(3) A large number of joints in a unit may be brazed simultaneously.

(4) Extremely thin gauge aluminum can be brazed.

(5) Section sizes from less than .002 to over 2.5 inches (.05 to 63.5 mm) thick can be brazed.

(6) Rapid and uniform heating rates are possible.

(7) Metal surfaces are protected from oxidation during the brazing operation by a film of salt.

(8) The rate of heating in a salt bath furnace is four to five times faster than in an atmosphere furnace. Shorter time cycles and faster heating reduce the risk of degradation.

(9) The buoyancy provided by the molten flux minimizes the distortion or collapse of fragile assemblies.

Limitations

(1) Parts that float cannot be brazed.

(2) Blind joints can trap air which will cause incomplete brazing or trap salt which could cause corrosion problems.

(3) Salt baths must be maintained at operating temperature and require periodic maintenance for successful brazing.

(4) Post braze cleaning of the brazement is required.

(5) All work must be preheated to prevent freezing of the salt and flushing of moisture when parts are immersed.

Some limitations can be overcome by proper fixturing, joint design, and control of process parameters.

EQUIPMENT REQUIREMENTS

A salt bath furnace consists of a metal or ceramic pot that holds the molten salt. Some salt bath furnaces have the salt containment vessel heated by gas, oil or electric resistance. Others are heated with immersion heaters heated by gas, oil or electrical resistance. A third type of salt furnace uses immersed metal or carbon electrodes which conduct current through the molten salt.

Externally Heated Furnaces

Externally heated furnaces are usually gas-fired, oil-fired, or heated by means of electrical resistance elements (less frequently used). This method heats the wall of the chamber which conducts the heat to the salt. Often the waste heat of fuel gases from fuel-fired furnaces may be fed to an adjacent chamber and used to preheat work pieces.

The size of the pot used to contain salt or molten metal is limited by the design strength of the material used, and the ability to maintain the bath at the operating temperature. Externally heated pots are normally fabricated from materials varying from aluminized low-carbon steel to high-nickel-chromium alloys.

Internally Heated Furnaces

Internally heated furnaces are of two types. One type is heated by immersion heaters which are hot and conduct the heat to the bath. The second type is heated by bath conducting electricity between two electrodes. With Type 2 furnace configuration the electrodes produce an electromagnetic stirring action of the salt that assists in maintaining temperature uniformity.

With salt baths, the heating medium contacts the work at all points. Heating is by conduction and is uniform. Distortion of the work is minimized.

Pot Construction

Materials used for construction of the pot in salt-bath furnaces depends on the type of salt to be contained. A steel alloy or magnesia-chrome refractory brick is used with cyanide salts and noncyanide carburizing salts that contain sodium carbonate. A ceramic (refractory) pot is usually preferred for neutral chloride salts and fluxing salts that consist of neutral chloride salts plus a fluxing agent such as borax or cryolite. Slight modifications can be made in the pot material when neutral salts and salts containing flux are used. When carburizing or cyaniding is to be performed in addition to brazing, a steel or heat-resistant alloy can be used; ceramic linings are also applicable.

SALTS AND FLUXING AGENTS

There is a variety of salts used in dip brazing. These include such materials as neutral chloride salts, fluxing agents (borax or cryolite) carburizing and cyaniding salts. Certain fluxing salts are also used with carburizing and cyaniding materials.

Though bath density is less than any of the metals, it does provide some support for the work. Molten salt's relative weight is two to three times that of water. Thus, aluminum work in the bath weighs 25 to 30% of its weight in air. This buoyancy feature helps to reduce distortion even further by minimizing the shock effect during immersion.

Neutral Salts

Neutral salts do not attack or modify the surface of the material being brazed. They protect the surface from attack by oxygen in the air. Because neutral salts do not attack oxides on the work piece, a flux must generally be provided.

The types of salts used in dip brazing of carbon and low-alloy steels are neutral chloride salts, neutral chloride salts with a fluxing agent such as borax or cryolite, and carburizing and cyaniding salts, which are also fluxing types of salts.

The neutral salts are mildly oxidizing to steel when they are used at recommended austenitizing temperatures. The oxides produced by heating steel in molten salt are largely soluble; hence, the steel is scale-free after heating. The accumulation of oxide in the molten salt, however, progressively makes the salt more decarburizing, and for this reason baths may require periodic replacement. Flux that is applied to the surface of the assembly and dried before the assembly is immersed in the neutral salt will be quickly dissipated by dissolving in the salt or gas. For this reason, there is generally no difficulty in removing flux from an assembly that has been brazed in a salt bath.

Fluxing Agents

Mixtures, which may be proprietary, are added to neutral chloride salts to produce a fluxing environment in the bath. A fluxing agent performs by (1) removing residual traces of oxide, (2) chemically etching the surface, and (3) providing a chemically clean surface, slightly mottled, for good filler metal wetting, flow, and penetration of the joint. Fluxing agents can be applied to the work or as an addition to the molten salt bath. When fluxing agents are used in the bath, periodic replenishment is required as the active ingredients are chemically depleted by the fluxing action. Above 1200°F (650°C), the fluxing materials will more rapidly decompose due to oxidation and other chemical reactions. In these cases, the fluxing agent must be replenished more frequently.

Carburizing and Cyaniding Salts

These salts provide their own fluxing action. In addition, they supply carbon or carbon and nitrogen to the surface of the steel assembly as it is being brazed. Although BAg filler metals have been used successfully, RBCuZn-A filler metal is generally preferred. A case depth up to 0.012 in. (0.30 mm) can be obtained in 20-30 minutes with excellent flow and filleting.

Aluminum Fluxes

For aluminum dip brazing there is a range of commercial fluxes. Some types are more suitable than others for a particular product or condition. The choice of a flux for aluminum dip brazing is based on fluxing ability, dragout, and required surface finish.

Commercial brazing fluxes for aluminum are generally similar to fluxes used in other brazing methods. Their main ingredients include sodium chloride, sodium fluoride, potassium chloride, aluminum fluoride, and lithium chloride. The chloride salts act as the carrier of the flux ingredients. High lithium-content fluxes melt and flow at lower temperatures reducing dragout. By properly proportioning the mixture NaCl, KCl, NaF, Alf and LiCl, the fluidity of the flux as well as the fluxing action can be tailored to the application. Salt manufacturers work closely with the users and often will develop special fluxes.

BASIC DESIGN CONSIDERATIONS

Simple, strong design normally begins with consideration of the application requirements. Full and careful study is required of the relationship between the brazed joints, the parts they are to join, and the dimensional criteria of the completed unit. During this study, the designer should consider the following brazing parameters:

(1) The distance between faying surfaces, i.e., joint clearance, may be estimated on the basis of past experience, but the final dimension, as ascertained by testing, may differ.

(2) The relationship between the coefficients of thermal expansion of the part within an assembly and the assembly to the fixture must be considered. For example, aluminum's coefficient is roughly one-third greater than that of the metals commonly used for fixturing and jigging. If the fixtures are not designed to allow for this expansion, parts may be damaged during brazing.

(3) At the brazing temperature, many base metals become very soft and are not self-supporting. This is particularly true for aluminum assemblies.

(4) Some distortion may be expected when an assembly is quenched after brazing.

Anticipation can reduce the above considerations to negligible factors through both design and process parameters.

Joint clearance is the distance separating the two base metal surfaces to be joined. The shape and orientation of one surface to the other is critical. Joint clearance influences capillary action. It is capillary action that makes brazing possible, drawing the molten filler metal deeply into the joint. Capillary force is controlled by surface tensions at the brazing interface. The smaller the clearance, the greater capillary force. However, the smaller the joint gap, the greater the possibility of flux, oxides, gas or foreign matter being trapped within the joint interfering with metal flow. In addition the smaller joint gap increases the possibility of filler metal stoppage by virtue of the filler metal alloying with the base metal to form a new alloy that is less fluid at the brazing temperature. Conversely, overly large clearances pose difficulties of their own. Most important, capillary action is reduced and the joint is not completely filled. More stress is placed on the fillet because of its larger cross-section. The strength of the joint will be closer to that of the filler metal than the base metal. Voids may appear in the joint. A smooth and even fillet may not form, and filler metal is wasted.

Fortunately, arriving at the correct or best joint clearance for a particular brazing job is not difficult. In practice, satisfactory joint clearance is quickly determined by means of a few test joints made under actual production conditions. Once gap dimensions have been established, they will hold true as long as other factors such as brazing method, time/temperature, flux and alloys are not changed.

Surface tension of the molten filler metal also plays a significant role in flow of the braze filler metal. One of the desirable characteristics of dip brazing is the tendency of the molten flux to reduce surface tension. The flux removes any residual oxide films on the surface of the molten filler metal and thereby enhances wetting. Since most filler metals have a melting range, a brazing temperature must be selected above the melting range for suitable flow and penetration.

JOINT PARAMETERS

All the following conditions should be met if sound brazed joints are to be realized:

(1) The faying surfaces are to be clean and free of discontinuities.

(2) The oxide surface shall be controlled to promote wetting. This may be accomplished either chemically or mechanically.

(3) The gap between mating surfaces is correct for the width of the joint, the type of filler and parent metals used and the time/temperature selected for the brazing method.

(4) The joint is designed to permit flux and filler metal to enter easily and to permit flux, filler, oxide, and heat-generated gases to exit just as easily as the molten filler metal flushes the joint.

(5) The correct quantity of filler metal shall be properly positioned prior to brazing. In instances where the braze filler metal is not preplaced in the joint, the filler metal shall be placed in such a manner where it melts after the joint has been fluxed.

(6) The assembly shall be designed and processed to permit the escape of gases generated during brazing and prevent trapped air.

(7) The joint is brought to proper temperature and held there long enough for brazing to be accomplished.

(8) Parts are maintained in fixed relationship during brazing and cooling.

FIXTURING

A self-supporting joint is preferred over a design requiring fixturing for support. By providing shoulders or locking devices, peening, staking, welding, knurling, etc., the assembly can be self-supporting. Fixturing is thus minimized to the use of a hook, a rod, a rack, or a basket which is used to carry the assembly freely into and out of the brazing operation.

If a fixture is used for supporting the assembly, complications are added to the process, such as the extra mass of material which must be heated. In some cases, the close proximity of the fixture to the braze joint may result in distortion of the assembly. However, there are many assemblies which must be held together by springs, clips, or special machined fixtures. Fixturing requires design attention and maintenance to preserve dimensional tolerances.

Joint Clearance

The dimensional relationships between mating surfaces often determine the success or failure of an application. For aluminum brazing the clearance is often specified within 0.002 in. (0.05 mm) and 0.010 in. (0.30 mm). For BAg or BCuZn brazing applications, a clearance between mating surfaces of 0.001 to 0.003 in. (0.025 to 0.076 mm) is common. When copper brazing ferrous assemblies, joint tolerances which range from a slight interference fit to approximately 0.001 to 0.002 in. (0.025 to 0.05 mm) are conventional. These dimensions are nominal and serve only as a guide in designing a joint for brazing.

Joint clearances and interferences are complicated when using dissimilar metals which have different rates of thermal expansion. It is possible to have an assembly where the joints at room temperature are loose but approach an interference fit at the brazing temperature. The opposite of this is also true. It is, therefore, necessary to consider expansion of the component parts so that the tolerance of the joint is maintained as constant as possible during the brazing operation. This precaution also applies to the selection of fixture material which can greatly influence dimensional stability of the assembly if the difference in thermal expansion is significant.

APPLICATION OF FILLER METAL

Filler Metal Placement

Sufficient filler metal, in the form of wire, washers, foil, powder, or paste shall be preplaced in close proximity to the joint. When placing filler metal external to the joint, it should be placed on one side only, and in sufficient quantity to produce visible fillets at the edges of the joint's faying surfaces. All binders and flux compounds used to coat and hold the filler metal powder in place must not leave a residue during or after the flow of the molten filler metal.

Joints, having one end inaccessible to visible inspection, shall have the filler metal placed at the blind end prior to assembly. Joints using clad brazing sheet will have the filler metal in contact over the entire faying area.

Brazing Filler Metal Types

Various brazing filler metals in the form of wire, washers, shim, sheets, cladding, powder, paste, etc. are available for salt bath brazing. Selection of the filler metal should be based on base metals to be brazed, design, service of finish product, process parameters and cost effectiveness. Chapter 3 on brazing filler metals aids in the selection of the proper material.

PROCESS DESCRIPTION

Precleaning

Because brazing involves capillary action of the filler metal, the surfaces of the workpieces must be clean, free of oil, paint, grease, oxide, scale, burrs or other contaminants that would prevent the filler metal from wetting the workpiece surfaces.

Preheating

Preheating in some form is always required prior to dip brazing. Preheating an assembly before brazing serves several purposes. If prefluxing is used, preheating dries the flux and vaporizes all moisture from the assembly and fixture. Even a slight amount of moisture can cause spattering in contact with molten salt.

The use of preheating will decrease the heat requirement of the salt bath, reduce brazing time, and reduce salt dragout. Preheating can prevent the premature melting of externally placed filler metal. In joining an assembly consisting of both heavy and light sections, adequate preheating reduces thermal gradients and subsequent distortion, as well as improving the wetting action on the heavier parts.

The preheating temperature must be below the melting temperature of the filler metal and below the temperature at which rapid oxidation occurs in the parent metal. It is frequently several hundred degrees lower than the brazing temperature for carbon and low-alloy steels; however aluminum may be preheated to within 100°F (55°C) for brazing with BA1Si filler metals.

Preheating is done by using a variety of methods. Ovens and furnaces are the most common. The temperature is selected based on the filler metals and parent metal used. Other less technical methods, such as placing parts on top of furnace cover or holding them over the surface of the salt, are also used.

Brazing

The time in the molten salt bath varies. For thin-section parts to be brazed only, the holding time may be less than one minute. For assemblies that are to be case hardened as well as brazed, the holding time is that required to produce the desired depth of case.

After the workpieces have been in the bath for the required time, they are carefully lifted from the salt bath. A uniform motion controlled removal rate is necessary. Jerky movements can cause the liquid filler metal to be displaced from the joint.

Brazing and Heat Treating

It is possible to combine the brazing operation with heat treating. The assembly can be air cooled to room temperature or quenched in a suitable medium, as required to obtain the heat treated properties.

If the assembly is to be water quenched, it must momentarily cool in air until the brazing filler metal has solidified. Otherwise, the molten filler metal will usually be forced out of the joint during quenching. Generally, it is possible to oil quench directly from the brazing temperature without air cooling.

Molten salt bath quenching is also used. However, it is imperative that the salt used for the quenching bath be compatible with the salt used for the brazing. Neutral chloride salts are recommended for brazing and austenitizing when a nitrate-nitrate salt is to be used as a quenching bath. The assembly must be air cooled until the filler metal solidifies before quenching in a nitrate-nitrate bath, as the molten salt will react with molten filler metal.

SALT BATH MAINTENANCE

A properly maintained salt bath will produce workpieces that are bright and shiny, with fillets well formed and complete. Such a bath will be relatively free of sludge and surface film, and the pH will be controlled based on the formulation of the salt. Its chemical composition will be relatively unchanged from its original formulation. A poorly working salt solution will produce porous, pitted fillets with voids along their length. Fillets and parts will emerge dull and gray.

The depleted salt solution may have a pH that is out of range. The bath may be filled with sludge or covered with floating debris and its chemical composition will have changed beyond the formulation limits. Any of these conditions may be severe enough to cause brazing difficulties.

Of the conditions described, variation in pH is by far the most frequent and troublesome. This is caused by one of two phenomena: water in the bath which decomposes into oxygen and hydrogen to form oxides and hydrates which turn the bath alkaline, or metals and their oxides which will turn the bath more acidic.

Water enters the bath in two ways. Dry salt, as received from the manufacturer, in air-tight packages, contains a small quantity of water. The flux as manufactured contains a small amount of water and flux is hydroscopic and will absorb water from the air. A pot of freshly melted salt may require days of dehydration or rectification, as it is often called, before its water content is sufficiently lowered to permit satisfactory brazing. Metals enter the bath through fixturing, the work, and maintenance tooling. In addition, the salt bath must be dehydrated or rectified on a periodic basis to control water absorbed from the air during operation.

In aluminum brazing excess water can be detected and removed with little difficulty by dipping clean aluminum into the hot flux. Water in the salt reacts with the aluminum to release hydrogen which rises to the surface and bursts into puffs of yellow flame. When the flames stop, the water is effectively gone. Gas bubbles alone should be ignored as various gases are always being generated at the aluminum/flux interface. Some shops do not depend upon hydrogen flames alone, but braze a series of test joints until joint quality indicates the absence of water. Raising flux temperature to 1300°F (704°C) or more is not recommended as a practical method of dehydration. The flux will fume excessively and there is a strong possibility some of the flux constituents will be destroyed.

The chemical composition or balance of a bath changes very slowly, most slowly when the bath is active and fresh flux is added constantly to replace flux lost by dragout. The rate of change, even in inactive baths, is so slow that a once-a-month bath test is enough to prevent brazing difficulties. Companies manufacturing flux and other products offer salt-bath testing services and also sell the chemicals necessary to correct bath composition changes. These are merely added as needed.

Some changes in bath composition are fairly obvious. For example, a bath that is low in fluorides would be fluid at its operating temperature, but would braze poorly. Without fluorides, the bath is very slow to remove oxide from the surface of aluminum. If the bath's fluoride content is high, or if its lithium content is low, the flux would be viscous at operating temperature, more salt would freeze out as skulls, but it would braze well.

The color of the molten flux is, however, only slightly useful in analyzing its composition. A desludged bath with carbon electrodes and sodium fluoride will be clear. The same bath with aluminum fluoride will be gray. A bath with nickel electrodes will be light blue in color when it is clean. As the quantity of metallic nickel in solution increases, the bath's color becomes increasingly blue.

The preventative maintenance schedule is specified by the manufacturer of the equipment and the molten salt. It may be modified based on the experience of the operator as required. The following comments are related to the operation of aluminum brazing salt baths, which are particularly unique and well documented.

Aluminum Salt Bath Maintenance

With aluminum brazing salts, on dry days, when a large volume of work is brazed, the need for dehydration is minimal; the clean aluminum workpieces remove water from the flux. On moist days or when the bath is inactive, considerable dehydration may be required. Idle baths should be tightly covered to prevent moisture pickup. It is also desirable to avoid drawing air across the bath surface. For best results, the furnace installation should be enclosed with exhaust take-off from the top of the enclosure.

In most instances of dehydration the removal of water (actually the hydrogen) is all that is necessary to turn the bath acidic. If the bath tests above 7.0 pH, anhydrous aluminum fluoride, or one of the proprietary concentrates manufactured for this purpose, may be added. Aluminum fluoride reacts unfavorably with lithium chloride and should be used sparingly. Aluminum fluoride should not exceed 1% of the total flux volume and should be combined with 20 times the amount of brazing salt for better bath mix.

If pH testing equipment is not available, a test joint can be used to judge the condition of the salts. This is generally made in the form of a tee with a line of contact a foot or more in length. Filler metal is placed at one end of the joint, and that end is lowered vertically and first into the hot flux. If the bath is working properly the filler will run up the joint and form satisfactory fillets on both sides of the tee.

Undissolved contaminants in and on top of the hot salt interfere with brazing far less frequently and far less severely than alkalinity. Contaminants that float to the surface are called scum and are composed mainly of iron oxide and carbon particles that scale from the electrodes (when carbon electrodes are used). Surface contaminants are readily removed by skimming the salt's surface with a sheet of aluminum.

A simple rule of thumb test by which surface scum concentration may be evaluated consists of the immersion of a clean sheet of metal into the salt. If the scum forms more than a 1/2 in. (1 cm) band around the top of the aluminum sheet, the bath should be desludged.

Contaminants that sink to the bottom of the pot are called sludge. Sludge is composed of the oxides of dehydration, oxides from the brazed assemblies, insoluble fluoride complexes, and the results of the reaction of the flux with the ceramic sides of the pot. Sludge is also the heavy metal impurities that find their way into the salt: nickel, iron, copper, zinc and lead. Depending on flux formula, flux condition and brazing volume, roughly 5 lb (2.27 Kg) of sludge may be expected daily for every 1000 lb (454.5 Kg) of salt in the pot.

Sludge cannot be prevented, but heavy metal contamination can be reduced by avoiding ferrous fixtures and springs, and steel-clad electric heating elements for initially melting the flux.

Some sludge is removed from the salt by brazing. The sludge adheres to the work in the form of dark stains. These stains may be removed by dipping the washed assemblies in nitric plus hydrofluoric acid. Neither hot water nor nitric acid by themselves are particularly effective. Some sludge is removed from the salt by dehydration treatment; the sludge adheres to the aluminum that is dipped into the pot.

The major portion of the sludge collects at the bottom of the pot and is removed by ladling with a perforated tool. Some shops prefer to remove sludge at the start of the work day, before brazing activity has stirred up the collected contaminants. Early desludging, however, tends to have salt lumps in the bath that froze during the preceding night. Sludging after work hours removes less sludge but reduces the incidence of salt freezing at the bottom of the tank.

The lumps of frozen salt on the bottom of the tank are called skulls. Ninety to 95% of the skull is composed of salt. The balance is sludge. The safest and fastest way to remove the skulls consists of melting the solid salt with the aid of a motor-driven impeller and an increase in bath temperature of 25 to 50°F (3.9 to 10.0°C). The larger furnaces have impellers and motors built into their side for this purpose. Small tanks can be stirred with stainless steel impellers mounted on portable electric motors.

In small quantities, neither sludge nor surface scum adversely affect brazing. Large quantities of contaminants can, however, completely stop joint wetting in portions of the brazed assembly.

The last and least frequent of the general causes of poor brazing, attributable to the salt itself, is a major change in flux formula.

POST BRAZE CLEANING

For post braze cleaning of dip brazed assemblies, boiling water is the best medium for removing salt. Agitated hot water, approximately 180°F (82°C), is also effective. The complexity of the assembly determines the amount of washing required. After washing, the assembly will be checked for residual salt followed by an optional cleaning process. The presence of residual chlorides can be checked by adding a few drops of silver nitrate reagent solution to the final draining of rinse water that has been acidified with a small amount of HN3. A white, milky precipitate indicates that chlorides are present and that washing should be repeated until the work is chloride free. Residual chlorides cause a frosty corrosion product which develops with time and may cause failure of the part in service.

NOTE FOR ALUMINUM: Extremely thin-walled tube or sheet brazements sometimes develop leaks during cleaning or thereafter in storage. This is a result of intergranular corrosion due to incomplete flux removal. Most acid brightening and cleaning solutions should not be used until the salt has been completely removed. If there is any doubt about the completeness of flux removal from complicated assemblies, a 10% solution of nitric acid can be employed in cleaning if the salt content of the acid does not exceed 0.5%. A solution of 5% phosphoric acid and 2% chromic acid in tap water at 180°F (82°C) is recommended for cleaning thin-walled work. After thorough cleaning and rinsing, assemblies should be dried before storage.

NOTE FOR STEEL AND COPPER: For post braze cleaning of steel and copper brazements, many of the procedures for cleaning salts from steels and copper are similar to those used for aluminum except the use of nitric acid for brazements containing silver, silver alloys, copper, or copper alloys. Nitric acid should not be used because of its rapid attack on these particular alloys.

SAFETY AND PRECAUTIONS

Steel and Copper Alloys

Cautions and references for the safe use of salt baths and accessories, and the need for all parts to be free of moisture before immersion in salt bath, are discussed below and in Chapter 6 on safety and health.

NOTE: SAFETY PRECAUTION. The introduction of cyanide salts or other reducing agents into a nitrate-nitrate quenching bath will cause violent explosions.

Since assemblies that are being brazed in salts containing cyanide must not be quenched in nitrate-nitrate salts because of the explosion hazard, and the filler metal must solidify before quenching, the following procedure has been used:

A brazing alloy such as RBCuZn-A is selected that will solidify above the transformation temperature range of the steel being brazed. The brazed assembly is quenched from the cyanide-containing bath into a neutral chloride rinse salt bath maintained below the freezing temperature range of the filler metal but within the austenitizing range of the steel. The assembly is then quenched from the neutral chloride rinse to the nitrate-nitrite bath. It is essential to control the amount of cyanide buildup in the neutral chloride rinse bath. When tests indicate more than 5% cyanide in the chloride

rinse, part of the salt should be discarded and the remainder diluted with new neutral chloride salt. After quenching in the nitrate-nitrite bath, the assemblies are air cooled, washed, and then tempered, if tempering is required. All fixtures must be thoroughly cleaned and dried after quenching, to prevent transfer of quenching salt to either cyanide or neutral chloride baths. Nitrate-nitrite salts will cause an explosion if they are mixed with cyanide. A chloride bath that is contaminated with nitrate-nitrite salts will produce pitted and decarburized parts.

Carburizing and cyaniding salts contain cyanide. If taken internally, cyanides are fatally poisonous; if allowed to come in contact with scratches or wounds, they are highly toxic. Fatally poisonous fumes are evolved when cyanides are brought into contact with acids. To avoid possible toxic effects, it is mandatory that an exhaust system be provided to remove fumes from salt baths containing cyanides.

Aluminum

Dip brazing has the inherent danger of very hot metal and fluxes. Aluminum does not change color when heated and other higher temperature alloys may not glow but be sufficiently hot to cause painful and dangerous burns. Aluminum's color cannot be used as a danger sign. Instead a system or pattern of metal handling must be established which eliminates the need for handling hot parts by the workman.

Safe Work Systems and Practices

Protective equipment for personnel would apply to all salt bath operations and would include long, heat resistant gloves, high-temperature safety apron, a face shield, and safety glasses.

Salt bath brazing may produce dusts, fumes, and gases hazardous to health. Therefore, adequate ventilation is necessary. Some silver brazing alloys contain cadmium. Cadmium oxide fumes are hazardous and inhalation of these fumes can be fatal. The fluorides in fluxes pose a dual problem; it is not only necessary to provide adequate ventilation to carry away fumes, but skin contact with these fluxes should be avoided and means should be provided to make it unnecessary for anyone to come in contact with the flux.

Mechanical aids for lowering and lifting parts in and out of hot equipment is advisable. The dangerous practice of having several people on one pole to handle a particularly heavy part should be avoided. Computer controlled work handling systems, with hooded enclosures, are custom designed to provide a safe, repetitive process.

Splatter produced by parts accidentally dropped into molten flux can be prevented with properly designed supports, hangers and fixtures, and by the avoidance of overloading and crowding. Properly designed hangers and supports are self-balancing and equipped with long handles. Indentations and hooks are provided to hold parts securely in place. Hooded enclosures and remote work handling systems are recommended.

Caustic Solution Precautions

Caustic precleaning solutions require two-fold precautions. Care must be exercised to prevent the caustic from dripping or splattering on personnel. Workers must wear protective clothing and goggles. And secondly, the room must be well ventilated. Caustic solutions combine with aluminum and liberate hydrogen which transports caustic vapors. Inhaled, the caustic vapors are painful and injurious to the respiratory system. If the hydrogen is permitted to accumulate, an explosive hydrogen/air mixture will form. Smoking should not be permitted in these areas.

Flux Precautions

Many brazing fluxes contain materials which are toxic. Flux must therefore be handled with caution. When a package or container of flux is first opened, care must be taken to not spill it. Flux is dry at this time and can be easily blown about.

Burns and Injuries

For burns and injuries, seek medical help immediately. For first aid information contact the Red Cross or other expert advisers for information.

A number of high-volume, low-pressure showers should be installed at strategic points throughout the brazing shop. They should be close enough to be of immediate value, but not so close that water can splash into a flux pot or onto a brazing furnace. The showers should be controlled by foot or elbow treadles so that cold water can be turned on quickly and easily.

A victim of serious burns should not be moved. No attempt should be made to remove his clothing; the skin may come off with it. Burns should be instantly flushed with volumes of cool water. A physician shall be consulted.

Ventilation

Salt pots must be force vented. Hot flux produces a constant stream of irritating, toxic gases. Maximum fuming occurs when the pot is first fired up and when fresh flux is added. The hot gases tend to condense on the nearest cool surface. If the gases and flux vapors are not removed, everything made of metal in the brazing room will corrode and the health of those present will be endangered. Venting can be accomplished by means of an overhead hood system or a long, narrow, horizontal vent placed close to the surface of the molten flux, or both.

The overhead hood system is preferred with top exhaust. The side vent is not nearly as effective, and it tends to draw more air over the surface of the flux, thereby increasing the moisture content of the molten salt.

Aluminum is recommended for hood ducting. Surprisingly, flux vapors attack the hood and leave the rest of the venting system virtually untouched.

Safety Sources

The reader is also referred to the following sources in which detailed information on safe practices, cautions, and hazards in brazing are discussed:

Safety in welding and cutting, American National Standard Z49.1

Specification for brazing filler metal, AWS A5.8, American Welding Society

Safe practices in welding and cutting, *Welding handbook*, 8th Edition, Volume 1, Chapter 16, American Welding Society

Molten Salt Baths, chapter 29, *Industrial fire hazards handbook*, Second Edition, Revised by Quentin O. Mehrkan, Ajax Electric Co. (Published by National Fire Protection Association)

REFERENCES

The Aluminum Association. *Aluminum brazing handbook*. Third Edition. The Aluminum Association, 1979.

ASM International. Dip brazing of copper and copper alloys. Revised by Martin Prager. *Metals handbook*, 9th Edition, Vol. 6.

ASM International. Dip brazing of aluminum alloys. Revised by A. Lentz. *Metals handbook*. 9th Edition, Vol. 6.

ASM International. Dip brazing of steels in molten salt. Revised by Q. Mehrkam. *Metals handbook*. 9th Edition, Vol. 6.

Gempler, E. B. Dip or vacuum - which is best for brazing heat exchangers. *Heat Treating*. April 1982.

Glen L. Martin Company. Dip brazing boosts output of aluminum assemblies. *Tool Engineer*. May 1956.

IBM Specification 28500912, 03B/10036/01. Brazing, aluminum molten flux.

Lynch, E. F. and Scott, J. K. Aluminum dip brazing for the electronic industry. Hughes Aircraft Company, *Western Machinery and Steel World*. Reprint No. 177.

Maston, J.E. High penetration produces strength brazed joints. Kearfoot Company. Modern Steel. 1958.

Mehrkam, Q. D. Never underestimate the power of a salt bath furnace. Ajax Electric Company. *Welding Design and Fabrication*. March 1968.

Mehrkam, Q. D. Salt bath brazing, original notes for publication in *ASM handbook*. Ajax Electric Company.

Mehrkam, Q. D. Salt bath brazing techniques. *Welding Journal*, April 1970.

Military specification. MIL-H-6088. Heat treatment of aluminum alloys.

Military specification. MIL-F-80113. Furnaces, vacuum, heat treatment and brazing.

Park Chemical Company. *Aluminum brazing in salt baths*. Technical Bulletin L-1 (H-26 Rev.).

Park Chemical Company. *Aluminum brazing in salt baths*. Technical Bulletin L-4.

Slotta, E. G. Dip brazing of aluminum pays with accurate assemblies. Raytheon Manufacturing Company. Waltham, Mass.

Keywords — resistance brazing, resistance brazing equipment, resistance welding transformers, manual resistance brazing, resistance brazing materials, resistance brazing processes, automated resistance brazing

Chapter 13

RESISTANCE BRAZING

PROCESS DESCRIPTION

Resistance brazing is a process in which heat is obtained from resistance to an electric current flowing in a circuit which includes the workpieces. The heat for resistance brazing is developed in either the workpieces or the electrodes that contact the workpieces, or both, when electric current is passed through them.

Electrical resistance can be a source of heat in conductors because crystalline imperfections (such as vacancies, dislocations, solute atoms, and grain boundaries) and phonons (which are thermally-induced quantized elastic waves in the crystal lattice) impede the movement of electrons within a conductor. These impediments to electron flow constitute the electrical resistivity of a conductor (an intrinsic physical property). Thus, their combined action converts some energy of each electron into heat.

Resistance heating is sometimes referred to as I^2R heating because the amount of heat in watt-seconds generated by the flow of electrons is directly proportional to the product of three variables:
(1) Square of electrical current, amperes
(2) Electrical resistance, ohms
(3) Time, seconds

In electronic devices and electrical machinery, resistance heating is normally undesirable because it reduces the efficiency and useful life of the equipment. In a joint to be resistance brazed, however, the heat has the beneficial effect of raising the brazing filler metal to its brazing temperature. The technical details of resistance heating can be found in the *Standard Handbook for Electrical Engineers*, published by McGraw-Hill.

Advantages

No brazing process is universally applicable; each has unique advantages. Resistance brazing is most useful for joints where heating must be highly localized, flameless, noncontaminating, rapid, and closely controlled. Specific advantages are

(1) Resistance brazing can be cost-effective. A large motor stator or rotor may contain many small copper-to-copper braze joints. It would be uneconomical to furnace heat the entire workpiece just to braze the small joints. Localized application of resistance heating is widely used to braze such joints.

(2) Resistance heating can be applied to a small area. In some electrical equipment, braze joints are situated close to temperature-sensitive components. The highly localized, flameless, and rapid application of resistance heating to only the joint area will avoid significant temperature rise in nearby delicate components.

(3) Resistance brazing can be extremely rapid, resulting in minimum oxidation of the workpieces and little or no absorption or diffusion of contaminates. For example, resistance brazing of two 1/8 in. (3.17 mm) square copper wires can be completed in a few seconds.

(4) Since the current level and duration of electrical current can be closely controlled,

the resistance brazing process itself can be controlled precisely.

Disadvantages

The disadvantages of resistance brazing are

(1) At least one workpiece, and preferably all of them, must be electrically conductive.

(2) Large workpieces (i.e., more than a few pounds or a few square inches) are impractical to resistance braze because they would require excessively high current to heat them to brazing temperature in a reasonable time. Also, uniform current distribution and thus, uniform heating, is difficult to achieve in large joints.

(3) Joints with nonuniform cross sections will not heat uniformly. (Within limits, electrode placement can compensate for this.)

(4) Joints with mechanically fragile components are difficult to hold in position during brazing because they may not have sufficient strength at the brazing temperature to withstand the electrode clamping pressure.

(5) Fixturing of the workpieces can be complicated by the high electrical current that flows through the joint during brazing. (Fixtures should be nonconductive or properly insulated from the workpieces.)

JOINT DESIGN

In addition to the general joint design principles discussed in Chapter 2, resistance brazing works best on geometrically simple workpieces. Electrical resistance is directly proportional to the resistivity of the conductor and the length of the current path. It is inversely proportional to the cross-sectional area of the current path. Therefore, heat distribution will be more nearly uniform in workpieces where the current path is through uniform cross sections and lengths. This is important because liquid brazing filler metal will flow to the region of highest temperature. Therefore, both workpieces must reach the required brazing temperature to facilitate good wetting and flow of the brazing filler metal.

Workpieces to be resistance brazed should be designed to be self-fixturing or self-nesting so that auxiliary fixtures cannot

extract heat from the joint or inhibit electrical contact.

Since electrical current flows through the joint, the workpieces must maintain electrical contact during thermal expansion while heating. Furthermore, interface resistance between the various parts and the electrodes in the conducting path can be the highest resistance in the brazing circuit. This should be considered when contemplating temperature distribution across the joint.

In many resistance brazing applications, the electrodes clamp the workpieces together. Therefore, the joint should not require clamping in more than two places.

Although it is estimated that over 90 percent of all resistance brazing applications involve copper-to-copper joints in electrical equipment, some exceptions are

(1) Monel eyeglass frames

(2) Carbon steel automotive parts

(3) Stainless steel orthodontic appliances

All of these use flux and silver filler metals.

RESISTANCE BRAZING EQUIPMENT

Transformers

Standard resistance welding transformers are well suited to provide the low voltage, high amperage ac current normally used for resistance brazing. The required electrical capacity of the transformer depends directly upon the mass of the joint components and their heat capacity; and to a lesser extent upon the contact area of the joint, and the amount and melting point of the brazing filler metal.

The size and electrical resistance of the electrodes, and the desired brazing time also influence transformer capacity. For example, a 10 kVA transformer is suitable for resistance brazing in a few seconds a copper-to-copper joint approximately 0.3 in. (7.62 mm) thick having a contact area of 0.5 in. (160 mm^2). The same transformer would be inadequate for a larger joint involving workpieces weighing perhaps a pound.

It is not possible to arrive at the necessary transformer capacity by performing exact calculations. Before purchasing a transformer, it is advisable to consult with the transformer manufacturer to determine the model and kVA rating that are suitable for a specific resistance brazing application. Transformer manufacturers normally have engineering departments that are knowledgeable in sizing transformers for resistance brazing.

Electrodes

Electrodes transfer electric current and, depending on the alloy being brazed, heat to the workpieces. Therefore, the workpieces totally determine all major aspects of the electrodes' material, size, and shape.

Although many materials are used for resistance brazing electrodes, chromium-copper, molybdenum, tungsten, and graphite are the predominate ones. For a given joint, a suitable electrode material usually is determined empirically. If there is a probability that molten brazing filler metal will contact an electrode, graphite is the preferred material because most filler metals do not wet graphite readily.

Electrodes with high electrical conductivity (such as chromium-copper, molybdenum, and tungsten) should be used for brazing low conductivity workpieces (steels) to avoid overheating the electrodes. Because high electrical conductivity alloys (such as those of copper, silver, aluminum, and gold) are difficult to heat with resistance heating, low conductivity electrodes (graphite) are best suited for resistance brazing them. The flow of electrical current heats the electrodes, which, in turn, heats the workpieces by thermal conduction.

Refractory metal electrodes (molybdenum and tungsten) survive resistance brazing conditions better than the other electrodes. However, they are relatively expensive and are difficult to fabricate into all but the simplest shapes. Sometimes simple disks or buttons of these refractory metals are used as inserts in chromium-copper holders.

Graphite electrodes are fabricated from mixtures of carbon or graphite particles and inorganic elements (such as sulfur or phosphorus) with coal tar pitch by common powder consolidation techniques, i.e., compaction and sintering. By carefully controlling the particle sizes, proportions of ingredients, and processing parameters, graphite electrodes are made in a wide variety of mechanical strengths, electrical con-

ductivity, and oxidation resistance. All grades oxidize with increasing rapidity at elevated temperature. Although graphite electrodes must be replaced more frequently than metal electrodes, they can be more economical for the following reasons:

(1) Availability
(2) Low initial cost
(3) Ease of machining
(4) Not readily wetted by brazing filler metals
(5) High temperature can be achieved at low current
(6) Uniform heating
(7) Suitability when using long cables and tongs
(8) Ability to distribute heat over a large joint area

For brazing parts of large equipment, the electrodes must be portable. In these cases the electrodes are usually attached to tongs (see the subsection below) for manual resistance brazing.

Spot Welding Machines

Spot welding equipment has been used successfully in many resistance brazing applications. Spot welding equipment normally includes a transformer, upper and lower electrodes, a mechanism for clamping the workpieces between the electrodes, and all associated controls for current, voltage, clamping pressure, and time. Spot welders are common in high volume resistance brazing that is fully or semi-automated. Generally, lower clamping pressure, lower current, and longer brazing time are used for brazing than are used for spot welding.

Tongs

Tongs are widely used for manual resistance brazing, with custom designed tongs being tailored to each specific application.

Figures 13.1 and 13.2 show "typical" brazing tongs. The hand-held tongs in Figure 13.1 weigh a few pounds and are about 15 in. long. Figure 13.1 indicates the major parts: cooling water lines (electrical cables are inside the water lines); parallel-acting pliers (for maintaining electrical contact between the tongs and the workpieces); electrode holders (made from a high-strength chromium-copper alloy, UNS-18200); insulating pads between the pliers and electrode holders (to eliminate the possibility of an electrical hazard to the brazer); and graphite electrodes. These tongs are powered by a 10 kVA transformer designed for intermittent duty. The brazer energizes the tongs with a foot switch that is connected to the transformer.

The tongs shown in Figure 13.2 are considerably larger and heavier. They weigh about 50 pounds and draw power from a 25 kVA transformer. These larger tongs have several features in common with the smaller ones in Figure 13.1 — but also differences. Since the longer tongs are used to braze larger workpieces, they require a transformer with higher capacity, a greater flow of cooling water, larger electrodes, and a stronger clamping mechanism. During use, these tongs are suspended from an overhead crane with an electric hoist. The handles are made from a nonconducting, high-strength plastic.

Although not considered tongs, a large resistance brazing unit is shown in Figure 13.3. It is anchored to the floor adjacent to a 75 kVA transformer. Cooling water passages are inside the electrode holders. The graphite electrodes measure about 5 in. x 3 in. x 2.5 in. (127 mm x 76.2 mm x 63.5 mm) and the top one is fastened to a slow-acting air cylinder for the purpose of clamping the large copper workpieces.

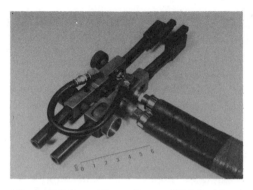

Figure 13.2 — These Tongs Weigh About 50 Pounds and Draw Power From a 25 KVA Transformer

RESISTANCE BRAZING MATERIALS

Base Metals

Resistance brazing can be used to join many alloys, but the widest application is the brazing of high electrical conductivity metals, such as copper and silver. High conductivity metals do not heat readily because their electrical resistance is low, even at brazing temperatures. For these metals, resistance brazers use electrode materials that have high resistance. The flow of electrical current develops the necessary temperature in high resistance electrodes, and they in

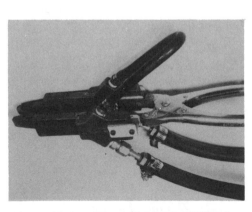

Figure 13.1 — Hand-Held Tongs Weigh a Few Pounds and Are About 15 in. Long. Electrical Cables Are Inside The Water Line

Figure 13.3 — Although Not Considered Tongs, a Large Resistance Brazing Unit is Shown Above For Comparison. The Vertical Gap Between the Electrodes is Three Inches.

turn heat the workpieces by thermal conduction.

Conversely, for resistance brazing low conductivity alloys, high conductivity electrodes should be used. The high electrical resistance of the alloys will develop sufficient internal heat to melt the braze filler metal. The electrodes should be of high conductivity material to minimize heating and maximize service life.

Filler Metals

Because resistance brazing is usually performed in air, excessive oxidation of the workpieces can lead to subsequent cleaning difficulties. To minimize oxidation, brazing should be done at the lowest practical temperature. Therefore, resistance brazers prefer filler metals with low melting points.

Brazing filler metals with high melting points, such as nickel base types, are rarely used for resistance brazing because of the rapid cooling effect of radiation heat loss. All hot metals will radiate to the immediate surroundings an amount of heat that is proportional to the fourth power of the temperature difference between the workpieces and the environment. Thus, a small increase in filler metal liquidus temperature leads to a large increase in the rate of heat loss. Also, high brazing temperatures result in rapid deterioration of the electrodes.

Selection of a resistance brazing filler metal is a simple matter because only low-melting filler metals are used in resistance brazing. Table 13.1 lists the AWS classification, nominal chemical composition, liquidus temperature, and solidus temperature for the most widely used resistance brazing filler metals. (See Chapter 3 for more detail about brazing filler metals.) In addition, a large body of information is available from filler metal manufacturers to assist in choosing the best filler metal for an application.

The form of the brazing filler metal should be selected in order to optimize the brazing process. For example, if the joint is a large, flat area, a strip or shim preform should be used. If the joint is cylindrical, ring-shaped preforms are recommended. If the workpieces are somewhat irregular, a paste or powder may be appropriate. Filler metal in rod form is used primarily for feeding additional metal into the joint to supplement preplaced filler metal.

Flux

The purposes of brazing flux are to remove thin oxide layers and to protect the workpieces from oxidation during brazing. Flux provides chemically clean surfaces that are essential for proper wetting and flow of the filler metal in the joint.

In resistance brazing, the flux is also part of the electrical circuit. Therefore, it must not be permitted to dry out and significantly increase its electrical resistance. This could cause excessive heating in the joint, and flux could be expelled rapidly from the joint. Powdered filler metal can be added to dry flux to improve electrical conductivity.

AWS Class BCuP-1, BCuP-2, and BCuP-5 filler metals are excellent for brazing copper alloys. The phosphorus addition renders them self-fluxing on copper. At brazing tem-

Table 13.1
Filler Metals Typically used for Resistance Brazing

AWS Class	Chemical Composition, percent						Liquidus		Solidus	
	Ag	Cu	Zn	Cd	P	Sn	°F	°C	°F	°C
BAg-1	45	15	16	24			1145	620	1125	605
BAg-1A	50	15.5	16.5	18			1175	635	1160	625
BAg-2	35	26	21	18			1295	700	1125	605
BAg-7	56	22	17			5	1205	650	1145	620
BAg-8	72	28					1435	780	1435	780
BAg-18	60	30				10	1325	720	1115	600
BCuP-1		95			5		1695	925	1310	710
BCuP-2		97			3		1460	795	1310	710
BCuP-5	15	80			5		1475	800	1190	645

perature, the phosphorus chemically reacts with any surface oxide to form phosphorus pentoxide (P_2O_5), which is given off as a colloidal smoke. Finished brazements that are resistance brazed in air using BCuP-1, BCuP-2, or BCuP-5 filler metals may exhibit some minor entrapped P_2O_5 in the form of porosity.

Atmospheres

Workpieces maintained at brazing temperature for a sufficiently long time will oxidize excessively. If an oxide layer is objectionable or if internal oxidation is harmful to the workpiece material, then a protective atmosphere should be employed (see Chapter 10). Nitrogen, argon, and sometimes helium are used to form a nonreactive, oxygen-excluding blanket of gas around the workpieces. Whichever gas is chosen, it should have a dew point sufficiently low to protect the base metals at brazing temperature (see Chapter 4 for metal/metal oxide equilibria data). The protective gas should be introduced as closely as practical to the joint, and allowed to flow long enough to purge reactive gases from the workpieces before starting the heating cycle (usually several seconds).

RESISTANCE BRAZING PROCESSES

Manual Operation

Manual resistance brazing is generally more suitable than automated resistance brazing when:

(1) Only a few joints are to brazed.
(2) Workpiece geometry is complex.
(3) The workpieces are part of a large equipment item.
(4) The joint configuration requires filler metal in addition to that which was preplaced to be hand-fed during brazing (this is called face feeding), in order to produce a sound part.

To make a manually resistance brazed joint, the brazer first cleans the parts that make up the joint and fluxes them if necessary. Depending upon the amounts and types of foreign matter to be removed, alkaline, acidic, and solvent cleaning solutions may sometimes be needed in combination with ultrasonic agitation.

Prebraze cleaning is critical to achieving an acceptable braze joint — all foreign material must be completely removed. Brazing flux is formulated to remove thin oxide layers only; it can never overcome poor cleaning. In addition, dirty workpieces can be a safety hazard because organic contaminants, such as cutting oil, paint, and insulation, will burn at brazing temperature.

When necessary, the brazer positions the braze filler metal preforms on the workpieces. Next, the workpieces are assembled in their prebrazed configuration and clamped with the electrodes. Clamping pressure depends upon the brazer's skill, experience, and close attention in order to achieve an acceptable brazement. Too little pressure will not provide sufficient electrical contact between the workpiece, and the result can be a low heating rate and arcing. Too much pressure, on the other hand, can (a) deform the workpieces at brazing temperature, (b) crack graphite electrodes, if used, or (c) squeeze the liquid filler metal out of the joint. In the following two cases it is necessary to exert sufficient pressure on the workpieces while at brazing temperature to squeeze excess brazing filler metal out of the joint.

Case 1: In brazements where there is appreciable alloying between the brazing filler metal and the base metal, the chemical composition of the filler metal changes during the brazing cycle. The filler metal/base metal alloy that forms initially has a low melting point. It usually includes those constituents which, due to their inadequate strength or poor corrosion resistance, would provide inadequate service. Brazement service life would be improved significantly if these low-melting constituents were removed. This can be accomplished by exerting added force on the workpieces at the appropriate time. Technically speaking, that initially molten filler metal/base metal alloy is a flux.

Case 2: Squeezing the workpieces together is also necessary when a thin bond is required — as in high precision or high-strength joints.

In addition, while closely observing the joint during brazing, the brazer may need to vary the clamping pressure.

The brazer initiates the flow of current, usually by a foot switch or a hand trigger on the tongs, and closely observes the workpieces as they heat. The brazer may also need to pulse the current on and off to achieve a uniform temperature throughout the joint. When preformed filler metal, if used, begins to melt, the brazer may, depending upon workpiece geometry, need to face-feed additional filler metal to fill the joint completely.

This part of the brazing cycle may require current pulsing because the liquid filler metal flows by capillary attraction and seeks the hottest part of the joint. It is important to refrain from force feeding more braze filler metal than the joint can absorb. Excess filler metal is not only a safety hazard and wasteful, but could also damage other areas of the component. Once the brazement contains the proper quantity and distribution of filler metal, the brazer stops current flow — but continues to maintain clamping pressure until the filler metal solidifies and develops sufficient strength to make the brazement self-supporting.

Automated Operation

In automated resistance brazing, process variables such as the type and amount of braze filler metal and flux, electrode pressure, current, and time, are usually determined empirically. As with manual resistance brazing, prebraze cleaning of all workpieces is essential.

Since automated resistance brazing usually requires more capital investment than manual resistance brazing, it is most advantageous when there are many identical joints to be brazed. Automated resistance brazing is also preferred when brazing parameters or finished workpiece dimensions must be closely controlled, and when face feeding is not required.

The filler metal (paste, powder, wire, strip, or ring) and flux are fed into or placed in the joint, and the components are assembled in the prebrazed configuration. The pressure of the electrodes or the design of the parts is used to hold the joint together. The current is then turned on for a controlled time. Resistance heating raises the joint area in the current path to brazing temperature. The flux melts first and removes surface oxides from the area to be brazed. The braze filler metal then melts and flows throughout the joint by capillary

action. When the current is shut off, the joint is cooled to below the brazing filler metal solidus. Cooling may be accelerated by gas or liquid quenching. Only then is the electrode pressure released. Finally, the solidified flux is removed, usually by water immersion or spraying.

Filler Metal Thickness

The thickness of filler metal depends upon the smoothness of the workpieces and the precision with which the workpieces fit together. Smooth workpieces may be brazed properly with a preform of 0.003 in. (0.08 mm) thick filler metal. Conversely, joints with large gaps or rough workpieces would require thicker preforms, perhaps as much as 0.010 in.(0.25 mm).

Strip preforms are available in a wide range of thicknesses. However, for ease of handling, most resistance brazing applications use 0.003 in. (0.08 mm), 0.005 in. (0.13 mm), or 0.010 in. (0.25 mm) thick filler metal.

SAFE PRACTICES

There is a potential for electrical shock during resistance brazing. The brazer, therefore, should exercise care and attention before, during, and after brazing.

Before brazing, the brazer should make sure that the workpieces are clean and free from all foreign materials, and are not in contact with any grounded metal object. The brazer should also visually inspect the brazing transformer, cooling water lines, tongs, and any other equipment to ensure that there are no frayed wires, loose connections, or other potentially hazardous conditions.

During brazing, the brazer should energize the tongs only when both legs of the tongs are in contact with the workpieces. In addition, the brazer should not contact any current-carrying part of the tongs or workpieces while grounding himself. After completing the brazed joint, the brazer should return the tongs to a safe location, preferably a receptacle that is designed to hold the tongs safely when they are not being used. Other, more detailed safe practices are addressed in Chapter 6.

Keywords — braze welding, braze welding equipment, oxyfuel gas process, braze welding procedure, braze welding technique, braze weld types, braze weld quality

Chapter 14

BRAZE WELDING

INTRODUCTION

Braze welding differs from brazing in that the filler metal is distributed by deposition rather than by capillary action. Welding processes utilized for braze welding include: oxyfuel gas welding (OFW), gas metal arc welding (GMAW), gas tungsten arc welding (GTAW), plasma arc welding (PAW), carbon arc braze welding (CABW) and shielded metal arc welding (SMAW). Base metals are not melted — only the filler metal. In all cases, bonding between the deposited filler metal and the hot unmelted base metals occurs in the same manner as in conventional brazing.

Joint designs for braze welding, shown in Figure 14.1, are similar to those used for fusion welding with the applicable welding process. However, sharp corners should be avoided.

Braze welding was originally developed for the repair of cracked or broken cast iron parts. Fusion welding of gray cast iron required extensive preheating and slow cooling to minimize the development of cracks and the formation of hard cementite. With braze welding, the lower joining temperature reduces the preheating time, virtually eliminates the formation of the brittle white cementite, and reduces the likelihood of thermal cracks.

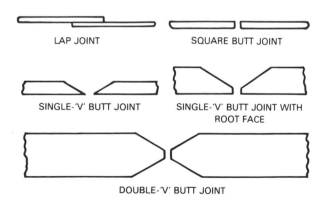

LAP JOINT SQUARE BUTT JOINT

SINGLE-'V' BUTT JOINT SINGLE-'V' BUTT JOINT WITH ROOT FACE

DOUBLE-'V' BUTT JOINT

Figure 14.1 — Typical Joint Designs for Braze Welding. With Large Jobs the Double V Design May be Braze Welded. Note: (1) Whenever Possible Braze Weld in an Uphill Direction Slightly Inclined to Allow Gases to Escape. (2) Break Sharp Edges to Prevent Overheating and Localized Melting

To obtain a strong bond between the filler metal and the unmelted base metal, the molten filler metal must wet the hot base metal. The base metal is heated to the required temperature with an oxyfuel gas torch or an electric arc. With oxyfuel gas braze welding, flux is required to clean both the base metal and previously deposited filler metal. Flux also aids in the process of wetting (precoating) the joint faces. During braze welding, filler metal is fed to the progressively heated joint surfaces in one or more passes until the joint is completely filled to the desired size.

Historically, braze welding has been performed using the oxyfuel welding process with a copper alloy brazing filler metal and a suitable flux. Recently, there has been extensive utilization of the carbon arc braze welding process, gas tungsten arc welding, gas metal arc welding, and plasma arc welding for braze welding. These processes can be used in the manual, semi-automatic or automatic modes, depending upon the process or the set up, or both, without a flux.

The gas tungsten arc, gas metal arc, and plasma arc welding processes normally employ inert shielding gases. Helium, argon, or a combination of these are suitable for braze welding with filler metals that have relatively high melting temperatures. The carbon arc braze welding process does not utilize a shielding gas. Filler metal selection, proper wetting of the base metal, and adequate shielding from the atmosphere are important considerations for effective use of these processes.

ADVANTAGES

Braze welding often offers the following advantages over conventional fusion welding processes:

(1) Less heat is required to accomplish bonding. Braze welding produces less distortion from thermal expansion and contraction than most fusion welding processes thereby reducing the potential for cracking.

(2) Deposited filler metal is relatively soft and ductile, making it machineable with low residual stresses.

(3) Braze welds can be produced with adequate strength for many applications.

(4) Equipment is simple and easy to use.

(5) Base metal with low ductility, such as gray cast iron, can be braze welded without extensively preheating the base metal.

(6) Braze welding provides a convenient way to join dissimilar metals. Examples are copper to steel or cast iron, and nickel-copper alloys to cast iron and steel.

DISADVANTAGES

The following disadvantages should be noted:

(1) Weld strength is limited to that of the deposited filler metal.

(2) Permissible performance temperatures of the product are lower than those of fusion welds because of the lower melting temperature of the filler metal. With copper alloy filler metal, service is limited to 500°F (260°C) or lower.

(3) Braze welded joints may be subject to galvanic corrosion and differential chemical attack.

(4) Brazing filler metal color may not match that of the base metal.

(5) Nonwetted areas are more difficult to detect than similar interfacial weld defects.

EQUIPMENT

Braze welding employs the same equipment as is used with each process for conventional fusion welding, as described in the *AWS Welding Handbook*.

MATERIALS

Base Metals

Braze welding is normally used for joining steel and gray cast iron. However, it can also join copper, nickel, and nickel alloys. Certain other metals can be braze welded by using suitable filler metals that will wet and form a strong metallurgical bond with the base metals. In addition, many dissimilar base metals can be joined by braze welding if suitable filler metals are used.

Weldability of the braze weld joint can be controlled, in many cases, by precoating (buttering) one or both faces of the braze weld joint. When braze welding a copper or copper nickel alloy to nickel, the copper base metal is usually buttered with a nickel filler metal. A high silicon bronze base metal can be buttered with ECuA1-A2 filler metal before braze welding to copper.

Filler Metals

The filler metals for oxyfuel braze welding are usually brass containing 60% copper and 40% zinc. Small additions of tin, iron and manganese may be added to improve flow characteristics, decrease volatilization of zinc, scavenge oxygen, and increase strength and hardness. Nickel (10%) can be added to whiten the color and increase weld strength. Chemical compositors of various braze welding filler metals are shown in Table 14.1 Properties and applications of these filler metals are listed in Table 11.2.

Braze welds made with bronze filler metals will display little ductility while those made with phosphor bronze and copper nickel filler metals will likely contain porosity.

Silicon bronze filler metal is used by the automotive industry and is applied by the gas metal arc welding process. It is used to close and seal the sheet metal body joints that are ground smooth prior to painting.

Chemical composition and properties of braze welding rods are given in Table 14.1. The suggested applications of these filler metals are tabulated in Table 14.2. Note that the minimum tensile strengths of these filler metals vary from 40 to 68 ksi (275-400 MPa). However, the strengths of these filler metals will decrease rapidly above 500°F (260°C).

A completed braze weldment joint may galvanically corrode in certain environments. Also, the filler metal may be less resistant to certain chemical solutions than the base metal.

Fluxes

Fluxes for braze welding are usually proprietary compounds designed for higher temperatures than those fluxes used for conventional brazing. The following types of flux are commonly used for braze welding of gray irons and steels:

(1) A basic flux of alkali fluoride and borax mixtures that cleans the base metal and weld deposits, assists in the precoating of the base metal, and has an active range over the full melting range of the filler metal.

(2) Proprietary fluxes that perform the same functions as the basic flux and also suppress the formation of zinc oxide fumes when galvanized sheet is braze welded.

(3) A flux that is formulated specifically for braze welding of gray or malleable cast iron. It contains iron oxide or manganese dioxide that combines with the free carbon on the cast iron surface to remove it.

Flux may be applied by one of the following methods:

(1) The heated filler rod may be dipped into the flux and transferred to the joint during braze welding.

(2) The flux may be brushed on the joint prior to brazing.

(3) The filler rod may be precoated with flux.

(4) The flux may be introduced through the oxyfuel gas flame.

METALLURGICAL CONSIDERATIONS

The bond that develops between the filler metal and the base metal is the same with braze welding as it is with conventional brazing.

Wetting is the first action needed to accomplish bonding. Following this, atomic diffusion may take place between the braze weld filler metal and the base metal in a narrow zone at the interface. Also, for some combinations, the braze welding filler metal may penetrate the grain boundaries of the base metal and contribute to higher bond strength. At times, however, this increases susceptibility to intergranular corrosion.

The ductility of most braze filler metal deposits is sufficient during solidification and subsequent cooling to minimize the shrinkage stresses and avoid cracks in the filler metal. Two phase alloys having a low melting grain boundary constituent should not be used as filler metals because they may crack during solidification.

Table 14.1
Chemical Composition Welding Rods for Braze Welding

AWS Classification	AWS Specification	UNS[a] Numbers	C	Cu	Zn	Sn	Fe	Ni+CO	Mn	Ti	Al	Si	S	Pb	P	Other Elements Total
ERCuAl-A2	A5.7	C61800	—	Rem	0.02	—	1.50	—	—	—	8.50-11.00	0.10	—	0.02	—	0.50
RCuSi-A	A5.27	C65600	—	Rem	1.0	1.0	0.50	—	1.5	—	0.01	2.8-4.0	—	0.02	—	0.50
ERCuNi	A5.7	C71580	—	Rem	—	—	0.40-0.75	29.0-32.0	1.0(max)	0.20-0.50	—	0.25	0.01	0.02	0.02	0.50
BNi-3[b]	A5.8	N99630	0.06-max	—	—	—	0.5-max	Rem	—	0.05-max	.05-max	4.0-5.0	0.02-max	—	0.02-max	0.50
ERNiCu-7	A5.14	N04060	0.15	Rem	—	—	2.5	62.0-69.0	4.0	1.5-3.0	1.25	1.25	0.015	—	0.02	0.50
RBCuZn-A	A5.27	C47000	—	57.0-61.0	Rem	0.25-1.00	—	—	—	—	0.01	—	—	0.05	—	0.50
RBCuZn-B	A5.27	C68000	—	56.0-60.0	Rem	0.80-1.10	0.25-1.2	0.20-0.80	0.01-0.50	—	0.01	0.04-0.15	—	0.05	—	0.50
RBCuZn-C	A5.27	068100	—	56.0-60.0	Rem	0.80-1.10	0.25-1.2	—	0.01-0.50	—	0.01	0.04-0.15	—	0.05	—	0.50
RBCuZn-D	A5.27	C77300	—	46.0-50.0	Rem	—	—	9.0-11.0	—	—	0.01	0.04-0.25	—	0.05	0.25	0.50
ENiCu-A	A5.15	W84001	0.35-0.55	35.0-45.0	—	—	3.0-6.0	50.0-60.0	2.30-max	—	—	0.75-max	0.025	—	—	1.00

a. ASTM/SAE Metals and Alloys in the Unified Numbering System

b. Boron is 2.75 to 3.50; Zr is 0.05 max; Co is 0.01 max; Se is 0.005 max.

Table 14.2
Properties and Applications of Braze Filler Metals

AWS Classification	Tensile Properties (Minimum) Ksi	MPA	Liquidus Temperatures °C	°F	Applications
ECuAl-A2	68	469	1925	956	Silicon bronze to steel, monel, and nickel alloys. Zinc coated steels (GMAW)
ENiCu-A ENiCu	53	365	-	-	Nickel alloys to copper and copper nickels
BNi3	—	—	—	—	Nickel alloys to copper and copper nickels
ERNiCu-7	50	344	—	—	Nickel alloys to copper and copper nickels
RBCuZn-A	40	275	1650	900	A silicon bronze used on copper sheet, mild tube steel, deep drawing steel and cast iron
RBCuZn-B	50	344	1630	890	Higher strength, used to surface overlay or bond copper, steel, cast iron or wrought iron
RBCuZn-C	50	344	1630	890	Better color match with mild steel, cast iron, and wrought iron
RBCuZn-D	60	413	1715	935	Best strength and color match with cast iron and wrought iron
RBCuSi-A	50	344	—	—	Iron and steel sheet metal joining and joint filler

GENERAL PROCESS APPLICATIONS

Historically, braze welding has primarily been performed with oxyfuel gas equipment for the welding and repair of carbon steels and the cast irons. For carbon steel applications, a neutral flame is employed. For the cast irons, a slightly oxidizing flame is used to aid in the removal of graphite from the joint surfaces. Currently, the arc welding processes are also being utilized for braze welding. Carbon arc braze welding (CABW), gas tungsten arc welding (GTAW), gas metal arc welding (GMAW) and the plasma arc welding (PAW) processes, using copper-silicon, copper, and aluminum filler metals, provide a significant contribution to braze welding. No flux is used with these processes. Except for the carbon arc braze welding process, a shielding gas is necessary to protect the molten weld pool from the atmosphere. The shielding gases normally used for braze welding are either argon or helium, or a combination of both. The absence of flux with these processes eliminates a potential harmful corrosion effect from the entrapment of flux in the braze weld. Typical condition for braze welding galvanized sheet steel with the gas metal arc welding process are shown in Table 14.3.

Braze welding enables large components to be repaired in place. Gas metal arc braze welding has a higher disposition rate than oxyfuel gas braze welding. The former process will generally be economical by reducing energy consumption, shortening braze-welding time, and reducing labor hours.

BRAZE WELDING PROCEDURE

Fixturing

Fixturing is usually required to hold parts in their proper location and alignment for braze welding. For repairing cracks and other defects in cast iron parts, no fixturing

Table 14.3
Typical conditions for braze welding galvanized steel sheet with the gas metal arc-short-circuiting transfer method

Joint Type	Weld Type	Sheet Thickness, in.	Electrode[a] Diam., in.	Electrode[a] Feed Rate, in./min.	Arc Voltage, V	Welding Current,[b] A	Travel Speed, in./min.
Butt	Square Groove	0.062	0.030	255	18	100	18
Tee	Fillet	0.062	0.030	280	19	130	18
		0.080	0.045	160	16	150	16
Corner	Fillet on Outside Corner	0.062	0.030	255	18	100	14
		0.080	0.045	150	16	140	18
Lap	Fillet	0.062	0.030	260	18	110	18

a. ERCuAl-A2—Electrode with argon shielding
b. Direct current, reverse polarity (DCRP)

may be necessary, unless the part is broken or separated.

Joint Preparation

Joint configurations are the same as those used for fusion welding with the applicable welding process. Fit-up tolerances are not as critical as those of brazing. Examples of braze weld joints are shown in Figure 14.1. For thicknesses over 3/32 in. (2 mm), single- and double- V grooves are prepared with an included angle of from 90 to 120 degrees. These grooves provide large bonded areas between the base metal and the filler metal. Square grooves may be used for thicknesses of 3/32 in. (2 mm) or less. Tubular joints are prepared with configurations shown in Figure 14.2. Sheet metal grooves can be one of the configurations shown in Figure 14.3. In large casting repairs, studs may be screwed into the faying surfaces to increase the interfacial strength of the repair as shown in Figure 14.4.

Precleaning

The prepared joint faces and adjacent surfaces of the base metal must be cleaned to remove all oxide, dirt, grease, oil and other foreign material that will inhibit wetting. When braze welding gray cast iron, the joint faces must also be free of graphite smears caused by prior machining. These graphite smears act as stop-off and inhibit wetting.

Malleable iron, because of its low total graphite content and its approximately spherical graphite nodules, can be easily cleaned by abrasive blasting. Ductile iron, however, is more difficult to braze weld. Its total carbon content is higher, but the graphite nodules are spherical. Gray cast iron contains carbon in flake form which makes it most difficult to braze weld and necessitates special surface preparation. One common way to remove graphite smears is to quickly heat the cast iron to a dull red color with a flame that is slightly oxidizing, and then wire brush it after the cast iron cools to a black heat. If the casting has been heavily soaked with oil, it should be heated in the range of 600°F (320°C) to 1200°F (650°C) to burn off the oil. These surfaces must also be wire brushed to remove any residue.

In production braze welding of cast iron components, the surfaces to be joined are usually cleaned by immersion in an electrolytic molten salt bath. This is the most effective method for removing free graphite from machined or broken surfaces.

In general, the joint surfaces must be free of contamination. Mechanical blasting material must be clean, and chemical cleaning agents must not leave a residue. Furthermore, braze welding should be accomplished as soon as possible after the joint has been cleaned.

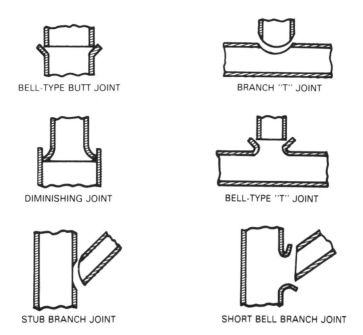

BELL-TYPE BUTT JOINT

BRANCH "T" JOINT

DIMINISHING JOINT

BELL-TYPE "T" JOINT

STUB BRANCH JOINT

SHORT BELL BRANCH JOINT

Figure 14.2 — Typical Tube Joint Designs for Braze Welding

Preheating Methods

The method and requirements for preheating parts to be joined will depend on the type of base metal and the size of the part. Special precautions may be needed to ensure that cracking does not result from thermal stresses in large cast iron parts. In the braze welding of a copper part, preheating may help to reduce the ultimate amount of heat required and the time needed to complete the joint.

Preheating may be either local or general, depending upon the size of the part and its thermal heat transfer characteristics. Preheat temperatures should range from 800 to 900°F (425° to 480°C) for cast iron. Higher temperatures are used for copper. After braze welding is completed, cast iron parts should be thermally insulated to ensure slow cooling to room temperature. Such insulation will minimize development of thermal stresses in the casting.

Technique

The joint to be braze welded must be properly aligned and fixtured in position. Braze welding flux, when required, is applied to

preheated filler rod (unless precoated), and also is spread on thick joints during heating with an oxyfuel gas torch.

Next, the base metal is heated with the flame until the temperature is sufficient to allow flow and wetting of the joint area as the filler metal rod is melted. The joint is filled with one or more passes using operating techniques similar to those for oxyfuel gas or arc welding. Braze welding requires a more diffused heat pattern than does fusion welding. The inner cone of an oxyfuel flame should not impinge on copper-zinc alloy filler metals nor on iron or steel base metal. The techniques used with arc welding processes are similar to those described for oxyfuel gas braze welding, except that flux is not used.

Quality of Braze Weld

Braze weld quality will depend on the care and control exercised during the preparation of the base metal and the braze welding operation. The strength and soundness of the joint will depend on the filler metal used, cleanliness of the base metal, proper oxyfuel gas flame adjustment, proper heat

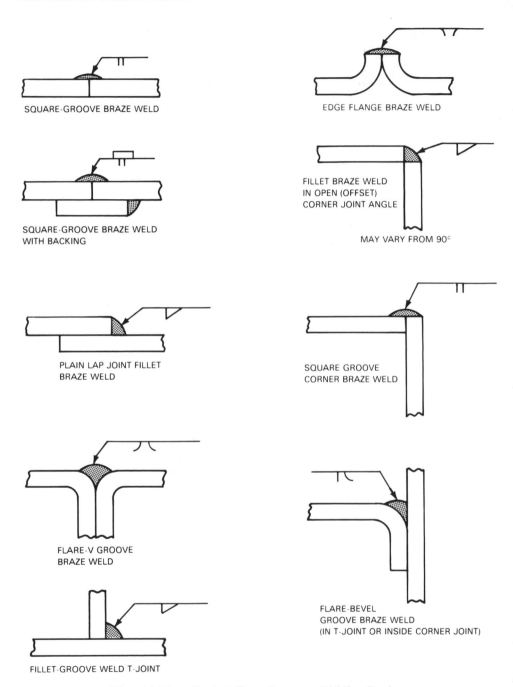

SQUARE-GROOVE BRAZE WELD

SQUARE-GROOVE BRAZE WELD
WITH BACKING

PLAIN LAP JOINT FILLET
BRAZE WELD

FLARE-V GROOVE
BRAZE WELD

FILLET-GROOVE WELD T-JOINT

EDGE FLANGE BRAZE WELD

FILLET BRAZE WELD
IN OPEN (OFFSET)
CORNER JOINT ANGLE

MAY VARY FROM 90°

SQUARE GROOVE
CORNER BRAZE WELD

FLARE-BEVEL
GROOVE BRAZE WELD
(IN T-JOINT OR INSIDE CORNER JOINT)

Figure 14.3 — Typical Sheet Braze — Welding Designs

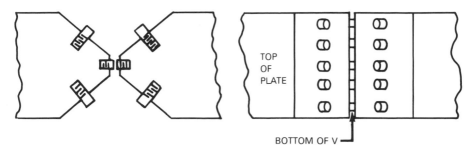

TOP
OF
PLATE

BOTTOM OF V

Figure 14.4 — Studding to Increase Bond Strength Along Faying Surface

input control and the adequacy of fluxing during the operation to ensure complete wetting of the base metal. Using correct joint design and procedures, braze welded joints can be produced that will be as strong as the adjacent base metal. The size of braze welded joints can vary from those on electronic chips to a repaired section of a huge cast iron structure weighing many tons.

Types of Braze Welds

A wide variety of typical weld joint designs can be braze welded. Groove, fillet, and edge welds can be used to join simple and complex assemblies made from sheet and plate, pipe and tubing, rods and bars, castings and forgings. To obtain good joint strength, an adequate bond area is required between the filler metal and the base metal. Weld groove geometry should provide adequate groove face area so that the joint will not fail along an interface. Selection of a proper joint design, filler metal, and welding procedure can produce a braze-weld joint strength which can meet or exceed the base metal tensile strength. With the arc braze welding processes that utilize a shielding gas rather than a flux, the potential for entrapped flux or oxides at the base metal/filler metal interface is greatly diminished. The result is higher joint strengths and improved corrosion resistance.

Keywords — brazing processes, arc brazing, block brazing, carbon arc brazing, diffusion brazing, electron beam brazing, exothermic brazing, flow brazing, infrared brazing, laser brazing, step brazing, twin carbon arc brazing.

Chapter 15

OTHER PROCESSES

BLOCK BRAZING

Block brazing is a brazing process that uses heat from heated blocks applied to the joint. The blocks also act as a jig to hold the parts in alignment.

This process, using heated blocks, requires proper heating of the blocks to obtain uniform and sufficient heat to obtain proper flowing of the filler metal. Filler metals are preplaced in the form of rings, washers, shims, or paste. Proper fluxing must be used.

If there is much difference in the cross-sections of the parts to be joined, preheating must be used to insure the uniform temperature required for brazing is obtained.

This process was used in earlier years but presently finds no commercial use and is considered obsolete.

DIFFUSION BRAZING

Diffusion brazing is a brazing process that forms liquid braze metal by diffusion between dissimilar base metals or between base metal and filler metal preplaced at the faying surfaces. The filler metal may be either preplaced in the joint or it may be formed at the faying surfaces during the heating cycle. The filler metal is diffused with the base metal to the extent that the joint properties approach those of the base metal. Pressure may or may not be applied.

The main difference between diffusion brazing and diffusion welding is that during diffusion brazing, there is a liquid filler metal produced that subsequently solidifies at the brazing temperature. The solidification of the filler metal takes place when one or more of the filler metal elements diffuses into the base metal, thus changing the chemistry of the filler metal. This change results in high joint strength, high remelt temperature, and improved ductility. Diffusion of elements from the base metal into the filler metal can also strengthen the joint.

During the brazing cycle, molten filler metal, when used, is distributed throughout the joint by capillary action. The filler metal is subsequently diffused with the base metal at an appropriate temperature and with sufficient time to produce the desired physical properties in the resultant joint. A diffusion brazed joint, when viewed metallographically, will show a substantial change in the microstructure of the filler metal within the joint. When diffusion is extensive, the joint may show no indication of a distinct filler metal layer, but may show base metal grain growth across the entire joint area. When complete diffusion takes place between base and filler metals with high mutual solubility, the resulting joint appears as all base metal.

The desired diffusion can be obtained and controlled by holding the brazement at a temperature near the liquidus temperature of the filler metal for an extended time. Diffusion can also be accomplished by holding the brazement at a much higher temperature for shorter periods of time.

The joint properties of the final brazement can be changed by varying the processing procedure and the filler metal selected.

The following are examples of diffusion brazing:

(1) A copper piece was brazed to a Type 304 stainless steel member by first plating the stainless steel member with 0.005 in. (0.13 mm) layer of pure silver. The assembly was then furnace brazed in pure dry hydrogen at 1700°F (930°C) for 60 minutes at temperature. The resulting brazement was suspended at one end and a weight applied to the other end so that the joint would fail if the filler metal remelted. The specimen was then heated in a controlled atmosphere furnace to 1900°F (1040°C). The copper member elongated substantially, but the copper-to-stainless steel diffusion brazed joint did not fail. Thus, the joint did not remelt when heated above the melting temperature of the copper-silver eutectic composition.

(2) A high-temperature cobalt-base alloy was diffusion brazed with a nickel-chromium-boron filler metal, in powder or foil form, between the faying surfaces of the joint. Pressure was applied to the joint. The parts were placed in a vacuum furnace and heated to 2050°F (1120°C) for from one to 24 hours. The mechanical properties of the joint can approach or be equal to those of the base metal. The remelt temperature for a diffusion brazed joint in a heat and corrosion resistant base metal can be above 2500°F (1370C), even though the original filler metal solidus was as low as 1780°F (970°C).

With the use of diffusion brazing, aircraft parts can develop a remelt temperature that will allow them to operate at 1900°F (1040°C) with excursions to about 2100°F (1150°C) without failure of the diffusion brazed joint.

ELECTRON BEAM BRAZING

Electron beam brazing is another highly specialized brazing procedure that is useful in limited applications. It is always done in a high vacuum (10^{-4} to 10^{-5} torr) and is very similar to electron beam welding (EBW). For brazing, the electron beam is focused to a relatively larger beam spot to avoid melting the base metal. Because brazing is done in a vacuum, flux is not needed to assure flow of the braze filler metal. This eliminates the need for post braze cleaning and any problems associated with flux entrapment.

Electron beam brazing is a very practical brazing method for small assemblies. It is relatively easy to control the process and affords the advantages of performing the operation in vacuum. This brazing method is especially useful for brazing products that require an internal vacuum.

EXOTHERMIC BRAZING

By definition, an exothermic process is one in which heat is generated in a chemical reaction. Exothermic brazing (EXB) is the process of using an exothermic compound as a heat source to melt the braze filler metal. These processes may use a highly exothermic material, such as compounds based on zirconium, aluminum, and oxides of chromium and iron. When ignited, this material is used to heat both the base metal and the braze filler metal to the required temperature where the filler metal will flow and fill the braze joint. Exothermic materials containing braze filler metal (as a product of an exothermic reaction) can be placed between the faying surfaces to produce not only the heat for brazing, but simultaneously, the braze filler metal. The use of an exothermic compound as the only source of heat when making a braze joint offers the following advantages:

(1) The heat entering the workpiece may be localized, thus eliminating undesirable conditions such as heat treatment, distortion, and oxidation of other areas of the components being brazed,

(2) The time from initiation to completion of the exothermic brazing reaction can be very short (20 to 30 seconds), further reducing undesirable time-temperature related reactions,

(3) Equipment is portable allowing braze joints to be made during actual construction in large and complex structures or me-

chanical systems and in areas otherwise inaccessible by other brazing processes,

(4) Equipment is low cost,

(5) Braze filler metal can be preplaced,

(6) Rapid heating to brazing temperature, particularly in brazements with extremely large masses, minimizes liquation of the braze filler metal,

(7) In many cases neither flux nor controlled gas atmosphere is required.

Disadvantages of exothermic brazing are:

(1) Although not highly complex, much experience is required to effectively apply the process,

(2) Safety is always a major consideration when working with exothermic materials,

(3) Filler metal selection is limited.

Externally applied compounds which produce the heat necessary to melt conventional braze filler metals have been developed for copper alloys, iron alloys, and refractory metals. Aluminum compounds that react and produce the braze filler metal within the joint can be used for joining aluminum alloys.

Using thermit compounds based on zirconium or aluminum and oxides of chromium and iron, it is possible to braze using most conventional braze filler metals. A problem with thermit compounds is that they are so extremely exothermic that considerable smoke, flame and noise is often produced during the brazing process. When using a 30g mixture of 19.90% Mg, 51.84% NiO, 18.24% MnO, 4.24% Al and 5.36% MnO_2 on a stainless steel tube of 25.4 mm OD and a 1 mm wall thickness, a satisfactory braze joint can be made without the characteristic thermit combustion. The Mg-NiO-MnO-Al-MnO_2 compound is less exothermic than the typical thermit compounds and produces temperatures that are in the 2000°F (1100°C) range when properly applied.

Refractory metals can be exothermically brazed in short times with a minimum of base metal recrystallization and diffusion with the filler metal. Temperatures in the range of 3200°F (1780°C) have been obtained using a compound consisting of 24.9% B, 75.1% V_2O_5, or 19.8% B, 60.2% V_2O_5, 16.5% Ta_2O_5, and 6.8% Al.

Several approaches can be applied to aluminum alloys using an exothermic reaction between two or more melting point depressant elements that react and form a compound with a higher melting point element. In such a reaction, the result is an intermetallic compound which forms a heterogeneous filler metal. This type of reaction can also be successfully applied without additional fluxing agents. During the reaction, surface oxides are displaced or disintegrated, which may eliminate the need for protective atmospheres. An example of this type of exothermic compound consists of a powdered mixture of 3.5g (50% Al-50% Mg) mixed with 10.0g (88% Al-12% Si) and 13.5g Al. The resulting filler metal is an intermetallic compound composed of 27g (89.1% Al, 10.2% Mg_2Si, 0.7% Si).

FLOW BRAZING

Flow brazing (FLB) is a process that uses heat from molten nonferrous filler metal poured over the joint until brazing temperature is attained. The filler metal is distributed in the joint by capillary attraction.

This process, must have all parts held in alignment by jigging. Preheating is advisable to shorten the brazing cycle and minimize stress cracking in some materials from thermal shock.

This process was used in earlier years but presently finds no commercial use and is considered obsolete.

INFRARED BRAZING

Infrared brazing (IRB) is a brazing process that uses heat from infrared radiation.

While infrared heat has been used for a long time as a heat source, infrared brazing did not develop as a brazing process until the development of the high-intensity quartz lamp and the need for light-weight honeycomb panels. With those two developments, infrared brazing became a commercially important process.

Infrared heat is radiant heat obtained below the red rays in the spectrum. While with every "black" source there is some visible light, the principal heating is done by the invisible radiation. Heat sources (lamps) capable of delivering up to 5000 watts of radiant energy are commercially available. The lamps do not necessarily need to fit the contour of the part to be heated even though the heat input varies inversely as the square of the distance from the source. Reflectors are used to concentrate the heat.

The assembly to be brazed is supported in a position that enables the energy to impinge on the part. In some applications, only the assembly itself is enclosed. There are, however, applications where the assembly and the lamps are placed in a bell jar or retort that can be evacuated, or in which an inert gas atmosphere can be maintained. The assembly is then heated to a controlled temperature, as indicated by thermocouples. Figure 15.1 shows an infrared brazing arrangement. The part is moved to the cooling platens after brazing.

Infrared radiation is comprised of electromagnetic energy of wavelengths greater than those of the visible spectrum. Supplied by high intensity lamps with power outputs up to 5000 watts, the infrared energy is absorbed by the workpiece. Although the heat input varies as the square of the distance between the part and the source, the lamps need not precisely follow the contour of the part. Thermal conduction and convection by the atmosphere will help to distribute heat away from hot spots. Placing the components and lamps in a bell jar or retort allows the option of vacuum or controlled atmosphere brazing. In some instances, the containment vessel is constructed so that the lamps can be situated outside of the heating zone.

LASER BRAZING

Laser brazing is a brazing method that is used only for specialized applications. This process uses the thermal energy created by laser beams to make localized braze joints on thin-walled critical joints. Its major advantage over other brazing procedures is that the joint area can be heated precisely without heating the entire workpiece to the liquidus of the brazing filler metal. This limits the flow of the braze filler metal to the area heated by the laser beam and precludes any flow due to capillary action. Because of high costs, laser brazing should be reserved for jobs that require the strength of a brazed joint but cannot tolerate the thermal consequences of conventional brazing techniques.

Most braze joints made by laser brazing are produced by a series of overlapping spot brazes. These are generally made by preplacing the braze alloy and moving the assembled joint beneath a fixed laser beam. Fluxes may or may not be required to produce a sound braze joint. Good atmospheric protection in the form of argon or vacuum is always required.

Laser brazing is expensive, but has been found to be useful in certain situations. These include (1) small parts that may be distorted from the heat of conventional brazing techniques, (2) brazing of very thin base metals (0.004 in. [0.01 mm] and less), (3) joints on assemblies that have heat sensitive parts, and (4) brazing of joints inside

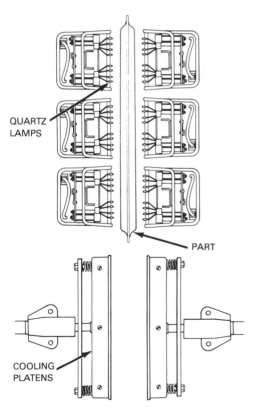

QUARTZ LAMPS

PART

COOLING PLATENS

Figure 15.1 — Infrared Brazing Apparatus

transparent solids, such as sealed glass or quartz tubes.

STEP BRAZING

Step brazing is the brazing of successive joints on a given part with filler metals of successively lower brazing temperature, so as to accomplish the joining without disturbing the joint previously brazed. A similar result can be achieved at a single temperature if the remelt temperature of the prior joint is increased by metallurgical interaction.

Step brazing is a technique which makes use of the differences of brazing temperature ranges of related types of brazing filler metals. By means of this technique, it is possible to braze one section of an assembly using the filler metal with the higher brazing temperature range. Supplemental operations may be performed followed by another brazing operation using a filler metal with a lower brazing temperature range. The brazing filler metals are selected so that the temperature employed for the second braze does not seriously impair the braze made at the higher temperature. Step brazing involving three or more sequences has also been employed.

Several brazing processes may be used. However, furnace, induction, and resistance brazing have proven the most successful.

There is a second procedure by which a number of sequential brazed joints can be made in an assembly. All of the joints can be made individually at the same brazing temperature. This process is called diffusion brazing (DFB), which was covered previously.

TWIN CARBON ARC BRAZING

Twin carbon arc brazing (TCAB) uses the heat from an arc between two carbon electrodes to accomplish brazing.

The equipment used for this process consists of either a welding transformer or a welding generator, an electrode holder, and the leads. The electrode holder is designed so that the two carbon electrodes can be brought together to initiate the arc and then separated to produce a suitable arc. The separation is manually controlled by squeezing the two handles of the torch.

The arc between the electrodes is the heat source used to heat the base metals and filler metals, which have previously been fluxed.

While this process found considerable usage in the earlier days, there is now little commercial application. For a small shop that has only a welding transformer or generator, the twin carbon arc torch takes the place of an oxyacetylene torch. For the shop that has an oxyacetylene torch, this torch is handier and easier to control during the brazing operation. Since most shops currently have oxyfuel gas torches, there are practically no twin carbon arc torches in commercial use today.

Keywords — inspection of brazed joints, acceptance limits, discontinuities in brazed joints, lack of fill, flux entrapment, noncontinuous fillets, base metal erosion, unsatisfactory surface conditions, cracks, inspection methods, nondestructive inspection, destructive inspection

Chapter 16

INSPECTION OF BRAZED JOINTS

GENERAL CONSIDERATIONS

Inspection of the completed assembly or subassembly is the last step in the brazing operation. Along with prior inspections during various processes, it is essential for assuring satisfactory and uniform quality. Final inspection also reveals how well previous steps were completed with respect to the overall integrity of the brazed joints.

The inspection procedure should be an important factor from the very beginning when a brazement is being designed. Care must be taken to ensure not only that the assembly can be manufactured to the required standard, but also that later it can also be properly inspected to assure that all quality requirements have actually been met and maintained.

The inspection method chosen to evaluate a final brazed component should depend on the service requirements. In many cases, the inspection methods are specified by the ultimate user or by regulatory codes. When establishing these codes or quality standards for brazed joints, an approach similar to the one used in establishing standards for any other phase of manufacturing should be used. These standards should be based, if possible, upon requirements that have been established by prior service tests or history. The brazing process should be validated by destructive testing to assure the ability to produce brazements to the required quality standards and dependability.

CRITICAL BRAZED COMPONENTS

Testing and examination of test specimens and process control samples may be an acceptable method for the control of noncritical components. They are not, however, an acceptable substitute for examination by nondestructive inspection of components intended for critical applications. A critical component is one where primary failure of the component would cause significant danger to persons or property, or would result in a significant operational penalty. Persons responsible for design, manufacture, and inspection of critical brazed components should refer to AWS C3.3, *Recommended Practices for Design, Manufacture, and Inspection of Critical Brazed Components*, available from the American Welding Society.

An organization that makes components for critical applications should have an adequate general quality control system. In addition to the documentation normally required, the system should assure that all required inspections are performed on every brazed joint and that the results are properly documented.

Nondestructive inspection methods are applicable to brazements made by all of the brazing processes, and should be used to control the quality of brazements for critical application. The size of the component, its complexity, and degree of critical application will dictate the most suitable inspection method or methods. If an accurate and

259

dependable method of inspecting a critical brazed joint cannot be developed, the part should be redesigned to permit inspection, or another more inspectable joining process should be used. Periodic destructive inspections or process qualification requirements are not satisfactory substitutes for nondestructive inspection of actual components for critical service.

ACCEPTANCE LIMITS

When defining the acceptance limit for any type of braze discontinuity, the following should be considered: (1) shape, (2) orientation, (3) location in the brazement (surface or subsurface), and (4) relationship to other discontinuities. Acceptance limits always should be stated as the minimum requirements for acceptability. Judgments for disposition of discrepant components should be made by persons competent in the fields of brazing metallurgy and quality assurance. These persons should also understand the function of the component. All such dispositions should be documented.

DISCONTINUITIES IN BRAZED JOINTS

Limits of acceptability should be specifically defined when nondestructive inspection is used to identify the following types of common discontinuities.

Lack of Fill

Lack of fill or voids in the joint can be the result of one or more of the following: (1) improper cleaning, (2) improper joint clearance, (3) insufficient filler metal, (4) entrapped gas, and (5) movement of the mating parts (or differential thermal expansion) while the filler metal is in the liquid or partially liquid state. Such discontinuities can cause low joint strength by reducing the load-carrying area. They frequently provide a path for leakage.

Flux Entrapment

Entrapped flux may be found in a brazed joint where a flux had been added to pre-

vent and remove oxidation during the heating cycle. Entrapped flux prevents flow of the filler metal into that particular area, and may result in low joint strength. It also may lead to failure during leak testing and, if corrosive, may severely reduce service life.

Noncontinuous Fillets

These discontinuities usually are found by visual inspection. They may or may not be acceptable depending upon the service requirements of the brazed joint.

Base Metal Erosion

This condition is caused by the filler metal alloying with the base metal during brazing. The base metal cross section may be partially or completely dissolved. Such conditions may reduce the strength of the brazement and reduce the load carrying area. The process of erosion involves dissolution of the base metal and flow of the filler metal.

Unsatisfactory Surface Conditions

Excessive brazing filler metal, rough surfaces, and flow of filler metal onto the base metal may be detrimental for several reasons. In addition to aesthetic considerations, these conditions may act as stress concentrations or corrosion sites, or both. They also may interfere with inspection of the brazement.

Cracks

Cracks reduce both the strength and the service life of brazed joints. They act as stress risers that can lower the static strength or cause premature fatigue failure of the braze.

INSPECTION METHODS

Brazed joints are inspected by nondestructive methods for quality and conformance to specifications. Brazing procedures should be qualified to meet specification requirements using both nondestructive and destructive inspection methods.

Nondestructive Inspection

Visual Inspection. Visual examination of a brazed joint is a widely used nondestructive method of inspection. It is also a convenient preliminary evaluation test where other inspection methods are used. The degree of magnification to be used, if any, should be specified.

A visual inspection is effective in evaluating external evidence of voids, porosity, surface cracks, fillet size and shape, noncontinuous fillets, base metal erosion, and general braze appearance. To assure adequate filling of a capillary joint using visual inspection, the braze filler metal must be applied to one side of the joint and visually inspected for continuous fillet on the other side of the joint. In some applications such as long sheet metal joints and long tube joints, inspection (witness) holes can be used to assure flow has reached the area where the holes are located. Continuous fillets in the holes and on the inspection side of the joint indicate adequate flow of braze filler metal has taken place.

When visual inspection of braze fillets is the primary method of evaluation, the first parts processed must be inspected nondestructively (i.e. X-ray or ultrasonic) or destructively (i.e. cross section or peel test), or both. This procedure is used to ensure that the brazing operations internally and externally are producing the required joint quality. After this evaluation is completed, strict control of process parameters is required to assure quality brazements. Such parameters include temperature, range, brazing temperature, time at temperature, cleanliness, and joint clearance. Periodic destructive or nondestructive sampling, or both, may be used to monitor the brazing process.

Internal imperfections in a brazed joint, such as trapped flux, porosity, lack of fill, and internal cracks cannot be detected by visual inspection. Fillets on both sides of a joint, even if continuous, do not guarantee complete filling of the joint with filler metal. However, with good statistical data and experience background, visual inspection can be used to assure that the part will meet the required service. To avoid misinterpretation, the inspector should be provided with samples, photos, or sketches showing the precise visual conditions that are acceptable and unacceptable.

Proof Testing. In proof testing, brazed joints are subjected to loads that exceed those that will be applied during service.

The proof loads can be applied by hydrostatic methods, tensile loading, spin testing, or other methods. Sometimes, it is not possible to assure a serviceable brazement by other nondestructive methods of inspection. Proof testing should not be specified as the primary method of inspection of brazed joints for critical applications. This test does not evaluate braze quality, but applies a one-time loading that may not closely simulate all the conditions encountered. It may not accurately predict service life, especially if cyclic loadings are encountered. A suitable inspection is required after proof testing to assure that the test itself did not cause cracks which could propagate in service.

Pressure Testing. This method is used when gas or liquid tightness is required of the brazement. Brazements that will be subjected to low pressure service, and are not covered by the ASME Boiler and Pressure Vessel Code requirements, may be tested with air. Brazements that will be subjected to high pressure service and those covered by the ASME Boiler and Pressure Vessel Code requirements should be tested hydrostatically in addition to testing with air. The hydrostatic test pressure should be at least 1.5 times the maximum pressure to which the brazement will be subjected. Test pressures are generally greater than service pressures, and are specified by Code or by purchasers' specifications.

Pressure testing with air may be accomplished by one of the following procedures:

(1) Close all openings, pressurize the brazement with air, submerge the assembly in water, and note any signs of leakage indicated by rising air bubbles.

(2) Close all openings, pressurize the assembly with air, brush a soap solution or commercially available indicator over the joint area, and note where bubbles are formed.

(3) Close all openings, pressurize the assembly with air, close the air inlet source, and note any change in air pressure over a period of time. (Corrections for temperature may be necessary.)

Note that pneumatic testing is a critical operation and that adequate safety precautions must be taken to protect life and property. The maximum test pressure and the equipment setup are extremely important for this type of test.

Pressure testing with helium frequently is preferred for finding very small leaks. A mass spectrometer leak detector generally is used to detect leakage of helium. Prior to testing, it is important to remove all liquid or vapors, or both, from the assembly by a suitable drying operation. Purging with dry gas while heating the assembly above the boiling point of the liquid is one method of drying.

Pressure testing with a halogen gas is often used in the refrigeration industry to determine very small volume leakage. Special halogen detection equipment is available, the simplest being a commercial halide torch. The torch samples the atmosphere from the area being checked, and any leak is indicated by a color change of the detector flame. Electronic halogen leak detectors can indicate very minute leaks.

Vacuum Testing. Evacuation of the brazed component may be employed in checking assemblies such as refrigeration equipment, electronic devices, and other high vacuum systems. A mass spectrometer leak detector is used for this testing method and helium is the sensing medium. The assembly being tested is either placed in a helium-filled container, or the test gas is flushed over the surface of the assembly. The mass spectrometer leak detector is connected to the vacuum side of the assembly, preferably between the auxiliary diffusion pump and the mechanical backing pump. If mass spectrometer helium leak testing is to be used, the brazed joints should not be exposed to liquids prior to testing because of the possibility of plugging any leaks.

Liquid Penetrant Inspection. This method is often used to detect discontinuities extending to the surfaces of brazed joints. Cracks and porosity in the braze fillet can be detected, but interpretation of indications is sometimes difficult when small surface irregularities are present. Incomplete flow and partial filleting also can be observed. This type of inspection should not be used if subsequent repairs are contemplated because the penetrant often is difficult or impossible to remove completely.

Radiographic Inspection. Brazed joints can be inspected by radiographic techniques but the applications are limited. In most cases, radiographic inspection may not detect discontinuities in braze joints where the joint clearance is 2% or less of the joint cross section. Joints may be radiographed satisfactorily only if thickness and x-ray absorption ratios permit delineation of the filler metal. Differences in x-ray absorption characteristics of base metal and filler metal should always be considered in interpreting the radiographs. A joint may be inspectable when brazed with one filler metal but not with another filler metal because of differences in x-ray characteristics of the two filler metals.

Special techniques are required to reliably inspect joints of varying thickness and part configuration. For example, radiographs of a cylindrical brazed joint shown in Figure 16.1 must be made with the radiation passing through two wall thicknesses. This reduces the sensitivity and complicates the interpretation of the radiographs because the projected sizes of discontinuities are dependent on their location relative to the film. Radiographs that show two sides of a cylindrical joint, such as those taken through a pipe fitting, should be interpreted with particular care.

Radiographic inspection may show the presence of filler metal in a joint, but it cannot verify a metallurgical bond between the base metal and filler metal. Metallurgical bonding must be assured by brazing process controls.

A radiograph showing no indications of discontinuities does not guarantee that filler metal has flowed into the joint. A joint may be completely void of filler metal as a result of a gross error in processing. Radiographic film readers should be aware of this, and be particularly suspicious of joints that have no indication of discontinuities on the film. Radiography is not recommended for inspecting braze joints where preplaced filler metal foils have been used.

Ultrasonic Inspection. Ultrasonic inspection of brazed joints depends upon the reflection of sound waves by surfaces. The principal method used is that of reflected waves from a single transducer. In commonly used procedures, a transducer emits

SHAFT RADIOGRAPHY

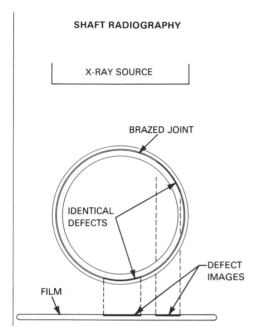

X-RAY SOURCE

BRAZED JOINT

IDENTICAL DEFECTS

FILM

DEFECT IMAGES

Figure 16.1 — Effect of Radiographic Technique on Defect Image on Film

an ultrasonic sound pulse and then receives back the pulse echoes. In a brazed lap joint, these pulse echoes will come from the front and rear surfaces of the base metal in areas of complete bonding. At defective areas, a third echo located between the first two will reflect from unbrazed faying surfaces of the joint.

The pulse echoes may be displayed on an oscilloscope. The echo signal may be used also to trigger the pen of a recording device to produce a facsimile of the joint showing bonded and unbonded areas. An immersion ultrasonic machine used to inspect the brazed joint in a turbine engine power shaft is shown in Figure 16.2. Figure 16.3 is an ultrasonic facsimile produced by this machine.

Ultrasonic inspection is sensitive to setup variables, part configuration, and materials. A reference standard identical to the part being inspected and containing defects of known configuration should be used to set up the equipment and then to calibrate it at specified intervals.

Figure 16.4 is a facsimile of the ultrasonic scan of a braze joint between a fibre metal

seal and a solid component. Four millimeter voids were intentionally produced by holes placed in the brazing filler metal foil. This procedure was used to provide a standard for instrument calibration because the sound attenuation properties of fibre metal may vary from lot to lot. Thus, each part has a built-in inspection calibration area.

Ultrasonic inspection can be used for many brazing applications, and it is one of the best methods of evaluating joint quality. The method is sensitive both to the presence of filler metal in the joint and to the bonding between the filler and base metals. Thus, ultrasonic inspection is the preferred method for evaluating joints in which preplaced filler metal foils are used.

Thermal Transfer Inspection. This method of inspection may be used only in certain cases. For example, brazed aircraft propeller blades have been photographed while still red hot after removal from the brazing furnace. The covering skins appear bright red in areas where they are brazed to the reinforcing ribs, but the skins are a darker red or black in areas where there are voids in the brazed joints.

Infrared heat lamps can be used to inspect brazed honeycomb panels. This procedure uses powder or liquid materials with low melting points to indicate differences in heat transfer characteristics when an assembly is placed under the lamp. Temperature variations cause the liquid to be repelled from warm areas and to accumulate in cool spots. The core partitions act as heat sinks, causing the fluid to flow to the brazed areas.

Other techniques use thermally sensitive phosphors, liquid crystals, and other temperature-sensitive materials. Infrared-sensitive electronic imaging devices with television readout are commercially available to monitor temperature differences produced by variations in heat transfer across a brazed joint. The resulting images, which can be recorded on video tape, show the brazed areas as light spots caused by the rapid conduction of heat through to the opposite part. The void areas are seen as dark spots which do not emit infrared as intensely. An example of infrared imaging is shown in Figure 16.5.

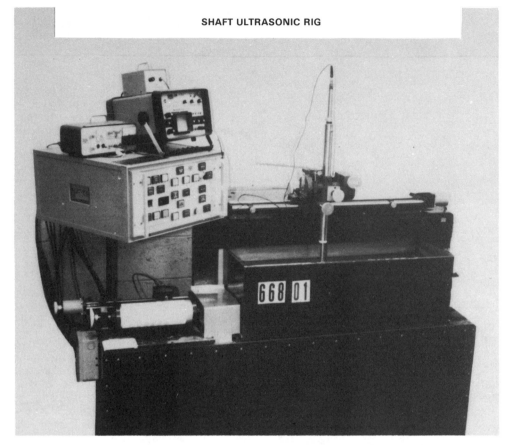

SHAFT ULTRASONIC RIG

668 01

Figure 16.2 — An Immersion Type Ultrasonic Inspection Machine

Other Methods. Laser holography, ultrasonic scanning microscopy, acoustic emission, and real time radiography show promise as inspection techniques for improving the ability to assure quality. In all cases, a rational decision based on experimental evidence should be the basis for process selection and application. When critical components are being inspected, careful verification of the procedure is absolutely crucial.

Destructive Inspection

Destructive testing methods are used for random or lot testing of brazed joints. In lot testing, a specified percentage of all production is tested to destruction. The results of these tests are assumed to apply to the entire production lot, and the various lots or batches are accepted or rejected accordingly. When used as a check on a nondestructive method of inspection, a production joint may be selected at regular intervals and tested to destruction so that rigid control of brazing procedures is maintained. Rules for selection, testing, and lot acceptance or rejection should be established and strictly followed.

Peel Tests. These tests are frequently used for evaluating lap joints. One member is held rigid, as in a vise, while the other is peeled away from the joint. This test may be used as a means of production quality control. It can determine the general quality of the bond as well as the presence of voids and

SHAFT ULTRASONIC FACSIMILE

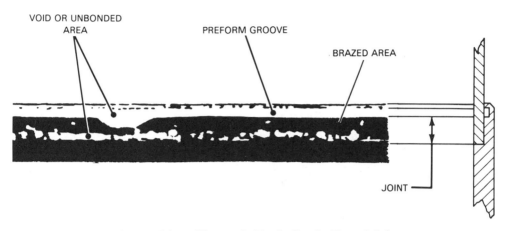

Figure 16.3 — Ultrasonic Facsimile of a Brazed Joint

flux inclusions in the joint. The permissible number, size, and distribution of these discontinuities will depend upon the service conditions of the joint and may be limited by applicable codes.

Tensile And Shear Tests. These tests are used to determine the strength of a joint in tension or in shear. They are generally used for development rather than for production quality control. However, selected production samples may be evaluated with these tests.

Fatigue Tests. Testing under cyclic loading is used to a limited extent. In most cases, they test the base metal as well as the brazed joint. As a general rule, fatigue tests require

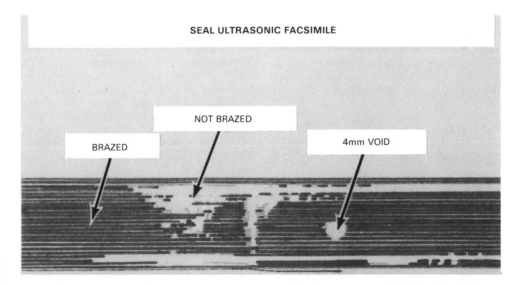

Figure 16.4 — Ultrasonic Facsimile of Brazed Joint With Intentional Void for Calibration

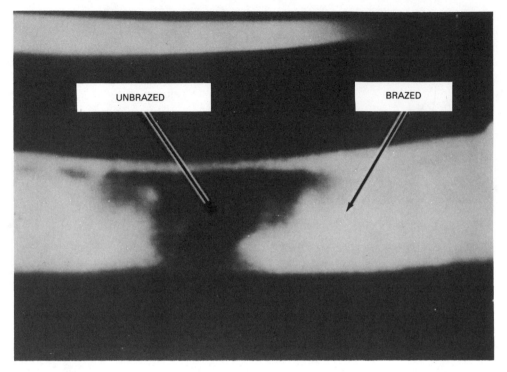

Figure 16.5 — Infrared Braze Image

long times to complete and, for this reason, are very seldom used for quality control.

Impact Tests. These tests, like fatigue tests, generally are limited to laboratory work in determining basic properties of brazed joints. As a general rule, the standard notch-type specimens are not suitable for brazed joints. Special types of joints may be required to obtain accurate results.

Torsion Tests. Occasionally, torsion tests are used on brazed joints in production quality control. Examples are studs, screws, or tubular members brazed to thick sections.

Metallographic Examination. Examination under a microscope frequently is used to determine the general quality of brazed joints. Discontinuities such as porosity, poor filler metal flow, excessive base metal erosion, and improper fit-up can be detected. Microscopic examination often provides a better understanding of the cause of various discontinuities than other inspection methods.

Keywords — codes, standards, technical requirements, recommended brazing practices, specifications, applications of codes and standards, sources of codes and standards, definitions and symbols, qualification and testing, Federal Government standards, Unified Numbering System

Chapter 17

CODES AND OTHER STANDARDS

DEFINITIONS

The purpose of this chapter is to familiarize fabricators and purchasers of brazed products with the basic documents that govern or guide brazing activities. These documents describe the requirements for (1) the production of safe and reliable brazed products and (2) necessary conditions for brazing in an environment free from safety and health hazards.

Publications relating only to the manufacture of brazing materials or equipment are not covered in this chapter. However, those publications may be referenced in the basic documents, and their contribution to safety and reliability should not be underestimated.

The American Welding Society uses the general term *standards* to refer to documents that govern and guide welding and brazing activities. Standards describe the technical requirements for a material, process, product, system or service. They also indicate procedures, methods, equipment, or tests used to determine that the requirements have been met.

Standards include codes, specifications, guides, methods, and recommended practices. These documents have many similarities, and the terms are often used interchangeably, but sometimes incorrectly. Codes and specifications are similar types of standards that use the verbs "shall" and "will" to indicate the mandatory use of certain materials or actions, or both. Codes differ from specifications in that their use may be mandated with the force of law by one or more governmental jurisdictions. The use of specifications becomes mandatory only when they are referenced by codes or contractual documents.

Guides and recommended practices are standards that are technology transfer documents. They use verbs such as "should" and "may" because their use is usually optional. However, if these documents are referenced by codes or contractual agreements, their use may become mandatory. If the codes or agreements contain nonmandatory sections or appendices, the use of referenced guides or recommended practices is at the user's discretion.

The user of a standard should become acquainted with its scope and intended use, both of which are usually included within the Scope or Introduction. It is equally important, but often more difficult, to recognize subjects that are not covered by the document. These omissions may require additional technical consideration. A document may cover the details of the product form without considering special conditions under which it will be used. Examples of special conditions would be corrosive atmospheres, elevated temperatures, and dynamic rather than static loading.

Standards vary in their method of achieving compliance. Some have specific requirements that do not allow for alternative actions. Others permit alternative actions or procedures as long as they result in properties that meet specified criteria. These criteria are often given as minimum requirements.

SOURCES

Private and governmental organizations develop, issue, and update Standards for their particular areas of interest. Table 17.1 lists those organizations of concern to the brazing industry. The interests of many of these groups overlap with regard to brazing, and some agreements have been made to reduce duplication of effort. Several Standards that are concerned with brazing and allied processes are prepared by the American Welding Society (AWS) because these subjects are of primary interest to its members. Standards that apply to a particular product are usually prepared by the group that has overall responsibility.

Table 17.1
Sources of Codes and Standards of Interest to the Welding Industry

---

American National Standards Institute (ANSI)
1430 Broadway
New York, NY 10018
(212) 354-3300

American Society of Mechanical Engineers (ASME)
345 East 47 Street
New York, NY 10017
(212) 705-7722

American Society for Testing and Materials (ASTM)
1916 Race Street
Philadelphia, PA 19103
(215) 299-5400

American Welding Society (AWS)
550 N.W. LeJeune Road
Miami, FL 33126
(305) 443-9353

Canadian Standards Association (CSA)
178 Rexdale Boulevard
Rexdale, Ontario
Canada M9W 1R3
(416) 744-4000

Compressed Gas Association (CGA)
1235 Jefferson Davis Highway
Arlington, VA 22202
(703) 979-0900

International Organization for Standardization (ISO)
(See American National Standards Institute)

(Continued)

National Board of Boiler and Pressure Vessel Inspectors (NBBPVI)
1055 Crupper Ave.
Columbus, OH 43229
(614) 888-8320

National Fire Protection Association (NFPA)
Batterymarch Park
Quincy, MA 02269
(617) 770-3000

Society of Automotive Engineers (SAE)
400 Commonwealth Drive
Warrendale, PA 15096
(412) 776-4841

Standardization Document Order Desk[a]
Building 4D
700 Robbins Ave.
Philadelphia, PA 19111-5094
(215) 697-3321

Superintendent of Documents[b]
U.S. Government Printing Office
Washington, D.C. 20402
(202) 783-3238

Uniform Boiler and Pressure Vessel Laws Society (UBPVLS)
2838 Long Beach Road
Oceanside, NY 11572
(516) 536-5485

---

[a] Source for Military Specifications
[b] Source for Federal Specifications

Each organization that prepares Standards has committees or task groups perform this function. Members of these committees or groups are specialists in their field. They prepare drafts of Standards that are reviewed and approved by a larger group. The review group is selected to include persons with diverse ranges of interests including, for example, producers, users and government representatives. To avoid control or undue influence by one interest group, agreement must be achieved by a high percentage of all members.

The federal government develops or adopts Standards for items and services that are in the public rather than the private domain. The mechanisms for developing federal or military documents are similar to those of private organizations. Standard writing committees usually exist within a federal department or agency that has responsibility for a particular item or service.

The American National Standards Institute (ANSI) is a private organization responsible for coordinating national standards for use within the United States. ANSI does not actually prepare standards. Instead, it forms national interest review groups to determine whether proposed standards are in the public interest. Each group is composed of persons from various organizations concerned with the scope and provisions of a particular document. If there is a consensus regarding the general value of a particular standard, then it may be adopted as an American National Standard. Adoption of a standard by ANSI does not, of itself, give it mandatory status. However, if the standard is cited by a governmental rule or regulation, it may then be backed by force of law.

Other industrial countries also develop and issue Standards on the subject of brazing. The following are examples of other national standards designations and the bodies responsible for them:

BS — BRITISH STANDARD. Issued by the British Standards Association

CSA — CANADIAN STANDARD. Issued by the Canadian Standards Association

DIN — WEST GERMAN STANDARD. Issued by the Deutsches Institute fur Normung

JIS — JAPANESE INDUSTRIAL STANDARD. Issued by the Japanese Standards Association

NF — FRENCH NORMS. Issued by the Association Francaise de Normalisation

There is also an International Organization for Standardization (ISO). Its goal is the establishment of uniform standards for use in international trade. This organization is discussed in a following section.

APPLICATIONS

The minimum requirements of a particular Standard may not satisfy the special needs of every user. Therefore, a user may find it necessary to invoke additional requirements to obtain desired quality.

There are various mechanisms by which most Standards may be revised. These are used when a Standard is found to be in error, unreasonably restrictive, or not applicable to new technological developments. Some Standards are updated on a regular basis, while others are revised as needed. The revisions may be in the form of addenda, or they may be incorporated in superseding documents.

If the user has a question about a particular Standard regarding either an interpretation or a possible error, he should contact the responsible organization.

When the use of a Standard is mandatory, whether as a result of a government regulation or a legal contract, it is essential to know the particular edition of the document to be used. It is unfortunate, but not uncommon, to find that an outdated edition of a referenced document has been specified, and must be followed to be in compliance. If there is a question concerning which edition or revision of a document is to be used, it should be resolved before commencement of work.

Organizations responsible for preparing Standards that relate to brazing are discussed in the following sections. The publications are listed without reference to date of publication, latest revision, or amendment. New publications relating to brazing may be issued, and current ones may be withdrawn or revised. The responsible organization should be contacted for current information on its Standards.

Some organizations cover many product categories while others may cover only one. The National Fire Protection Association is not listed in the table because its standards are concerned with safe practices and equipment rather than with products. The American Welding Society also publishes standards concerned with brazing and welding safety.

AMERICAN NATIONAL STANDARDS INSTITUTE

The American National Standards Institute provides means for determining the need for standards, and ensures that organizations competent to fill these needs undertake the development work. The approval procedures for American National Standards ensure that all concerned national interests have an opportunity to participate in the development of a standard or to comment on its provisions prior to publication. ANSI is the U.S. member of nontreaty international standards organizations, such as the International Organization for Standardization (ISO) and the International Electrotechnical Commission (IEC).

American National Standards, which now number approximately 10 000 documents, encompass virtually every field. They deal with dimensions, ratings, terminology and symbols, test methods, and performance and safety specifications for materials, equipment, components, and products. These fields include construction; electrical and electronics; heating, air conditioning, and refrigeration; information systems; medical devices; mechanical; nuclear; physical distribution; piping and processing; photography and motion pictures; textiles; welding; and brazing.

The ANSI federation consists of not only companies, large and small, but also trade, technical, professional, labor, and consumer organizations. The standards are primarily developed by the member organizations.

AMERICAN SOCIETY OF MECHANICAL ENGINEERS

Two standing committees of the American Society of Mechanical Engineers (ASME) are actively involved in the formulation, revision, and interpretation of Standards covering products that may be fabricated by welding and brazing. These committees are responsible for preparing the ASME Boiler and Pressure Vessel Code and Code for Pressure Piping, which are American National Standards.

Boiler and Pressure Vessel Code

The *ASME Boiler and Pressure Vessel Code* is designed to establish rules of safety governing design, fabrication, and inspection of boilers and unfired pressure vessels during construction. The ASME Boiler and Pressure Vessel Code Committee is charged with the preparation of these rules and their interpretations when questions arise regarding intent. In formulating the rules, the committee considers the needs of users, manufacturers, and inspectors of pressure vessels. The Code is published in several sections as follows:

Section I — *Power Boilers*
Section II — *Material Specifications*
Part A — *Ferrous Materials*
Part B — *Nonferrous Materials*
Part C — *Welding Rods, Electrodes, and Filler Metals* (includes brazing filler metals)
Section III — *Nuclear Power Plant Components*
Section IV — *Heating Boilers*
Section V — *Nondestructive Examination*
Section VI — *Recommended Rules for Care and Operation of Heating Boilers*
Section VII — *Recommended Guidelines for the Care of Power Boilers*
Section VIII — *Pressure Vessels*
Division 1
Division 2 — *Alternative Rules*
Section IX — *Welding and Brazing Qualifications*
Section X — *Fiber-Reinforced Plastic Pressure Vessels*
Section XI — *Rules for Inservice Inspection of Nuclear Power Plant Components*

The individual sections were prepared by subcommittees. Specific rules with respect to welding and brazing may appear in the individual sections as deemed necessary by the various subcommittees. Welding and brazing qualifications are set forth in Section IX and are applicable to all sections of the Code.

Provisions for brazing of unfired pressure vessels are in Section VIII of the Code. The entire Code is republished every three years. In the intervals, corrections, revisions, and additions appear in semi-annual addenda. New materials, processes, and applications are covered by the "Code Case" procedures, as needed. Each new edition incorporates all the changes accomplished by the addenda and the Code Cases of the intervening period.

Base metals used in brazed construction must conform to Section II, *Material Specifications*: Part A for *Ferrous Materials* and Part B for *Nonferrous Materials*. While these specifications are similar to the ASTM specifications carrying the same numbers, there are, occasionally, significant differences. Such differences are noted under the titles of the individual specifications in Section II; therefore, it is important to work with the Code versions. A further limitation is imposed on the use of base metals: Only those base metals for which allowable stress values have been established may be used.

In the tabulation of the Code sections just given, it will be noted that Section II, Part C, relates to *Welding Rods, Electrodes, and Filler Metals* (including brazing filler metals). Applicable AWS Filler Metal Specifications are included in Section II, as Part C. The Code versions of these specifications are essentially identical with the AWS versions. As is the case with the ASTM specifications in Parts A and B, any deviations are noted under the title.

There are no specifications for fluxes or atmospheres under the Code requirements; however, the successful qualification of a brazing procedure in turn qualifies the specific flux or atmosphere used. Responsibility for the strength of brazed joints rests with the contractor, who must determine from suitable tests or from past experience that the specific brazing filler metal selected can produce a joint which will meet the design requirement for strength over the operating temperature range.

The *ASME Boiler and Pressure Vessel Code* is referenced in the safety regulations of most states and major cities of the USA, and also the Provinces of Canada. A number of federal agencies include the Code as part of their respective regulations.

The Uniform Boiler and Pressure Vessel Laws Society (UBPVLS) has, as its objective, uniformity of laws, rules, and regulations that affect boiler and pressure vessel fabricators, inspection agencies, and users. The Society believes that such laws, rules, and regulations should follow nationally accepted codes and standards. It recommends the *ASME Boiler and Pressure Vessel Code* as the standard for construction and the *Inspection Code of the National Boiler and Pressure Vessel Inspectors* (discussed in a following section) as the standard for inspection and repair. The *ASME Boiler and Pressure Vessel Code* is unique in that it requires third party inspection independent of the fabricator and the user. The National Board of Boiler and Pressure Vessel Inspector (NBBPVI) commissions inspectors by examination. These inspectors are employed either by authorized inspection agencies (usually insurance companies) or by jurisdictional authorities.

Prior to building a boiler or pressure vessel, a contractor must have a quality control system in place and a manual that describes it. The system must be acceptable to the authorized inspection agency and to either the jurisdictional authority or the NBBPVI. Based on the recommendations to ASME, a code symbol stamp and Certificate of Authorization may be issued by ASME to the contractor. The authorized inspection agency is also involved in monitoring the fabrication and field erection of boilers and pressure vessels. An authorized inspector must be satisfied that all applicable provisions of the *ASME Boiler and Pressure Vessel Code* have been followed before allowing the fabricator to apply its code symbol stamp to the unit.

AMERICAN SOCIETY FOR TESTING AND MATERIALS

The American Society for Testing and Materials (ASTM) develops and publishes specifications for use in the production and testing of materials. The committees that develop the specifications are comprised of producers and users as well as others who have an interest in the subject materials. The specifications cover virtually all materials used in industry and commerce with the exception of welding consumables, which are covered by AWS specifications.

ASTM publishes an *Annual Book of ASTM Standards* that incorporates new and revised standards. It is currently composed of 15 sections comprising 66 volumes and an index. Specifications for metal products, test methods, and analytical procedures of interest to the welding industry are found in the first three sections, comprising 17 volumes. Section 1 covers iron and steel products; Section 2, nonferrous metal products; and Section 3, metal test methods and analytical procedures. Copies of single specifications are also available from ASTM.

Prefix letters, which are part of the alphanumeric designation of each specification, provide a general idea of the specification content. They include *A* for ferrous metals, *B* for nonferrous metals, and *E* for miscellaneous subjects including examination and testing. When ASME adopts an ASTM specification for certain applications, either in its entirety or in a revised form, it adds an *S* in front of the ASTM letter prefix.

Many ASTM specifications include supplementary requirements that must be specified by the purchaser if they are desired. These may include vacuum treatment, additional tension testing, impact testing and ultrasonic examinations.

The producer of a material or product is responsible for compliance with all mandatory and specified supplementary requirements of the appropriate ASTM specification. The user of the material is responsible for verifying that the producer has complied with all requirements.

Some codes permit the user to perform the tests required by an ASTM or other specification to verify that a material meets requirements. If the results of the tests conform to the requirements of the designated specification, the material can be used for the application.

AMERICAN WELDING SOCIETY

The American Welding Society (AWS) publishes documents covering the use and quality control of welding and brazing. The general subject areas covered are:
  (1) Definitions and symbols
  (2) Filler metals
  (3) Qualifications and testing
  (4) Processes
  (5) Safety

Definitions and Symbols

ANSI/AWS A2.4, *Standard Symbols for Welding, Brazing, and Nondestructive Examination.* This publication describes the standard symbols used to convey welding, brazing, and nondestructive testing requirements on drawings. Symbols in this publication are intended to facilitate communications between designers and fabrication personnel. Typical information that can be conveyed with welding symbols include type of weld or braze, joint geometry, weld size or effective throat, extent of welding, contour and surface finish, and any requirements for nondestructive examination of the weldment or brazement.

ANSI/AWS A3.0, *Standard Welding Terms and Definitions.* This publication lists and defines the standard terms that should be used in oral and written communications conveying welding, brazing, soldering, thermal spraying, and thermal cutting information. Nonstandard terms are also included; these are defined by reference to the standard terms.

Filler Metals

The AWS filler metal specifications cover most types of consumables used with the various welding and brazing processes. The specifications include both mandatory and nonmandatory provisions. The mandatory provisions cover such subjects as chemical properties, mechanical properties, manufacture, testing, and packaging. The nonmandatory provisions, included in an appendix, are provided as a source of information for the user on the classification, description, and intended use of the filler metals covered.

ANSI/AWS A5.8, *Specifications for Filler Metals for Brazing* is the only brazing filler metal specification. Most AWS filler metal specifications have been approved by ANSI as American National Standards and adopted by ASME. When ASME adopts an AWS filler metal specification, either in its entirety or with revisions, it adds the letters SF to the AWS alphanumeric designation. Thus, specification ASME SFA5.8 would be similar, if not identical, to the AWS A5.8 document.

AWS also publishes two documents to aid users with the purchase of filler metals. AWS A5.01, *Filler Metal Procurement Guidelines* provides methods for identification of filler metal components, classification of lots of filler metals, and specification of the testing schedule in procurement documents.

The AWS *Filler Metal Comparison Charts,* assist in determining the manufacturers that supply filler metals in accordance with the various AWS specifications and the brand names of the filler metals. Conversely, the AWS specification, classification, and manufacturer of a filler metal can be determined from the brand name. The Filler Metal Comparison Chart is not a standard.

Qualification and Testing

ANSI/AWS B2.2, *Standard For Brazing Procedure and Performance Qualification.* The requirements for qualification of brazing procedures, brazers, and brazing operators for furnace, machine, and automatic brazing are covered by this publication. It is to be used when required by other documents, such as codes, specifications, or procurement contracts. Those documents must specify certain requirements applicable to the production brazement. Applicable base metals are carbon and alloy steels, cast iron, aluminum, copper, nickel, titanium, zirconium, magnesium, and cobalt alloys.

ANSI/AWS C3.2, *Standard Method for Evaluating the Strength of Brazed Joints in Shear.* This standard describes a test method used to obtain reliable brazed joint shear strengths. Specimen preparation, brazing practices, and testing procedures must be consistent for purpose of comparison. Production brazed joint strength may not be the same as test joint strength if the brazing practices are different. With production furnace brazing, for example, the actual part temperature or time at temperature, or both, may vary from those used to determine joint strength.

Processes

AWS publishes recommended practices and guides for brazing. The applicable documents are listed below:

ANSI/AWS C3.3, *Recommended Practices for Brazing, Design, Manufacture, and Inspection of Critical Brazed Components* covers those procedures that should be followed to assure component reliability in service. The recommended procedures represent the best current practice, in the opinion of the AWS C3 Committee on Brazing

and Soldering, and are necessary to the control of brazed joint quality. All practices may not be applicable to all products or all brazing processes. However, when some of these practices are omitted on critical components, it should be the result of a rational decision, not the result of a lack of knowledge of the best practice.

ANSI/AWS Z49.1, *Safety in Welding and Cutting* was developed by the ANSI Accredited Standards Committee Z49, *Safety in Welding and Cutting,* and then published by AWS. The purpose of the Standard is the protection of persons from injury and illness, and the protection of property from damage by fire and explosions arising from welding, cutting, and allied processes. It specifically covers arc, oxyfuel gas, and resistance welding, and also thermal cutting, but the requirements are generally applicable to other welding processes as well. The provisions of this standard are backed by the force of law since they are included in the General Industry Standards of the U.S. Department of Labor, Occupational Safety and Health Administration.

Additional brazing specifications include:

ANSI/AWS C3.4, *Specification for Torch Brazing*

ANSI/AWS C3.5, *Specification for Induction Brazing*

ANSI/AWS C3.6, *Specification for Furnace Brazing*

ANSI/AWS C3.7, *Specification for Aluminum Brazing*

ANSI/AWS C3.8, *Recommended Practices for the Ultrasonic Inspection of Brazed Joints*

COMPRESSED GAS ASSOCIATION

The Compressed Gas Association (CGA) prepares and coordinates Standards in the compressed gas industries, including end uses of products.

The *Handbook of Compressed Gases,* published by CGA, is a source of basic information about compressed gases, their transportation, uses, and safety considerations, and also the rules and regulations pertaining to them.

CGA C-3, *Standards for Welding and Brazing on Thin Walled Containers* is directly related to the use of welding and brazing in the manufacture of DOT (Department of Transportation) compressed gas cylinders. It covers procedure and operator qualifications, inspection, and container repair.

The following CGA publications contain information on the properties, manufacture, transportation, storage, handling, and use of gases commonly used in brazing operations:

G-1, *Acetylene*

G-1.1, *Commodity Specification for Acetylene*

G-4, *Oxygen*

G-4.3, *Commodity Specification for Oxygen*

G-5, *Hydrogen*

G-10.1, *Commodity Specification for Nitrogen*

G-11.1, *Commodity Specification for Argon*

P-9, *The Inert Gases Argon, Nitrogen, and Helium*

Safety considerations related to the gases commonly used in operations are discussed in the following CGA pamphlets:

P-1, *Safe Handling of Compressed Gases in Containers*

SB-2, *Oxygen-Deficient Atmospheres*

SB-4, *Handling Acetylene Cylinders in Fire Situations*

FEDERAL GOVERNMENT

Several departments of the Federal Government, including the General Services Administration and the Department of Defense, are responsible for either developing Standards or adopting existing Standards, or both.

Consensus Standards

The U.S. Departments of Labor, Transportation, and Energy are primarily concerned with adopting existing national consensus Standards. However, they also make amendments to these Standards or create separate Standards, as necessary. For example, the Occupational Safety and Health Administration (OSHA) of the Department of Labor issues regulations covering occupational safety and health protection. The brazing portions of Standards adopted or established by OSHA are published under Title 29 of the *United States Code of Federal Regulations*. Part 1910 covers general industry. These regulations were derived primarily from national consensus Standards of ANSI and of the National Fire Protection Association (NFPA).

Military and Federal Specifications

Military (MIL) specifications are prepared by the Department of Defense (DOD). They cover materials, products, or services specifically for military use, and commercial items modified to meet military requirements.

Military specifications have document designations beginning with the prefixes MIL or DOD. They are issued as either coordinated or limited-coordination documents. Coordinated documents cover items or services required by more than one branch of the military. Limited-coordination documents cover items or services of interest to a single branch. If a document is of limited coordination, the branch of the military which uses the document will appear in parentheses in the document designation.

Federal Specifications are developed for materials, products, and services that are used by two or more federal agencies, one of which is not a Defense agency. Federal Specifications are classified into broad categories. The QQ group, for example, covers metals and most welding and brazing specifications. Soldering and brazing fluxes are in the O-F group.

Some military and federal specifications include requirements for testing and approval of a material, process, or piece of equipment before its submission for use under the specification. In such cases, the testing procedure and acceptance criteria are given in the specification. If the acceptance tests pass one specification requirement, the material or equipment could be included in the applicable Qualified Products List (QPL).

In other specifications, the supplier is responsible for product conformance. This is often the case for brazed fabrications. The supplier must show evidence that the brazing procedures and the brazing shop are qualified in accordance with the requirements of the specification, and must certify the test report.

The following Military and Federal Standards (currently listed in the Department of

Defense Index) address brazing. Those Standards covering base metals and brazing equipment are not included.

Braze-Welding, Oxyacetylene, of Built-Up Metal Structures, MIL-B-12672

Brazing Alloy, Gold, QQ-B-653

Brazing Alloy, Silver, QQ-B-654

Brazing Alloys, Aluminum and Magnesium, Filler Metal, QQ-B-655

Brazing Alloys, Copper, Copper-Zinc, and Copper-Phosphorus, QQ-B-650

Brazing of Steels, Copper, Copper Alloys, Nickel Alloys, Aluminum, and Aluminum Alloys, MIL-B-7883

Brazing Sheet, Aluminum Alloy, MIL-B-20148

Brazing Silver, General Process for, MIL-STD-1881

Brazing, Aluminum, Process for, MIL-B-47292

Brazing, Nickel Alloy, General Specification for, MIL-B-9972

Brazing, Nickel, High Temperature Vacuum, MIL-STD-1877

Brazing, Oxyacetylene, of Built-Up Metal Structures, MIL-B-12673

Flux, Brazing, Silver Alloy, Low Melting Point, O-F-499

Welding and Brazing Procedure and Performance Qualification, MIL-STD-248

Welding Symbols (ABCA-323), Q-STD-323

Welding Terms and Definitions (ABCA-324), Q-STD-324

MIL-STD-248, *Welding and Brazing Procedure and Performance Qualification*, covers the requirements for the qualification of (1) welding and brazing procedures for non-aerospace applications; (2)welders; (3) brazers; and (4) welding and brazing operator performance qualifications. It allows the fabricator to submit for approval certified records of qualification tests prepared in conformance with the Standards of other government agencies, American Bureau of Shipping, ASME, and other organizations. Its use is mandatory when referenced by other specifications or contractual documents.

INTERNATIONAL ORGANIZATION FOR STANDARDIZATION

The International Organization for Standardization (ISO) promotes the development of Standards to facilitate the international exchange of goods and services. It is comprised of the Standards-writing bodies of more than 80 countries, and has adopted or developed over 4000 Standards.

The American National Standards Institute is the designated U.S. representative to ISO. ISO Standards and publications are available from ANSI.

NATIONAL BOARD OF BOILER AND PRESSURE VESSEL INSPECTORS

The National Board of Boiler and Pressure Vessel Inspectors (NBBPVI), often referred to as the National Board, represents the enforcement agencies empowered to assure adherence to the *ASME Boiler and Pressure Vessel Code*. Its members are the chief inspectors or other jurisdictional authorities who administer the boiler and pressure vessel safety laws in the various jurisdictions of the United States and provinces of Canada.

The National Board is involved in the inspection of new boilers and pressure vessels. It maintains a registration system for use by manufacturers who desire or are required by law to register the boilers or pressure vessels that they have constructed. The National Board is also responsible for investigating possible violations of the *ASME Boiler and Pressure Vessel Code* by either commissioned inspectors or manufacturers.

The National Board publishes a number of pamphlets and forms concerning the manufacture and inspection of boilers, pressure vessels, and safety valves. It also publishes the *National Board Inspection Code* for the guidance of its members, commissioned inspectors, and others. The purpose of this code is to maintain the integrity of boilers and pressure vessels, after they have been placed in service, by providing rules and guidelines for inspection after installation, repair, alteration, or rerating. In addition, it provides inspection guidelines for authorized inspectors during fabrication of boilers and pressure vessels.

In some states, organizations that desire to repair boilers and pressure vessels must obtain the National Board Repair (R) stamp by application to the National Board.

The firm must qualify all welding procedures and welders in accordance with the *ASME Boiler and Pressure Vessel Code*, Section IX, and the results must be accepted by the inspection agency. The firm must also have and demonstrate a quality control system similar to, but not so comprehensive as that required for an ASME code symbol stamp.

NATIONAL FIRE PROTECTION ASSOCIATION

The mission of the National Fire Protection Association (NFPA) is the safeguarding of the environment from destructive fire through the use of scientific and engineering techniques and education. NFPA Standards are widely used as the basis of legislation and regulation at all levels of government. Many are referenced in the regulations of the Occupational Safety and Health Administration (OSHA). The Standards are also used by insurance authorities for risk evaluation and premium rating.

Installation of Gas Systems

NFPA publishes several Standards that present general principles for the installation of gas supply systems and the storage and handling of gases commonly used in brazing. These are:

NFPA 50, *Bulk Oxygen Systems at Consumer Sites*

NFPA 51, *Design and Installation of Oxygen-Fuel Gas Systems for Welding and Cutting and Allied Processes*

NFPA 54, *National Fuel Gas Code*

Users should check each standard to see if it applies to their particular situation. For example, NFPA 51 does not apply to a system comprised of a torch, regulators, hoses, and single cylinders of oxygen and fuel gas. Such a system is covered by ANSI/AWS Z49.1, *Safety in Welding and Cutting.*

Safety

NFPA publishes several Standards which relate to the safe use of welding and cutting processes. These are:

NFPA 51B, *Fire Prevention in Use of Cutting and Welding Processes*

NFPA 91, *Installation of Blower and Exhaust Systems for Dust, Stock, and Vapor Removal of Conveying*

NFPA 306, *Control of Gas Hazards on Vessels to be Repaired*

NFPA 327, *Cleaning Small Tanks and Containers*

NFPA 410, *Standard on Aircraft Maintenance*

Again, the user should check the Standards to determine those that apply to the particular situation.

SOCIETY OF AUTOMOTIVE ENGINEERS

The Society of Automotive Engineers (SAE) is concerned with the research, development, design, manufacture, and operation of all type of self-propelled machinery. Such machinery includes automobiles, trucks, buses, farm machines, construction equipment, airplanes, helicopters, and space vehicles. Related areas of interest to SAE are fuels, lubricants, and engineering materials.

Aerospace Materials Specifications

Material specifications are published by SAE for use by the aerospace industry. The *Aerospace Material Specifications* (AMS or MAM) cover fabricated parts, tolerances, quality control procedures, and processes. Brazing-related specifications are listed below. The appropriate AWS filler metal classification or a common trade name follows some of the specifications, in parentheses, for clarification.

Processes

2664 Brazing-Silver, for Use up to 800°F (425°C)

2665 Brazing-Silver, for Use Up to 400°F (205°C)

2667 Brazing-Silver, For Flexible Metal Hose-600°F (315°C) Max Operating Temperature

2668 Brazing-Silver, For Flexible Metal Hose-400°F (200°C) Max Operating Temperature

2669 Brazing-Silver, For Flexible Metal Hose-800°F (425°C) Max Operating Temperature

2670 Brazing-Copper Furnace, Carbon and Low Alloy Steels

2671 Brazing-Copper Furnace, Corrosion and Heat Resistant Steels and Alloys

2672 Brazing-Aluminum

2673 Brazing-Aluminum Molten Flux (Dip)

2675 Brazing-Nickel Alloy Flux

3410 Flux, Silver Brazing

3411 Flux, Silver Brazing, High Temperature

3412 Flux-Brazing, Aluminum

3414 Flux-Welding, Aluminum

3415 Flux-Aluminum Dip Brazing, 1030°F (555°C) or Lower Liquidus

3416 Flux-Aluminum Dip Brazing, 1090°F Fusion Point

3430 Paste, Copper Brazing Filler Metal, Water Thinning Aluminum-Alloys

4184 Wire, Brazing-10 Si, 4 Cu (4145)

4185 Wire, Brazing-12 Si (4047)

Brazing and Soldering Filler Metals

4750 Solder-Tin-Lead; 45 Sn, 55 Pb

4751 Solder-Tin-Lead, Eutectic, 63 Sn, 37 Pb

4764 Brazing Filler Metal-Copper, 52.5 Cu, 38 Mn, 9.5 Ni, 1615-1700°F (880-925°C) Solidus-Liquidus Range

4765 Brazing Filler Metal-Silver; 56 Ag, 42 Cu, 2.0 Ni, 1420-1690°F (770-895°C) Solidus-Liquidus Range

4766 Brazing Filler Metal-Silver; 85 Ag, 15 Mn, 1760-1780°F (960-970°C)

4767 Brazing Filler Metal-Silver; 92.5 Ag, 7.2 Cu, 0.22 Li, 1435-1635°F (780-890°C) Solidus-Liquidus Range

4768 Brazing Filler Metal-Silver; 35 Ag, 26 Cu, 21 Zn, 18 Cd, 1125-1295°F (605-700°C) Solidus-Liquidus Range

4769 Brazing Filler Metal-Silver; 45 Ag, 24 Cd, 16 Zn, 15 Cu, 1125-1145°F (605-620°C) Solidus-Liquidus Range

4770 Brazing Filler Metal-Silver; 50 Ag, 18 Cd, 16.5 Zn, 15.5 Cu, 1160-1175°F (625-635°C) Solidus-Liquidus Range

4771 Brazing Filler Metal-Silver; 50 Ag, 16 Cd, 15.5 Zn, 15.5 Cu, 3.0 Ni, 1170-1270°F (630-690°C) Solidus-Liquidus Range

4772 Brazing Filler Metal-Silver; 54 Ag, 40 Cu, 5.0 Zn, 1.0 Ni, 1325-1575°F (720-855°C) Solidus-Liquidus Range

4773 Brazing Filler Metal-Silver; 60 Ag, 30 Cu, 10 Sn, 1115-1325°F (600-720°C) Solidus-Liquidus Range

4774 Brazing Filler Metal-Silver; 63 Ag, 28.5 Cu, 6.0 Sn, 2.5 Ni, 1275-1475°F (690-800°C) Solidus-Liquidus Range

4775 Brazing Filler Metal-Nickel; 73 Ni, 4.5 Si, 14 Cr, 3.1 B, 4.5 Fe, 1790-1970°F (975-1075°C) Solidus-Liquidus Range

4776 Brazing Filler Metal-Nickel; 73 Ni, 4.5 Si, 14 Cr, 3.1 B, 4.5 Fe, (Low Carbon) 1790-1970°F (975-1075°C) Solidus-Liquidus Range

4777 Brazing Filler Metal-Nickel; 82 Ni, 4.5 Si, 7.0 Cr, 3.1 B, 3.0 Fe, 1780-1830°F (970-1000°C) Solidus-Liquidus Range

4778 Brazing Filler Metal-Nickel; 92 Ni, 4.5 Si, 3.1 B, 1800-1900°F (980-1040°C) Solidus-Liquidus Range

4779 Brazing Filler Metal-Nickel; 94 Ni, 3.5 Si, 1.8 B, 1800-1950°F (980-1065°C) Solidus-Liquidus Range

4780 Brazing Filler Metal-Manganese; 66 Mn, 16 Ni, 16 Co, 0.80 B, 1770-1875°F (965-1025°C) Solidus-Liquidus Range

4782 Brazing Filler Metal-Nickel; 71 Ni, 10 Si, 19 Cr, 1975-2075°F (1080-1135°C) Solidus-Liquidus Range

4783 Brazing Filler Metal-High Temperature; 50 Co, 8.0 Si, 19 Cr, 17 Ni, 4.0 W, 0.80 B, 2050-2100°F (1120-1150°C) Solidus-Liquidus Range

4784 Brazing Filler Metal-High Temperature; 50 Au, 25 Pd, 25 Ni,

2015-2050°F (1100-1120°C) Solidus-Liquidus Range

4785 Brazing Filler Metal-High Temperature; 30 Au, 34 Pd, 36 Ni, 2075-2130°F (1135-1165°C) Solidus-Liquidus Range

4786 Brazing Filler Metal-High Temperature; 70 Au, 8 Pd, 22 Ni, 1845-1915°F (1005-1045°C) Solidus Liquidus Range

4787 Brazing Filler Metal-High Temperature; 82 Au, 18 Ni, 1740°F (950°C) Solidus-Liquidus Temperature

Unified Numbering System

The Unified Numbering System (UNS) provides a method for cross referencing the different numbering systems used to identify metals, alloys, and filler metals. With UNS, it is possible to correlate over 3500 metals and alloys used in a variety of specifications, regardless of the identifying number used by a society, trade association, producer, or user.

UNS is produced jointly by SAE and ASTM, and designated SAE HSJ1086/ ASTM DS56. It cross references the metal and alloy designations of the following organizations and systems:

AA (Aluminum Association)
SFSA (Steel Founders Society of America)
AISI (American Iron and Steel Institute)
SAE (Society of Automotive Engineers)
AMS (SAE Aerospace Materials Specifications)
ASME (American Society of Mechanical Engineers)
ASTM (American Society for Testing and Materials)

AWS (American Welding Society)
CDA (Copper Development Association)
Federal Specification Numbers
MIL (Military Specifications) Numbers

Over 500 of the listed numbers are for welding and brazing filler metals. Numbers with the prefix W are assigned to welding and brazing filler metals that are classified by deposited metal composition.

MANUFACTURER'S ASSOCIATIONS

The following organizations publish literature which relates to brazing. The committees that write the literature are comprised of representatives of equipment or material manufacturers. They do not generally include users of the products. Although some bias may exist, there is much useful information that can be obtained from this literature. The organization should be contacted for further information.

The Aluminum Association
818 Connecticut Avenue N.W.
Washington, D.C. 20006

American Iron and Steel Institute
1000 16th Street N.W.
Washington, D.C. 20036

Copper Development Association, Inc.
Greenwich Office Park 2
Box 1840
Greenwich, CT 06836

Electronic Industries Association
2001 Eye Street, N.W.
Washington, DC 20006

Keywords — copper, copper alloys, oxygen-bearing coppers, deoxidized and oxygen-free coppers, special coppers, high coppers, copper-zinc alloys, leaded brasses, copper-tin alloys, copper-aluminum alloys, copper silicon alloys, copper-nickel alloys

Chapter 18

COPPER AND COPPER ALLOYS

Copper and its alloys are selected for many applications because of their wide range of properties. This whole family of metals normally has excellent corrosion resistance and formability. Brazing can be performed readily on most copper alloys with proper precautions. Electrical and thermal conductivities vary widely depending on composition.

The classes of wrought and cast copper alloys are shown in Tables 18.1 and 18.2. In comparison with steels, copper and its alloys have higher thermal expansion. This is important when brazing dissimilar metals.

If proper precautions are not taken when copper and copper alloys are brazed, cracking, distortion, and unacceptable softening may occur. However, these potential problems should not discourage the use of these alloys in brazed assemblies. Knowledge about the factors that cause such problems makes it possible to predict the properties of the resulting brazement.

Softening of the base metal occurs frequently during brazing because many copper base alloys derive their properties from low temperature heat treatment or cold work, or both. The degree of softening increases with temperature and the length of exposure to high temperatures. Softening of areas close to the braze can be minimized by (1) cooling the assembly, except for the area to be brazed, (2) immersion in water, (3) packing with wet rags, or (4) providing a heat sink to keep the overall temperature of the part as low as possible. Brazing with a

low melting filler metal for a minimum time will also minimize softening.

Residual stresses arising from cold working, casting, or machining operations can cause cracking of some alloys during brazing. Heating and cooling can induce additional stress as a result of uneven expansion or contraction. Relatively low stresses from all of these factors may be sufficient to cause cracking at elevated temperatures in the presence of liquid brazing filler metal.

Uniform and controlled heating is very important, especially with brasses, cold worked phosphor bronzes, and cold worked silicon bronzes. They are especially susceptible to cracking when subject to tensile stress in areas exposed to liquid braze filler metal as a result of integranular penetration.

BASE METALS

Oxygen-Bearing Coppers

This group includes the fire-refined and electrolytic tough pitch grades of copper and silver-bearing copper [10-20 oz./ton (300-600 g/tonne)]. Fire-refined copper contains very small percentages of impurities, as well as from 0.02 to 0.05% oxygen in the form of copper oxide which constitutes part of a copper-copper oxide eutectic. The latter is scattered as globules throughout the wrought metal, and appears as an interdentritic structure in cast products. The cop-

Table 18.1
Wrought Coppers and Copper Alloys

Range of UNS Numbers	Description	Typical Composition Ranges	Approximate Electrical Conductivity Range, % IACS
C10100 to 10700	Oxygen free coppers	99.5% Cu or better	100
C10400 to 14200	Tough pitch and deoxidized coppers	Contain oxygen or deoxidizers	80 - 100
C14000 to 14700	Free machining coppers	Small additions of S, Te, etc.	45 - 95
C15000 to 19400	High copper alloys	of 1 or 2% additions of Be, Cr, Co, Fe, Ni, Zn, and/or Sn	20 - 90
C20500 to 24000	Red brasses	Up to 20% Zn	35 - 60
C25000 to 29800	Yellow brasses	From 25 to 50% Zn	25 - 35
C31000 to 38500	Leaded brasses	From 10 to 45% Zn and up to 4.5% Pb	25 - 45
C40500 to 45500	Tin brasses	To 5.5% Sn, to 48% Zn	25 - 30
C50200 to 52900	Copper-tin alloys (phosphor bronzes)	From 1% to 11% Sn	10 - 50
C53200 to 54600	Leaded phosphor bronzes	1 to 4% Pb, about 5% Sn, some with additions of Zn	10 - 20
C60600 to 64200	Aluminum bronzes	From 2.6 to 13% Al, to 5% Fe, some with additions of Si or Ni	10 - 20
C64700 to 66100	Silicon bronzes	From 1 to 3.5% Si, some with Mn, Si, or Sn	7 - 12
C66800 to 69700	Alloy brasses	Zinc-containing alloys with additions of Ni, Sn, Mn, Al, and Si	20 - 25
C70100 to 82000	Copper nickels	From 2 to 40% Ni, additions of Fe, Be, Mn, or Cr	4 - 10
C73200 to 79800	Nickel silvers	From about 43 to 73% Cu, from 7 to 23% Ni, some with Pb or Mn, balance Zn	5 - 10

Note: For specific compositions and properties see Standards Handbook, Part 2, Copper Development Association, NY.

per-copper oxide eutectic has no serious effects on mechanical properties or electrical conductivity. However, it does make the copper susceptible to embrittlement when heated in a hydrogen atmosphere.

Hydrogen rapidly diffuses into the metal and reacts with oxides to form steam. The steam then expands and creates porosity at the grain boundaries of the copper. Carbon monoxide, which may be present in oxyacetylene flames or reducing environments, can contribute to weakness if moisture is also present. That happens because carbon monoxide can reduce water vapor to hydrogen, which then diffuses into the metal. A final note on embrittlement: When oxygen-bearing coppers are heated above 1690°F (920°C) for prolonged periods, as during some brazing operations, copper oxide concentrates in the grain boundaries and reduces strength and ductility.

Oxygen-bearing coppers are of medium strength and low hardness. When the copper oxide is uniformly distributed, these coppers are tough, ductile, and highly malleable. Their electrical and thermal conduc-

Table 18.2
Cast Coppers and Copper Alloys

Range of UNS Numbers	Description	Composition Ranges	Approximate Electrical Conductivity Range, % IACS
C80100 to C81100	Coppers	Minimum of 99.70% Cu and remainder Ag	92 - 100
C81300 to C82800	High copper alloys	Additions of up to about 2.5% Be, Co, Si, Ni, and Cr	20 - 80
C83300 to C83800	Red brasses	83 to 93% Cu, to 12% Zn with lesser amounts of Sn, Pb	15 - 40
C84200 to C84800	Semi-red brasses	76 to 80% Cu, 8 to 15% Zn with lesser amounts of Sn, Pb	15 - 20
C85200 to C85800	Yellow brasses	57 to 72% Cu, balance primarily Zn, 1 to 2% Sn, Pb, Ni, or Al	18 - 28
C86100 to C86800	High-strength yellow brasses	55 to 67% Cu, additions of Fe, Ni, Mn, Al, balance Zn	7 - 22
C87200 to C87900	Silicon brasses and silicon bronzes	65 to 90% Cu, about 3 to 5% Si, some with large amounts of Zn	6 - 15
C90200 to C94500	Tin bronzes	3 to 19% Sn, some with large amounts of Pb, less Zn, Ni	7 - 15
C94700 to C94900	Nickel-tin bronzes	About 5% Sn and 5% Ni, to 2.5% Zn, Alloy C94800 has 1% Pb	12
C95200 to C95800	Aluminum bronzes	7 to 11% Al, at least 71% Cu, balance Ni, Fe, Mn and Si	3 - 13
C96200 to C96600	Copper nickels	10 to 31% Ni, about 1% additions of Fe,Cb, Si, Mn and Si	4 - 11
C97300 to C97800	Nickel silvers	55 to 65% Cu, Pb and Sn additions, 12 to 25% Ni, balance Zn	4 - 5

NOTE: For specific compositions and properties see Standards Handbook, Part 7, Copper Development Association, NY.

tivities are the highest of all copper alloys except for the oxygen-free copper grades. They may be softened at temperatures of 450 to 1500°F (230 to 815°C), depending on the time and temperature of exposure. Annealing in a hydrogen-containing reducing atmosphere should be avoided because of the potential for creating voids or fissures, which weaken the material. It is advisable to carry out an annealing operation in vacuum, pure nitrogen, or inert atmosphere.

Deoxidized and Oxygen-Free Coppers

This group includes phosphorus-deoxidized copper and oxygen-free (OF) copper. Phosphorus-deoxidized copper is usually copper from which oxygen has been re-moved by the addition of 0.01 to 0.04% phosphorus before casting. In some cases, phosphorus is added to oxygen-free copper to provide a greater margin of protection against hydrogen embrittlement. If the phosphorus addition is controlled closely to obtain less than 0.01% residual, such copper has a high conductivity and is called deoxidized low phosphorus (DLP) copper. Phosphorus-deoxidized copper with 0.01 to 0.04% phosphorus (DHP) has lower electrical and thermal conductivities than the low phosphorus (DLP) grade.

Oxygen-free copper has an oxygen content that has been deliberately minimized. This is accomplished by melting and casting copper cathodes under atmospheres that reduce oxygen and prevent the formation of oxides during the casting. Since no deoxidant is introduced into this cast copper, it can absorb some oxygen from the air during relatively long heating times at high temperatures.

Phosphorus-deoxidized and oxygen-free coppers are very ductile. The absence of the copper-copper oxide eutectic improves the cold-working properties of these coppers over those of oxygen bearing coppers. This is especially true for deep drawing and spinning operations. These coppers can be annealed between 450 and 1500°F (230 and 815°C) in reducing atmospheres, since they will not be embrittled by hydrogen. Corrosion resistance is essentially the same as that of oxygen-bearing coppers.

Special Coppers

This group includes some alloys that offer high electrical conductivity plus special properties such as machinability. Lead, tellurium, selenium, and sulfur-bearing coppers have machinability ratings of about 80, compared to about 20 for ordinary coppers (based on a rating of 100 for free-cutting brass). However, ductility and workability are reduced somewhat by the presence of the inclusions that impart free-machining characteristics. These free-machining coppers can be supplied with a matrix of deoxidized or oxygen-free copper to prevent embrittlement or gassing when they are brazed. They are used widely in the manufacture of electrical connectors.

High Coppers

The high copper group includes those with additional small amounts of alloying elements that enhance mechanical properties. Chromium copper has a tensile strength of about 75 ksi (517 MPa) and an electrical conductivity of about 80% IACS after cold working and age hardening. Zirconium copper develops somewhat lower strength than chromium copper, but it has higher electrical conductivity of about 90% IACS.

Copper-beryllium[1] alloys are of two types: One has 1.5 to 2% beryllium, while the other has about 0.4% beryllium. Cobalt or nickel additions to these alloys restrict

grain growth during annealing. Alloys with about 0.4% beryllium have higher conductivity but lower strength. Beryllium-containing alloys have a moderate tensile strength in the cold-rolled condition. When age hardened by heat treatment at temperatures in the range of 500° to 900°F (315° to 480°C), they have the highest tensile strength and hardness of all copper alloys. Cold-worked and age-hardened alloys are softened and made malleable by annealing in an exothermic atmosphere at temperatures above 1450°F (788°C) depending on composition, followed by water quenching.

Copper-Zinc Alloys (Brasses)

Copper-zinc alloys are produced with varying ratios of these two elements to provide the desired properties and casting characteristics. Other elements occasionally are added to enhance particular mechanical or corrosion properties. These special brasses are identified by the added element — for example, aluminum brass and aluminum bronze. Additions of manganese, tin, iron, silicon, nickel, lead, and aluminum, either singly or collectively, rarely exceed 4%. These additions, however, have important effects on the brazing characteristics of brasses.

In general, brasses can be divided into three classifications: low brasses (20% Zn max.), high brasses (more than 20% Zn), and alloy brasses. On sudden exposure to elevated temperatures, brasses that are stressed in tension are susceptible to cracking. To avoid cracking, stressed material should be uniformly brought up to brazing or heat treating temperature.

Brasses can be softened after cold-working operations by annealing in the temperature range of 700° to 1400°F (370° to 760°C). When quenched from the annealing temperatures, the alpha-beta brasses show a higher hardness than would be found after air or furnace cooling. This is a result of a higher proportion of martensitic beta phase in quenched material. Stress relieving at 500° to 700°F (260° to 370°C) will reduce residual stresses.

Leaded Brasses

Lead is commonly added to brass in amounts up to 5% to improve machinability. This addition usually has no effect on

---

1. Beryllium has been identified as a toxic element. Brazing beryllium-containing alloys requires adequate ventilation and cleanliness. See Chapter 27.

the tensile strength, particularly of cold worked materials, but it does increase the possibility of cracking in materials that are under tensile stress and subjected to heat. Lead has no effect on corrosion resistance, but it impairs brazeability.

Copper-Tin Alloys (Phosphor Bronzes)

Alloys of copper and tin are termed tin bronzes. During casting, 0.03 to 0.5% phosphorus is added as a deoxidizing agent and may be found as a residual in the alloys. Thus, the tin bronzes are known commercially as phosphor bronzes. Phosphor bronzes have moderate to high tensile strength, depending on tin content and degree of cold work.

Commercial alloys are available with from 1.0 to 10% tin. In the completely homogenized condition, they are single-phase alloys with a structure similar to alpha brass. Alloys with tin content over 5% are difficult to cast without dendritic segregation. During cooling, they give rise to a brittle delta phase. The low tin phosphor bronzes have low electrical and thermal conductivities. Those with high tin content have particularly low values of these properties.

All the phosphor bronzes have good cold working properties with high strength and hardness in the cold-rolled tempers. After cold working, these alloys can be rendered soft and malleable by annealing at temperature between 900° and 1400°F (482° and 760°C), depending on the properties desired. In a stressed condition, the phosphor bronzes are subject to cracking during brazing. Sudden application of high heat must be avoided.

Some wrought phosphor bronzes contain up to 4.5% lead to improve machinability. Many cast bronzes also contain lead up to 6% to improve both casting and machining properties. Segregation of lead to the surface of these metals at brazing temperatures tends to prevent wetting by the filler metals. It also tends to produce brittle joints. The degree of lead segregation increases with temperature and lead content.

The lead-bearing phosphor bronzes also tend to crack readily when stressed at brazing temperatures. It is often necessary to avoid brazing of components and assemblies that have sharp corners or notches that

concentrate stresses. Also, abrupt changes in cross section can result in high thermal stresses during heating.

Copper-Aluminum Alloys (Aluminum Bronzes)

Copper-aluminum alloys (aluminum bronzes) are high copper alloys that contain 3 to 13% aluminum and varying amounts of iron, nickel, manganese, and silicon. The aluminum bronze alloys are divided into two types. The first includes the alpha or single-phase alloys that do not undergo phase changes on heat treatment. The second type includes the duplex or multiphase alloys that can be hardened by heat treatment. The properties of the latter type depend on composition and heat treatment. Both types of aluminum bronzes have low electrical and thermal conductivities.

The alpha phase aluminum bronzes are characterized by relatively high tensile strength, high toughness, good ductility, and moderate hardness. They have a reasonably wide plastic range and can be worked at temperatures between 1450° and 1650°F (788° to 900°C). They can be softened by annealing between 900° and 1400°F (482° to 760°C) to achieve specific properties desired.

The duplex alloys have relatively high tensile strengths. As the aluminum content is increased, ductility decreases and hardness increases. The harder grades are useful for metal-to-metal wear applications and for service under abrasive conditions. The physical properties of the duplex alloys can be improved by quenching in water from 1550° to 1750°F (843° to 954°C), followed by heat treating at 800° to 1200°F (427° to 649°C), depending on the specific composition of the alloy.

The formation of aluminum oxide and other compounds creates difficulty when brazing these alloys. Highly reactive brazing fluxes are required with copper-aluminum alloys.

Copper-Silicon Alloys (Silicon Bronzes)

Copper-silicon alloys commonly called silicon bronzes, generally contain 1.5 to 3.5% silicon and 1.25% or less of zinc, tin, manganese, or iron. The addition of iron increases tensile strength and hardness. Toughness and shear strength increase with silicon content, but electrical and thermal conductivities decrease.

The high silicon bronzes work-harden rapidly. High tensile strength and hardness can be obtained in cold-worked products. All these alloys can be softened after cold working by annealing between 900° and 1400°F (482° to 760°C). The silicon bronzes are subject to cracking during brazing, and their crack susceptibility increases as silicon content increases. The alloys are all susceptible to intergranular penetration by molten filler metal and to hot-shortness. Stress relieving of these alloys and uniform heating for brazing are important considerations.

Copper-Nickel Alloys

The copper-nickel alloys are available commercially with nickel content ranging from 5 to 30%. Those most commonly used in fabrication contain 10%, 20%, or 30% nickel. These alloys may also contain other elements including iron, manganese, or chromium.

The copper-nickel alloys have moderately high tensile strength, good ductility, and are relatively tough. Electrical and thermal conductivities are low, and they decrease as nickel content increases. The alloys have good hot-working characteristics but are not well-suited to hot forging. Their cold-working properties are excellent, and the degree of malleability with low nickel content approaches that of copper. These alloys can withstand severe drawing, stamping, and spinning operations because they do not work-harden rapidly. They are annealed at temperatures from 1200° to 1600°F (649° to 871°C), depending on composition. Stress relief heat treatment is advisable prior to joining.

Copper-Nickel-Zinc Alloys (Nickel Silvers)

Nickel is added to copper-zinc alloys to make them silvery in appearance for decorative purposes and to increase strength and corrosion resistance. The resulting alloys are called *nickel silvers*. These alloys are of two general types: (1) alloys containing about 65% copper plus nickel and consisting of a single alpha phase, and (2) other alloys containing 55 to 60% copper plus nickel and consisting of two phases, alpha and beta. Manganese and magnesium are added to these alloys to control free sulfur. In the hard (37% cold rolled) temper, the yield strength of these alloys is high, exceed-

ing 70 ksi (480 MPa). Electrical and thermal conductivities are quite low.

Single-phase nickel silvers are satisfactory for deep drawing, stamping, and spinning. Because of their good plasticity over a wide temperature range, two-phase nickel silvers may be hot worked by any of the commercial processes. The two-phase alloys are difficult to cold work, and therefore seldom are. However, they can be softened by annealing at 900° to 1300°F (480° to 705°C), depending upon the properties desired. Brazing characteristics are good. The alloys should be stress relieved and heated evenly to brazing temperature.

FILLER METALS

Silver base filler metals with brazing temperatures from 1145° to 1600°F (620° to 870°C) and copper-phosphorus filler metals with brazing temperatures of 1300° to 1500°F (704° to 816°C) are used most commonly. All RBCuZn, BCuP, BAu, and BAg filler metals may be used for brazing most coppers and copper alloys, provided the brazing temperatures are sufficiently lower than the melting range of the base metal. BCu filler metal may be used to braze the copper-nickel alloys, but its liquidus temperature is too high for use with the other copper-base metals.

RBCuZn filler metals may be used to braze the coppers and the copper-nickel, copper-silicon, and copper-tin alloys. The liquidus temperatures of these filler metals are too high for brazing the brasses and nickel silvers. RBCuZn filler metals are not useful for brazing aluminum bronzes because the required brazing temperatures destroy the effectiveness of the fluxes these base metals.

BCuP filler metals are useful for most of the copper base metals because fluxes are not needed. BCuP filler metals have been used for brazing 90Cu-10Ni alloy, but their usefulness for copper nickels should be established by suitable testing for each application. These filler metals are useful for brazing cast high-leaded brass pipe fittings when precautions are taken to flux properly and avoid overheating.

The BAg filler metals may be used on all copper base metals. The BAu filler metals are used primarily in electronic applications when a low vapor pressure filler metal is necessary.

Corrosion resistance is an important factor in selecting a brazing filler metal, since quite often a copper-base metal is used for that same reason. In many cases with copper-nickel, copper-silicon, and copper-tin alloys, the RBCuZn filler metals will not have adequate corrosion resistance when in contact with these alloys. For example, it is not advisable to use BCuP filler metals for service in high temperature water systems or in sulfurous atmospheres.

Other factors in the selection of a filler metal include service temperature and the sequential brazing operations in step brazing procedures.

JOINT DESIGN

The temperatures required to braze copper base metals are high enough to anneal them. Therefore, joint designs must be based only on the strength of annealed metals. All the joint designs discussed in Chapter 2 are feasible. Joint clearances of 0.001 to 0.005 in. (0.03 to 0.13 mm) are best suited for maximum joint strength and soundness. However, somewhat larger clearances may be tolerated where it is impossible or impractical to maintain the suggested values. If lap joints can be used, they can compensate for lower strength necessitated by wide clearances. In this type of joint, a brazed area at least three times the cross-sectional area of the thinner member will develop the full strength of that member.

The coefficients of thermal expansion of copper alloys vary over a wide range. They are generally higher than those of nickel alloys, carbon steel, and cast iron. However, they are much lower than those of aluminum alloys. The thermal expansion characteristics of austenitic stainless steels are similar to those of copper alloys.

PROCESSES AND EQUIPMENT

Copper and copper-base alloys can be brazed using a wide variety of processes and techniques. In general, all the processes and equipment described in other Chapters are applicable with the following precautions:

(1) Brazing may not be suitable if the base metal and the filler metal have similar melting ranges or if they are readily soluble in one another. Prolonged heating may cause excessive base metal erosion, resulting in a decrease in the base metal melting range. One example is the brazing 90Cu-10Ni alloy with copper (BCu) filler metal. Time at brazing temperature has to be limited, in this instance, since dissolution mainly occurs above the solidus of the filler metal as soon as a liquid phase appears.

(2) Heating should be uniform when brazing phosphor bronze, silicon bronze, and nickel silvers to prevent cracking in cold-worked materials. With resistance, induction, dip, and oxyfuel gas brazing processes, thermal stresses may be induced if heating is too rapid.

(3) Exposure of oxygen-bearing coppers to hydrogen-containing atmospheres can result in embrittlement of the copper. Both torch brazing of large assemblies and furnace brazing in reducing atmospheres should be avoided to preserve the integrity of the assembly. Embrittlement by hydrogen is temperature and time sensitive; higher temperatures and longer times increase the damage.

(4) Copper alloys containing lead are susceptible to lead segregation. Consequently, torch brazing of large assemblies and furnace brazing may prove troublesome. Excessive segregation of lead from leaded alloys, especially those containing more than about 2.5% lead, can result in defective brazed joints due to embrittlement and incomplete bonding.

(5) Furnace brazing of sulfur-bearing coppers may prove troublesome when done in hydrogen or inert gases because sulfur vapor can inhibit wetting of the copper surface.

PRECLEANING AND SURFACE PREPARATION

Consistent and sound brazed joints cannot be achieved unless the joint surfaces are free of oxides, dirt, and other foreign substances. Conventional solvent or alkaline degreasing procedures are suitable for cleaning copper base metals. Mechanical methods, such as wire brushing, and sand blasting may be used to remove oxides.

Table 18.3
Guide to Brazing Copper and Copper Alloys

Material	Commonly Used Brazing Filler Metals	AWS Brazing Atmospheres*	AWS Brazing Flux No.	Remarks
Coppers	BCuP-2**, BCuP-3II, BCuP-5** RBCuZn, BAg-1a, BAg-1, BAg-2, BAg-5, BAg-6, BAg-18	1 or 2 or 5	FB3-A,C,D,E,I,J	Oxygen-bearing coppers should not be brazed in hydrogen-containing atmospheres.
High coppers	BAg-8, BAg-1		FB3-A	
Red brasses	BAg-1a, BAg-1, BAg-2 BCuP-5, BCuP-3, BAg-5, BAg-6, RBCuZn	1 or 2 or 5	FB3-a,C,D,E,I,J	
Yellow brasses	BCuP-4, BAg-1a, BAg-1, BAg-5, BAg-6 BCuP-5, BCuP-3	3 or 4 or 5	FB3-A,C,E	Keep brazing cycle short.
Leaded brasses	BAg-1a, GAg-1, BAg-2 BAg-7, BAg-18 BCuP-5	3 or 4 or 5	FB3-A,C,E	Keep brazing cycle short and stress relieve before brazing.
Tin brasses	BAg-1a, BAg-1, BAg-2 BAg-5, BAg-6 BCuP-5, BCuP-3 (RBC7Zn for low tin)	3 or 4 or 5	FB3-A,C,E	
Phosphor bronzes	BAg-1a, BAg-1, BAg-2 BCuP-5, BCuP-3 BAg-5, BAg-6	1 or 2 or 5	FB3-A,C,E	Stress relieve before brazing.
Silicon bronzes	BAg-1a, BAg-1, BAg-2	4 or 5	FB3-A,C,E	Stress relieve before brazing. Abrasive cleaning may be helpful.
Aluminum bronzes	BAg-3, BAg-1a, BAg-1 BAg-2	4 or 5	FB4-A	Flux required with fuhrnace brazing.

Table 18.3 (Continued)

Material	Commonly Used Brazing Filler Metals	AWS Brazing Atmospheres*	AWS Brazing Flux No.	Remarks
Copper nickel	BAg-1a, BAg-1, BAg-2 BAg-18, BAg-5, BCuP-5, BCuP-3	1 or 2 or 5	FB3-A,C,E	Stress relieve before brazing.
Nickel silvers	BAg-1a, BAg-1, BAg-2 BAg-5, BAg-6 GCuP-5, BCuP-3	3 or 4 or 5	GB3-A,C,E	Stress relieve before brazing and heat uniformly.

* Inert gas, including hydrogen, or vacuum atmospheres are usually acceptable (AWS Type 6 or 9 or 10). See Chapter XX.

** Protective atmosphere or flux is not required for brazing copper.

Complete chemical removal of oxides requires proper choice of pickling solution. Typical procedures used for chemical cleaning of copper alloys are as follows:

Aluminum Bronzes

Successive immersions are needed in the following two solutions:

(1) Cold 2% hydrofluoric acid and 3% sulfuric acid mixture by volume in water, followed by

(2) A solution of 5% by volume sulfuric acid at 80 to 120°F (27 to 49°C).

This procedure may be repeated until the work is clean.

Copper-Silicon Alloys

Cleaning procedure for this group of base metals consists of immersing in hot 5% sulfuric acid, then in cold 2% hydrofluoric and 5% sulfuric acid mixtures, all volume percent aqueous solutions.

Brass and Nickel-Silver Alloys

The cleaning solution in this case is cold 5% by volume sulfuric acid.

Coppers

Immerse in cold 5 to 15% by volume sulfuric acid.

Prior to brazing it is often desirable to copper plate those alloys containing strong oxide-forming elements to simplify brazing and fluxing requirements. Plating with about 0.001 in. (0.03 mm) of copper is used on chromium coppers, and 0.0005 in. (0.013 mm) on beryllium coppers, silicon bronzes, and aluminum bronzes.

FLUXES AND ATMOSPHERES

Fluxes

The AWS flux Types FB3-A, -C, and -E are suitable for use with BCuP and BAg filler metals when brazing all the copper-base metals except the aluminum bronzes. Refractory oxides form easily on the latter, and the more active Type FB4-A fluxes are needed to cope with them (see Table 18.3). The effectiveness of Types FB3-A, -C, and -E fluxes may be reduced rapidly at the brazing temperatures needed for RBCuZn filler metals, while it is completely destroyed when brazing with BCu filler metal. Types

FB3-D, -I and -J fluxes may be used with RBCuZn and BCu filler metals except for brazing aluminum bronze and beryllium copper where more active fluxes are needed. Mixtures of Types FB4-A and FB3-D, -I, and -J fluxes may be found satisfactory for applications of this kind.

Atmospheres

Combustion fuel gases are economical and useful atmospheres for brazing most copper base alloys. The obvious exceptions are atmospheres with high hydrogen content, namely Types 2, 3, and 4, which may not be used on oxygen-bearing coppers because of their embrittling action. Dissociated ammonia and hydrogen are also useful, except for oxygen-bearing coppers. Inert-gas atmospheres, such as argon and helium, and also nitrogen may be used for all copper base alloys without harmful effects.

Vacuum atmosphere is suitable for copper and copper alloys when the alloys are essentially free of elements having high vapor pressures at the brazing temperature (lead, zinc, etc.). The brazing filler metal should be restricted to vacuum grades containing low amounts of high vapor pressure elements, such as zinc and cadmium. For high-vacuum brazing, the vapor pressure-temperature curves for copper and any alloying elements should be consulted to determine maximum permissible brazing temperatures. Introduction of inert gases to create a partial pressure in a furnace may help to avoid the volatilization of components of the filler metal/base metal or the base metal, or both.

ASSEMBLY, PROCEDURE, AND TECHNIQUE

The methods described in Chapter 8 may be used in assembling components of copper-base alloys for brazing, and the requirements outlined therein should be observed in the design of fixtures. Assembly procedures should start with thorough cleaning of all joint areas. When flux is to be used, it should be applied to all joint areas, preferably immediately after the cleaning operations.

Procedures and techniques are determined by the brazing process to be used. Additional steps may be incorporated as needed for certain applications. For instance, it is possible to braze heat-treatable base metals in combination with suitable heat treatments to obtain desired mechanical properties. This procedure can be applied to the brazing of beryllium and chromium coppers, but not to zirconium coppers. Some increase in strength properties can be obtained by applying a suitable aging heat treatment.

This combined operation necessitates the use of brazing filler metals appropriate for the required solution annealing temperatures. It is also practical only when the brazed assembly can be cooled rapidly to maintain the microstructure and properties achieved in the solutioned treated condition. In order to avoid damaging the brazed joints, the assembly must be supported properly during brazing, quenching, and subsequent aging heat treatment. Before proceeding with the operation, information should be checked on the required temperatures for solution annealing and aging.

POSTBRAZE OPERATIONS

Flux residues can be a source of corrosion, and their removal is recommended. These residues are loosened or dissolved by immersion in hot water. Removal of oxides requires mechanical cleaning, such as wire brushing, or pickling with the appropriate solutions. The procedures described previously should be used.

INSPECTION

Most of the inspection procedures described in Chapter 16 are feasible for brazed assemblies of copper-base alloys. Selection depends on the design of the assemblies and requirements of the application.

PROCEDURES AND PROPERTIES

Copper

Prior to brazing, the copper oxide surface film can be removed easily by abrading. Where large areas are to be brazed, pickling

in hot 5 to 15% by volume sulfuric acid is recommended.

Oxygen-free, high-conductivity copper and deoxidized coppers are brazed readily by furnace or torch methods. Oxygen-bearing (tough-pitch) copper is susceptible to oxide migration or hydrogen embrittlement, or both, at elevated temperatures. Therefore, oxygen-bearing copper should be furnace brazed in an inert atmosphere, nitrogen, or vacuum. Torch brazed should be done with a neutral or slightly oxidizing flame. The copper-phosphorus and copper-silver-phosphorus filler metals are self-fluxing on copper. Flux is beneficial, however, for heavy assemblies where prolonged heating would otherwise cause excessive oxidation. Due to the danger of corrosive attack, joints brazed with filler metal containing phosphorus should not be exposed for long periods at elevated temperatures in water or sulfurous atmospheres. Microstructures of copper brazed with BCup-3 are shown in Figures 18.1 and 18.2.

With copper-zinc filler metals, care should be taken not to overheat them since volatilization of zinc causes voids in the joint. When torch brazing, an oxidizing flame will reduce zinc fuming. The AWS brazing fluxes Type FB3-D, -I and -J should be used. Corrosion resistance of RBCuZn-type filler metals is inferior to that of copper.

There is a rule of thumb for design that is widely used: A lap joint assembly will develop the full strength of annealed copper with an overlap of three times the thickness of the thinner member. Actually, a lap joint

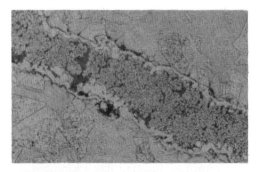

Figure 18.1 — Copper Joint Brazed With BCuP-3 Filler Metal, 120x (Reduced by 20% on Reproduction).

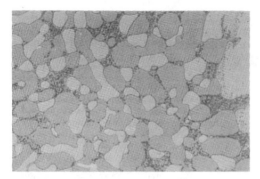

Figure 18.2 — Phases Present in Copper Brazed With BCuP-3 Filler Metal: Copper (Light Gray); Cu₃P Compound (Dark Gray); and Ternary Eutectic (Fine Structure), 1000x (Reduced by 20% on Reproduction).

in deoxidized copper develops the full strength of the base metal with a lesser overlap, as shown in Figure 18.3. The knee in the tensile strength-overlap distance relationship at an overlap of twice the thickness reflects the change in the location of the fracture from the joint to the base metal. At an overlap greater than 2T, short-time tensile fracture generally occurs in the base metal.

Even when specimens are deliberately made to fail in the joint area, the location of failure generally is partly or completely through the base metal in a plane parallel to the joint interface. This is explained by the fact that the room temperature strengths of the BAg and BCuP filler metals are greater than those of annealed base metal. At elevated temperatures, the strength of the brazing filler metal decreases more rapidly than that of the copper, and eventually failure occurs through the filler metal. Maximum operating temperatures for some filler metals are listed in Table 18.4. The properties of joints of electrolytic tough-pitch and phosphorus-deoxidized copper are shown in Table 18.5 Stress-rupture and fatigue strength properties for brazed tough-pitch copper are shown in Figures 18.4 and 18.5.

Special Copper Alloys

The coppers that contain small additions of silver, lead, tellurium, selenium, or sulfur (generally no more than 1%) are brazed readily with the self-fluxing BCuP filler metals. Wetting action is improved when a flux is used.

High Coppers

Beryllium copper surfaces must be clean for brazing. Oxide scale can be removed using the following pickling procedure: (1) immersion in 20% by volume sulfuric acid solution at 160-180°F (71-82°C) followed by water rinse; (2) quick dip (less than 30 seconds) in cold 30% by volume nitric acid solution finally followed by immediate and thorough rinsing. Chromium copper can be pickled by immersing in hot 5% by volume sulfuric acid, then in a cold mixture of 1-5 oz./gal (15 to 37 g/l) sodium bichromate with 3-5% by volume sulfuric acid. Copper plating to a thickness of 0.001 in. (0.03 mm) prior to brazing facilitates the operation.

The high-strength beryllium-copper alloys (2% beryllium) can be furnace brazed and simultaneously solution treated at 1450°F (788°C). Temperature then is lowered to 1400°F (760°C) to solidify the braze filler metal. The assembly is quenched in water, and then aged at 600 to 650°F (316 to 343°C). The silver-copper eutectic filler metal BAg-8 is generally used with AWS Type FB3-A and FB3-E fluxes.

Another method suitable for sections that can be heated rapidly (less than 1 min. is desirable) is the brazing of solution annealed material at a temperature below the solution annealing temperature. Brazing is followed by aging without reannealing. Sufficiently fast heating rates can be attained with induction brazing.

High-conductivity beryllium-copper alloys (0.5 percent beryllium), which are annealed at 1700°F (927°C) and precipitation-hardened between 850° and 900°F (454° to 482°C), can be silver brazed readily with BAg-1 or BAg-1a filler metal in the aged condition. However, the mechanical properties of hardened base metal will be reduced.

Chromium copper and zirconium copper are age hardened at about 900°F (482°C), following solution annealing at 1650° to 1850°F (899° to 1010°C) and cold working. Brazing with silver-bearing filler metal and fluoride-containing flux is best performed after solution annealing and cold working, but before age hardening. Heat treated

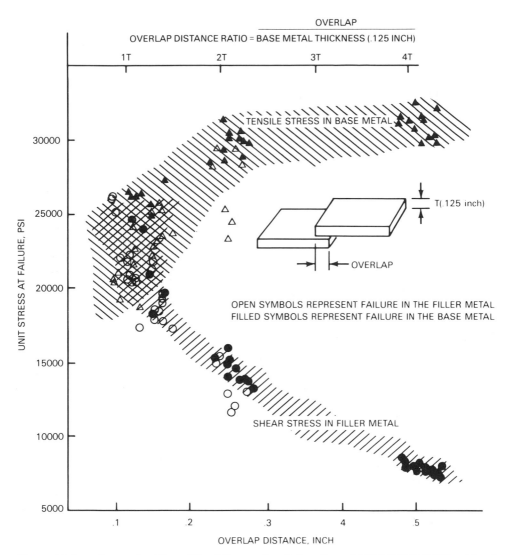

Figure 18.3 — Strength of Deoxidized Copper Lap Joints Brazed With BCuP-3 Filler Metal and Various Overlap Ratios.

properties of base metals after this procedure are lower than normal values. Properties of chromium and zirconium coppers after brazing at temperatures lower than solution heat treating temperatures are given in Tables 18.6 and 18.7. Brazing at the solution annealing temperature followed by rapid cooling can give enhanced mechanical properties for chromium coppers; zirconium coppers require an intermediate cold-

working step, which is not practical for most brazed structures.

Copper-Zinc Alloys (Brasses)

All the brasses can be brazed with the BAg and BCuP filler metals, and the higher melting point (low zinc) brasses can also be brazed with RBCuZn filler metal. The AWS Type FB3-A, -C, and -E fluxes should be used with BAg and BCuP; AWS Types

Table 18.4
Maximum Operating Temperatures for Brazing Filler Metals

Filler Metal Classification	Service Temperature			
	Continuous Service		Short - Time Service	
	°F	°C	°F	°C
BCuP	300	150	400	200
BAg	400	200	600	315
BAu	800	425	1000	540
RBCuZn	400	200	600	315

FB3-D, -I, and -J fluxs should be used with RBCuZn filler metal.

Chapter 4 lists the protective atmospheres commonly used in furnace brazing. Even in a protective atmosphere, flux should be used to promote good wetting by the brazing filler metal. Flux should also be used to reduce zinc fuming. When heated above 750°F (400°C), brass tends to lose zinc by vaporization. This loss can be reduced by fluxing the parts during furnace brazing, or by using an oxidizing flame during torch brazing. Brasses are also subject to cracking and should therefore be heated carefully and uniformly. Sharp corners and changes in cross section that concentrate stresses and produce thermal strains should be avoided. It is good practice to bring parts to temperature gradually.

Brasses containing aluminum or silicon require treatment similar to aluminum or silicon bronzes. Lead added to brass to improve machinability may alloy with the brazing filler metal and cause brittleness. Major brazing difficulties occur when the lead content is over 2 to 3%. To maintain good flow and wetting during brazing, leaded brasses require complete flux coverage to prevent the formation of lead oxide or dross. Rapid heating of high-leaded brasses may cause cracking. Stress relief annealing before brazing and also slower uniform heating and cooling will minimize this cracking tendency.

Copper-Tin Alloys (Phosphor Bronzes)

In a stressed condition, these alloys are subject to cracking. It is good practice to stress relieve or anneal the parts before brazing. It is also advisable to support them in a stressfree condition during brazing, and to avoid thermal shock by using slow heating cycles. Adequate flux protection during brazing is necessary when the tin content is high or when there are appreciable lead additives. All the phosphor bronzes can be brazed with the BAg and BCuP filler metals. The low tin varieties also can be brazed with the RBCuZn filler metals.

The phosphor bronzes are sometimes in the form of powder metal compacts. Before brazing, these compacts require pretreating

Table 18.5
Tensile Strength of Single-lap Brazed Joints in Copper

Brazing Filler Metal	Tensile Strength			
	Tough Pitch Copper		Deoxidized Ccopper	
	70°F (21°C)	-321°F (-196°C)	70°F (21°C)	-321°F (-196°C)
	KSI Mpa	KSI Mpa	KSI Mpa	KSI Mpa
Type BAg,ksi	18 - 20	24 - 30	17 - 19	10 - 11
MPa	124 - 138	165 - 207	117 - 131	69 - 76
Type BCuP,ksi	18 - 20	17 - 22	18 - 20	10 - 11
MPa	124 - 138	117 - 152	124 - 138	69 - 76

NOTES: Specimens were made from 0.25 in. (6.4 mm) thick sheet; joints had an overlap of 0.15 (3.8 mm) and no fillet.

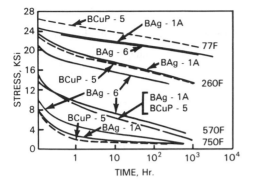

Figure 18.4 — Stress-Rupture Strength of Copper Brazed With Several Filler Metals Using Modified Plug and Ring Creep Specimens as a Function of Annealing Time. At 77 and 260°F (25 and 127°C) Failures Were Largely in Copper (Ref. 3).

by painting the surface with a water base or oil base colloidal graphite suspension, followed by baking at a low temperature, cleaning, and degreasing. This procedure seals the pores so that brazing can be performed.

Copper-Aluminum Alloys (Aluminum Bronzes)

Aluminum bronzes can be brazed with silver-bearing brazing filler metals and AWS Types FB3-D, -I, and -J fluxes. Formation of refractory aluminum oxide at brazing temperature in alloys containing more than 8% aluminum presents difficulties. However, this problem can be avoided by electroplating at least 0.0005 in. (0.013 mm) of copper

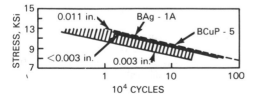

Figure 18.5 — Comparison of Torsional Fatigue Strengths of Two Brazing Filler Metals. Joint Clearances From 0.003 to 0.011 in. (0.076 to 0.279 mm) Give Little Difference in Fatigue Strength of Joint (Ref. 3.)

on surfaces to be brazed. To furnace braze unplated surfaces, flux should be used with a protective atmosphere (e.g., Type 4 and 5).

Copper-Silicon Alloys (Silicon Bronzes)

Copper-silicon alloys should be cleaned and then flux coated or copper plated before brazing to prevent the formation of refractory silicon oxide. Mechanical cleaning is recommended. For light oxidation, the material can be pickled, as noted earlier. Silver-bearing brazing filler metals and AWS Types FB3-A, -C, and -E fluxes are generally used.

Silicon bronzes are subject to intergranular penetration by the filler metal and to hot-shortness under stress. They should be stress relieved before brazing, and then brazed below 1400°F (760°C).

Copper-Nickel Alloys

Copper-nickel alloys generally are brazed with BAg filler metals. The base metal must be thoroughly evaluated for performance and microstructure before using filler metal of the BCuP class for joining. It is possible for a brittle nickel phosphide to form during brazing. The base metal must be free of sulfur or lead, which might cause cracking during the brazing cycle. Standard solvent or alkaline degreasing procedures may be used. Oxides can be removed by abrading or by pickling in hot 5% by volume sulfuric acid and immediately and thoroughly rinsing. AWS Types FB3-A, -C, and -E fluxes are suitable for most applications.

Copper-nickel alloys in the stressed condition are susceptible to intergranular penetration by molten filler metal. To prevent cracking, they should be stress relieved before brazing. Stresses should not be introduced during brazing.

Copper-Nickel-Zinc Alloys (Nickel-Silvers)

Nickel silvers can be brazed readily with the same procedures used for brazing brass. When RBCuZn filler metals are used, however, great care is required because of the relatively high brazing temperatures. These alloys are subject to intergranular penetration by the filler metals unless they are stress relieved before brazing. The poor thermal conductivity of these alloys tends to lead to local overheating. Slow and uni-

Table 18.6
Mechanical Properties of 0.83% Chromium Copper After Simulated Brazing Cycles

Brazing Cycle	Tensile Strength				0.2% Yield Strength				Elongation in 1 in. %		Vickers Hardness No.	
	As Cooled		Aged*		As Cooled		Aged*					
	ksi	MPa	ksi	MPa	ksi	MPa	ksi	MPa	As-Cooled	Aged*	As-Cooled	Aged*
	1110°-1190° (599°-643°C)											
1 min.	68.1	469	66.1	456	61.8	426	56.2	387	17	17	139	132
4 min.	56.4	389	56.7	391	42.5	293	43.9	303	19	19	122	113
16 min.	44.8	309	44.6	308	23.1	159	21.7	150	34	45	86	85
	1200°-1290° (649°-599°C)											
1 min.	60.3	416	58.7	405	47.0	324	47.0	334	23	21	125	116
4 min.	44.6	308	45.2	312	22.4	154	22.8	157	—	31	94	85
16 min.	42.3	292	42.3	292	17.9	123	17.9	123	35	38	85	78
	1290°-1380° (699°-719°C)											
1 min.	43.0	246	43.9	303	21.3	147	22.0	152	31	36	86	86
4 min.	42.8	295	42.1	290	17.7	122	17.5	121	37	39	76	79
16 min.	41.4	285	41.2	284	15.9	110	15.0	103	41	39	77	74
	1380°-1470° (749°-799°C)											
1 min.	42.8	295	43.5	300	18.8	130	18.1	125	36	39	75	78
4 min.	41.2	284	42.1	290	15.2	105	15.4	106	42	42	72	74
16 min.	41.2	284	41.2	284	13.4	92	13.4	92	46	45	69	70
	1470°-1560° (799°-849°C)											
1 min.	41.2	284	45.0	310	14.6	101	20.6	142	42	38	70	88
4 min.	40.8	281	47.5	328	12.5	86	25.8	178	41	35	67	97
16 min.	40.5	279	46.4	320	11.4	79	21.9	151	47	36	64	91
	NONE (solution heat treated and aged*)											
	76.6 ksi		528 MPa		70.3 ksi		485 MPa		16		152	

* Aged at 840° (449°C) for 8 hours.

form heating is recommended and sufficient flux should be used to prevent oxidation.

Dissimilar Metals

Dissimilar combinations of the copper alloys can be brazed readily. Copper alloys in combination with steel, austenitic stainless steel, or nickel also can be brazed using BAg filler metals. Suggested brazing filler metals for dissimilar metal systems are given in Table 18.8. Brazing copper alloys to aluminum and magnesium alloys is not practical. In cases where base metals have widely dissimilar of thermal expansion characteristics, the design and production of satisfactory brazed joints may be extremely difficult or impractical.

SAFETY

Some brazing filler metals contain zinc and cadmium, some fluxes contain fluorides,

Table 18.7
Mechanical Properties of 0.16% Zirconium Copper After Simulated Brazing Cycles

Brazing Cycle	Tensile Strength				0.2% Yield Strength				Elongation in 1 in. %		Vickers Hardness No.	
	As Cooled		Aged*		As Cooled		Aged*					
	ksi	MPa	ksi	MPa	ksi	MPa	ksi	MPa	As-Cooled	Aged*	As-Cooled	Aged*
					1110°-1190° (599°-643°C)							
1 min.	56.4	389	56.2	387	50.8	350	48.6	335	17	17	123	126
4 min.	40.3	278	39.4	272	19.4	134	18.8	130	33	42	74	74
16 min.	39.0	269	38.1	263	15.4	106	14.1	97	42	46	72	67
					1200°-1290° (649°-699°C)							
1 min.	41.0	283	44.1	304	21.5	148	26.0	179		39	74	76
4 min.	38.5	265	37.4	258	15.0	105	13.0	90	44	46	68	66
16 min.	39.9	275	37.2	256	13.4	92	12.8	88	49	46	70	64
					1290°-1380° (699°-749°C)							
1 min.	38.5	265	37.4	258	14.3	99	13.9	96	46	44	70	69
4 min.	37.6	259	37.4	258	13.6	94	12.8	88	46	47	67	59
16 min.	37.9	261	37.0	255	11.6	80	11.6	80	46	46	63	60
					1380°-1470° (749°-799°C)							
1 min.	38.3	264	37.0	255	12.1	83	12.5	86	-	45	68	66
4 min.	37.6	259	36.5	252	10.5	72	9.4	65	44	44	64	61
16 min.	37.4	258	36.7	253	9.6	66	9.8	68	44	49	62	55
					1470°-1560° (799°-849°C)							
1 min.	37.2	256	37.0	255	9.2	63	9.2	63	44	46	59	61
4 min.	37.4	258	36.7	253	8.5	59	8.7	60	43	48	60	55
16 min.	38.3	264	37.0	255	8.5	59	8.7	60	45	47	59	51
					NONE (heat treated and aged*)							
	68.3 ksi		471 MPa		63.1 ksi		435 MPa		17		155	

* Aged at 840° (449°C) for 8 hours.

and some base metals contain beryllium, lead, or zinc. When these materials are used, adequate ventilation must be provided to protect personnel (Chapter 6.) Do not attempt to braze without consulting *Safety in Welding and Cutting*, American National Standard Z49.1 (latest edition). In addition, *Hygiene Guide Series*, American Industrial Hygiene Association, is a useful data source.

Table 18.8
Braze filler metal selection chart for joining dissimilar metals

	Cu and Cu Alloys	Carbon and low alloy Steels	Cast Iron	Stainless Steel	Ni and Ni Alloys	Ti and Ti Alloys	Other Reactive Metals	Refractory metals	Tool Steels
Cu and Cu Alloys	BAg, BAu BCuP, R-BCuZn	BAg, BAu, RBCuZn	BAg, BAu, RBCuZn	BAg, BAu	BAg, BAu, RBCuZn	BAg	BAg	BAg	BAg, BAu, RBCuZn, BNi

Keywords — precious metals, gold and gold alloys, platinum group metals, silver and silver alloys, inspection of brazed precious metals

Chapter 19

PRECIOUS METALS

Pure and alloyed precious metals are desirable in industrial applications for electrical and electronic contacts because they have (1) good electrical and thermal conductivity, (2) resistance to corrosion, (3) high melting temperatures, (4) suitable behavior under arcing conditions and (5) freedom from sticking. In dental applications, precious metals offer resistance to corrosion, strength, bio-compatibility, workability, and in some cases, aesthetics. The jewelry industry uses precious metals as an art medium of high intrinsic value primarily because of aesthetics. Corrosion resistance, however, is another consideration.

Brazing is the most common method of joining precious metals to precious metals and to other base metals. The brazing process, which utilizes the flow of the filler metal between the fitted joint surface, is ideal for joints between electrical contacts and the support conductors. Such brazements provide good thermal and electrical conductivity plus adaptability for joining many combinations of contacts and base metals. Production brazing of electrical contacts is a highly specialized field where the quantity produced may greatly influence the selection of materials and techniques. Examples of finished electrical contact parts are shown in Figure 19.1.

Although it is usually called *soldering* when employed in the jewelry or dental fields, brazing can provide invisible joints free from surface imperfections in which foreign materials can become lodged.

MATERIALS

Gold and Gold Alloys

Commonly used karat gold alloys contain additions of copper, silver, zinc and sometimes nickel. If special electronic, physical, or chemical properties are required, platinum, palladium, iridium, rhodium, ruthenium, indium or other metals may be added to gold.

Platinum Group Metals

Well known for their corrosion resistance, the platinum group metals and alloys have numerous industrial uses. These include catalysts for chemical reactions, glass handling tools and equipment, heater elements and thermocouples. Combinations of platinum, palladium, iridium, rhodium and ruthenium are widely used, as are gold, silver, copper and nickel for alloying constituents.

Silver and Silver Alloys

Alloying elements may be palladium, copper, nickel, iridium, or indium. The jewelry industry most often uses sterling silver (92.5% silver and 7.5% copper). For strength, silver is sintered with nickel, iron, tungsten, molybdenum, and with certain nonmetallic materials such as cadmium oxide or graphite for use as electrical contacts.

Plated Materials

Materials plated with precious metals require special considerations for brazing ap-

Figure 19.1 — Brazed Electrical Contacts

plications. Plating thickness must be adequate to prevent excessive diffusion into the base metal. Plating discontinuities may result in joint failure in service.

BRAZING FILLER METALS

Silver (BAg), gold (BAu), copper-phosphur (BCuP) filler metals are used for brazing the contacts to the holders. Relatively low brazing temperatures are preferable unless elevated service temperatures are to be encountered. The filler metal is usually preplaced in the form of washers or discs between the contact and the holder. Wire forms, rings, or pastes are used similarly. Contacts are sometimes supplied preclad with brazing filler metal for convenience in assembly.

A filler metal containing phosphorus will provide better wetting of contact materials containing molybdenum and those of the silver-cadmium oxide type. Care must be taken, however, to prevent the flow of these filler metals onto the face of the contact. This is done by controlling the heating pattern and by avoiding the usage of excess flux and filler metal. Stopoff materials also may be used.

Filler metals for dental and jewelry applications are very similar in composition. They usually contain gold, silver, copper, zinc, and tin, and sometimes small amounts of other metals, such as germanium, to lower the melting points or to improve wetability. The common gold filler metals for dental applications are listed in Table 19.1. Filler metals usually contain a slightly lower percentage of gold (karat) than the materials to be joined. For example, a 75% gold (18 Kt) alloy may be joined with a 65.4% gold filler metal. Sterling silver jewelry alloys are usually joined with filler metals from the BAg group.

Platinum group alloys commonly are brazed with filler metals from the BAu group. Filler metals of the platinum-copper type are also available. In jewelry applications, platinum or palladium alloy parts are frequently joined with karat gold filler metals.

JOINT DESIGN

Joint design for precious metal assemblies follows the same principles used for other braze joints (see Chapter 2). Typically, dental parts employ butt joints. Jewelry parts may be joined using all types of joint designs, generally with loose clearances. Electrical contacts require joints which will provide either good contact with flat surfaces on the support, or matching curved surfaces to provide the proper capillary action for brazing. Optimum electrical conductivity of the braze joint is obtained with minimum joint clearance.

PROCESSES AND EQUIPMENT

Most methods of heating may be used for brazing precious metal parts (see Chapter 1). Manual torch brazing frequently is used for large assemblies or where production requirements are limited, such as in dentistry and jewelry.

Induction heating is adaptable to many production applications. Induction brazing procedures are described in Chapter 11.

Resistance brazing can be accomplished using resistance welding equipment, although lower currents and longer times are required than for resistance welding. For example, heating times may vary from one to six seconds for brazing. Resistance brazing may also include use of carbon block electrodes and low voltage transformers.

Furnace brazing is employed in the dental laboratory when the parts being joined have been porcelain coated. Electrical contact assemblies are also furnace brazed, particularly where one or more contacts are to be brazed on each assembly. A contact assembly that is self-positioning is desirable for furnace brazing, otherwise suitable fixtures are required to position the work. Volatile elements such as zinc should be avoided to prevent contamination of the surface of precious metal contacts during long brazing cycles.

PRECLEANING AND SURFACE PREPARATION

As in all brazing, best results are obtained when the parts are precleaned and this cleanliness is maintained throughout all subsequent operations. Jewelry and dental materials are usually degreased and sometimes etched in dilute acid solutions. Nitric acid is never to be used with silver alloys.

Most silver-nonmetallic contacts, such as those of silver-cadmium oxide, generally are supplied with a backing of pure silver to facilitate brazing. In an alternative method,

Table 19.1
Composition and Melting Points of Dental Gold Filler Metals

Solder No.	Gold (%)	Silver (%)	Copper (%)	Zinc (%)	Tin (%)	Melting Range (°C)	(°F)
A	65.4	15.4	12.4	3.9	3.1	745-785	1375-1445
B	66.1	12.4	16.4	3.4	2.0	750-805	1385-1480
C	65.0	16.3	13.1	3.9	1.7	765-800	1410-1470
D	72.9	12.1	10.0	3.0	2.0	755-835	1390-1535
E	80.9	8.1	6.8	2.1	2.0	745-870	1375-1595

from SKINNER'S SCIENCE OF DENTAL MATERIALS, RALPH W. PHILLIPS, MS, D.Sc, WB SAUNDERS COMPANY, PHILA. SEVENTH EDITION 1973

the cadmium oxide is removed from the faying surfaces by chemical leeching. For silver-graphite contacts, the graphite may be removed by firing the contact at elevated temperatures [about 1500°F (816°C)] in an oxidizing atmosphere (air). The surfaces that will serve as the contacts should be protected during cleaning to avoid removing nonmetallic constituents. The treated surface then is suitable for brazing, although a coining operation may be advisable to compact the area from which the nonmetallics have been removed. Silver-backed contacts are easy to clean and braze; they are preferred over other types for economical reasons.

FLUXES AND ATMOSPHERES

Controlled atmosphere brazing without flux is used extensively in production because it eliminates the need for post-braze cleaning. The controlled atmosphere to be used is determined by the base materials to be joined (see Chapter 4).

Fluxes are used when brazing in air and in an atmosphere furnace if either zinc or cadmium is present in the precious metal, base metal, or filler metal. Zinc and cadmium generate toxic fumes (see Chapter 6). Any tarnish or surface oxide is much easier to remove from precious metals than from nonprecious base or filler metals. A wide selection of flux choices is available because precious metals do not readily oxidize. For example, when brazing silver-cadmium oxide to steel, any flux or atmosphere suitable for the steel will be more than adequate for the precious metal. This holds true for most other applications where precious metals are to be joined to other materials.

ASSEMBLY PROCEDURES AND TECHNIQUES

A problem encountered in many brazing operations is heating of both members to the brazing temperature simultaneously. In some cases, especially when joining a small precious metal contact to a large holder, the objective may be to avoid directly heating the contact and allow it to be heated by conduction. Pressure applied to the contact will assist seating as the brazing filler metal melts and flows. To maintain rigidity of the contact holders or supports, it may be necessary to avoid annealing those members. For such applications, heating must be localized in the joint area.

Dissolution of the precious metal by the brazing filler metal should be avoided. This action can be minimized by controlling the amount of filler metal used, avoiding excessive heating, limiting time at the brazing temperature, and striving for uniform heat distribution.

Because of the variable shapes of dental parts, fixtures must conform to the parts. The fixturing method commonly used for joining parts of a dental apparatus is called "investing". The parts to be joined are first aligned and temporarily held in place with a wax at the joint. The attached parts are then covered by a plaster material, or investment, leaving the joints exposed. After the investment has solidified, the wax and any extra investment is removed. Then the parts are ready for brazing. An example of invested parts being readied for the brazing operation is shown in Figure 19.2. Figure 19.3 shows examples of dental bridge work.

POST-BRAZE OPERATIONS

Heat treatments of brazed assemblies are often performed during the brazing or cooling cycle by selecting an appropriate filler metal with a suitable brazing temperature range. In some cases, however, heat treatments are performed after brazing at temperatures below the solidus for the filler metals.

When flux is employed in the brazing of precious metal parts, it is necessary to remove the residue after brazing. Any of the flux removal methods discussed in Chapter 4 may be used. It is important to avoid solutions containing nitric acid during all phases of the operation when dealing with materials which contain silver or copper.

Stopoff materials may be removed by mechanical means (brushing or blasting), by water or steam flushing, or by acid pickling. The method depends on the type of stopoff used. Oxides resulting from the brazing process are usually removed by a

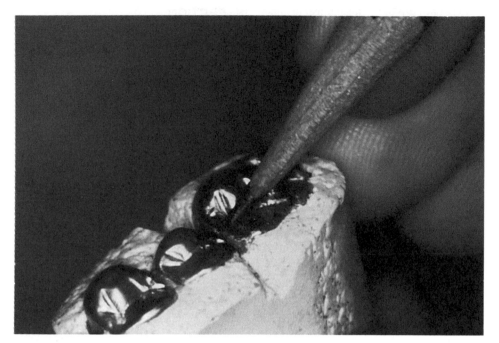

Figure 19.2 — Application of Graphite As Stop-Off To Invested Dental Bridge Before Brazing

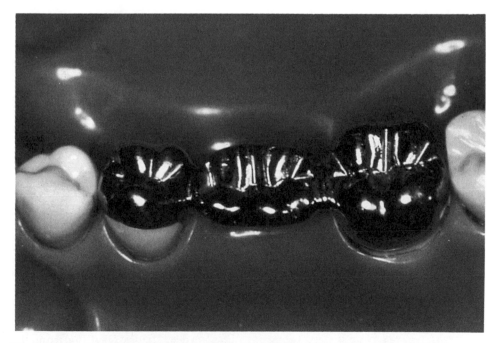

Figure 19.3 — Finished Three Unit Bridge. (Photo Courtesy of Dr. Richard Oliva)

chemical bath dip or by mechanical means. Typically, dental and jewelry assemblies are burnished or polished to enhance the appearance.

INSPECTION

In most cases, the inspection method for brazed precious metal parts is visual (as discussed in Chapter 16). In some cases, inspection has been successfully achieved with ultrasonic techniques and radiography.

Keywords — low carbon, low alloy steels, tool steel, applications for low carbon low alloy steels

Chapter 20

LOW CARBON, LOW ALLOY, AND TOOL STEELS

Brazing of low carbon, low alloy, and tool steels is a widely used and highly developed production process. The brazing filler metals often are used with flux, and manually or automatically applied or placed in the joint. Heat treating characteristics of these base metals help determine the specific filler metal and brazing temperature to be used. The parts normally are brazed using one of several heat sources, including torch, induction, furnace, infrared, and resistance heating.

LOW CARBON AND LOW ALLOY STEELS

By definition, low carbon steel and low alloy steel fall into the steel group that contains about 0.10% carbon with a fractional percentage of alloying elements such as nickel, manganese, copper, and chromium. The earliest recorded uses of these materials indicates they were brazed with copper, silver, and gold alloys, often to produce a tool or weapon. Today, low carbon and low alloy steels are used in a wide variety of applications.

Base Metals

Low carbon steels (less than 0.30% carbon) and low alloy steels (less than 5% total alloy content) are used for various fabricated shapes as well as for castings and forgings.

Free-machining steels can be brazed using filler metals and flux combinations normally recommended for steel. Low-carbon free-machining steel containing lead can be torch brazed with Type BAg filler metals and AWS Type FB3 flux (Chapter 15). Satisfactory furnace brazed joints can be made in leaded steels using BCu and BNi filler metals. Free machining steels containing manganese or bismuth can be furnace brazed using special techniques such as preplating with copper, nickel, or iron. Consideration must be given, however, to such metallurgical aspects as the vaporization and melting temperatures of the lubricant compounds. Consequently, the heat source and heating parameters must be selected carefully.

Braze Filler Metals

All the BAg classifications can be used for brazing ferrous metals. Silver-base filler metals containing nickel usually provide better wettability, and are preferred for brazing certain low alloy steels where joint strength is most important. The BCu group is used primarily for preplacement in controlled-atmosphere furnace brazing. Types RBCuZnA and RBCuZnD filler metals often are used for face feeding applications, but can be preplaced for furnace and induction brazing. Selecting braze filler metal with a high solidus temperature may permit simultaneous brazing and heat treating of some low alloy steels. The BNi brazing filler metals are used for joining ferrous metals when special joint requirements are required. The selection of braze filler metals will ultimately depend on the end-use of the product as well as metallurgical, mechanical, and economic factors.

Joint Design

Joints should be close-fitting and properly designed (refer to Chapter 2). Joint clearances of 0.002 to 0.005 in. (0.05 to 0.13 mm) produce the best mechanical properties with most filler metals when using a mineral flux. Light press fits are recommended with the BCu filler metals when furnace brazing. Joint clearances for the BNi group and BAg group range between 0.000 to 0.005 in. (0.0 to 0.13 mm) depending upon the alloy type. (See Table 7 in Chapter 2.)

It is important to consider the thermal expansion characteristics of the parts being joined to determine the joint clearance at brazing temperature for the selected filler metal. Filler metals with relatively narrow melting ranges are required for close-fitting joints. Conversely, filler metals with wide melting ranges have good bridging characteristics when wide clearances are involved. Furnace dew point can be used to control the fluidity of BCu filler metal when brazing joints with wide gaps. Furnace variables, such as time, temperatures, dew point, and heating rates, can be used to control the fluidity of BCu and some BNi and BAg braze filler metals.

Processes and Equipment

Low carbon and low alloy steels can be brazed using most heating methods. Torch, furnace, and induction heating techniques are the most common. Filler metals in the form of continuous wire or strip can be applied automatically using electro-mechanical wire feeders; powder filler metals blended with flux and paste-forming ingredients are applied automatically with pressurized dispensing equipment. For torch brazing, the equipment would include standard oxyacetylene or propane torches. Furnaces of the batch or conveyor type, with or without control of the atmosphere, can be used. They can be electric, gas, or oil-fired, and should be equipped with accurate temperature controls. (Refer to Chapter 4 for specific information on fluxes and atmospheres.)

Precleaning

For best results, parts should be cleaned thoroughly prior to brazing. Organic and inorganic contaminants must be removed from areas that are to be locally brazed. For furnace brazing entire parts must be thoroughly cleaned.

The procedures used for cleaning will depend upon the extent and types of contaminants present on the part. Many different cleaning procedures are used, including mechanical, chemical, and electro-chemical processes. Often combinations of the three are used. Neither fluxes nor furnace atmospheres can be considered as the "cleaners" of dirty parts.[1]

There are several recommended techniques for determining the degree of cleanliness. They include solvent wipe tests, where evidence of soil may indicate the need for additional cleaning, and water-break testing. Water will form a continuous film on surfaces that are free of organic contaminants, and will bead on surfaces that have organic residue. Many braze problems can be attributed to lack of cleanliness.

Fluxes and Atmospheres

Flux or a proper atmosphere is required when brazing steel. Selection will depend upon the filler metal type. For example, AWS brazing flux Types FB3A, FB3B, and FB4 are suitable for BAg filler metals. Types FB4 and FB5 normally are used with the RBCuZn filler metals. Fluxes and some atmospheres may be used together for certain applications.

The flux can be in either paste or powder form, or combined with the filler metal. In face-fed operations, hand-held filler metal may be coated with the appropriate flux. In furnace and induction brazing, the filler metal is preplaced in or near the joint. Maximum furnace temperature as well as time at temperature must be controlled to ensure proper melting and flow of the filler metal into the joint. (Refer to Chapter 4 for com-

---

1. For additional information regarding cleaning of steels for brazing, refer to Volume 5 of the *Metals Handbook*, 9th Edision, and Volume 2 of the *Metals Handbook*, 8th Edition published by American Society for Metals.

prehensive data on fluxes and atmospheres.)

Procedure and Technique

In torch brazing, a neutral or slightly reducing flame is usually preferred. The filler metal is face-fed into the prefluxed joint. Flux-coated filler metal often is used. As in all brazing processes, it is important not to overheat during brazing to prevent undesirable metallurgical effects on the base metal, the filler metal, or the flux. Time at temperature is an important consideration, especially when the filler metal contains volatile elements, such as zinc and cadmium.

Excess heat may also affect the integrity of the braze and mechanical properties in the joint. Automated torch and burner-type production equipment is available for high production applications. These units usually use brazing fluxes and BAg filler metals.

In production applications, the filler metal (often BCu or BNi) is preplaced in, or adjacent to, the joint before the preassembled parts are moved into a controlled-atmosphere, batch or conveyor type furnace for brazing. Induction heating for brazing can be used to control the maximum temperature to the immediate joint area. This is done by selective coil design and the choice of an appropriate current frequency for the induction heating circuit.

In joining certain hardenable low alloy steels, it may be desirable to use low-melting BAg filler metals. Such steels can be brazed below their lower transformation temperatures. However, some local annealing may occur. When postheat treatments are required, higher melting filler metals are necessary to preclude the possibility of braze remelt. Low alloy steels are often brazed and heat-treated simultaneously using filler metals of the RBCuZn and BAg classifications. The solidus temperature of the selected filler metal must be above the austenitizing temperature recommended for the base metal. In an application of this type, the joint is made at normal brazing temperature, removed from the heat source to permit a drop to the hardening temperature, and then quenched. The procedure is satisfactory only for base metals compatible with rapid cooling.

Braze components are best designed to be self-fixturing — that is, capable of being self-positioning and self-supporting without the use of extraneous fixturing. Such techniques as staking, crimping, knurling, swaging, and tack welding usually can ensure adequate positioning.

Always observe safety precautions specified by the supplier of braze filler metal and equipment. (Also, see Chapter 6.)

Postbraze Operations

When flux has been used, all residue must be removed. Quenching a brazement in water while still hot produces sufficient thermal shock to accelerate removal of the flux residue, providing the base metals are compatible with such treatment. Washing and brushing in warm or hot water should be followed by a mild chemical dip, such as a dilute sulfuric acid solution, followed by a water rinse. If the brazement is to be painted or plated, specific cleaning procedures might be required for good adhesion. When protective atmospheres without the use of flux are used properly in furnace brazing, no postbraze cleaning operation is needed if the parts come out bright and clean. For brazements requiring a postbraze heat treatment, the effects of heating and quenching on the joint integrity must be considered.

Inspection

Many inspections of brazed joints are visual only. Inspecting at low magnification usually will be satisfactory for detecting surface voids, flux inclusions, and general wetting and flow of filler metal. There are also a number of nondestructive tests for brazed joints (Chapter 16). For certain production brazing applications, destructive tests are recommended in order to establish final manufacturing parameters.

Preproduction workmanship samples and braze schedules can provide criteria for accepting or rejecting parts.

Typical Applications

A variety of applications for the brazing of low carbon and low alloy steels in everyday production include many components for automobiles, trucks, bicycles, motorcycles, snowmobiles, all-terrain vehicles, and the like. Other common brazements include window and door frames, ducts, tanks, containers of all types, perforated and expanded steel panels, steel partitions, and shelving. A great variety of tubular steel furniture is also manufactured using brazing. Cutting tools and knives, hydraulic tanks, reservoirs, electronic chassis and supports, hand tools, honing appliances, instruments, and steel assemblies of all types also are routinely brazed.

TOOL STEELS

Tool steels are used in the fabrication of cutting tools and dies where high hardness and resistance to wear are prime requisites. Frequently, a cutting or wearing surface may be only a small section that must be joined to the tool assembly. Brazing is one commonly used method for doing this.

Base Metals

A classification system that groups tool steels of similar properties has been developed by the American Iron and Steel Institute (AISI) and the Society of Automotive Engineers. Compositions of several typical tool steels with their AISI designations are given in Table 20.1.

For brazing, it is convenient to group tool steels in two broad classifications: carbon tool steels and high-speed tool steels. Carbon tool steels depend primarily on high-carbon content (0.60-1.40%) for hardness. Except for thin sections, these steels must be quenched rapidly during heat treatment to achieve optimum properties. Alloying elements may be added to the carbon tool steels to impart special properties, such as reduced distortion on heat treatment, greater wear resistance, improved toughness, and better high-temperature properties. Such steels are referred to as alloy tool steels. They are known by various trade names and grades, and their properties are well documented in manufacturers' published information and in various handbooks.

High-speed tool steels, although logically falling in the alloy steel group, are classified separately. Their properties depend on relatively high percentages of certain alloy elements including tungsten, molybdenum, chromium, and vanadium. The carbon content of these steels is normally much lower than that of the carbon tool steels. High-speed tool steels are widely used in industry as metal cutting tools. A common type

Table 20.1
Commonly Used Tool Steels

AISI Designation	Type	Typical Composition, % (Balance Iron)						
		C	Mn	Si	Cr	V	W	Mo
W1	Water hardening	0.60-1.40	....	....	....	...	....	....
W2	Water hardening	0.60-1.40	....	....	....	0.25	....	....
S1	Shock resisting	0.60	....	....	1.50	....	2.50	....
S5	Shock resisting	0.50	0.80	2.00	....	....	....	0.40
01	Oil hardening	0.90	1.00	....	0.50	....	0.50	....
A2	Air hardening Med. alloy	1.00	....	....	5.0	....	....	1.00
D2	High carbon-high chromium	1.50	....	....	12.0	....	....	1.00
H12	Chromium hot work	0.35	....	....	5.0	0.40	1.50	1.50
H21	Tungsten hot work	0.35	....	....	3.5	....	9.50	....
T1	Tungsten high speed	0.70	....	....	4.0	1.00	18.0	....
M1	Molybdenum high speed	0.80	....	....	4.0	1.00	1.50	8.50

known as 18-4-1 or T1 contains 18% tungsten, 4% chromium, and 1% vanadium.

Hard sintered carbides also are used extensively as cutting tools but do not fall under the classification of steels. Brazing of carbides is discussed in Chapter 30.

Filler Metals

The choice of brazing filler metal to be used depends on the properties of the tool steel being brazed and the heat treatment required to develop its optimum properties. Most brazing filler metals of the BAg, BCu, and RBCuZn classifications are used to braze tool steels. The best filler metal for the specific application should be determined during development of the initial braze schedule and preproduction testing.

Joint Design

Lap or sleeve type joints generally are used in brazing tool steels. With BAg or RBCuZn filler metals, clearances of 0.002 to 0.005 in. (0.05 to 0.13 mm) at the brazing temperature are preferred. Preplacing the filler metal where it is protected from direct contact with the heat source is recommended. Methods of preplacing are described in Chapter 2. Where brazing and hardening operations are combined, the joint should be designed to maintain the brazing filler metal under compression during quenching. Tensile stress on the joint at temperatures near the melting point of filler metal may cause cracking.

Process and Equipment

Tool steels can be brazed using most brazing process but torch, furnace, and induction brazing are most commonly used. The available equipment and the quantity and size of parts are frequently the main factors in selecting the brazing process.

Surface Preparation

Base metals must be clean and free from oxides and organic contaminants to achieve proper flow of the filler metal. A machined or roughened surface is preferable to a smoothly ground or polished surface. Flux and braze filler metal do not flow well on very smooth surfaces. Capillary flow is enhanced greatly when the surface is roughened. Abrasive blasting is often used to prepare a finish machined surface for brazing.

Fluxes and Atmospheres

In general, Types FB3A and FB3B fluxes are used for brazing tool steels. Some modification of the flux may be required depending upon the particular tool steel used and the brazing temperature required.

Controlled atmosphere furnaces are used extensively to prevent oxidation during heating and to avoid the necessity of postbraze cleaning. With a controlled atmosphere, steps must be taken to guard against either carburizing or decarburizing the tool steel.

Brazing Technique

Brazing of tool steel is best done prior to or in conjunction with the hardening operation. The hardening temperature for carbon steels is normally in the range of 1400° to 1500°F (760° to 815°C). When brazing is done prior to the hardening operation, a filler metal is customarily used which has a solidus temperature well above this range. With this procedure, the brazed assembly will have sufficient strength when reheated to the hardening temperature to avoid joint failure. Frequently, a BCu filler metal is used for this purpose. At times, however, the high temperature required for copper brazing may adversely affect the structure of the steel. Silver (BAg) and copper-zinc (RBCuZn) filler metals are available that can be used at brazing temperatures of 1700° to 1800°F (925° to 980°C).

When brazing and hardening are combined, a filler metal is often used with a solidus close to the quenching temperature. In this procedures, attention should be focused on the brazement the design and handling procedures to avoid cracking. Joint strength is extremely low at the quenching temperature.

The steel being brazed should be studied carefully to determine (1) the proper heat treating cycle, (2) the kind of quench necessary (water, oil, or air), (3) the optimum brazing filler metal, and (4) the proper technique for the combined heat treating and brazing operation to achieve desired properties and service life.

Quenching can set up severe temperature gradients which produce differential expansions and contractions that may rupture brazed joints. Austenite to martensite transformations in certain steels, that can take place during the brazing cycle or heat treatment, or both, may result in the parts contracting, expanding, finally contracting again. When the brazed joint is supported properly during quenching, a filler metal which will melt below the transformation temperatures of the steel can be used successfully.

The high-speed tool steels require hardening treatments at temperatures above the usual BAg brazing temperatures. It is common practice to harden these steels prior to brazing, and then braze during or after the secondary tempering treatment. Tempering is usually done in the range of 1000 to 1200°F (540 to 650°C). Filler metals, such as BAg-1 or BAg-1a, can be used for brazing at temperatures above 1150°F (620°C). If short brazing cycles are used, hardened high-speed tool steel may be brazed without over-tempering.

Broken high-speed tools, such as broaches, circular saws, and milling cutters, may be repaired by brazing. This can minimize long delays in production while seeking a replacement.

The fundamentals of brazing outlined elsewhere apply to the tool steels and should be observed. Since each application will require a knowledge of the service requirements as well as the particular mechanical and metallurgical needs, it is not practical to outline specific procedures here.

Always observe safety conditions outlined by suppliers of braze filler metal and equipment.

Typical Applications

Figure 20.1 shows a furnace brazed assembly of 1.5% carbon, 12% chromium tool steel. The joint is at the juncture of the rod and boss on the ribbed disc. A copper brazing filler metal was used in a hydrogen atmosphere. Figure 20.2 shows a furnace brazed assembly of 1% carbon, 5% chromium, and 1% molybdenum tool steel. A BAg filler metal was used in a hydrogen at-

mosphere. Hardening was accomplished in conjunction with the brazing operation. Figure 20.3 shows a photomicrograph of the brazed joint in the assembly in Figure 20.2. Figure 20.4 shows the torch brazing of a tool steel assembly.

Figure 20.1 — Furnance Brazed Alloy Tool Steel

Figure 20.2 — Alloy Tool Steel Assembly

Figure 20.3 — Photomicrograph Of Joint In Figure 20.2

Figure 20.4 — Torch Brazing High Alloy Tool Steel

Keywords — cast iron brazing, preparation of cast iron for brazing, metallurgical considerations in brazing cast iron, brazing of gray cast iron to steel, brazing of cast iron to brass

Chapter 21

BRAZING OF CAST IRON

INTRODUCTION

The brazing of gray, ductile, and malleable cast irons differs from brazing of steel in two principal respects: a special precleaning method is necessary to remove graphite from the surface of the iron, and the brazing temperature should be kept as low as feasible to avoid reduction in the hardness and strength of the iron.

The processes used for the brazing of cast irons are the same as those used for the brazing of steel, including furnace, torch, induction, and dip brazing. Like other metals, selection of the brazing process largely depends on the size and shape of the assembly, the quantity of assemblies to be brazed, and the equipment available. There are many applications in which it is desirable to braze gray, malleable, and ductile cast irons either to themselves or to other metals. White cast irons are seldom brazed.

PREPARATION OF CAST IRON FOR BRAZING

Gray cast iron is characterized by the presence of graphite flakes in its microstructure. For many service applications, the graphite flakes are beneficial in that they enable the cast iron to dampen mechanical vibrations, to produce small chips when machined, and to have self-lubricating sliding surfaces. In other instances, however, the presence of graphite flakes is harmful because they cause the alloy to be weak in tension and to be brittle. The machining of gray cast iron surfaces smears the surfaces with graphite, making it difficult to wet the surfaces with brazing filler metal. During brazing, silver and copper filler metals do not wet the graphite flakes or nodules. Graphite must be removed from the surfaces prior to brazing. Filler metals that contain chromium, titanium, or other carbide formers will wet and bond to the graphite. Filler metals such as BNi-2 can be used without the normal cleaning process to remove the graphite.

Some high silicon cast irons may have a silicon oxide on the surface of the casting which will also prevent wetting by the filler metal. These surfaces must be cleaned by one of the methods described below. When using metals that do not contain carbide formers, smeared graphite and graphite flakes must be removed from the surfaces to be brazed. There are several processes that may be used, depending on the parts application and the cost.

Treatment by Flame

An oxidizing oxyacetylene flame is normally applied to heat the surface and, at the same time, to oxidize surface graphite. Heating is then continued with a slightly reducing flame to reduce any iron oxide formed during the first heating. The treatment by flame, if conducted properly, will permit satisfactory brazing of a joint.

Chemical Treatment

In this process, components are generally immersed for 15 minutes in a fused 50:50 solution of sodium and potassium nitrate salts. The temperature of the salt bath is kept at 662° to 752°F (350° to 450°C) by external heating. The surface graphite is oxidized by the salts, leaving the metal with a light film of iron oxide over it. The samples are then cooled and washed. This is followed by pickling for about one minute in a 10% by volume solution of commercial hydrofluoric acid (50-60% strength) to remove the iron oxide film.

Electrolytic Treatment

This is the most effective cleaning treatment, despite its high cost, and should be used in applications where high quality is most important. In this process, a cast iron part is suspended in a steel crucible filled with a salt bath having the composition: NaOH-75%, NaC1-5%, NaF-5%, Na_2CO_3-14%, and K_2CO_3-1%. The bath is operated at 860° to 914°F (460° to 490°C) by controlling the heating current supplied to the crucible furnace. Direct current is applied between the crucible and part, and by changing polarity, the action of the molten bath is changed from oxidizing to reducing. When the part is made the anode, its surface is oxidized and CO that forms from the graphite is absorbed by the NaOH in the bath. Then when the part is made the cathode, iron oxide on its surface is reduced to iron. Different samples are electrolyzed at different current densities in order to determine the optimum conditions for surface preparation. The total electrolyzing time is chosen by trial and error. The part is finally removed from the molten salt bath, immersed in hot water to rinse away the salt, and then dried.

In the case of ductile irons where the flakes are in nodular form, the wetting problem exists to a lesser extent than with gray cast iron, and the cleaning procedure is essentially the same.

METALLURGICAL CONSIDERATIONS

In the brazing of ductile and malleable irons, certain precautions are imperative. If ductile or malleable irons are heated above 1400°F (760°C), the metallurgical structure may be damaged; brazing should thus be performed below this temperature. Figure 21.1 shows the relationship of time at temperature to the changes in malleable iron microstructure; ductile iron behaves in a similar manner.

Filler Metals

With proper surface preparation, any filler metal suitable for use with iron or steel can be used for brazing cast iron. However, the lower melting BAg filler metals are preferred, since they produce minimum distortion of the assembly during brazing. Those containing nickel, such as BAg-3 and BAg-4, have better wettability on cast irons and therefore produce higher strength joints. Copper and copper-zinc filler metal can be used, but great care must be exercised because of their high brazing temperature ranges. Sometimes a nickel base filler metal is used, particularly for brazing dissimilar metals such as cast iron to steel. Brazing Flux Type FB3 is normally used with the recommended BAg filler metals.

Preheating

Preheating of the assembly can minimize brazing time and dries the flux when brazing in a furnace or salt bath. Because cast irons have relatively low thermal conductivities, temperature variations may be reduced by preheating the entire assembly to a temperature between 400 to 800°F (205 to 425°C). The formation of deleterious carbides is also minimized by preheating.

Joint Clearances

Normally joint clearances of 0.002 in. to 0.007 in. (0.05 to 0.18 mm) is recommended for brazing cast iron. Joint clearances should be determined for specific applications while considering the thermal expansion characteristics of the metals being joined, the methods of heating, type of filler metals, and other important factors.

Surface Plating

In spite of effective cleaning, graphite flakes sometimes remain on the cast iron surfaces to be brazed. Nickel plating the surface improves wetting and minimizes surface oxidation.

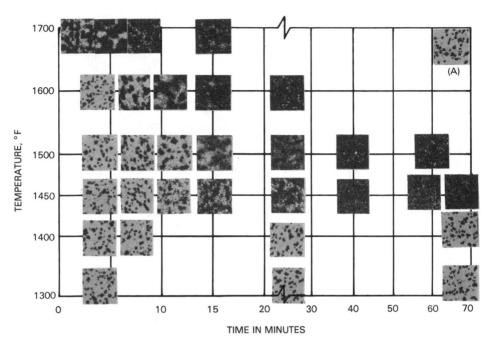

Figure 21.1 — Relationship of Time at Temperature to the Changes in Malleable Iron. Inset (A) (Top Right) Shows Typical Original Microstructure

Heating and Cooling

When a gray, malleable, or ductile cast iron is heated above its transformation temperature, the structure normally present begins to change into austenite. Upon rapid cooling, this phase changes either to martensite or to a fine pearlite structure having a cementite network. In either case, the metallurgical structure in the heat-affected zone becomes extremely unfavorable. The cast irons expand rather readily and conduct heat quite poorly. Thus, the heating and cooling cycles require careful consideration.

BRAZING PROCESSES

Any of the processes described in other chapters are applicable to the brazing of cast irons. The choice of brazing process will depend upon the (1) metals being joined, (2) brazing filler metals used, (3) design of the joint, and (4) relative masses of the parts. Those processes that can be used

with automatic temperature control are desired, and overheating should be avoided. Vacuum furnace brazing is particularly suited for the production of high quality parts and the brazing of dissimilar metals. Sometimes normalizing following brazing is done to restore the original properties of the cast iron. For applications in which little or no decrease in strength can be tolerated, it is mandatory to use a brazing filler metal with the lowest permissible flow temperature. This permits the use of a low brazing temperature and a minimum brazing time at temperature. Brazing with copper and nickel filler metals is generally accomplished in a furnace.

DISSIMILAR METALS BRAZING

Joint clearance is critical to the properties and performance of brazed joints between cast iron and a dissimilar metal. This is true for all types of loading (static, fatigue, and impact) and for all joint designs. With similar metals of about equal mass, the clearance at room temperature before brazing is a satisfactory guide. When brazing dissimilar metals, the differences in thermal expansion characteristics may tend to increase or decrease the joint clearance, depending on the relative positions and configurations of the base metals. Therefore, when brazing dissimilar metals, consideration must be given to both the required clearance at brazing temperature and the clearance at room temperature, which must be designed to provide the desired clearance at brazing temperature.

The joint clearance at the brazing temperature is calculated from thermal expansion data. With high differential thermal expansion between two materials, the brazing filler metal must be strong enough to resist fracture and the base metal must yield during cooling. Whenever possible, the brazement should be designed so that the cast iron is in compression after brazing.

Brazing of Gray Cast Iron to Steel

In a specific example of brazing dissimilar metals, Class 35 cast iron and a steel casting (0.2% carbon, 1.35% manganese, and 0.25% molybdenum) were chosen. The brazing filler metal used was Type BNi-2. The differential coefficients of thermal expansion between the cast iron and the steel is approximately 1.5 x 10-6 inch/inch/°F (2.7 x 10-6 m/m/°C) assuming the brazing is done at 1975°F (1080°C)). For a mean diameter of about 9.6 in. (244 mm) at 75°F (24°C) for the assembled parts, the joint clearance was calculated as 1.5 x 10-6 in/in/°F x 9.6 in. x 1900°F (2.7 x 10-6 m/m/°C x 245 mm x 1038°C) or 0.027 in. (0.69 mm) on the diameter or 0.014 in. (0.35 mm) of radial clearance. Since grey iron will expand more than the cast steel on heating to 1975°F (1080°C), the initial room temperature radial clearance between them was 0.015 to 0.016 in. (0.39 to 0.41 mm). The inner part was made of grey cast iron. On heating, the gap would close to about 0.001 to 0.002 in. (0.03 to 0.05 mm) which is desirable for the filler metal used. The arrangement for brazing is shown in Figure 21.2.

The assembly was electrolytically cleaned and the cast iron part was nickel plated. They were then furnace brazed in a reducing hydrogen atmosphere at 1975°F (1080°C) for 15 minutes. The resultant brazed joint is shown in Figure 21.3. The photomicrograph in Figure 21.3 shows a metallurgically sound joint with minimum dissolution. Additionally, tensile testing showed good results.

Cast Iron to Brass

This assembly is part of a bellows-type seal for automotive air-conditioner compressors. A cast iron washer is joined to a half-hard 0.008 in. (0.20 mm) thick brass bellows. Operating requirements: 100% gastight under 250 to 300 psi (1.7 to 2.1 MPa) pressure; withstand vibration, twist and refrigerant corrosive action. BAg-1 filler metal rings are placed on bellows coated with AWS Type FB3-A flux, with preform BAg-1a filler metal between. Assemblies are then jigged and induction-brazed four at a time. Brazed units are then water-sprayed to hold bellows temper.

Other Applications

The assembly in Figure 21.4 is made of two pieces of cast iron cut from an automotive engine block and one piece of 1010 carbon steel that is to be brazed between the two castings so that unused passages in the cast iron block would be sealed. The brazed assembly is used for testing automotive gasketing materials, using sections of the actual engine block and flanges. The test was run for many hundreds of cycles from -40°F to 300°F (-40°C to 150°C) over a 10-minute time period with no detrimental effects on the brazed joint. The part was brazed with BNi-2 filler metal at 1950°F (1065°C) for 30 minutes at temperature in a vacuum of 10-3 torr. No pretreatment of the machined surfaces was required prior to brazing with this filler metal because the chromium in the filler metal dissolves any free carbon on the surface of the cast iron.

Figure 21.5 shows a steel sprocket brazed to a cast iron hub.

Figure 21.6 shows an assembly of steel tubes brazed to a malleable iron header.

Figure 21.7 shows several diesel engine subassemblies made of cast iron fittings brazed to steel tubing.

Figure 21.8 shows a tripod section made of steel tubing brazed to a cast-iron base.

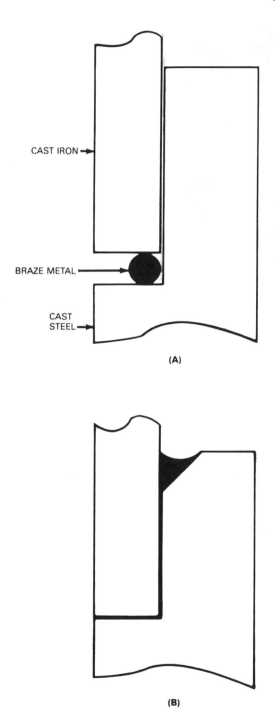

CAST IRON →

BRAZE METAL →

CAST
STEEL →

(A)

(B)

Figure 21.2 — Sketch Showing (A) Preplaced Filler Metal, Class 35 Cast Iron, and Steel Casting; and (B) Cross Sectional View of the As-Brazed Joint.

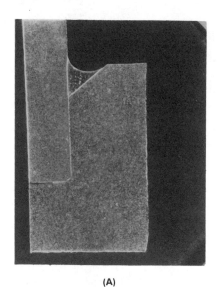

(A)

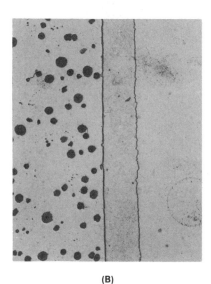

(B)

Figure 21.3 — (A) Shows a Cross Section of the As-Brazed Joint Between Cast Iron and Steel. Photomicrograph in 21.3(B) Shows Minimum Dissolution of the Base Metal

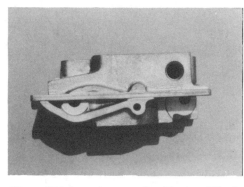

Figure 21.4 — Automotive Engine Block Flanges Brazed With BNi-2 Without Pre-cleaning

Figure 21.5 — Steel Sprocket Brazed to Cast Iron Hub

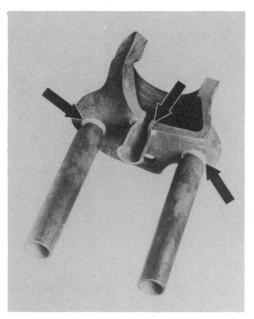

Figure 21.8 — Steel Tubing Brazed to Cast Iron Base

Figure 21.6 — Steel Tubes Brazed to Malleable Iron Header

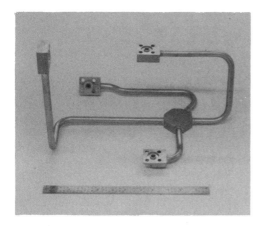

Figure 21.7 — Cast Iron Fittings Brazed to Steel Tubing

Keywords — pipe and tube, cutting or sizing pipe and tubing, fluxes for brazing tube or pipe, filler metals used in brazing of pipe and tubing

Chapter 22

PIPE AND TUBE

It would be difficult to say when man first used metal tubular goods to convey materials from one location to another. Remnants of copper pipe show that metal was used to convey water as early as 3000 B.C. Certainly by the time of the Roman Empire the use of metals to convey water and other materials was well established. At this time, too, tubular members were probably first joined to various alloys. Basic joining methods have changed little since then and for the most part they are still manual operations. However, the demands and service conditions of today's technology have created the need for a wide variety of material requirements and the development of new alloys, as well as advanced joining techniques.

Brazed joints are economical and can easily be made by skilled personnel with the correct equipment and materials. Brazed joints can readily be made in aluminum alloys, copper and copper alloys, steel, stainless steel, and most other metals. Most of pipe and tube materials that can be soldered can also be brazed, and when service conditions are too severe for soldered joints, brazed joints can very often be used satisfactorily.

A wide variety of brazing filler metals and fluxes are available, and most of them are readily applicable to pipe and tube. A filler metal for use with tube or pipe material should be chosen from the charts in the chapters concerned with the various metals. In situations where joint clearances cannot be held to close tolerance, the brazing filler metal should have a melting range wide enough to be easily workable, and be fluid enough to flow readily into the capillary space with good wetting action.

Most brazed joints in pipe and tubing do not need as much depth of socket insertion (overlap) as do soldered joints. Ordinarily, a depth of tube insertion three times the thickness of the tube will result in a joint as strong or stronger than the base metal. This is illustrated by the test results shown in Figure 22.1. Attempting to draw brazing filler metal to the base of a socket designed for soldering will often result in serious overheating.

CUTTING AND SIZING

Pipe or tubing should be cut square to seat evenly on the shoulder at the bottom of the socket. Hacksaws or abrasive cutoff discs are suitable for cutting lengths of pipe and tubing. Smaller size tubing, such as copper water service tubing, can be cut with a roller cutter. During any cutting operation, clamping pressure should not cause distortion of the wall. Care should be exercised to ensure that the ends to be joined are true and round.

Tables 22.1 shows the standard dimensions of copper water tube; Table 22.2 gives the dimensions of copper and brass socket-type fittings. Fittings and tubing both are sized to

319

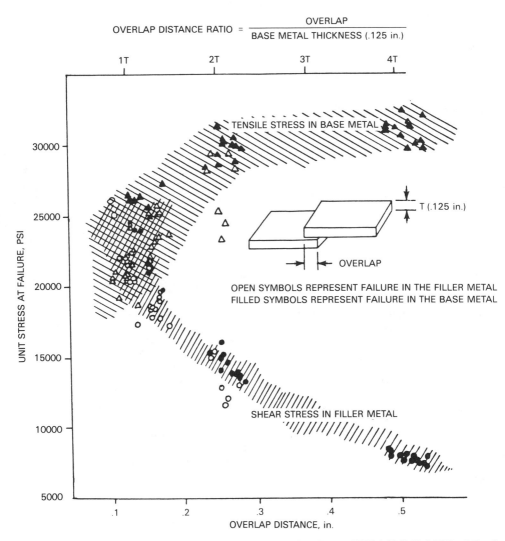

Figure 22.1 — Strength of Deoxidized Copper Lap Joints Brazed With BCuP-3 Filler Metal And Various Overlap Ratios (Ref. 1)

provide proper joint clearance for capillary flow of the filler metals and also adequate joint strength. Tables 22.3 and 22.4 provide the dimensions for steel and stainless steel pipe, and Table 22.5 provides the dimensions for threadless copper pipe. Roundness tolerances are given in Table 22.6.

The dimensions in these tables generally allow about 0.004 in. (0.10 mm) clearance between the outside wall of the tubing and the inner wall of the solder (braze) cup. The

overall difference in diameter between the male end (fitting connection) and the female end (solder cup) is about 0.009 in. (0.23 mm). Sound brazed joints can be produced using this joint clearance and proper brazing procedures. Clearances that are larger than about 0.012 in. (0.30 mm) or are too tight increase the probability of discontinuities in the joints.

After cutting, the tube should be reamed to remove burrs or uneven edges that would

Table 22.1
Sizes and Weights of Copper Water Tube

Nominal Size		Outside Diameter, Types K-L-M-DWV		Inside Diameter							
				Type K		Type L		Type M		Type DWV	
in.	mm	in.	mm	in.	mm	in.	mm	in.	mm	in.	mm
1/4	6.4	0.375	9.5	0.305	7.75	0.315	8.00	0.325	8.00	—	—
3/8	9.5	0.500	12.7	0.402	10.21	0.430	10.92	0.450	11.43	—	—
1/2	12.7	0.625	15.9	0.527	13.39	0.545	13.84	0.569	14.45	—	—
5/8	15.9	0.750	19.1	0.652	16.56	0.666	16.92	0.690	17.53	—	—
3/4	19.1	0.875	22.2	0.745	18.92	0.785	19.94	0.811	20.60	—	—
1	25.4	1.125	28.58	0.995	25.27	1.025	25.04	1.055	26.80	—	—
1 1/4	31.8	1.375	34.93	1.245	31.62	1.265	32.13	1.291	32.79	1.295	32.89
1 1/2	38.2	1.625	41.28	1.481	38.62	1.505	38.23	1.527	38.79	1.541	39.14
2	50.8	2.125	53.98	1.959	49.76	1.985	50.42	2.09	51.03	2.041	51.84
2 1/2	63.5	2.625	66.68	2.435	61.85	2.465	62.61	2.495	63.37	—	—
3	76.2	3.125	79.38	2.907	73.84	2.945	74.80	2.981	75.72	3.035	77.09
3 1/2	88.9	3.625	92.08	3.385	85.98	3.425	87.00	3.459	87.86	—	—
4	101.6	4.125	104.78	3.857	97.97	3.905	99.19	3.935	99.94	4.009	101.83
5	127.0	5.125	130.18	4.805	122.05	4.875	123.83	4.907	124.64	4.981	126.52
6	152.0	6.125	155.58	5.741	145.82	5.845	148.46	5.881	149.38	5.959	151.36
8	203.0	8.125	206.38	7.583	192.61	7.725	196.22	7.785	197.74	7.907	200.84
10	254.0	10.125	257.18	9.449	240.00	9.625	244.48	9.701	246.41	—	—
12	305.0	12.125	307.98	11.315	287.40	11.565	293.75	11.617	295.07	—	—

Wall Thickness								Weight*							
Type K		Type L		Type M		Type DWV		Type K		Type L		Type M		Type DWV	
in.	mm	in.	mm	in.	mm	in.	mm	lb/ft	kg/m	lb/ft	kg/m	lb/ft	kg/m	lb/ft	kg/m
0.035	0.89	0.030	0.76	0.025	0.064	—	—	0.145	0.216	0.126	0.187	0.106	0.158	—	—
0.049	1.24	0.035	0.89	0.025	0.064	—	—	0.269	0.400	0.198	0.295	0.145	0.216	—	—
0.049	1.24	0.040	1.02	0.028	0.71	—	—	0.344	0.512	0.285	0.424	0.204	0.304	—	—
0.049	1.24	0.042	1.07	0.030	0.76	—	—	0.418	0.622	0.362	0.539	0.263	0.391	—	—
0.065	1.65	0.045	1.14	0.032	0.81	—	—	0.641	0.954	0.455	0.677	0.328	0.488	—	—
0.065	1.65	0.050	1.27	0.035	0.89	—	—	0.839	1.25	0.665	0.975	0.465	0.692	—	—
0.065	1.65	0.055	1.40	0.042	1.07	0.040	1.02	1.04	1.55	0.884	1.32	0.682	1.01	0.650	0.96
0.072	1.83	0.060	1.52	0.049	1.24	0.042	1.07	1.36	2.02	1.14	1.70	0.940	1.40	0.809	1.20
0.083	2.11	0.070	1.78	0.058	1.47	0.42	1.07	2.06	3.07	1.79	2.60	1.46	2.17	1.07	1.59
0.095	2.41	0.080	2.03	0.065	1.65	—	—	2.93	4.36	2.48	3.70	2.03	3.02	—	—
0.109	2.77	0.090	2.29	0.072	1.83	0.045	1.14	4.00	5.95	3.33	4.96	2.68	3.99	1.69	2.51
0.12	3.05	0.100	2.54	0.083	2.11	—	—	5.12	7.62	4.29	6.38	3.58	5.33	—	—
0.134	3.4	0.110	2.79	0.095	2.41	0.058	1.47	6.51	9.69	5.38	8.01	4.66	6.93	2.87	4.26
0.16	4.06	0.125	3.18	0.109	2.77	0.072	1.83	9.67	14.4	7.61	11.3	6.66	9.91	4.43	6.57
0.192	4.88	0.140	3.56	0.122	3.10	0.083	2.11	13.9	20.7	10.2	15.2	8.92	13.3	6.10	9.05
0.271	6.88	0.200	5.08	0.170	4.32	0.109	2.77	25.9	38.5	19.3	28.7	16.5	24.6	10.6	15.7
0.338	8.59	0.250	6.35	0.212	5.38	—	—	40.3	60.0	30.1	44.8	25.6	38.1	—	—
0.405	10.29	0.280	7.11	0.254	6.45	—	—	57.8	86.0	40.4	60.1	36.7	54.6	—	—

* Slight variations from these weights must be expected in practice.

Table 22.2
Dimensional Data for Braze and Solder Joint Fitting Ends

Nominal Size		Male End (fitting connection)				Female End (solder cup)			
		O D Max		O D Min		I D Max		I D Min	
in.	mm	in.	mm	in.	mm	in.	mm	in.	mm
1/4	6.4	0.376	9.55	0.374	9.50	0.380	9.65	0.378	9.60
3/8	9.5	0.501	12.73	0.499	12.67	0.505	12.83	0.503	12.78
1/2	12.7	0.626	15.90	0.624	15.85	0.630	16.00	0.628	15.95
5/8	15.9	0.751	19.08	0.749	20.19	0.755	19.18	0.753	19.13
3/4	19.1	0.876	22.25	0.874	22.20	0.880	22.35	0.878	22.30
1	25.4	1.1265	28.61	1.1235	28.54	0.1305	28.71	1.1285	28.66
1 1/4	31.8	1.3765	34.96	1.3735	34.89	1.3805	35.06	1.3785	35.01
1 1/2	38.2	1.627	41.33	1.623	41.22	1.6315	41.44	1.629	41.38
2	50.8	2.127	54.03	2.123	53.92	2.1315	54.14	2.129	54.08
2 1/2	63.5	2.627	66.73	2.623	66.62	2.6315	66.84	2.629	66.78
3	76.2	3.217	79.43	3.123	79.32	3.1315	79.54	3.129	79.48
3 1/2	88.9	3.627	92.13	3.263	92.02	3.632	92.25	3.629	92.18
4	101.6	4.127	104.8	4.123	104.7	5.132	130.4	4.129	104.9
5	127.0	5.127	130.2	5.123	130.1	5.132	130.4	5.129	130.3
6	152.0	6.127	155.6	6.123	155.5	6.132	155.8	6.129	155.7
8	203.0	8.127	206.4	8.123	206.3	8.132	206.6	8.129	206.5
10	254.0	10.127	257.2	10.119	257.0	10.132	257.4	10.129	257.3
12	305.0	12.127	308.0	12.119	307.8	12.132	308.2	12.129	308.1

Solder Joint Fittings Cast Bronze, Wrought Copper				Solder Joint Fittings**							
				Fitting End Length				Solder Cup Depth			
Fitting End Length		Solder Cup Length		Wrought Drainage		Cast Drainage		Wrought Drainage		Cast Drainage	
in.	mm	in.	mm	in.	mm	in.	mm	in.	mm	in.	mm
3/8	9.5	5/16	7.9	—	—	—	—	—	—	—	—
7/16	11.1	3/8	9.5	—	—	—	—	—	—	—	
9/16	14.3	1/2	12.7	—	—	—	—	—	—	—	—
11/16	17.5	5/8	15.9	—	—	—	—	—	—	—	—
13/16	20.6	3/4	19.1	—	—	—	—	—	—	—	—
31/32	24.6	29/32	23.0	—	—	—	—	—	—	—	—
1 1/32	26.2	31/32	24.6	11/16	17.5	0.44	11.2	5/8	15.9	0.38	9.7
1 5/32	29.4	1 3/32	27.8	3/4	19.1	0.50	12.7	11/16	17.5	0.44	11.2
1 13/32	35.7	1 11/32	34.1	13/16	20.6	0.56	14.2	3/4	19.1	0.50	12.7
1 17/32	38.9	1 15/32	37.3	—	—	—	—	—	—	—	—
1 23/32	43.7	1 21/32	42.1	1	25.4	0.81	20.6	15/16	23.8	0.75	19.1
1 31/32	50.0	1 29/32	48.4	—	—	—	—	—	—	—	—
2 7/32	56.4	2 5/32	54.8	1 1/4	31.8	1.06	26.9	1 3/16	30.2	1.00	25.4
23/32	69.1	2 21/32	67.5	—	—	1.31	33.3	—	—	1.25	31.8
3 5/32	80.2	3 3/32	78.6	—	—	1.62	41.1	—	—	1.50	38.1
4 1/32	102.4	3 31/32	100.8	—	—	2.12	53.8	—	—	2.00	50.8
4 1/8	104.8	4	101.6	—	—	—	—	—	—	—	—
4 5/8	117.5	4 1/2	114.3	—	—	—	—	—	—	—	—

* For general plumbing use.

** For copper drainage use.

Table 22.3
Steel Pipe, Schedule 40

Pipe Size		Outside Diameter		Inside Diameter		Wall Thickness	
in.	mm	in.	mm	in.	mm	in.	mm
1/8	3.2	0.405	10.29	0.307	7.80	0.049	1.24
3/4	6.4	0.540	13.72	0.410	10.41	0.065	1.65
3/8	9.5	0.675	17.15	0.545	13.84	0.065	1.65
1/2	12.7	0.840	21.34	0.675	17.12	0.083	2.11
3/4	19.1	1.050	26.67	0.884	22.45	0.083	2.11
1	25.4	1.315	33.40	1.097	27.86	0.109	2.77
1 1/4	31.8	1.660	42.16	1.442	36.63	0.109	2.77
1 1/2	38.2	1.900	48.26	1.682	42.72	0.109	2.77
2	50.8	2.375	60.33	2.157	54.79	0.109	2.77
2 1/2	63.5	2.875	73.03	2.635	66.93	0.120	3.05
3	76.2	3.500	88.90	3.260	82.80	0.120	3.05
3 1/2	88.9	4.000	101.6	3.760	95.50	0.120	3.05
4	101.0	4.500	113.4	4.260	108.2	0.120	3.05
5	127.0	5.563	141.3	5.295	134.5	0.134	3.40
6	152.0	6.625	168.3	6.357	161.5	0.134	3.40
8	203.0	8.625	219.1	8.239	211.6	0.148	3.76

interfere with the flow of the filler metal or the alignment of the mating parts. A half-round file or other device can be used to remove these surface burrs and irregularities. Care should be exercised to prevent distortion or increased joint clearance by over-reaming of the fitting sockets.

CLEANING

The ends of tubing or pipe as well as the internal surfaces of fittings must be thoroughly cleaned to permit good for wetting and distribution of the filler metals at the faying surfaces. All traces of dirt, grease, lacquers, and oxides on the base metals must be removed. Degreasing with organic solvents will often remove various oils. However, these solvents will not ordinarily be effective in removing oxides or organic coatings that have been applied to provide oxidation resistant surfaces. Cleaning should be performed so that surface contaminants are removed without excessive loss of the base metals. That is best accomplished by lightly abrading the tube ends

Table 22.4
Stainless Steel Pipe, Schedule 5

Pipe Size		Outside Diameter		Inside Diameter		Wall Thickness	
in.	mm	in.	mm	in.	mm	in.	mm
1/2	12.7	0.840	21.34	0.710	18.03	0.065	1.65
3/4	19.1	1.050	26.67	0.920	23.37	0.065	1.65
1	25.4	1.315	33.40	1.185	30.10	0.065	1.65
1 1/4	31.8	1.660	42.16	1.530	38.86	0.065	1.65
1 1/2	38.2	1.900	48.26	1.770	44.96	0.065	1.65
2	50.8	2.375	60.33	2.245	57.02	0.065	1.65
2 1/2	63.5	2.875	73.03	2.709	68.81	0.083	2.11
3	76.2	3.500	88.90	3.334	84.68	0.083	2.11
3 1/2	88.9	4.000	101.6	3.834	97.38	0.083	2.11
4	101.6	4.500	114.3	4.334	110.1	0.083	2.11

Table 22.5

Dimensions, Weights and Tolerances in Diameter and Wall Thickness for Copper Threadless Pipe (TP) Sizes

Standard Pipe Size	NOMINAL DIMENSIONS										TOLERANCES			
	Outside Diameter*		Inside Diameter		Wall Thickness		Cross-Sectional Area of Bore		Nominal Weight lb/ft (kg/m)		Average Outside Diameter, Minus Valves		Wall Thickness, Plus or Minus	
in.	in.	mm	in.	mm	in.	mm	in.²	cm²	lb/ft	kg/m	in.	mm	in.	mm
1/4	0.540	13.7	0.410	10.4	0.065	1.65	0.132	0.852	0.376	0.559	0.004	0.10	0.0035	0.089
3/8	0.675	17.1	0.545	13.8	0.065	1.65	0.233	1.50	0.483	0.719	0.004	0.10	0.004	0.10
1/2	0.810	21.3	0.710	18.0	0.065	1.65	0.396	2.55	0.613	0.912	0.005	0.13	0.004	0.10
3/4	1.050	26.7	0.920	23.4	0.065	1.65	0.665	4.29	0.780	1.16	0.005	0.13	0.004	0.10
1	1.315	33.4	1.185	30.1	0.065	1.65	1.10	7.10	0.989	1.47	0.005	0.13	0.004	0.10
1 1/4	1.660	42.2	1.530	38.9	0.065	1.65	1.84	11.9	1.26	1.87	0.006	0.15	0.004	0.10
1 1/2	1.900	48.3	1.770	45.0	0.065	1.65	2.46	15.9	1.45	2.16	0.006	0.15	0.004	0.10
2	2.375	60.3	2.245	57.0	0.065	1.65	3.96	25.5	1.83	2.72	0.007	0.18	0.006	0.15
2 1/2	2.875	73.0	2.745	69.7	0.065	1.65	5.92	38.2	2.22	3.30	0.007	0.18	0.006	0.15
3	3.500	88.9	3.334	84.7	0.083	2.11	8.73	56.3	3.45	5.13	0.008	0.20	0.007	0.18
3 1/2	4.000	102.0	3.810	96.8	0.095	2.41	11.4	73.5	4.52	6.73	0.008	0.20	0.007	0.18
4	4.500	114.0	4.286	109.0	0.107	2.72	14.4	92.9	5.72	8.51	0.010	0.25	0.009	0.23
5	5.562	141.0	5.298	135.0	0.132	3.40	22.0	142.0	8.73	13.0	0.012	0.30	0.010	0.25
6	6.625	168.0	6.309	160.0	0.158	4.01	31.3	202.0	12.4	18.5	0.014	0.36	0.010	0.25
8	8.625	219.0	8.215	209.0	0.205	5.21	53.0	342.0	21.0	31.2	0.018	0.46	0.014	0.36
10	10.750	273.0	10.238	260.0	0.256	6.50	82.3	531.0	32.7	48.7	0.018	0.46	0.016	0.41
12	12.750	324.0	12.124	308.0	0.313	7.95	115.0	742.0	47.4	70.5	0.018	0.46	0.020	0.51

* The average outside diameter of a tube is the average of the maximum and minimum outside diameters, determined at any one cross-section of the tube.

and the sockets with small compatible wire brushes, steel wool, or fine grades (00) of abrasive papers or cloth. Embedding of abrasive particles in the joint surfaces should be avoided, and any residue remaining after mechanical cleaning must be removed.

Although mechanical cleaning of steel or copper pipe and tubing is usually adequate, the use of mineral acids or alkaline cleaners is often required to prepare surfaces for brazing. Nickel and aluminum alloys, as well as stainless steel and cast iron, usually require active chemical cleaning before joining. Thorough washing of chemically cleaned parts is necessary to remove any residues resulting from the operation. Suitable chemical cleaning procedures are given in Chapter 7.

Table 22.6

Roundness Tolerances

Ratio of Nominal Wall Thickness to Nominal Outside Diameter (t/d)	Diameter Roundness Tolerances, % of Nominal Outside Diameter*
0.01 to 0.03, incl	1.5
Over 0.03 to 0.05, incl	1.0
Over 0.05 to 0.10, incl	0.8
Over 0.10	0.7

* The deviation from roundness is measured as the difference between major and minor outside diameters as determined at any one cross section of the tube.

FLUXES AND FLUXING OPERATIONS

Fluxes are usually necessary for all manual brazing of tubing and pipe except when brazing copper tubing with copper phosphorus (BCuP) filler metals, or when brazing in controlled atmospheres and vacuum. The phosphorus in BCuP filler metals gives them a self-fluxing capacity when used on copper; however, it is recommended that a flux be used on other copper alloys. Flux must be used if the copper tube is being joined to a cast brass or bronze component.

One purpose of flux is to prevent oxidation of joint surfaces both after cleaning and during the brazing process. Fluxes are not intended to be used as cleaning compounds to remove heavy oxide films, but they should be sufficiently active to remove thin oxide films that may develop after cleaning. In addition, fluxes should dissolve or absorb any metallic oxide formed on joint surfaces during the joining operation.

It is important that flux be applied to the surfaces to be joined as soon as possible after cleaning. A small brush or a clean cloth is suitable for coating the joint surfaces with flux. It is essential that joints be brazed soon after fluxing. One hour is the usual limit to wait before brazing after fluxing, but a shorter interval is desirable.

Fluxes used in brazing pipe or tubing usually fall under AWS brazing flux classifications FB3-A, -C, -D, and -E. When choosing flux, the proprietary mixtures recommended by the filler metal manufacturer are usually satisfactory. Flux activity and stability depend upon the extent and type of oxides present, the heating method employed, and the temperature of the heated parts. Fluxes are characterized by their high fluidity at brazing temperatures; prolonged heating or over-heating of fluxes will usually destroy their effectiveness. All surfaces to be wetted by the filler metals should be coated with flux, but application of flux to the inside of surfaces of pipe and tubing should be avoided. Flux inclusions in the joint can often be minimized by preplacing a ring of the brazing filler metal at the seat of the pipe joint. Upon heating, the filler metal from the ring will flow outward into the joint, displacing the flux ahead of it.

ASSEMBLY

Before any joining operation, the assembly should be carefully aligned and adequately supported. Misalignment will affect the joint by altering the clearance between the tube and the fittings. Supports and fixtures should allow for expansion and contraction of the assembly during brazing and in service. Rings, straps, and clamps can be used to help maintain alignment of the joints. It is an good practice to verify the clearance with a feeler gauge after the assembly has been supported.

APPLYING HEAT AND FILLER METALS

All the preliminary steps for preparing the joint surfaces must be completed before any attempt is made to braze the joint. Since pipe and tubing assemblies normally are brazed with oxyfuel gas torches, the heating of joints by torches fueled with acetylene, propane, natural gas, or other gases are emphasized here. Figure 22.2 illustrates this operation.

The amount of heat required for the size of the pipe or tubing being joined governs the size of the torch tip. The selected tip should provide a low velocity, bulbous flame for heating the joint rapidly and evenly. This flame should be adjusted neutral or slightly reducing. A neutral flame has a smooth and even inner cone, without a feathered appearance. To adjust an acetylene flame to its neutral state, light a flame that has an excess of acetylene. Then decrease the acetylene flow, or increase the oxygen flow very slowly, until the feather just disappears.

The flame should be directed so that it small diameter pipe or fittings. If the pipe is larger than 4 in. (101. mm), multiple tip torches additional heating torches may be required. Usually the pipe or tube is heated before the fittings and, after a short heating period, the flame is directed alternately from the pipe and then to the fitting. This procedure should be used to heat the fittings and pipe to the same temperature. Holding the flame too long at one location on the fit-

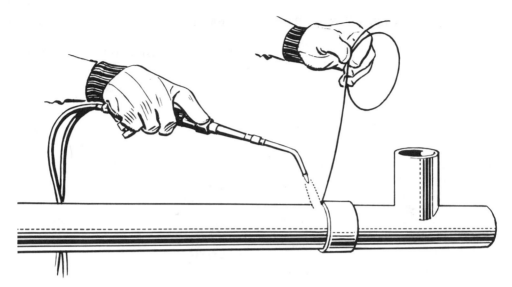

Figure 22.2 — Manually Brazing Pipe Using Oxyfuel Gas and Face Fed Filler Metal

ting or pipe may result in localized over-heating, excessive drying of the flux, distortion of the pipe, and possible cracking of the fitting.

Overheating is the most common cause of poorly brazed joints. The inner cone of the flame should not impinge directly on the tube or the shoulder of the fitting, or into the socket of the fitting. The flame should be kept in motion so that heat is distributed evenly on the joint.

The correct temperature for making brazed pipe joints can be tested by touching the filler metal to the heated junction between the pipe and fitting and observing the flow of the filler metal. The flux should be very fluid on both fitting and pipe. Fluxes usually melt into clear liquid at brazing temperature.

When brazing a horizontal joint, it is a common mistake to begin feeding the filler metal into the top of the joint. The correct procedure, applicable to both small and large diameter pipes, is to begin heating on the tube or pipe to start the conduction of heat into the fitting socket. Then both the fitting and the tube are heated as the torch is moved slowly back and forth to keep the temperature uniform. When the tube and fitting are uniformly heated, the torch is moved to the lower quarter of the tube to

heat the joint to the melting temperature of the filler metal. When the filler metal begins to melt, capillarity draws it into the joint clearance. At the same time, the torch is moved around towards the top of the fitting with the filler metal being fed to the joint just behind the path of flame. For large diameters, a second torch may be used to simultaneously braze the opposite side from the bottom to the top of the joint. Heat should not be applied directly to the filler metal. If the torch flames are directed to the heel or bottom of the socket of the fitting, the filler metal will flow throughout the joint.

Opinions differ as to whether or not a fillet of filler metal is necessary or desirable on pipe joints. Often, a full annular fillet indicates good filler metal flow and distribution in the joint. However, generous fillets at the bottom of the joint may be caused by accumulation of filler metal that has solidified after flowing over relatively cold metal. Heating from the bottom of the fitting will generally prevent this condition.

Induction or furnace heating is sometimes used to braze pipe and tubing assemblies. Usually the filler metal is preplaced in, or adjacent to, the fluxed joint. Alternately, brazing pastes may be used to paint the joint surfaces with mixtures of flux and powdered filler metals before joining. Heat should be applied to the joint in such a manner to evenly distribute the filler metal within the joint.

FILLER METALS

Filler metals used in brazing of pipe and tubing must be metallurgically compatible with the base metals they are intended to join (see Chapter 3). In addition, they must provide adequate strength to the joints in service. Normally when brazing copper, BCuP filler metals are good general purpose filler metals. They should not, however, be used on copper-nickel alloys with a nickel content of more than 10 percent. Brazing filler metals commonly used in joining copper alloy pipe and tubing are given in Table 18.3. Filler metals with a broad melting range should be selected for brazing pipe joints because of the difficulty in maintaining the workpieces in a narrow temperature range.

POST BRAZE CLEANING

After the filler metal solidifies, the remaining flux and residues can be removed by wiping with a wet cloth or by wet brushing. Cast fittings should be allowed to cool slowly to below 300°F (150°C) before applying swabs to the joints. Since most brazing fluxes are hygroscopic and may contain active corrosive agents, they should be removed. Where aluminum alloys are brazed, it is essential to remove flux residues.

INSPECTION

Though brazed joints are not commonly examined by nondestructive testing methods other than visual, it is possible to conduct both radiographic and ultrasonic inspections. Ultrasonic inspection can be done in a short time, but it require a highly skilled and experienced operator, and it usually does not provide a permanent record of the test. Radiography also requires a trained operator and may require a relatively long time to obtain the results. However, it does give a permanent record. Figures 22.3, 22.4, and 22.5 show pipe joints that were inspected by radiograph.

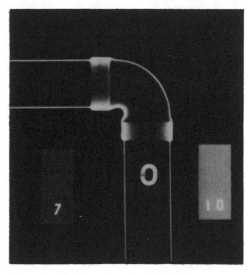

Figure 22.3 — Copper Tube Brazed To Wrought Copper Fitting With 15Ag-5P-80Cu Filler Metal. Shows Minor Voids

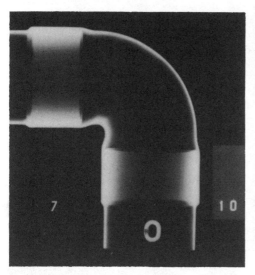

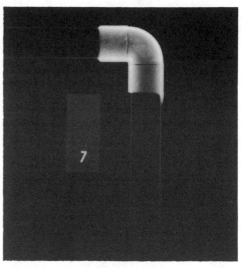

Figure 22.4 — 2.5 in. (63.5 mm) Diameter Copper Tube Brazed To Wrought Copper Fitting With 50Ag-Cu-Cd-Zn Filler Metal. Shows A Nearly Perfect Joint

Figure 22.5 — Copper Tube (0.75 in. [19.05 mm] Diameter Brazed To Cast Brass Fitting With 50Ag-Cu-Cd-Zn Filler Metal. This Radiograph Sho0ws a Very Good Joint And Indicates The Difficulty Of Radiographing Brazed Joints In Castings As Contrasted With Wrought Fittings (BAg-1a)

Keywords — nickel base alloy, cobalt-containing alloys, filler metals suitable for joining nickel and high nickel alloys, cobalt used as an alloying element

Chapter 23

NICKEL BASE AND COBALT-CONTAINING ALLOYS

Brazing is a widely used joining process for nickel and high-nickel alloys. The brazing process and technique used for a specific application will depend on the chemical and metallurgical properties of the alloy and anticipated service conditions. These factors can vary widely within this group of alloys. An important precaution is the susceptibility of nickel and high-nickel alloys to attack by both sulfur and low-melting-point metals, and to liquid-metal embrittlement in contact with molten brazing filler metals.

Generally, cobalt-containing alloys are brazed in the same manner as the high-nickel alloys. The same equipment and procedures and filler metals can be used. A brief discussion of these alloys is presented below.

NICKEL BASE ALLOYS

From a brazing standpoint, the important categories of nickel and high-nickel alloys are (1) commercially pure nickel, (2) nickel-copper alloys, (3) solid-solution-strengthened super- alloys, (4) precipitation-hardenable superalloys, and (5) oxide dispersion-strengthened (ODS) alloys.

Unalloyed nickel and nickel-copper alloys are used primarily for applications where corrosion resistance is important, or in processing equipment where product purity must be maintained. Any of the common brazing processes can be used and the choice of process, equipment, and filler metal depends on design and service conditions.

The superalloys and dispersion strengthened alloys are designed for applications requiring high strength and resistance to corrosion at high temperatures. The alloy compositions, service conditions, and designs usually require furnace brazing with nickel-base filler metals or, in some cases, with precious metal alloys having relatively high melting ranges.

The precipitation-hardenable superalloys of require special consideration for brazing. The aluminum and titanium contained in these alloys oxidize readily to form films that deter wetting and flow of brazing filler metals. Assemblies with clean surfaces can be brazed successfully in tight, clean vacuum systems with pumping capacity to maintain pressures below 0.001 torr (0.15 Pa) at temperature. (See Chapter 4 on Fluxes and Atmospheres, and Chapter 10 on Furnace Brazing.)

Strategic use of oxygen and nitrogen getters, such as hot titanium chips, can enhance the quality of the vacuum and thus promote brazing. Metals such as nickel, that form low-melting eutectics with titanium, must be avoided in making containers for such getters.

Nickel plating is recommended for complex assemblies that may require a rebraze cycle in vacuum, or for parts that are to be brazed in hydrogen. The required plating thickness depends on the titanium aluminum content, or both, of the base metal.

Plating of 0.0002 in. (0.005 mm) thick is adequate for alloys with trace amounts of these elements, while alloys with up to 4% of these elements require 0.0008 to 0.0012 in. (0.02 to 0.03 mm) thick plating.

The heat treating cycles are another factor to be considered when alloys of this group are brazed. Generally, brazing of these alloys in the precipitation-hardened condition is not recommended. The 1100° to 1500°F (590°-815°C) aging temperature range restricts the choice of filler metals to the silver or silver-copper based systems. The inherent stresses present in the precipitation-hardened alloys make them particularly susceptible to liquid metal embrittlement in contact with molten silver and silver-copper alloy filler metals.

General practice is to combine the solution annealing and brazing cycles, then age the brazed assemblies. The brazing filler metal must flow at the solution annealing temperature and provide brazed joints of adequate strength to withstand stresses during the aging cycle. Some nickel-base alloys are susceptible to grain growth with consequent degradation of mechanical properties if the brazing temperature is too high [above approximately 1850°F (1010°C)]. For critical applications, gold or palladium base filler metals may be used in spite of their relative high cost. Nickel filler metals BNi-2 and BNi-3 are typical ones that can be used.

The dispersion-strengthened alloys are powder metallurgy materials; their high strength at high temperature is derived from the fine, uniform dispersion of stable thorium or yttrium oxides in the matrix. These oxides are not soluble in the base metal even at the melting temperature. However, melting can cause rejection of the oxide particles from the matrix and loss of strength. Brazing is, therefore, one of the preferred joining processes for these alloys. Alloys such as Inconel MA754, TD-Nickel, and TD-NiCr are fairly easy to braze. After machining the surfaces to be brazed and washing them with a clean, residue-free solvent, these alloys can be brazed in vacuum, hydrogen or inert gas atmospheres. Some oxide dispersion strengthened (ODS) alloys, like Inconel MA 6000, are also gamma prime strengthened, and require vacuum brazing after mechanical cleaning. When brazing with a filler metal that may alloy

with the base metal and erode the surface of the base metal, controlled brazing cycles should be used to avoid the erosion. Melting of the surface can cause agglomeration of the oxides at the joint surface with a resulting decrease in joint strength.

FILLER METALS

The brazing filler metals normally used for ferrous metals are suitable for joining nickel and high-nickel alloys. The selection of the brazing filler metal depends on the service conditions of the finished assembly. Any heat treatments that may be required for the base metal must be considered to be certain that the brazed joints will withstand the temperatures involved. The brazing filler metals are covered in detail in Chapter 3.

The BAg brazing filler metals may be used to join nickel and high-nickel alloys to themselves and to many other metals and alloys. With proper design and brazing techniques, the brazed joints should be as strong as the base metal. The low-melting filler metals BAG-1, BAg-1a, and BAg-3 are commonly used, but for many corrosive environments filler metals containing at least 50% silver are preferred. The BAg-7 filler metal is useful where liquid metal embrittlement (sometimes called stress cracking or stress corrosion) might occur.

High-nickel alloys are capable of being brazed with BCu filler metal using the same equipment that is used for steels with only minor changes in the brazing procedures to account for the different characteristics of these alloys. The BCu filler metals will alloy more rapidly with nickel alloys than with steel; therefore, filler metal will not flow far before it has alloyed with sufficient base metal to raise its liquidus and reduce its fluidity. To avoid this problem, the filler metal should be placed as close to the joint as possible, and there should be a sufficient reservoir to fill the joint. Slightly roughened or lightly etched surfaces improve the capillary flow of copper filler metal. Such surfaces can be wet for relatively long distances, but polished surfaces inhibit the flow of the copper filler metal. Excessive copper filler metal can alloy with and perforate a thin member. In the presence of liquid copper and certain other filler metals, cracking can take place in nickel alloys when tensile stresses are present. Thermal stress relieving or annealing prior to brazing is recommended to avoid cracking during brazing.

Phosphorus combines with many metals to form brittle compounds known as phosphides. For this reason, BCuP filler metals are not usually used with iron-base and nickel-base alloys. However, BNi-6 and BNi-7 filler metals that contain about 10 to 12% phosphorus are used for brazing nickel base alloys. The BNi filler metals have been widely used for brazing nickel-base and high-nickel heat-resistant alloys. Joint strength of these BNi filler metals increases directly with the increasing brazing temperature and time at temperature, and inversely with the quantity of filler metal in the joint.

Joints with high corrosion resistance and elevated-temperature strength can be obtained when properly brazed with these filler metals. The BNi filler metals will interact with the base metal at the brazing temperature. This property has the advantage of increasing the joint remelt temperature and joint strength, as well or corrosion resistance. Caution should be used, however, to apply the nickel-base filler metals so that erosion and excessive interaction with the base metal do not occur.

Nickel-base filler metals containing palladium or platinum additions as well as gold base, palladium base, and platinum base filler metals have been used successfully to braze high-nickel alloys. These filler metals generally have good wetting and flow characteristics. They have a low interaction rate with most nickel-base metals and are used to advantage in many special applications because they brazed joints having good ductility, high strength, and good oxidation resistance. Filler metals such as BNi-1 through -5, consisting of the base metal with additions of silicon and boron to supress the melting temperature have been used successfully for brazing certain high-strength heat-resistant alloys.

PROCESSES AND EQUIPMENT

Most commercial brazing processes may be used for nickel and high-nickel alloy assem-

blies. These include torch, furnace, induction, and resistance brazing. Salt bath and metal bath dip brazing have limited application. While the BAg filler metals may be used in torch brazing, the BCu and BNi filler metals are usually used in controlled atmosphere furnace brazing. The various types of brazing furnaces are described in Chapter 10. In furnace brazing, careful control of the temperature is important. The furnace should be designed to permit a suitable heating and cooling rate, and accurate control of the brazing temperature. The furnace must be leak tight and clean so that the proper atmosphere or vacuum can be maintained. When a vacuum furnace is used, the walls and other internal parts of the furnace should be designed so that there will be minimum outgassing.

PRECLEANING AND SURFACE PREPARATION

Since the primary purpose of a flux or atmosphere is not to remove oxides, oil, grease, dirt, or other foreign materials, precleaning and surface preparation are important steps of the brazing procedure. The usual requirements for precleaning and surface preparation, outlined in Chapter 7, apply to the nickel alloys. Precleaning the nickel alloys just prior to brazing is particularly important since they are subject to attack by low melting point elements, particularly lead and sulfur, at elevated temperatures. Material which has been embrittled by these elements cannot be salvaged. Grease, oil, paint, shop dirt, and other foreign materials must be entirely removed before brazing because they usually contain these harmful elements.

The oxide films formed on nickel alloys are tenacious and wire brushing alone may not remove them. However, they may be removed by emery cloth or grinding. Uniform oxide removal by pickling cannot be expected unless the high-nickel alloy is first thoroughly cleaned of all other foreign material. The cleaning processes for the nickel alloys must be thorough, and they are similar to those used for other metals. Soaps can be removed with hot water. Soluble oils, tallow, fats and fatty acid combinations are removed with a hot [180°-200°F (82°-93°C)] 10 to 20% solution of equal parts of sodium carbonate and sodium hydroxide. (Soda ash may be used instead of sodium carbonate.) Parts are immersed in this solution for up to one-half hour and then rinsed in water. Mineral oils and greases are readily soluble in trichloroethylene and other solvents.

Fluoride-ion cleaning is a procedure now used for gamma prime strengthened nickel-base alloys. Some details are proprietary and several different procedures are used. However, the basic reaction forms volatile fluoride compounds of titanium and aluminum by reaction with both the oxides and the metals themselves. The latter serves to deplete a thin surface layer of these elements to prevent reformation of the refractory oxides during the brazing cycle. Fluoride-ion cleaning is generally used in conjunction with diffusion brazing with special brazing filler metals, which in some cases may be derived by adding boron to the base metal composition.

FLUXES AND ATMOSPHERES

When using BAg filler metals, AWS types FB3-A, -C and -D fluxes are suitable for most nickel alloys that do not contain aluminum. An AWS Type FB4-A flux may be used with aluminum-containing nickel alloys. The standard oxide-reducing atmospheres listed in Chapter 4 may be used when furnace brazing nickel and high-nickel alloys.

Elements such as aluminum and titanium form refractory oxides which are not reduced during brazing in normal furnace atmospheres. Therefore, age-hardenable nickel alloys are more difficult to braze than nonheat treatable alloys. A procedure is necessary which will either prevent the formation of the oxide or will flux away any oxide formed during heating. Some methods of handling such a situation are as follows:

(1) The parts may be copper plated to prevent the formation of harmful oxides. The plating procedure is somewhat special in that oxides must not be present on the surface of the parts under the plating. Oxides can be avoided by preparing the surface with the normal procedure and then using a nickel chloride strike followed by copper plating in the usual manner. About 0.0003 in. (0.008 mm) of copper is desired. This copperplate will become part of the brazing filler metal by alloying.

(2) Another method is to nickel plate the parts to prevent the formation of refractory oxides. Nickel plating is generally used in conjunction with BNi-1 filler metal and the age-hardenable alloys containing aluminum and titanium. The thickness of the plating required depends on the time and temperature of the brazing cycle. The general range of plating thickness is 0.0002 to 0.002 in. (0.005 to 0.05 mm). A nominal thickness of 0.0005 in. (0.02 mm) is preferred to minimize the effect of alloying.

(3) The use of dry, oxygen and nitrogen free atmospheres can be used to prevent the formation of most oxides and in most cases will reduce those present. However, if the oxides of titanium or aluminum are involved, it is best to perform the brazing in a vacuum at a pressure of 0.001 torr (0.15 Pa) or less. When using inert gase atmospheress, such as argon or helium, the parts must be free of surface oxides when loaded in the furnace. In some cases, pure dry hydrogen may be used as the atmosphere. The required dew point of the hydrogen to be used is determined by the oxides which could form or must be reduced during the brazing cycle.

Vacuum brazing in a clean, leak-free furnace is widely used for the age-hardenable alloys.

ASSEMBLY, PROCEDURES, AND TECHNIQUES

Silver (BAg) Filler Metals

In general, when using BAg filler metals, the brazing procedures outlined in Chapters 1 through 11 apply. The joint clearance should be maintained between 0.002 and 0.005 in. (0.05 - 0.13 mm) when flux is used. When brazing in a controlled atmosphere, clearances as low as 0.0005 in. (0.01 mm) are feasible. As a rule, wider clearances will result in lower joint strengths.

Copper (BCu) Filler Metals

When brazing the nickel alloys with BCu filler metals, the steps are similar in many respects to those used on steel. The tolerances for assembly range between a light press fit to 0.002 in. (0.05 mm) maximum clearance. Brazed butt joints can have strengths approaching those of annealed base metals if proper joint clearances are maintained. However, the multiplicity of stresses that may be applied in service may cause a joint to fail prematurely.

Nickel (BNi) Filler Metals

These filler metals follow the same general requirements as the BAg group. While joint clearances between 0.002 and 0.005 in. (0.05-0.13 mm) are feasible, better strengths are obtained if the joint clearance is less than 0.002 in. (0.005 mm). Filler metals are available for brazing joint clearances to 0.010 in. (0.25 mm) (wide gap alloys). In general, when using BNi filler metals, the brazing is performed in a furnace using a suitable atmosphere. These filler metals are commonly applied in the form of a powder mixed with a volatile vehicle, such as an acrylic resin, for application by brush, spray, or extrusion. Precautions must be taken in using the vehicle so that no harmful residue will remain. When properly used, the vehicle should completely disappear during the heating cycle. Filler metal is also available in tape and foil forms that permit the application of a controlled quantity of braze filler metal to joints. Adjustments for density must be made when using tape filler metals.

POSTBRAZE OPERATIONS

For many applications, no postbraze treatment is required. If oxidation has occurred, a pickling operation may be used to clean the assembly. If flux has been used, a flux removal operation may be necessary. Flux removal is particularly important on parts that are to operate at elevated temperatures or in corrosive environments.

The age-hardenable nickel alloys may be aged after brazing. The brazing filler metal must have a remelting temperature that is higher than the temperature to which the base metal must be heated.

If excess copper brazing filler metal is present, it can be removed in a cold bath of 20 parts of ammonia and one part hydrogen peroxide or in a solution of 15 gal. of water to 1 gal. of sulfuric acid at room temperature. The length of time required in the bath depends on the thickness of copper to be removed.

INSPECTION

The general techniques outlined in Chapter 16 should be followed when inspecting brazed joints in any of the nickel and high-nickel alloys. The inspection requirements of components for elevated temperature and corrosion environments will be more rigid because of the service conditions involved. X-ray and ultrasonic inspection techniques have been widely used.

COBALT-CONTAINING ALLOYS

Cobalt, which is similar to nickel in most of its properties, is used primarily as an alloying element. In general, these alloys are used for their corrosion and heat-resistant properties. Some iron base alloys containing cobalt are used in metal-to-glass seal applications. There are three popular groups of alloys in which cobalt is used: iron base, nickel base, and cobalt base. A few cobalt-containing alloys are listed in the appendix section of this book. In general, the same principles as outlined for the nickel and high-nickel alloys apply to the cobalt-containing alloys. However, when using the BAg and BCu filler metals, some consideration should be given to strength and oxidation resistance as the service temperature is increased. The use of the BNi filler metals is recommended for applications where the service temperatures are high.

The oxide film that forms on these alloys during heat treatment strongly adheres to the base metal. This oxide can be removed by molten caustic baths and acid solutions. It is also important to remove foreign materials before heating to avoid embrittlement of the base metal.

Keywords — brazing stainless steels, austenitic nonhardenable stainless steels, ferritic nonhardenable stainless steels, martensitic hardenable stainless steels, precipitation hardening stainless steels, duplex stainless steels, stainless steel brazing processes

Chapter 24

STAINLESS STEELS

The term "stainless steels" describes a wide variety of chromium-containing iron base alloys used primarily in applications demanding heat or corrosion resistance. This class of materials is brazeable by all processes, but with tighter process controls than required to braze carbon steels.

The most rigorous requirements are imposed by inherent chemical characteristics of the stainless steels and the generally more arduous service environments. Success in brazing stainless steel components depends on a knowledge of the properties of stainless steels and rigid adherence to the appropriate process controls.

BASE METALS

Stainless steels may be grouped into five categories: (1) austenitic (nonhardenable) steels, (2) ferritic (nonhardenable) steels, (3) martensitic (hardenable) steels, (4) precipitation hardening steels, and (5) duplex stainless steels. All these alloys are iron based and contain chromium, the basic element that imparts corrosion resistance. The corrosion resistance of the stainless steels varies from one alloy to another, and for any given alloy varies from one corrosive medium to another. If doubt exists about the proper stainless steel to use in a given environment, standard reference works (References 1-4) or manufacturers' representatives should be consulted.

AUSTENITIC (NONHARDENABLE) STAINLESS STEELS

These steels contain sufficient nickel or nickel plus manganese additions to (1) stabilize austenite down to room temperature, and (2) cause these alloys to be nonmagnetic and nonhardenable by heat treatment. Stainless steels of this class possess the highest heat and corrosion resistance. The Cr-Ni steels are designated by an AISI (American Iron and Steel Institute) type number in the 300 series, and the Cr-Ni-Mn steels by a 200 series number. One commonly used alloy is Type 302 which contains nominally 18% (by weight) chromium, 8% nickel.

In the 200 series stainless steels, some of the nickel is replaced with manganese on a ratio of approximately 2% of manganese for each percent of replaced nickel. Type 202, the parallel to Type 302, contains 18% chromium, 5% nickel and 9% manganese.

Both torch brazing and furnace brazing processes are used extensively to braze 300 series (austenitic) stainless steels. These alloys exhibit relatively high thermal expansions and low thermal conductivities. Those properties make thermal distortion a major concern in furnace brazing large, complex assemblies, or assemblies in which dissimilar materials are brazed to stainless steel. The designers of fixtures, heat shields, and thermal cycles must consider the requirement of providing uniform heating and cooling.

In brazed assemblies where corrosion resistance is important, precautions should be taken to avoid sensitization to intergranular corrosion. This problem is discussed more fully in Chapter 2. Briefly, it occurs when unstabilized grades of austenitic stainless steel, such as Type 302 or Type 304, are held at temperatures in the range from 800°F to 1600°F (427° to 815°C), and slowly cooled through this range. The excess carbon combines with chromium and precipitates as chromium carbide along grain boundaries of the austenite. The region around the precipitate is depleted of chromium and thus becomes susceptible to corrosion.

There are several ways to prevent or minimize the harmful effects of carbide precipitation. First, because the reaction is time dependent, carbide precipitation can be minimized by holding the braze thermal cycle as short as possible. With short time cycles, such as would result from torch or induction brazing of small parts, even the unstabilized grades can be brazed without serious loss of corrosion resistance.

Secondly, the susceptibility to carbide precipitation also increases with the carbon content. Thus, Type 304 would be an improvement over Type 302, and the extra low carbon grades, such as Type 304L, are relatively insensitive to carbide precipitation.

For critical applications, Type 347, the columbium stabilized grade, is recommended. It has good high temperature strength and can be brazed without danger of impaired corrosion resistance. Type 321 is also a stabilized grade, but it has slightly lower general corrosion resistance than Type 347 and is more difficult to braze because titanium is used as the carbide stabilizing element. When high melting point filler metals are used, precipitated carbides can be redissolved by heat treatment after brazing. Alternatively, corrosion resistance can be restored by diffusing chromium back into the depleted area around the carbide precipitates in the process called stabilization. The recommended stabilization heat treatment temperature is 1750°F to 2150°F (954°C to 1177°C).

While the 300 series stainless steels have been widely used for brazed components, the 200 series have been used primarily in such applications as railroad car sidings, tank cars, sinks, hospital wares, or countertops, which do not require brazing. More recently, these Cr-Ni-Mn steels have been found promising for use in structural components of superconducting magnet assemblies. When the 200 series steels are used in furnace brazed assemblies, special care should be devoted to precleaning and maintaining good atmospheres. The relatively high manganese content makes these alloys more difficult to furnace braze in hydrogen atmospheres than the Cr-Ni stainless steels. Manganese forms an oxide that is not reduced readily by dry hydrogen at the furnace brazing temperatures normally used for stainless steel.

All the chromium-nickel steels are subject to stress corrosion cracking when exposed to molten brazing filler metals in a stressed condition. Molten filler metal penetrates the base metal along grain boundaries at the points of stress, producing a greatly weakened structure. The stresses can be residual from cold forming operations on the assembly details or can result from loads applied while the braze is being made. To avoid this problem, the parts should be stress relieved, either prior to assembly or during the brazing cycle. In the latter procedure, stress relief must be accomplished below the solidus temperature of the brazing filler metal. The parts also must be assembled and supported in a manner to avoid thermal or mechanical stresses during brazing. For further discussion of this subject, see Chapter 2.

FERRITIC (NONHARDENABLE) STAINLESS STEELS

These stainless steels are basically low carbon alloys of iron and chromium, where sufficient chromium has been added to the iron to stabilize ferrite, the low temperature phase in steels, over a wide temperature range. The more common grades in this category are AISI Types 405, 430, and 446. Type 430 is a widely used grade that is particularly subject to a form of interfacial corrosion when brazed with some BAg filler metals. The corrosion apparently is caused by electrochemical action whereby the bond between the base metal and the filler metal is destroyed. This action has been found to occur in many cases even in the presence of tap water. The addition of small percentages of nickel to silver base filler metals prevents interface corrosion of brazed joints in most stainless steels. Nickel-containing silver base filler metal is not completely effective, however, with Type

430, even though its use greatly reduces the rate of attack. A special silver base filler metal BAg-21 (63Ag-28.5Cu-6Sn-2.5Ni) has been developed for Type 430.

The high temperature strength of the ferritic stainless steels decreases drastically above 1500°F. Good fixturing is essential for furnace brazing with higher melting filler metals such as the BNi classifications. Caution must be exercised to avoid differential expansion between the part and the fixture. Where possible, parts should be designed for brazing without the need for fixturing.

MARTENSITIC (HARDENABLE) STAINLESS STEELS

These are iron-carbon-chromium alloys of two basic types: the low chromium, low carbon grades (Types 403, 410, and 416) and the high chromium, high carbon grades (Types 440 A, B, and C). These steels are related closely to the ferritic nonhardenable grades, but have their alloy compositions balanced so that they are capable of being hardened by heat treatment.

A primary requirement in brazing components made of hardenable stainless steel alloys is to use a brazing thermal cycle compatible with the heat treatment required by the alloys. Selecting brazing filler metals with brazing temperatures high enough to allow austenization of the base metal to take place at the brazing temperature is the first step in creating a brazing cycle compatible with the required heat treatment. The furnace cycle must then provide rapid cooling from the brazing temperature to allow transformation of the austenite to martensite. If high temperature nickel-base filler metals are used, it is possible to reaustenize the assembly by heat treatment after brazing. Although this procedure increases costs, it may be desirable to develop optimum properties in critical components.

The thermal expansion coefficients of the martensitic stainless steels are relatively low, similar to those of the ferritic alloys. One factor to consider is the dimensional change that occurs at the austenite-to-martensite transformation. If the cooling rate within an assembly is not uniform, significant stresses can be created as one section transforms before another.

PRECIPITATION HARDENING STAINLESS STEELS

Precipitation hardening stainless steels have been developed for applications where high strength plus heat and corrosion resistance are required. These alloys are austenitic stainless steels to which alloying elements such as aluminum, titanium, copper, and molybdenum have been added. The addition of these elements makes it possible to precipitation-harden the alloys by special heat treatments. The precipitation hardening alloys are not completely austenitic, since the hardening reactions are quite complex and sometimes involve some martensite formation.

Some of the designations of the precipitation hardening stainless steels are 17-7 PH, 17-4 PH, 15-5 PH, PH 15-7 Mo, PH 14-8 Mo, and AM350.[1] As in the case of the martensitic hardenable stainless steels, brazing thermal cycles used in joining these alloys must be compatible with their heat treatments. As the heat treatments vary widely, specific brazing procedures are required for each alloy. Suppliers of these alloys and of brazing filler metals should be consulted. In some cases, it may be necessary to compromise between optimum heat treat and brazing cycles, particularly regarding cooling rates of complex parts. The properties that will actually be achieved should be determined for critical applications.

Alloys such as 17-7 PH or PH 15-7 Mo which contain aluminum or titanium are difficult to wet in the usual furnace brazing atmospheres. Nickel plating is generally used as a surface treatment for furnace brazing. Brazing under high vacuum may eliminate the necessity for plating.

DUPLEX STAINLESS STEELS

The duplex stainless steels are a duplex of austenite and ferrite. They are characterized by high chromium contents, up to 28%. At the time of publication, these alloys have been available for about 30 years. Limited application, by comparison to other stainless steels, has caused the brazing experi-

---

1. Some of these may be tradenames

ence with these alloys to be very limited. Technical data for the specific alloy should be reviewed to determine such properties as thermal expansion rates, response to heat treatment, precipitation of various phases, and chemical compositions. Many of the newer duplex stainless steels contain additional nitrogen which can cause wetting problems during brazing.

FILLER METALS

A wide variety of filler metals is available commercially for the brazing of stainless steel. Compositions and properties of brazing filler metals are presented in Chapter 3. The factors to consider in selecting a filler metal for a particular application include the following:

(1) Service conditions, including operating temperature, stresses, and environment

(2) Heat treatment requirements for martensitic or precipitation hardening steels

(3) Brazing process

(4) Cost

(5) Special precautions, such as sensitization of unstabilized austenitic stainless steels at certain temperatures.

Commercially available brazing filler metals used for joining stainless steels are commonly the copper, silver, nickel, cobalt, platinum, palladium, and gold based alloys. These may be conveniently grouped according to service temperature as listed in the chart below.

Of the BAg filler metals, BAg-1, BAg-1a, and BAg-2 are general purpose filler metals, with good brazing characteristics, used where corrosion resistance is not vital. BAg-2 requires a somewhat higher brazing temperature. BAg-2 is particularly useful when joint fitup cannot be controlled as closely as is required for BAg-1 and BAg-1a. However, BAg-2 filler metal should not be used where the heating time is excessive due to its tendency to liquate.

Where improved corrosion resistance is needed, BAg-3 and BAg-24 are recommended over nickel-free silver base filler metals. BAg-5 and BAg-6 are general purpose filler metals for higher brazing temperatures and are used where cadmium is prohibited, along with BAg-24. Cadmium-free BAg-7 and BAg-21, because of their white color, are useful for fabricating food handling equipment. BAg-8, BVAg-86, BAg-18, and BVAg-29 are often used for vacuum and atmosphere brazing where freedom from volatile cadmium or zinc is required. Two filler metals with 0.2-0.5% lithium (BAg-8a and BAg-19) show increased wettability on stainless steels. BAg-19 is modified sterling silver and is used under the high temperature required to combine base metal heat treatment and the brazing process. Silver-based filler metals for service above 800°F (427°C) are generally furnace brazed in high purity reducing atmospheres, inert atmospheres, or in vacuum.

The BCu classification is the only copper base filler metal recommended for brazing stainless steels. Brazing is usually performed in a high purity reducing atmosphere of low moisture content. RBCuZn and BCuP filler metals are not recommended for use on stainless steels.

Nickel base filler metals are used primarily where extreme heat and corrosion resistance are required. They are commonly used in the manufacture of components for jet and rocket engines, chemical processing equipment, and nuclear reactors. These filler metals normally are supplied in paste form (atomized powder suspended in a polymeric binder). They can also be obtained as sintered or cast rods, preforms,

Service Temperature		Filler Metals
°F	°C	
400	204	BAg-1 through -8a, and BAg-24
700	371	BAg-13, -19, -21
800	427	BCu
800-1000	427-538	Copper-Manganese-Nickel
1000 and above	538 and above	BNi, BCo, BAu, BPd

plastic bonded sheet, plastic bonded wire, and tape. Many of these alloys are now are available as metallic foils produced by rapid solidification technology. Some compositions also are produced by boronizing nickel or nickel-chromium foil and wire.

BNi filler metals commonly are used on stainless steels for oxidation resistance to temperatures up to 1800-2000°F (982°-1093°C). Filler metals BNi-1, BNi-2, BNi-3, and BNi-4, which contain boron, tend to erode thin sheet metal because of their reaction with many base metal compositions. Therefore, time at braze temperature and the quantity of filler metal should be carefully controlled when these filler metals are used. Figure 24.1 shows a Type 347 stainless steel vacuum fitting brazed with BNi-2. The boron-free filler metals BNi-5, BNi-6, and BNi-7 are suitable for use in nuclear reactor components where boron cannot be tolerated because it absorbs neutrons. Oxidation resistance of BNi-5 is good up to 2000°F (1093°C). BNi-5 has the highest melting point of all the nickel filler metals.

Filler metals based on gold, platinum, palladium, and their alloys, such as gold-nickel, gold-nickel-chromium, gold-nickel-palladium, copper-platinum, and silver-palladium-manganese, are useful for brazing heat and corrosion resistant components. BAu-4 is used where maximum corrosion resistance is necessary, such as resistance to sulphur bearing gases and compounds. Figure 24.2 shows a Type 304

stainless steel microwave guide assembly brazed with BAu-4.

The advance of brazing technology in such industries as jet and rocket propulsion and nuclear energy mark the introduction of still more new and unique filler metals. These have not become sufficiently standardized to be classified. Suppliers of brazing filler metals should be consulted where unique requirements must be met.

PROCESS AND EQUIPMENT

Stainless steels can be brazed with any brazing process. Specific information on each process can be found in other chapters in this book. Much controlled atmosphere brazing is performed on stainless steels, and the acceptability of this technique is attributed to the ready availability of reliable atmospheres and vacuum furnaces. The primary requirements are that the furnaces have good temperature control at brazing temperature (plus or minus 15°F [8°C] is desirable) and be capable of fast heating and cooling. All gases used in atmosphere furnaces must be of high purity (>99.995 percent pure). Commercial vacuum brazing equipment operates at pressures varying from 10^{-5} to 10^{-1} torr (0.0015 to 13.5 Pa). The necessary vacuum level depends upon the particular grade of stainless steel, with those containing titanium or aluminum requiring better vacuums.

PRECLEANING AND SURFACE PREPARATION

Stainless steels require more stringent precleaning than do carbon steels. During the heating cycle, residual contaminants often form tenacious films which are difficult to remove by fluxes or reducing atmospheres. These films form as a direct reaction between the contaminant and stainless steel surface.

Precleaning of stainless steels for brazing should include a degreasing operation to remove any grease or oil films. The joint surfaces to be brazed also should be cleaned mechanically or with an acid pickling solu-

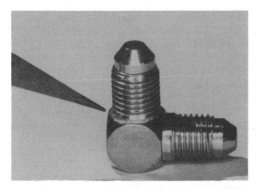

Figure 24.1 — Brazed Type 347 Stainless Steel Fitting

Figure 24.2 — Brazed Type 304 Stainless Steel Microwave Guide Assembly

tion. Wire brushing should be avoided, but if necessary, stainless steel brushes should be used. Cleaned surfaces should be protected to prevent soiling by dirt, oil, or fingerprints. For best results, parts to be brazed should be brazed immediately after cleaning. When this is not practical, the cleaned parts should be enclosed in containers such as sealed polyethylene bags or dessicator jars to exclude moisture and other contaminants until the part can be brazed.

FLUXES AND ATMOSPHERES

Stainless steel assemblies are routinely furnace brazed in atmospheres of dry hydrogen, argon, helium, dissociated ammonia, or vacuum, without the aid of flux. When fluxes are required, there are a number of special compositions available for use with stainless steels. Chapter 4 covers fluxes and atmospheres in detail including the special requirements for brazing stainless steel.

Dissociated ammonia atmospheres should be used with caution. Certain stainless steels can be nitrided at some brazing temperatures. Nitriding produces a hard surface that can be either beneficial or detrimental depending on the service requirements of the component. Nitriding can be detected by an increase in surface hardness or by metallographic examination.

Oxides of aluminum and titanium cannot be reduced in atmosphere furnaces at ordinary brazing temperatures. If these oxides are present in small amounts, satisfactory brazes can be obtained by the use of high purity gas atmospheres and vaporizing flux. When these elements are present in quantities exceeding one or two percent by weight, the metal surface should be cleaned and nickel-plated instead of using fluxes or vacuum atmospheres. Nickel plating will prevent the formation of harmful oxides and can be an effective means of limiting embrittlement or erosion of base metals. Thickness of the electrolytic nickel plating should be kept in the range of 0.0002 to 0.002 in. (0.005 to 0.05 mm). The thickness of plating should be controlled so that it is dissolved into the brazing filler metal, preventing the possibility of failure occurring in the remaining plating layer.

POSTBRAZE OPERATIONS

Major stainless steel postbraze operations that may be necessary are removal of flux or stopoff residues and any required postbraze heat treatment.

Depending on the materials used, flux or stopoff residues can be removed by water rinsing, chemical cleaning, or mechanical means. With abrasive cleaning, the grit should be sand or other nonmetallic material. Metallic shot, other than stainless steel, should be avoided because particles may become embedded in the stainless steel surface and cause rusting or pitting corrosion in service.

Unless the brazing cycle is compatible with the heat treating requirements of the base metal, heat treatment after brazing will be required for assemblies which are made with martensitic or precipitation hardening stainless steels. Since treatments vary so widely, no general rules can be made except that supplier recommendations should be followed.

REPAIR METHODS

When furnace brazed assemblies contain many joints, minor defects may occur that are beyond acceptance limits — but which are not economically or technically feasible to repair by rebrazing the entire assembly. In some cases, repairs can be made by localized rebrazing using oxyacetylene or gas tungsten arc torches.

The oxyacetylene method requires use of a high temperature flux. With martensitic type alloys, care must be taken to avoid stress cracking.

The manual gas tungsten arc method is useful for braze fillet repairs on applications like turbine engine stators. The defective area should be cleaned properly to remove surface oxides. Brazing is performed under an inert gas envelope, usually argon, with approximately 20 amps power input. Filler metal is added, as required. The plasma needle arc process also can be used, but requires a skillful operator. Prototype work on a mock-up with proper evaluation prior to actual repair work is recommended.

REFERENCES

American Society for Metals. *Metals handbook*, Vol. 3. Metals Park, OH: American Society for Metals, 1980.

———. *Metals handbook*, Vol. 6 Metals Park, OH: American Society for Metals, 1983.

———. *Sourcebook on brazing and brazing technology*, Metals Park, OH: American Society for Metals.

Peckner, D. and Bernstein, I.M. *Handbook of stainless steels*. New York, NY: McGraw-Hill, 1977.

Pickering. *The metallurgical evolution of stainless steels*. Metals Park, OH: American Society for Metals, 1979.

Keywords — aluminum and aluminum alloys, filler metals for brazing aluminum, aluminum brazing fluxes, safety in handling brazing flux, joint types and fixtures for brazing aluminum

Chapter 25

ALUMINUM AND ALUMINUM ALLOYS

INTRODUCTION

Aluminum and aluminum alloys are brazed using practices similar to those used for brazing other metals except that different fluxes, different filler metals, and generally lower brazing temperatures are used. Aluminum brazing can be divided into two areas: flux and fluxless. Aluminum can be brazed by most of the standard processes including torch, dip, and furnace processes. Furnace brazing may be done in air or a controlled atmosphere, including vacuum. Other methods, including induction, radiant lamp, and resistance brazing may be used for specific applications. Regardless of the method used, close control of the process parameters is required for successful brazing.

MATERIALS

Base Metals

Aluminum alloys that can be brazed are listed in Table 25.1. These include a number of nonheat treatable and heat treatable wrought alloys and casting alloys. The nonheat treatable alloys include the high-purity aluminums and those with low additions of alloy elements. The mechanical properties of these alloys are improved by cold working. Aluminum alloy 3003, the most commonly brazed alloy, is in this group. The heat treatable alloys generally contain magnesium and silicon and are described as the magnesium silicon types or 6000 series. These alloys begin to melt at lower temperatures than 3000 series aluminum alloys because of their higher total alloy content. Consequently, they are generally brazed at lower temperatures than the nonheat treatable alloys. The heat treatable alloys, which include the commonly used 6061 alloy, are thermally treated during post-braze fabrication or during the braze cycle to obtain their high mechanical properties. The cast aluminum alloys that are most easily brazed generally are low in silicon and magnesium, which may limit castability and weldability.

Not all aluminum alloys can be brazed. The high-strength wrought aluminum alloys and certain casting alloys contain high amounts of alloying ingredients. These alloying ingredients often prevent adequate wetting by filler metal due to their unique oxide film combination. These alloys also melt at temperatures below those of commercially available filler metals. (See Chapter 3 for a listing of filler metals for brazing aluminum.) Some experimental low-melting filler metals have been used to successfully braze the difficult-to-braze alloys such as 2024, 2219, and 7075. These low-melting filler metals are basically aluminum with zinc, copper, and sometimes silicon added to reduce the melting point. Although some of these combinations have been successful, none possess joint qualities recommended for commercial use.

Table 25.1
Nominal Composition and Melting Range of Common Brazeable Aluminum Alloys

Aluminum Assoc. No.	Brazeability Rating[2]	Nominal Composition[1]						Approximate Melting Range	
		Cu	Si	Mn	Mg	Zn	Cr	F°	C°
1100	A	99% Al Min.	—	—	—	—	—	1190-1215	643-657
1350	A	99.5% Al Min.	—	—	—	—	—	1195-1215	646-657
3003	A	—	—	1.2	—	—	—	1190-1210	643-654
3004	B	—	—	1.2	1.0	—	—	1165-1205	629-651
3005	A	0.3	0.6	1.2	0.4	0.25	0.1	1180-1215	638-657
5005	B	—	—	—	0.8	—	—	1170-1210	632-654
5050	B	—	—	—	1.2	—	—	1090-1200	588-649
5052	C	—	—	—	2.5	—	—	1100-1200	593-649
6151	C	—	1.0	—	0.6	—	0.25	1190-1200	643-649
6951	A	0.25	0.35	—	0.65	—	—	1140-1210	615-654
6053	A	—	0.7	—	1.3	—	—	1105-1205	596-651
6061	A	0.25	0.6	—	1.0	—	0.25	1100-1205	593-651
6063	A	—	0.4	—	0.7	—	—	1140-1205	615-651
7005	B	0.1	0.35	0.45	1.4	4.5	0.13	1125-1195	607-646
7072	A	—	—	—	—	1.0	—	1125-1195	607-646
Cast 443.0	A	—	5.0	—	—	—	—	1065-1170	629-632
Cast 356.0	C	—	7.0	—	0.3	—	—	1035-1135	557-613
Cast 710.0[3]	B	—	—	—	0.7	6.5	—	1105-1195	596-646
Cast C712.0[4]	A	—	—	—	0.35	6.5	—	1120-1190	604-643

1. Percent of alloying elements; aluminum and normal impurities constitute remainder.
2. Brazeability ratings: A = Alloys readily brazed by all commercial methods and procedures.
   B = Alloys that can be brazed by all techniques with a little care.
   C = Alloys that require special care to braze.
3. Sand Castings.
4. Mold Castings.

Filler Metal

Filler metals for brazing aluminum are based primarily on the aluminum-silicon alloy system. All of these alloy combinations are listed under the BAlSi classification in Chapter 3.

Filler metals for brazing aluminum alloys are supplied in the form of powder, paste, wire, or thin-gauge shim stock which is either face-fed or preplaced in the joint area. Round wire of various diameters is used in the form of rings or in straight lengths as required for preplacement. Preforms are also made from flat wire or from shim stock and the geometry is adapted to the shape of the part to be brazed.

Another convenient method of supplying filler metal is to use brazing sheet which consists of a core of aluminum clad with a lower melting filler metal. The cladding is an aluminum-silicon alloy and may be applied to one or both sides of the core sheet. The cladding is roll bonded to the core by hot rolling. Thus, the brazing sheet is a product that can be used in fabrication methods such as drawing, bending, and other normal metal working processes without removing this coating. The formed part can be assembled and brazed without placing additional filler metal in the joint. Brazing sheet is frequently used as the member of an assembly with the mating piece made from unclad brazeable alloy. The filler metal on the brazing sheet flows by capillary action and gravity to fill the joint. In addition, clad thin wall tubing is made by continuous seam welding of brazing sheets.

Other products such as wire, rod, bar, extruded shapes, rolled shapes, forgings, or castings are not currently available with clad brazing filler metals on their surface.

FLUX

Brazing flux must be used for every brazing process except fluxless inert gas or fluxless vacuum brazing. Aluminum brazing fluxes are mixtures of inorganic chloride and fluoride salts supplied in powder form. Brazing flux is applied dry or after with distilled water, deionized water, or alcohol for application by brushing, spraying, or dipping. In torch brazing, dry flux may be sprinkled on the joint area of the assembly being brazed, or applied to the end of a heated filler rod. Generally, brazing flux is used in slurry form: a mixture of three parts flux and one part liquid is satisfactory for brush application. A thinner mixture is used for dipping or spraying, and the amount of liquid needed is established by trial to suit the application, furnace conditions, or spray gun used. Because the performance of these fluxes degrades over time after being exposed to moisture, they should be used within a few hours of mixing. See Chapter 4 for more information on flux.

Care should be exercised in handling brazing flux. Flux mixtures are strong chemical agents and can be injurious to the body and corrosive to equipment. Flux should be sprayed in a ventilated or water-curtain spray booth. After flux spraying, equipment must be rinsed to retard corrosion or clogging of the gun. See Chapter 6 for further information on safety and health in the handling of brazing fluxes.

MATERIAL PREPARATION

Precleaning requirements vary widely depending on the brazing method, condition of the metal surfaces, and the type of contaminant present. Degreasing is recommended for fluxless brazing, and is desirable for flux brazing when contamination and oxide films are light.

Degreasing by vapor cleaning is usually adequate for nonheat treatable aluminum alloys in a clean condition. For most heat treatable alloys, however, chemical cleaning is necessary to remove the thicker oxide films. Solvent vapor degreasing is desirable when cleaning thin metals such as those used for heat exchanger applications. With thin-gauge clad material, a light etch can also be used, but care must be exercised to prevent removal of an excessive amount of cladding.

In certain applications, chemical cleaning is required in addition to degreasing. Chemical etching is used to remove heavy oxides which have been embedded by spinning operations or caused by severe metal working. The same is true when prototype parts are produced in small quantities by hand hammering or similar methods. In these cases, and when 6000 series alloys are used, a conventional caustic etch or nitric hydrofluoric etch is desirable. Many proprietary cleaning solutions on the market today have proven satisfactory. Two generic cleaning processes and cleaner compositions are as follows: Caustic-acid cleaner:

(1) Degrease

(2) Dip: 5 percent by weight sodium hydroxide at 140°F (60°C) up to 60 seconds

(3) Rinse: ambient tap water

(4) Dip: 50 percent by volume nitric acid at room temperature for 10 seconds.

(5) Rinse: hot or cold water and dry. Acid clean:

(1) Degrease: solvent or vapor.

(2) Dip: 10 percent volume nitric acid plus 0.25 percent by volume hydrofluoric acid at room temperature up to five minutes.

(3) Rinse: hot or cold water and dry. Type of rinse water required depends on condition of water.

Caustic etches are normally used for removal of heavy surface contaminants while acid etching works best for removing light oxide films.

Chemical cleaning of filler metals before preplacement is often desirable. Filler metals made from wire, or shims made from flat wire, may flow more uniformly if the oxide coating is removed by a light etch or mechanical abrasion.

JOINT TYPES AND FIXTURES

Lap joints rather than butt joints are preferred for brazing aluminum and aluminum alloys (See Chapter 2). With flux brazing type operations, press fit or tightly fitting joints must be avoided in assembling aluminum parts to facilitate fluxing and filler metal flow and to minimize flux entrapment. The exact opposite is required for fluxless brazing where mating parts of the joint must be in contact throughout the braze cycle and the braze alloy is normally placed in the joint. Figure 25.1 illustrates recommended joint designs.

With flux, joint clearance affects the capillary action which draws the filler material into place. Consequently, the joint clearance must be carefully controlled to provide good flow. As a rule of thumb when dip brazing, clearance of 0.002 to 0.004 in. (0.05 to 0.10 mm) is suitable. When furnace, mechanized torch, and induction brazing with 0.25 in. (6.4 mm) of overlap or less, similar clearances can be used. Larger clearances of up to 0.010 in. (0.6 mm) are used for longer lap joints because the filler

metal changes composition by dissolving the base metal and becomes sluggish as it flows through long lap joints. The correct clearance for any given joint is best determined by trial.

The joint design should permit easy assembly and inspection of the parts. Closed assemblies should be designed to provide for escape of the gases during brazing, both from the parts and the fixture. It is preferable in flux brazing to design the parts to be self-jigging by tabs or to be held in alignment by rivets or other projections that remain on the part after brazing. Typical self-jigging joints are shown in Figure 25.2.

Fixtures and jigs must be designed carefully (see Chapter 8). Differences in thermal expansion should not force the parts out of alignment. Nickel-base alloys, stainless steels, special alloy steels, and low carbon steels are commonly used for jigs and fix-

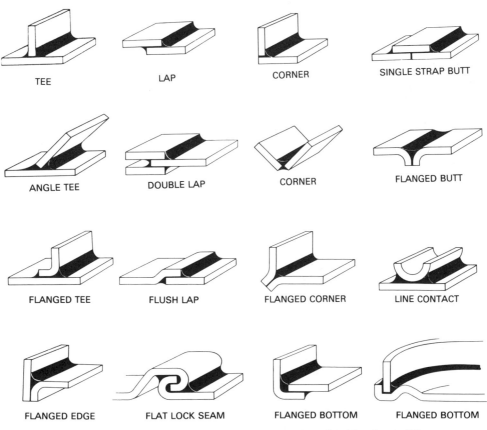

TEE LAP CORNER SINGLE STRAP BUTT

ANGLE TEE DOUBLE LAP CORNER FLANGED BUTT

FLANGED TEE FLUSH LAP FLANGED CORNER LINE CONTACT

FLANGED EDGE FLAT LOCK SEAM FLANGED BOTTOM FLANGED BOTTOM

Figure 25.1 — Recommended Brazed Joint Designs for Aluminum Alloys

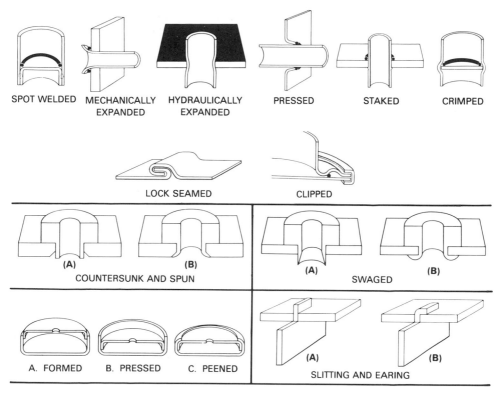

Figure 25.2 — Typical Self-Jigging Joints for Aluminum alloys

tures. Low carbon steel will have a short life in the presence of flux. However, coating the fixture with aluminum by dipping, followed by a thermal treatment to diffuse the coating, will increase its useful life.

Fixtures for fluxless brazing must control the joint to keep it closed as the part brazes. Hot strength is required but corrosion is not a factor in choosing a material. Oxidation of the fixture or coating with stop-off before use may be required to prevent metallurgical interaction between the fixture and the workpieces.

PERFORMANCE OF JOINTS

Brazing temperatures approach the melting points of most heat treatable base metals. Control of time in the critical temperature range is necessary to prevent excessive diffusion. Exposure to brazing temperatures will cause dissolution if the temperature is too high. If the dwell time at brazing temperature is too long, the same deleterious result may occur. In all cases, annealing takes place at brazing temperature. Brazed parts made of nonheat treatable aluminum alloys have mechanical properties corresponding to the annealed temper of these alloys.

The heat-treatable base metals are also annealed by brazing, but their strength and corrosion resistance can be substantially increased either by a heat treating operation applied after the parts are brazed or by quenching the parts from the brazing temperature. The later procedure is not always feasible, depending on the geometry of the parts. The heat treatable base metals can be quenched with an air blast, a water spray, or submersion in a tank of hot or cold water. An air quench is the slowest of the three. A slight delay after removal from the brazing heat source is necessary to permit the filler metal to solidify before the part is quenched. Otherwise, the dimensional changes that occur on quenching will rupture the joints.

In some complicated parts, cracking of the joints may occur even though the filler metal has solidified. Complicated assemblies should be cooled prior to heat treating. Proper heat treating times and temperature depend on the aluminum alloy involved.

The corrosion performance of brazed aluminum assemblies depends on two conditions: (1) removal of corrosives, such as flux, and (2) retaining the metallurgical structure. If the postbraze cleaning operation does not remove all of the flux residues, the combination of flux with moisture can cause corrosion attack of the base metal. Excessive time at brazing temperature may compromise the corrosion resistance due to changes in grain structure. Resistance to corrosion by specific chemicals cannot be described in general terms, but where aluminum welds have performed satisfactorily, brazed joints will perform similarly. In certain cases, exposure tests under specific conditions are required for performance proof.

BRAZING PROCESSES

Furnace Brazing

Aluminum may be furnace brazed in air using flux, but benefits can be gained by brazing in a controlled atmosphere. As the moisture content of air is reduced by dehumidification, less flux is required. In air or a nitrogen atmosphere with a dew point of about -60°F (-51°C) special noncorrosive fluxes can be used. With argon and vacuum atmospheres, no flux is required.

Atmosphere Furnace Brazing. Equipment for furnace brazing aluminum alloys is much the same as that used for heat treating these materials. Such equipment is designed to operate at temperatures up to 1200°F (650°C) with a temperature control of ±5°F (650°C ±3°C). Furnaces may be oil or gas fired or heated electrically. The furnace atmosphere is circulated to achieve uniformity of temperature by improved heating conduction rates. The recommended procedure is to adjust the furnace cycle time from 30 seconds to 2 minutes longer than the time required to bring the assembly to brazing temperature. Excessive

time at temperature may result in detrimental diffusion, penetration, or other undesirable filler metal or flux reactions.

Care to prevent contact of the molten brazing flux with metallic furnace parts is desirable because flux will corrode such parts. Protection often is provided by using shallow pans of low carbon steel to catch any flux that may drop. Such pans can be replaced at a much lower cost than that of the furnace lining or hearth parts. Brazing aluminum in furnaces used for brazing steel at high temperatures is not recommended because aluminum flux residues will attack the heating elements when used at high temperatures.

Inert Gas Brazing. Furnace chambers of various designs are used for brazing in inert gas. For fluxless brazing, the majority of water molecules must be removed from the base metal surface by using a combination of heat and vacuum before flushing the system with inert gas.

For flux brazing, an inert gas atmosphere is sufficient. The inert gas must have a dew point of -60°F (-51°C) or less. Nitrogen is commonly used because of its low cost, but helium or argon also can be used. Hydrogen may be used but the hazards far outweigh the advantages. Conduction or convection are the methods of heating. If flux is used, the amount required is directly proportional to the amount of water vapor in the system at temperatures below 400°F (204°C). Above that temperature oxides that form begin to seriously interfere with wetting by the filler metal.

Continuous-belt-type furnaces and batch furnaces both have been used successfully. In some furnace designs, a fast heating rate is obtained in a high temperature zone at the entrance of the furnace. Here, the part is brought up to within 50° to 100°F (10° to 38°C) of the brazing temperature and then moved to the brazing zone which is controlled to within ±5°F (±3°C) of the brazing temperature.

Vacuum Furnace Brazing. Aluminum alloys can be successfully vacuum brazed without flux. Furnaces operating in the 10^{-5} to 10^{-6} torr range are used in combination with magnesium which acts both as a getter of gaseous oxygen and modifier of the oxide film or the aluminum. Magnesium may be introduced into the process in three ways:

(1) As elemental magnesium in a tray

(2) By use of magnesium-bearing base metals

(3) By use of magnesium in the braze filler metal Furnace brazing in vacuum is primarily being used for production of heat exchangers. Vacuum brazing of vertical joints in primary components is not recommended.

Dip Brazing

Dip brazing has been used widely in the manufacture of wave guide and heat exchanger assemblies. This process consists of dipping a preheated assembly into a molten brazing flux. In this manner, fluxing and raising the part to a brazing temperature are accomplished simultaneously in rapid and uniform manner.

Several salt pot designs are commercially available (see Chapter 12 on Dip Brazing). Dip pots can be cylindrical but are more commonly rectangular in shape with either over-the-wall nickel electrodes or through the wall carbon electrodes. Heat is provided by the resistance of the molten flux to the current flowing between the electrodes. Thermal stirring of the bath provides temperature uniformity. Externally heated gas-fired pots have been occasionally used. Brick lining of high alumina content is the only material with sufficient flux resistance to provide long life. Metal pots made from pure nickel have exhibited longer flux resistance than other metals, but not more than six months of service can be expected due to intergranular corrosion.

Parts to be dip brazed are preheated prior to immersion in the molten brazing flux. Preheating temperatures range from 900°-1065°F (482°-575°C) depending on the size, alloy, and complexity of the assembly. Preheating is employed to dry the part and eliminate possible rapid steam generation, and also to prevent solidification of flux on a cold assembly when submersed. Solidification of flux would insulate the parts and seal small openings so that molten flux cannot reach the inside. After brazing, the assembly must be suspended over the salt pot for a sufficient length of time to allow all possible flux to drain into the pot. This draining operation not only reduces flux consumption, but also reduces the amount of flux that must be removed from

the brazed unit. It also provides for brazed alloy solidification before moving the part.

Dip brazing is an excellent method for consistently producing high quality brazed joints in complicated assemblies. Because the ample supply of flux, incomplete joints sometimes caused by lack of flux coverage in furnace brazing are nearly nonexistent. The major concern in dip brazing is flux flow. Air traps, such as blind holes, must be avoided.

Molten flux must be dehydrated before use to remove chemically combined moisture. Dehydration is accomplished by immersing clean aluminum sheets in the flux. Hydrogen will bubble up to the surface as long as moisture is present. The bubbles will ignite spontaneously, showing the characteristic small orange flames. Several hours of dehydration are usually necessary to remove moisture from a freshly melted flux bath. Regular dehydration of the flux and control of bath chemistry are required for continuously successful operation.

Torch Brazing

Torch brazing is generally used for small parts, short production runs, and the attachment of fittings to previously welded or brazed aluminum assemblies. Torch brazing is sometimes used for joining tubes to headers, return bends on heat exchangers, and other similar joint configurations. Air fuel gas or oxyfuel gas torches may be used. Brazing flux is applied to both work and filler metal by brushing, dipping, or sprinkling. The torch should be adjusted so the flame is slightly reducing. Heat is applied locally to the area to be joined until the flux and filler metal melt and wet the surfaces of the base metal. Extreme care must be taken to avoid overheating the base alloys because the melting point of the filler metal is very close to that of the base alloy.

DISSIMILAR METAL BRAZING

Aluminum can be brazed to certain other metals as noted in Table 25.2. Titanium, nickel, cobalt, and beryllium can be brazed directly to aluminum. Other alloys may be precoated with nickel to improve alloy flow and protect the surfaces from oxidation. Copper and brass cannot be brazed directly to aluminum.

**Table 25.2
Brazing aluminum to other metals.**

Metal or alloy joined to aluminum	Brazeability *	Remarks
Ferrous alloys	A	Best results obtained by electroplating or aluminum coating the ferrous part.
Nickel, Inconel	A	Can be brazed directly or precoated with aluminum.
Titanium, Kovar[1]	A	Aluminum coat the parts before brazing.
Beryllium	B	Can be wetted directly by aluminum brazing alloys.
Monel	B	Can be wetted directly but may tend to be brittle.
Copper, brass	C	Difficult, requires special techniques to avoid brittleness; can use a steel transition section.
Magnesium	D	Too brittle to be of use.

* Brazeability ratings:

A - generally brazeable by all commercial brazing procedures and methods.
B - brazeable with special techniques or on specific applications that justify preliminary trials or testing to develop brazing procedures and performance or brazed joints.
C - limited brazeability.
D - not recommended for brazing.

[1] Kovar is a trademark of Westinghouse Corporation.

Bare steel and titanium parts can be brazed to aluminum in vacuum or in inert gas atmospheres with very low dew points. Steel and titanium parts can be electroplated with nickel or hot dip coated with aluminum for torch, furnace, or dip brazing applications. During brazing, aluminum and steel will alloy to form a brittle intermetallic layer at the interface. Therefore, preheating and brazing times must be kept to a minimum. In addition, the designer must take into consideration the differential thermal expansion of these materials.

Aluminum cannot be brazed to magnesium because of their mutual solubility and the formation of an extremely brittle aluminum-magnesium intermetallic. Bimetal joints require special treatments to ensure adequate corrosion resistance because of the difference in electrochemical potential. Corrosion resistance can be enhanced by painting, encapsulation, or coating with a moisture impervious material.

POST BRAZE CLEANING

After brazing with flux, the corrosive flux residues must be removed. Flux residues form a hard, brittle layer which can best be removed in hot water followed by chemical cleaning. Mechanical cleaning such as wire brushing or grinding is not adequate. It breaks up the crust of flux into fine particles that become embedded in the surface and, on subsequent exposure to moisture, may cause corrosive attack.

Several procedures are suitable for the removal of brazing flux. Immersion in hot water before the part has entirely cooled is effective in removing the major portion of the flux. For torch brazed joints particularly, the application of hot running water combined with brushing is a good practice. A five minute immersion in a mixture of 10 percent by volume nitric acid with 0.25 percent by volume hydrofluoric acid bath, or 2 to 5 minute dip in a 1.5 percent by volume hydrofluoric acid bath applied after the hot water dip, will also remove the flux residue. An acid dip of this type is always required if the joint is inaccessible for brushing. Good results can be obtained by a number of commercial cleaners. All these methods require clean water rinses to remove the cleaning agents. Subsequent to the water rinsing the parts must be thoroughly dried. There are also commercially available noncorrosive fluxes which do not require postbraze cleaning.

Keywords — magnesium, magnesium alloys, combustibility of magnesium, prebraze assembly of magnesium components

Chapter 26

MAGNESIUM AND MAGNESIUM ALLOYS

The brazing techniques used for magnesium alloys are similar to those used for aluminum. Furnace, torch, and dip brazing can be employed, although the latter process is the most widely used.

BASE METALS

The composition and physical properties of brazeable magnesium alloys are given in Table 26.1. Furnace and torch brazing experience is limited to M1A alloy whereas dip brazing has been applied to AZ10A, AZ31B, K1A, M1A, ZE10A, and ZK21A alloys. Because of their low solidus temperatures, other magnesium alloys cannot be brazed with present brazing filler metals and techniques.

FILLER METAL

The filler metal used for magnesium brazing is a magnesium base alloy shown in Table 26.1. Brazing filler metal BMg-1 is suitable for the torch, furnace, or dip brazing processes. Standard forms and sizes of brazing filler metals are listed in Table 26.2.

JOINT DESIGN

Joints should be designed to take advantage of capillary action, and to allow the flux to be displaced by the brazing filler metal as it flows into the joint. Because of the corrosive nature of the flux, care should be taken in the design of the joints to minimize flux entrapment. Both lap and butt joints can be used for magnesium assemblies.

Joint clearances of 0.004 to 0.010 in. (0.10 to 0.25 mm) at the brazing temperature are satisfactory. However, designing joints with the smallest clearance that will permit good flow of filler metal is recommended to advantage of good capillary action. In dip brazing, the joint should be designed with slots or recessed grooves to accommodate the filler metal and prevent it from being washed into the flux bath (see Chapter 2).

COMBUSTIBILITY OF MAGNESIUM

Ignition of magnesium depends primarily upon the size and shape of the material, as well as the size or intensity of the source of ignition. If magnesium is in the form of ribbon, shavings, or chips with thin, feather-like edges, a spark or a match flame may be sufficient to start the metal burning.

Heavier sections, such as fabricated assemblies, ingots, or castings, are more difficult to ignite because heat is conducted rapidly away from the source of ignition. However, self-sustained burning will occur if the entire piece of metal is raised to its ignition temperature (approximately the solidus temperature of the alloy).

351

Table 26.1
Composition and Physical Properties of Brazeable Magnesium Alloys and Filler Metal

AWS A5.8 Classification	ASTM alloy Designation	Avail Forms	Nominal Composition, % (Bal. mg.)						Specific Gravity	Density		Solidus		Liquidus		Brazing Range	
			Al	Zn	Mn*	Zr	Rare Earth	Be		lbs/in	kg/m	°F	°C	°F	°C	°F	°C
—	AZ10A	E	1.2	0.4	0.20	—	—	—	1.75	0.063	1760	1170	632	1190	643	1080-1140	582-616
—	AZ231B	E,S	3.0	1.0	0.20	—	—	—	1.77	0.064	1790	1050	566	1160	627	1080-1140	582-593
—	K1A	C	—	—	—	0.70	—	—	1.74	0.063	1760	1200	649	1202	650	1080-1140	582-616
—	M1A	E,S	—	—	1.2	—	—	—	1.76	0.063	1760	1198	648	1202	650	1080-1140	582-616
—	ZE10A	S	—	1.2	—	—	0.17	—	1.76	0.063	1760	1100	593	1195	646	1080-1100	582-593
—	ZK21A	E	—	2.3	—	0.60	—	—	1.79	0.064	1790	1159	626	1187	642	1080-1140	582-616
FILLER METAL																	
BMg-1	AZ92A		9.0	2.0	0.1	—	—	0.0005	1.83	0.066	1850	830	443	1110	599	1120-1140	604-616

E = Extruded shapes and structural sections.
S = Sheet and Plate
C = Castings
* = Minimum

Table 26.2
Standard Forms and Sizes of Brazing Filler Metal

AWS-ASTM Classification	Standard Form	Size, in.
BMg-1	wire in coils rod, 36-in. lenghts	round 1/16, 3/32, 1/8, 5/32, 3/16

The probability of combustion occurring during the torch brazing of magnesium is extremely remote. In that instance, only the joint area is heated (not the entire structure), and the flux tends to prevent burning in the heated joint area. The dip brazing process does not present combustion problems with magnesium because the entire operation is performed in a flux bath. However, during the 850° to 900°F (454-482°C) preheat of the assembly prior to dip brazing, two precautions are in order: (1) make sure that proper temperature controls are used, and (2) determine that preheat time is the minimum needed to bring the assembly to temperature.

Furnace brazing has the greatest potential for magnesium combustion because the entire structure is heated to the brazing temperature. However, the risk of combustion is still minimal provided the furnace temperature is controlled within the prescribed brazing range and brazing time is held to the minimum necessary to achieve filler metal flow.

BRAZING PROCESSES

Furnace Brazing

When furnace brazing AZ31B alloy, electric or gas heating equipment should be used that has automatic controls capable of holding the furnace temperature within ± 10°F (6°C) of the brazing temperature. Such controls are needed to minimize incipient melting of the base metal and to reduce the potential of a magnesium fire. No special furnace atmosphere is required. However, atmospheres produced by the products of combustion in gas-fired furnaces and sulfur dioxide (used in heat treating magnesium alloys) reduce filler metal flow, and thus should be avoided.

Parts to be brazed should be assembled with the filler metal preplaced in or around the joints. Joint clearances of 0.004 to 0.010 in. (0.10 to 0.25 mm) should be used to obtain good capillary flow of the filler metal. Best results are obtained when dry powdered flux is sprinkled on the faying surfaces of the joint. Flux should not be mixed with water or alcohol because they will retard the flow of filler metal. Flux pastes made with benzol, toluene, or chlorbenzol may be used, but they are not smooth and thus more difficult to apply. Flux pastes should be dried by heating at 350° to 400°F (177-204°C) for 5 to 15 minutes in drying ovens or circulating air furnaces. Flame drying is not desirable because it leaves a heavy soot deposit.

Brazing time will depend somewhat on the thickness of the materials and the amount of fixturing necessary to position the parts. The brazing time should be the minimum necessary to obtain complete filler metal flow. The objective is to avoid excessive diffusion of the filler metal and minimize the potential for combustion of the magnesium. Normally, one or two minutes at the brazing temperature is sufficient to make the braze. After brazing, the parts should be allowed to air cool away from drafts to minimize distortion.2

Torch Brazing

Torch brazing is done with a neutral oxyfuel or air-fuel gas flame. Manual torch brazing is difficult with BMg1 filler metal, and requires great attention and care. Considerable skill is necessary for face feeding of filler metal because of the solidus temperature of the base metal and the flow point of the brazing filler metal are very close together.

The filler metal should be placed on the joint and fluxed before heating. Flux pastes made with alcohol give the best results, and are preferred over pastes made with water. The procedure is to apply heat to the part until the flux melts, and then to continue applying heat until the brazing filler metal melts, wets the surface, and flows into the joint by capillary action. Operators should avoid overheating the base metal because rapid diffusion and drop-through of the metal may take place. Natural gas with its relatively low flame temperature is well suited for torch brazing because it reduces the danger of overheating.

Dip Brazing

Dip brazing involves immersing the assembly in a molten brazing flux held at the desired brazing temperature. The flux serves the dual function of both heating and fluxing. When brazing AZ31B alloy, temperature control should be to within $\pm 10°F$ (6°C) of the desired brazing temperature to minimize incipient melting.

Joint clearance in dip brazing should be from 0.004 to 0.010 in. (0.10 to 0.25 mm). After preplacing the filler metal, the parts to be brazed are assembled in a fixture, preferably of stainless steel to resist the corrosive action of the flux. They are then preheated in a furnace to the 850°-900°F (455-482°C) range. Preheating is necessary to drive off any moisture on the assembly, as well as to minimize thermal shock and flux entrapment. Immersion time in the flux bath should be relatively short [30 to 45 seconds for 0.063 in. (1.6 mm) base metal] because the bath rapidly heats the parts. Large assemblies with massive fixturing may require from one to three minutes immersion. Because of the large volume of flux and uniform heating, more consistent results are achieved with dip brazing than with other brazing processes.

PRECLEANING AND SURFACE PREPARATION

Parts to be brazed should be thoroughly clean and free from burrs. Oil, dirt, and grease should be removed either with hot alkaline cleaning baths or by vapor or solvent degreasing. Surface films, such as chromate conversion coatings or oxides, must be removed by mechanical or chemical means immediately prior to brazing. In mechanical cleaning, abrading with aluminum oxide cloth or steel wool has proved very satisfactory. Chemical cleaning should consist of a 5 to 10 minute dip in a hot alkaline cleaner followed by a 2 minute dip in ferric-nitrate bright pickle solution which is described in Table 26.3.

FLUXES

AWS brazing flux, Type FB2-A (see Table 4.1), is used to when brazing magnesium alloys. Because of the corrosive nature of the flux, complete removal is extremely important if good corrosion resistance is to be maintained in brazed joints.

Table 26.3
Chemical Treatment Solutions

Treatment	Composition	Method of Application
Ferritic nitrate bright pickle	Chromic acid 1.5 lb (0.68kg) Ferritic nitrate, 5.33 oz (0.15 kg) Potassium fluoride, 0.5 oz (0.014 kg) Water to make one gallon Temperature 60-100 F (16-38C)	0.25 to 3 min immersion followed by cold and hot water rinses and air dry.
Chrome pickle	Sodium dichromate, 1.5 lb (0.68 kg) Conc. nitric acid, 1.5 pt (0.71 1k) Water to make one gallon Temperature 70-90F (21-32C)	1 to 2 min immersion, hold in air 5 s, followed by cold and hot water rinses and air or forced dry 250 F max (121 C max)
Postbraze cleaner	Sodium dichromate, 0.5 lb (0.23 kg) Water to make one gallon Temperature 180-212F (82-100C)	2 HR immersion in boiling bath followed by cold and hot water rinses and air dry.

ASSEMBLY

Staking, self-positioning, spring-loaded fixtures, or tack welding are methods commonly used for prebraze assembling of magnesium components. Mild steel or stainless steels fixtures generally are used in magnesium brazing. In dip brazing, fixtures should be small and of simple design to minimize heat losses in the flux bath and to allow complete drainage of the flux from the fixture. Tack welding may be desirable to eliminate massive or complex fixtures in short production runs. If tack welding is used, Type M1A welding rod is recommended because its melting point is above the brazing temperature of magnesium alloys. BMg-1 filler metal usually is hot-formed to fit the joint by heating it to between 500° and 600°F (260°-316°C). The filler metal can be formed over a heated steel mandrel to the desired contour.

POSTBRAZE CLEANING

Regardless of the brazing process used, complete removal of all traces of flux is of utmost importance. Brazed parts should be rinsed thoroughly in flowing hot water to remove flux. A stiff-bristled brush should be used to scrub the surface and speed up the cleaning process. The parts should then be immersed for one to two minutes in chrome pickle, followed by a two hour boil in postbraze flux remover cleaner. These solutions are described in Table 26.3.

CORROSION RESISTANCE

The corrosion resistance of brazed joints depends primarily on (1) the thoroughness of flux removal, and (2) the adequacy of joint design to prevent flux entrapment. Since the brazing filler metal is a magnesium base alloy, galvanic corrosion of brazed joints is minimized.

INSPECTION

After cleaning, visual inspection of the assembly may reveal incomplete filler metal flow. That problem might be attributed to inadequate cleanliness, insufficient flux, or too low a temperature. If this and other possible discontinuities may be present, certain other tests can be employed to discover them. These include nondestructive tests, radiography, liquid penetrant, pressure tests and electrical resistance. Flux inclusions also can be detected by exposing the joint to a high humidity atmosphere (95%) for several days and then inspecting for indications of flux residue.

The adequacy of test brazements can be determined by etching a cross section of the joint and examining it under a microscope. Optimum detail of a brazed joint microstructure is obtained when conventional magnesium metallographic etchants are used. However, sufficient joint detail can be revealed by etching the cross section in a 5 to 10% solution of acetic acid. A typical joint with a magnesium filler metal is shown in Figure 26.1.

STRENGTH OF BRAZEABLE MAGNESIUM ALLOYS

Typical mechanical properties of brazeable magnesium alloys are shown in Table 26.4. The temperatures involved in brazing will reduce the properties of work-hardened (tempered) magnesium sheet alloys to the

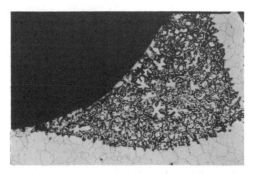

Figure 26.1 — Brazed Joint in AZ31B Base Metal (65x)

Table 26.4
Typical Mechanical Properties of Brazeable Magnesium Alloys

ASTM Alloy Designation	Form	Temper	% E	TYS Ksi	TYS MPa	UTS Ksi	UTS MPa	CYS* Ksi	CYS* MPa
AZ10A	E	F	10	21	145	35	241	10	69
AZ31B	E	F	14	28	193	38	262	15	103
AZ31B	S	H24	15	32	221	42	290	24	165
AZ31B	S	O	21	22	152	37	255	16	110
KIA	C	F	14	8	55	23	159	—	—
MIA	E	F	9	20	138	34	234	11	76
MIA	S	H24	9	27	186	37	255	—	—
MIA	S	O	15	16	110	32	221	—	—
ZE10A	S	H24	12	26	179	37	255	25	172
ZE10A	S	O	23	20	138	33	228	16	110
ZK21A	E	F	10	33	228	42	290	—	—

E = Extruded shapes and structural forms
S = Rolled sheet
C = Sand casting
F = As fabricated

O = Annealed condition
H24 = 0.5 Hard, work hardened condition
*Compressive yield strength

0 temper (annealed) level. Torch brazing will reduce properties only in those areas heated for brazing; furnace and dip brazing will reduce properties of the entire structure. Properties of cast or extruded alloys or of annealed sheet alloys are not greatly affected by the heat of brazing. Table 26.5 illustrates the minor reduction in strength of annealed sheet alloys after exposure to various brazing temperatures.

Table 26.5
Tensile Properties of Magnesium Sheet Alloys After Heating

Time, min.	Temp, F	C	UTS[1] Ksi	UTS[1] MPa	TYS[2] Ksi	TYS[2] MPa	% E	UTS Ksi	UTS MPa	TYS Ksi	TYS MPa	% E
	Heating				AZ31B-0*					ZE10A-0		
None	None		36.2	250	20.7	143	23.5	32.9	227	20.2	139	22.5
0.5	1075	579	34.4	237	17.0	117	19.0	27.3	188	10.8	74	16.5
1	1075	579	33.9	234	16.5	114	17.0	27.3	188	10.6	73	14.5
2	1075	579	34.0	234	16.9	117	17.5	26.6	183	10.0	70	12.0
3	1075	579	33.4	230	15.8	109	15.5	26.2	181	9.9	68	10.0
5	1075	579	32.9	227	16.0	110	13.0	25.2	174	9.0	62	10.5
0.5	1100	593	33.8	233	16.5	114	17.0	27.6	190	11.5	79	15.0
1	1100	593	33.2	228	15.9	110	16.0	27.0	186	11.1	77	14.0
2	1100	593	32.6	225	15.4	106	14.5	26.7	184	11.1	77	13.0
3	1100	593	32.5	224	14.6	101	15.0	26.8	185	11.0	76	12.0
5	1100	593	31.5	217	14.4	099	14.0	25.7	177	10.4	72	10.5
0.5	1125	607	32.5	224	16.1	111	14.0	27.5	190	10.6	73	16.0
1	1125	607	31.4	217	12.1	083	12.0	27.2	188	11.2	77	15.5
2	1125	607	29.2	201	10.8	074	9.0	26.5	183	11.0	76	16.5

* O = Annealed Condition
[1]Ultimate Tensile Strength
[2]Tensile Yield Strength

TYPICAL APPLICATIONS

Figure 26.2 shows the torch brazing of M1A magnesium alloy hydraulic lift floats. Filler metal BMg-1 was face fed into the annular groove and upon melting was distributed through the joint by capillary action. The flame was directed against the plug so that tube walls were heated largely by conduction. Figure 26.3 shows a longitudinal section of a completed float.

Figure 26.4 shows a ZWI (British magnesium alloy nearly equivalent to ASTM ZK21A alloy) battery container made of 0.040 to 0.625 in. (1.0 to 15.9 mm) sheet and plate. This assembly was dip brazed using BMg-2a filler metal, which is no longer

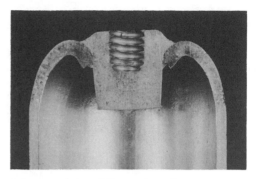

Figure 26.4 — Magnesium Battery Container. (A) and (B) Show Details of Assembly. (C) Shows Container After Brazing and Flux Removal

Figure 26.2 — Torch Brazing Magnesium Hydraulic Lift Float

Figure 26.3 — Longitudinal Section of a Completed Float

available. Views (A) and (B) show details of the magnesium assembly and the stainless steel brazing fixture. Spring loading of the front and back supports was used to accommodate differences in thermal expansion between magnesium and stainless steel. Photo (C) shows the battery container after brazing and flux removal.

Figure 26.5 shows a dip brazed M1A magnesium alloy microwave antenna. View (A) shows details of the assembly without feeder and end caps in position. View (B) shows the assembly in the stainless steel brazing fixture. An expendable section of magnesium wave guide is used in three locations along the length of the assembly to assist in positioning the parts during brazing. View (C) shows the antenna after brazing and flux removal.

SELECTION OF BRAZING TEMPERATURE

The melting temperature of the filler metal must be lower than that of the base metal. Usually, the lowest permissible brazing temperature is preferred in order to:

(1) Minimize heat effects on base metal (annealing, grain growth, or warpage, for example)

(2) Minimize base metal-filler metal interactions

(3) Increase the life of fixtures, jigs, or other tools

(A)

(B)

(C)

Figure 26.5 — Magnesium Microwave Antenna Is Shown in (A). Details of Assembly and Stainless Steel Brazing Fixture Are Shown in (B). Assembly After Flux Removal Is Shown in (C)

(4) Economize on heat energy required

Higher brazing temperatures may be desirable to:

(1) Combine annealing, stress relief, or heat treatment of the base metal with brazing

(2) Permit subsequent processing to elevated temperatures

(3) Promote base metal-filler metal interactions in order to modify the brazing filler metal (this technique is usually used to increase the remelt temperature of the joint)

(4) Effectively remove surface contaminants and oxides with vacuum or atmosphere brazing

(5) Avoid stress cracking

Keywords — titanium, zirconium, beryllium, beryllium as a health hazard, applications for titanium and beryllium

Chapter 27

TITANIUM, ZIRCONIUM, AND BERYLLIUM

INTRODUCTION

Titanium, zirconium, beryllium and their alloys are three of the several metals that react readily with oxygen to form stable oxides. Several commercial titanium alloys are being used in aircraft and corrosion applications. Titanium is often used in heat exchangers where both light weight and high strength are required.

Zirconium and beryllium have been used primarily in nuclear applications. Wrought beryllium is sometimes used for lightweight components in aircraft and aerospace applications.

Beryllium, titanium, and zirconium have similar brazeability. That is because they share two common characteristics: rapid reaction with oxygen to form very stable oxides, and high solubility for oxygen, nitrogen, and hydrogen at elevated temperatures. These three elements dissolve interstitially in the metals. Small amounts of dissolved oxygen and nitrogen significantly increase the hardness of the metals; dissolved hydrogen reduces toughness and increases notch sensitivity. Therefore, these metals must be brazed in a high-purity inert gas or in high vacuum to avoid embrittlement.

Titanium and zirconium also react with carbon at elevated temperatures to form carbides. Carbon is sometimes added intentionally as an alloying element to increase strength and hardness. However, excess car-

bides in these metals lowers ductility. The metals should be free of oil, grease, and other hydrocarbons before brazing to prevent carburization. Graphite should not be used for fixturing or for dams to control the flow of molten brazing filler metal.

Beryllium differs from titanium and zirconium in that it has low ductility at room temperature. Beryllium mill products are normally made by powder metallurgy techniques using several consolidation methods. Wrought products are produced from either cast or powder metallurgy billets. Cold-worked material may have good ductility in only one direction, and low ductility perpendicular to that direction (anisotropy). The physical properties of beryllium may vary greatly, depending on the manner of processing.

An adherent refractory oxide film will rapidly form on exposed beryllium, as with aluminum and magnesium. This oxide film will inhibit wetting, flow, and fusion during brazing. Therefore, beryllium parts must be adequately cleaned of surface oxides and protected from reoxidation prior to brazing. The brazing process and procedures must prevent oxidation during the operation by incorporating appropriate shielding medium and techniques, such as the use of inert gas and vacuum.

Titanium, zirconium, and beryllium also react with nitrogen to form a nitride film that presents problems when wetting with brazing filler metals. Filler metal selection for brazing each of the three metals is critical to avoid formation of undesirable intermetallic compounds. A study of phase diagrams will guide the user in the selection of suitable filler metals.

TITANIUM ALLOYS

General Alloy Classifications

Titanium and titanium alloys are usually classified into groups on the basis of the phases in the microstructure. The groups (and typical alloys in each group) are listed in Table 27.1. There are other alloys, not listed, that are either being evaluated for specific applications or are seldom used. Some alloys are produced with extra-low interstitial impurities (ELI) for applications where good ductility and toughness are needed at cryogenic temperatures. Titanium alloys can also be classified according to their corrosion or heat resistant properties.

Commercially Pure Titanium

Various grades of commercially pure titanium are classified on the basis of minimum mechanical properties and maximum interstitial impurities. In general, variations in strength are produced by differences in impurity levels, primarily the

Table 27.1
Classification of titanium alloys

Alloy	UNS No.
Commercially pure	
ASTM Grade 1	R50250
ASTM Grade 2	R50400
ASTM Grade 3	R50550
ASTM Grade 4	R50700
Alpha alloys	
Ti-0.2Pd (ASTM Grades 7 and 11)	R52250,R52400
Ti-0.8Ni-0.3Mo (ASTM Grade 12)	R53400
Ti-5Al-2.5Sn (ASTM Grade 6)	R54520,R54521
Near-alpha alloys	
Ti-8Al-1Mo-1V (AMS 4915)	R54810
Ti-2.25 Al-11Sn-5Zr-1Mo (AMS No. 4974)	R54790
Ti-5Al-5Sn-2Mo-2Zr	R54560
Ti-6Al-2Cb-1Ta-0.8Mo	R56210
Ti-6Al-4Zr-2Mo-2Sn (AMS 4975)	R54620
Alpha-bet alloys	
Ti-3Al-2.5V (AMS 4943, ASTM Grade 9)	R56320
Ti-6Al-4V (ASTM Grade 5)	R56400, R56401
Ti-7Al-4Mo (AMS 4970)	R56740
Ti-6Al-6V-2Sn (AMS 4918)	R56620
Ti-6Al-6Mo-4Zr-2Sn (AMS 4981)	R56260
Ti-8Mn (AMS 4908)	R56080
Ti-10V-2Fe-3Al	—
Beta alloys	
Ti-13V-11Cr-3Al (AMS 4917)	R58010
Ti-11.5Mo-6Zr-4.5Sn (ASTM Grade 10)	R58030
Ti-8Mo-8V-3A1-2Fe	R58820
Ti-3Al-8V-6Cr-4Zr-4Mo	R58640
Ti-15V-3Cr-3Sn-3Al (AMS 4914)	—

interstitial elements oxygen, nitrogen, carbon, and also iron. As the amounts of these elements in titanium increase, strength also increases. While strengthening by cold work is possible, it is seldom used.

Alpha and Near-Alpha Alloys

Alpha alloys are not normally heat-treated to increase strength. They always contain a high percentage of alpha phase in their microstructure and, in some cases, they are totally alpha phase. The alloys are commonly used where moderate elevated temperature strength or creep resistance is required. Near-alpha alloys are designed for outstanding creep strength and elevated temperature stability by the addition of small amounts of beta-stabilizing elements such as Mo, Va, Zr and Cb.

Alpha-Beta Alloys

Alpha-beta alloys contain mixtures of alpha and beta phases in their micro structures. These alloys can be strengthened by solution treating and aging heat treatments. Some alpha-beta alloys are used also in the annealed condition. They have excellent fracture toughness when annealed, and outstanding strength-to-density ratios in the heat-treated condition. The Ti-6Al-4V alloy is the most widely used alpha-beta titanium alloy.

Beta Alloys

Commercial beta alloys contain a high percentage of beta phase stabilizing elements, but they are not truly single phase. The transformation of beta to alpha phase is very sluggish. During normal processing, the microstructure is nearly all beta phase at room temperature. They can be aged by a long-time, low-temperature heat treatment to greatly increase tensile strength. The strengthening mechanism is the precipitation of alpha or a compound in the beta matrix. Beta alloys in general are characterized by excellent formability in the all beta-phase condition, and are often selected because their formability aids in the manufacture of complex shapes. When heat-treated, they exhibit high strength-to-weight ratios, but have relatively low ductility and fracture toughness. Beta alloys show exceptional work-hardening characteristics, and are used in fasteners and springs.

Heat Treatment

Annealing. Titanium and titanium alloys are annealed to improve ductility, dimensional or thermal stability, fracture toughness, and creep resistance.[1] Improvement in some properties is usually obtained at the expense of other properties. Therefore, the titanium producer should be consulted concerning a recommended heat treatment to provide desired properties. Suggested annealing treatments for several titanium alloys are given in Table 27.2.

Solution Heat Treatment And Aging. Alpha-beta and beta titanium alloys can be strengthened by heat treatment. A basic heat treatment consists of solution heating at a temperature high in the single-phase, alpha-beta field followed by quenching in air, water, or other suitable media. The beta phase present at the solutioning temperature may be completely retained during cooling, or a portion may transform to alpha phase. The specific response depends upon the alloy composition, the solutioning temperature, the cooling rate, and the section size.

The solution-treated alloy is then aged at a suitable temperature in the 900°, to 1200°F (480° to 650°C) range. During aging, fine alpha particles precipitate in the retained beta phase. This duplex structure is harder and stronger than an annealed structure. Solution treating and aging can increase the strength of these alloys up to 50 percent over that of annealed material. Typical solutioning and aging heat treatments for several titanium alloys are given in Table 27.3.

As with annealing, specific mechanical properties of titanium alloys are influenced by solutioning and aging heat treatments. The recommendations of the producers should be followed to obtain desired properties.

Thermal Effects

Brazing operations have very little effect on the properties of commercially-pure titanium

---

1. For additional information on heat treating titanium alloys, 763-74 refer to the *Metals Handbook*, Vol. 4, *Heat Treating*, 9th Ed., Metals Park, OH, American Society for Metals, 1981.

Table 27.2
Annealing Heat Treatments for Titanium Alloys

Alloy	Temperature °F	°C	Time, h	Cooling Medium
Commercially Pure Grades	1200-1400	650-760	0.1-2	Air
Alpha Alloys and Near Alpha Alloys				
Ti-5Al-2.5Sn	1325-1550	720-845	0.2-4	Air
Ti-8Al-1Mo-1V	(1) 1450a	790	1-8	Furnace
	(2) 1450	790	0.25	Air
Ti-5Al-5Sn-2Mo-2Zr	1200-1450	650-790	0.5-2	Air
Ti-6Al-2Co-1Ta-0.8Mo	1450-1650	790-900	1-4	Air
Ti-6Al-4Zr-2Mo-2Sn	(1) 1650a	900	0.5-1	Air
	(2) 1450	790	0.25	Air
Alpha-Beta alloys				
Ti-3Al-2.5V	1200-1400	650-760	0.5-2	Air
Ti-6A-4	1300-1450	705-790	1-4	Air or furnace
Ti-7Al-4Mo	1300-1450	705-790	1-8	Air
Ti-6Al-6V-2Sn	1300-1500	705-815	1-4	Air or furnace
Ti-6Al-6Mo-4Zr-2Sn	1300-1350	705-730	2	Air
Ti-8Mn	1200-1400	650-760	0.5-1	Furnace to 1000°F (540°C) then air
Ti-10V-2Fe-3Al	1450-1500	790-815	0.1-0.25	Air
Beta Alloys				
Ti-13V-11Cr-3Al	1300-1450	705-790	0.2-1	Air or water
Ti-11.5Mo-6Zr-4.5Sn	1300-1400	705-760	0.2-1	Air or water
Ti-8Mo-8V-3Al-2Fe	1425-1450	774-790	0.1-0.25	Air or water
Ti-3Al-8V-6Cr-4Zr-4Mo	1450-1500	705-815	0.3-0.5	Air or water
Ti-15V-3Cr-3Sn-3Al	1450	790	0.1-0.33	Air or equivalent

and the alpha titanium alloys. When used in the annealed condition, beta alloys are virtually unaffected by a brazing thermal cycle. If they are to be used in a heat-treated condition, the brazing temperature can have an important effect on mechanical properties. The best ductility is obtained when a beta titanium alloy is brazed at its solution treating temperature. As the brazing temperature is increased above this temperature, the ductility of the alloy decreases.

Depending on composition, the mechanical properties of alpha-beta titanium alloys may be altered greatly by both the starting micro-structure as well as changes to it caused by heat treatment. Wrought alpha-beta titanium alloys generally are fabricated to obtain a fine-grained, equiaxed, duplex microstructure that provides maximum ductility. When brazing an alpha-beta alloy, it is desirable to maintain this microstructure by brazing at a temperature that does not exceed the beta phase transformation temperature (beta transus). That temperature may vary from 1650 to 1900°F (900 to 1038°C) depending upon specific base metal composition. Alpha-beta alloys may be designated for use in either the annealed condition or the solution treated and aged condition.

Table 27.3
Solutioning and Aging Heat Treatments for Titanium Alloys

Alloy	Solutioning Temperature °F	Solutioning Temperature °C	Time, h	Cooling Medium	Aging Temperature °F	Aging Temperature °C	Time, h
Alpha-Beta Alloys							
Ti-3Al-2.5V	1600-1700	870-930	0.25-0.3	Water	900-950	480-510	2-8
Ti-6Al-4V	1650-1775	900-970	0.2-1	Water	900-950	480-510	4-24
Ti-7Al-4Mo	1675-1775	915-970	0.2-2	Water	950-1200	510-650	4-24
Ti-6Al-6V-2Sn	1550-1650	840-900	0.2-1	Water	875-1150	470-620	2-8
Ti-6Al-6Mo-4Zr-2Sn	1550-1700	840-930	0.2-1	Air	1050-1150	565-620	2-8
Ti-10V-2Fe-3Al	1450-1500	790-815	1	Water	925-975	495-525	8
Beta Alloys							
Ti-13V-11Cr-3Al	1400-1500	760-815	0.2-1	Air or water	825-1000	440-540	2-60
Ti-11.5Mo-6Zr-4.5Sn	1350-1450	730-790	0.2-1	Air or water	900-1100	480-590	8
Ti-8Mo-8V-3Al-2Fe	1450-1475	790-800	0.1-1	Air or water	900-950	480-510	8
Ti-3Al-8V-6Cr-4Zr-4Mo	1500-1700	815-930	0.5	Air or water	900-1100	480-590	6-12
Ti-15V-3Cr-3Sn-3Al	1450	790	0.1-0.33	air or equivalent	900-1150	480-620	4-16

a. The time at temperature depends upon the temperature and section thickness. Longer times are used for low temperatures and thicker sections.

If an annealed microstructure is desired after brazing, the following alternatives are possible:

(1) Anneal, then braze at or below the annealing temperature.

(2) Braze above the annealing temperature and incorporate a step-cooling operation in the brazing cycle to obtain an annealed structure.

(3) Braze above the annealing temperature and anneal after the brazing operation is completed.

Many commercial brazing filler metals flow at temperatures above the annealing temperature of a titanium alloy. Therefore, the latter two alternatives are normally used for brazing. Problems can be anticipated in selecting filler metals and brazing cycles that are compatible with the heat treatment required for alpha-beta and beta titanium alloys. Ideally, brazing should be performed at a temperature that is 100° to 150°F (38° to 66°C) below the beta transus tempera-

ture because the ductility of alpha-beta alloys may be impaired if this temperature is exceeded. The beta transus temperature can be exceeded when beta titanium alloys are brazed. However, if the brazing temperature is too high, base metal ductility may be impaired by considerable diffusion and alloying of the filler metal and base metal. The beta transus temperatures for selected titanium alloys are listed in Table 27.4.

The mechanical properties of heat treatable titanium alloys also may be affected adversely by brazing unless the brazement can be heat-treated afterwards. For example, the alpha-beta titanium alloys must be solution-treated, quenched, and aged to develop optimum properties (Table 27.3).

Brazing filler metals that permit brazing and solution treating in a single operation are not always readily available. Similarly, it is not always possible to quench a brazed assembly at the desired cooling rates. Also, certain configurations (e.g., honeycomb

Table 27.4
Beta Transus Temperatures of Titanium Alloys

Alloy	Beta Transus	
	°F, +/- 25°	°C,+/- 15°
Commercially pure Ti, 0.25 max O^2	1675	910
Commercially pure Ti, 0.40 max O^2	1735	945
Alpha and Near-Alpha Alloys		
Ti-5Al-2.5Sn	1925	1050
Ti-8Al-1Mo-1V	1900	1040
Ti-6Al-4Zr-2Mo-2Sn	1820	995
Ti-6Al-2Cb-1Ta-0.8Mo	1860	1015
Ti-0.8Ni-0.3Mo	1615	880
Alpha-Beta Alloys		
Ti-6Al-4V	1830[b]	1000[b]
Ti-6Al-6V-2Sn	1735	945
Ti-3Al-2.5V	1715	935
Ti-6Al-6Mo-4Zr-2Sm	1720	940
Ti-7Al-4Mo	1840	1000
Ti-8Mn	1475[d]	800[c]
Beta or Near-Beta Alloys		
Ti-13V-11Cr-3Al	1330	720
Ti-11.5Mo-6Zr-4.5Sn	1400	760
Ti-3Al-8V-6Cr-4Zr-4Mo	1460	795

(a) +/- 20°.   (b) +/- 30°.   (c) +/- 35°.   (d) +/- 50°.

sandwich structures) do not lend themselves to rapid quenching without distortion. Brazing at the aging temperature is impractical because few brazing filler metals melt and flow at this temperature.

The possibility of galvanic corrosion must be considered when filler metals are selected for brazing titanium alloys. The chemical activity of titanium tends to decrease in an oxidizing environment because the surface undergoes anodic polarization in a manner similar to that of aluminum. In many environments, titanium becomes more noble than most structural metals. The corrosion resistance of titanium generally is not affected by contact with structural metals. However, other metals (e.g., copper) will corrode rapidly when they are in contact with titanium in an oxidizing environment. Thus, filler metals must be chosen carefully to avoid preferential corrosion of brazed joints.

When titanium is brazed, extreme care must be taken to ensure that the brazing retort or chamber is free of contaminants from previous brazing operations. Experience has shown that the properties of titanium may be deteriorated by gaseous contamination from a brazing furnace. To avoid contamination during brazing in a furnace, a protective shield of titanium foil (AMS 4900) or other suitable material, 0.001 to 0.003 in. (0.003 to 0.08 mm) thick, should be put over or around the assembly. This shield will act as a "getter" of any gaseous residual contaminants. Precautions should also be taken to ensure that the foil does not directly contact the workpieces.

The choice of materials to be used in brazing fixtures also must be considered carefully to avoid contamination. For example, nickel and titanium will alloy together at about 1730°F (940°C) to form a low-melting eutectic (28.4% Ni). If a titanium workpiece and a nickel-base alloy fixture are in contact, they may fuse together if the brazing temperature is in excess of 1730°F (940°C). In addition, solid state diffusion may result if contact is made above 1500°F (816°C). This also can occur when a retort made of nickel-base alloy is used for brazing, and inadvertent contact takes place between the workpieces and the retort.

ZIRCONIUM ALLOYS

Zirconium base metals of commercial importance are pure zirconium and several alloys containing small percentages of tin, columbium, iron, chromium, and nickel. Their chemical compositions are listed in Table 27.5. The alloys were developed for corrosion resistance in nuclear applications, notably pressurized water nuclear power reactors.

Zirconium alloys are useful in constructing various components in nuclear reactors

Table 27.5
Chemical Composition of Commercial Zirconium and Zirconium Alloys

ASTM Grade	Zr + Hf, min[b]	Composition, Percent[a]			
		Sn	Cb	Fe + Cr	O, max
R60702[c]	99.2	—	—	0.2 max	0.16
R60704[d]	97.5	1.0-2.0	—	0.2-0.4	0.18
R60705[e]	95.5	—	2.0-3.0	0.2 max	0.18
R60706[e]	95.5	—	2.0-3.0	0.2 max	0.16

a. H - 0.0005 percent max.
   N - 0.025 percent max.
   C - 0.05 percent max.

b. Hf - 4.5 percent max.

c. Grade R60001 is nuclear grade unalloyed zirconium.

d. A similar nuclear Grade R60802 is commonly called Zircaloy-2, and Grade R60804 is called Zirealoy-4.

e. The similar nuclear Grade is R60901.

because of (1) low-absorption cross section for thermal neutrons, (2) excellent corrosion resistance in water, sodium, and potassium, and (3) good mechanical properties. They are used frequently as a cladding material for fuel elements.

Like titanium, zirconium will react readily with oxygen, nitrogen, and hydrogen, and the effects are similar. The metal will alloy with many other metals and alloys to form brittle intermetallic compounds.

Oxidation of zirconium occurs as low as 400°F (205°C), and the rate of oxidation increases with temperature. The reaction of zirconium with nitrogen occurs slowly at about 750°F (400°C). Above about 1500°F (815°C), the reaction rate increases rapidly. Zirconium is embrittled by the absorption of hydrogen, which occurs rapidly at temperatures between 600° and 1800°F (315° and 980°C). As a result of this reactivity, zirconium must be brazed in a vacuum of less than 10^{-4} torr with a maximum leak rate[2] of 1 micron/hr, or a dry atmosphere of argon or helium to prevent re-oxidation of the surface. Zirconium components must be cleaned carefully before brazing to remove oxides and other surface contaminants. Brazing should be done within 4 hours after cleaning to prevent the re-oxidation of the substrate.

BERYLLIUM

Much of the early development work on beryllium was done for nuclear energy application where high purity was vital. In addition to premium nuclear grades, beryllium is now available as massive hot-pressed block, large wrought plate, and sheet stock. From a modulus-to-density standpoint, beryllium is about four to six times as efficient as conventional structural metals. Beryllium has been long used for windows in X-ray tubes and similar devices because of the high permeability to X-rays. Beryllium is used for nuclear reactor moderators and reflectors because of the low

thermal-neutron-absorption cross section and the large scattering cross section. Beryllium is sometimes used for aerospace structures because of low density, high stiffness-to-weight ratio, and excellent dimensional stability.

The metal has some disadvantages. Its limited ductility presents problems in design and fabrication. Toxicity of beryllium and its compounds are well known and extreme precautions must be observed in fabrication.

Beryllium reacts with carbon, oxygen, and nitrogen at conventional brazing temperatures. High temperature brazing usually is done in an argon atmosphere or a vacuum of less than 10^{-4} torr with a leak rate of 1 micron/hr because oxidized or nitrided surfaces impair the wetting and flow of brazing filler metals. These surface contaminants cannot be removed by a brazing atmosphere.

Bare metal surfaces must be thoroughly cleaned before brazing. Fluxes have been used to prevent oxidation during low temperature brazing of beryllium in air. Sometimes, beryllium surfaces are plated with silver or painted with titanium hydride to improve the wetting and flow of a brazing filler metal. Reoxidation of the precleaned surface be prevented by the application of a thin film of amylacetate which readily burns off at brazing temperatures.

Joint design, filler metal selection, and choice of brazing cycle are complicated by low ductility and reactivity with many of the filler metals used for brazing. Low ductility of beryllium is aggravated by the presence of structural discontinuities, such as surface scratches and notches, and the asymmetrical stress patterns produced by single-lap joints. Beryllium parts must be handled carefully. Butt, scarf, step, and double-lap joint configurations should be considered when structures are designed.

Since beryllium reacts with the constituents of most brazing filler metals, brazing should take place under conditions that minimize the formation of intermetallic compounds. These conditions include rapid heating and cooling cycles, low brazing temperatures, minimum time at brazing temperature, and minimum amounts of filler metal. However, sufficient amounts of filler metals frequently must be preplaced between joint

---

2. Leak rate is a measure of the capability of a tight vacuum chamber to hold vacuum. After having been pumped down to a high vacuum, and with the furnace pumps valved off, a tight furnace should not lose more than 1 micron/hr. of vacuum.

members to fill the joint. Induction brazing in vacuum or argon with short heating cycles, as compared to furnace brazing, often is better for these applications.

SAFE PRACTICES

Titanium and Zirconium

The possibility of spontaneous ignition of titanium and zirconium components during brazing is extremely remote. Like magnesium and aluminum, the occurrence of fires is usually encountered where an accumulation of grinding dust or machining chips exists. Even in extremely high surface-to-volume ratios, accumulations of clean particles do not ignite at any temperature below incipient fusion temperature in air.

However, spontaneous ignition of fine grinding dust or lathe chips saturated with oil under hot and humid conditions has been reported. Water or water-based coolants should be used for all machining operations. Carbon dioxide is also satisfactory for cooling. Large accumulations of chips, turnings, and metal powders should be removed and stored in enclosed metal containers. Dry grinding should be accomplished in a manner that will provide adequate heat dissipation.

Dry fire extinguishing agents such as graphite powder or dry sand are effective for titanium and zirconium fires. Ordinary extinguishing agents such as water, carbon tetrachloride, and carbon dioxide foam are ineffective.

Violent oxidation reactions can occur between titanium or zirconium and liquid oxygen or red-fuming nitric acid.

Beryllium

Beryllium and its compounds in the form of dust, fume, or vapors are toxic and a serious health hazard. Inhalation or ingestion of these materials in any form must be avoided. They can cause acute or chromic lung disease and other health problems.

Hazardous concentrations of beryllium vapor may build up in the atmosphere when the metallic beryllium is heated above 1200°F (650°C) or when beryllium oxide is heated above 2800°F (1540°C). Therefore, all brazing operations should be done in a brazing furnace that is exhausted through a system of high-efficiency filters. Appropriate precautions must be taken when cleaning the furnace and exhaust system to avoid inhalation or ingestion of beryllium dust that may have collected during use on interior walls, components, or equipment. Users should comply with appropriate safety and health standards applicable to beryllium.

FILLER METALS

Titanium

Pure silver and certain silver alloys, such as 95Ag-5Al and 92.5Ag-7.5Cu (AWS BAg-19), can be used to braze titanium. These brazing filler metals can provide useful joint strengths up to about 800°F (425°C). The Ti-Ag intermetallic compound is reasonably ductile, but the Ti-Cu intermetallic is not. Therefore, the copper content of silver brazing filler metals should be kept low. Lithium additions tend to accelerate the diffusion and alloying of silver base filler metals with titanium.

The Ag-5Al-0.5Mn brazing filler metal has good resistance to salt-spray stress corrosion and excellent shear strengths in titanium joints. Its brazing temperature range is 1600 to 1650°F (870 to 900°C). The diffusion layer at the filler metal-titanium interface is a relatively ductile Ag-Al-Ti solid solution. Shear strengths in the 20 to 30 ksi range can be obtained when brazing Ti-6Al-4V alloy. Brazed joints show good fatigue life in shear and axial tension-tension modes.

Types 3003 and 4043 aluminum alloys and unalloyed aluminum as brazing filler metals for titanium can provide good shear strengths up to about 500°F (260°C). The brazing temperature should be in the range of 1200° to 1275°F (650 to 690°C), depending upon the melting temperature of the filler metal.

For applications requiring a high degree of corrosion resistance or high strength up to about 1000°F (540°C), 48Ti-48Zr-4Be or 43Ti-43Zr-12Ni-2Be filler metal is recommended. The Ti-Zr-Be filler metal can also be used to braze titanium to carbon steel, austenitic stainless steel, refractory metals, and other reactive metals. One disadvantage of these filler metals is their high brazing temperatures, in the range of 1600° to 2000°F (870° to 1095°C). In addition, the filler metals are not available commercially because of toxicity from the beryllium in them.

The two brazing filler metals, 82Ag-9Pd-9Ga alloy and 81Pd-14.3Ag-4.6Si alloys, flow well on titanium, provide good strength, and have good corrosion and oxidation resistance. The Ag-Pd-Ga filler metal has a brazing range of 1615° to 1690°F (879° to 921°C), and is available as foil, wire, and powder. These alloys must be brazed in either a high-purity argon atmosphere or a vacuum with partial pressure of argon gas because of the high vapor pressure of silver and gallium at the brazing temperature. Lap-shear strengths with this filler metal average 26 ksi.

Brazing alloy of Ti-Ni in a composite foil is used for brazing Ti alloys. This product is available in wire and powder forms. Copper and nickel-base filler metals are not recommended for conventional brazing of titanium because they form brittle intermetallic compounds and low-melting eutectic compositions. However, a thin (.002 in.) layer of preplaced copper (or other) filler metal can be used with diffusion brazing to form an in-situ brazing filler metal, formed by diffusion and alloying at brazing temperature. The braze layer diffuses into the titanium during a diffusion heat treatment. Copper film can be diffusion brazed at 1425 to 1450°F (775 to 790°C) for 15 minutes or longer, depending on the film thickness.

In the specialized fields of electronic and high vacuum devices, copper, ceramics, and magnetic alloys are commonly brazed to titanium. A binary titanium-beryllium filler metal is often used.

In all applications, the time at brazing temperature should be held to a minimum to limit diffusion and alloying in the joint.

Zirconium

The development of brazing filler metals for zirconium has been limited to those with suitable corrosion resistance to high-temperature water in a nuclear reactor environment. Suggested brazing filler metals for zirconium are given in Table 27.6.

The 95Zr-5Be filler metal, with a melting range of 1780° to 1815°F (970° to 990°C), can be used to braze zirconium to itself and other metals, such as stainless steel. This

Table 27.6
Suggested Brazing Filler Metals for Joining Zirconium

Composition, Percent	Brazing Temperature	
	°F	°C
95Zr-5Be	1840	1005
50Zr-50Ag	2770	1520
71Zr-29Mn	2500	1370
76Zr-24Sn	3150	1715

filler metal, which is usually applied in powder form, exhibits good wetting and flow properties. There is no noticeable erosion or attack of the base metal during brazing, and it forms a smooth braze fillet. Brazements show good corrosion resistance to water at 680°F (360°C).

Because of its ability to wet ceramic surfaces, the 95Zr-5Be filler metal can be used to braze zirconium to uranium oxide and beryllium oxide.

Beryllium

Filler metals normally considered for brazing beryllium are aluminum-silicon filler metals containing 7.5 or 12 percent silicon (AWS BAlSi-2 or BAlSi-4). They can provide high joint strengths up to about 300°F (150°C). However, these filler metals exhibit poor flow in capillary joints, and must be preplaced in the joint.

Silver and silver-based filler metals are normally used for brazing beryllium for high temperature applications. Types BAg-19 (Ag-7Cu-0.2Li) filler metal and BAg-8a (Ag-28Cu-72.3Li) filler metal are examples. Lithium is added to pure silver in small amounts to improve wettability. But even so, capillary flow on beryllium is poor. Another filler metal suitable for Be, is 64Ag-34Cu-2Ti. Preplacement of the filler metal is recommended, but outside corners and edges of the joint sometimes will show a lack of adequate filler metal flow. This leaves a notch at the joint. In this case, it is often necessary to machine the joint to a smooth surface to avoid the severe notch sensitivity of beryllium.

Selection of a proper combination of brazing technique and filler metal depends upon the service temperature, joint geometry, and required joint strength. Good capillary flow of filler metal is difficult wherever beryllium oxide is present on the surfaces to be brazed. Therefore, a brazing filler metal that has a low melting temperature and also provides the desired mechanical properties is generally recommended.

Alloying and grain boundary penetration into the beryllium can also be a problem. Brazing times should be selected to minimize this condition.

Silver-copper filler metals that melt at relatively low temperatures may exhibit less grain boundary penetration. However, both copper and silver can form brittle intermetallic compounds with beryllium at temperatures below the melting point of the filler metal. This will result in low joint strengths and erosion of the beryllium. Fast heating, short brazing times, and rapid cooling will minimize formation of such compounds in the joint. These metallurgical problems can be avoided by using copper-silver filler metals that have small percent additives of tin or lithium, both of which are commercially available.

PRECLEANING AND SURFACE PREPARATION

Titanium and Zirconium

Prior to brazing or subsequent heat treating, titanium and zirconium components must be cleaned of surface contaminants and dried. Oil, fingerprints, grease, surface coatings, and other foreign matter must be removed using a suitable nonchlorinated solvent cleaning method. Ordinary tap water should not be used to rinse titanium and zirconium parts. Chlorides and other cleaning residues on titanium and zirconium surfaces can lead to stress-corrosion cracking when the components are heated above about 550°F (290°C) during brazing or heat treating. Hydrocarbon residues can result in contamination and embrittlement of titanium and zirconium.

If the parts to be brazed have lightly oxidized surfaces in the vicinity of the joint, the oxide can be removed by pickling the parts in an aqueous solution of two to four percent hydrofluoric acid and 20 to 40 per-

cent nitric acid until clean.[3] The parts are then rinsed in deionized water and dried. Hydrogen absorption by titanium and zirconium alloys is generally not a problem when the temperature of the acid solution is kept below 140°F (60°C). The parts should be handled, at this point, with lint-free gloves during assembly for brazing.

Scale formed at temperatures above 1100°F (595°C) is difficult to remove chemically. Mechanical methods, such as vapor blasting and grit blasting, are used for scale removal. Mechanical operations are usually followed by a pickling operation to insure complete removal of scale and any contaminated metal on the surfaces.

Where extended storage time is required before brazing, it may be necessary to place the parts in sealed bags containing a desiccant or in a controlled humidity storage room. When this is not possible, thorough degreasing and light pickling of the parts just prior to brazing is recommended. Mechanical abrasion of the faying surfaces followed by washing with a suitable solvent, is sometimes used in lieu of the pickling treatment. Fixturing should be thoroughly cleaned and degreased prior to loading.

Beryllium

Prior to brazing, beryllium components should be degreased, and then pickled in an aqueous solution of 10 percent hydrofluoric acid or 40 percent nitric-5 percent hydrofluoric acid. Subsequently, the components are ultrasonically rinsed with deionized water and dried. Precoating of the joint faying surfaces by special vacuum metallizing techniques or electroplating may be effective for promoting wetting by various braze filler metals. A thin coating of silver, titanium, copper, aluminum, or AWS BAlSi-4 filler metal may be satisfactory.

A vacuum-deposited titanium coating may be very effective in promoting wetting on beryllium. This coating drastically changes the surface of the beryllium and promotes extensive spreading and wetting when pure aluminum filler metal is used. Tests have indicated that the ultimate tensile strength at room temperature of butt

---

3. Special safe practices are required when handling any acid, particularly hydrofluoric acid. Suitable protective clothing and equipment must be used.

joints brazed with commercially pure aluminum compares very favorably with that for the beryllium base metal.

Preplacement of titanium hydride powder on the beryllium joint faces may be an acceptable substitute for vapor deposited titanium. The powder is mixed with standard commercial brazing cement (polyethyl methacrylate and 1, 1, 1-trichloroethane). The cement will decompose during brazing in vacuum at about 1000°F (540°C). The titanium hydride will completely dissociate at 1220°F (660°C), leaving a layer of titanium metal on the surface.

FLUXES AND ATMOSPHERES

Brazing with high-purity inert gas atmospheres such as argon or helium, and in vacuum with extremely low partial pressures of oxygen, nitrogen, and water vapor are essential for titanium, zirconium, and beryllium. Vacuum and inert gas atmospheres will protect titanium during furnace and induction brazing operations. A pressure of 10^{-4} torr (0.013 Pa) or lower, or a dew point of -70°F (-54°C) or lower is necessary to prevent discoloration of titanium when brazing in a temperature range of 1400 to 1700°F (760 to 927°C). Discoloration during cooling indicates a chamber leak. Discolored Ti becomes bright when heated in vacuum to about 1650°F (900°C). Generally, argon obtained from the liquid, which is stored in a cryogenic vessel, is preferred because of the inherent purity.

Although information on the brazing of zirconium is limited as compared to titanium, the two metals are very similar chemically. In general, the brazing techniques used are applicable to both metals.

Beryllium is most commonly brazed in an argon atmosphere but helium or a vacuum of less than 10^{-3} torr (0.133 Pa) is also suitable. Two fluxes have been used for furnace and induction brazing of beryllium with good results: a 60% LiF-40% LiCl mixture and a tin-chloride compound. These fluxes should not be used when brazing in vacuum. Beryllium and beryllium compounds as flux residues are toxic. Only OSHA approved installations should be considered for machining, cleaning, assembling, and

brazing of beryllium, regardless of the methods used.

BRAZING PROCESSES

Induction Brazing

Induction heating is well suited for brazing titanium. It is especially recommended for use with filler metals that alloy readily with titanium to take advantage of short brazing cycles that minimize alloying in the joints. Inert gas protection is normally used for induction brazing operations although vacuum with partial pressures of argon can be used with brazing filler metals that have high vapor pressures when molten.

Furnace Brazing

Furnace brazing is easily adapted to various joint designs and large assemblies, and this process is frequently used for brazing titanium, zirconium, and beryllium. Furnace heating is a relatively slow method of heating and under certain conditions, the base material and braze filler metal will be in the brazing range for a longer period of time than with induction heating. Therefore, selection of a filler metal that does not alloy excessively with the base metal is important.

Titanium, zirconium, and beryllium can be brazed in a high-purity argon atmosphere in a sealed retort or bell. These alloys are normally brazed in a cold-wall vacuum furnace that can be evacuated to a high vacuum. It is desirable that the furnace have the capability of introducing partial pressures of argon to prevent vaporization of any high vapor pressure elements in the base metal or the braze filler metals. The use of a retort generally requires longer heating and cooling cycles than when brazing in a vacuum furnace. Longer heating and cooling cycles may result in excessive dissolution in the joint due to the inability to raise and lower the temperature of the components rapidly from the solidus to the brazing temperature.

Diffusion Brazing

Titanium And Titanium Alloys. Titanium and titanium alloys are readily joined by diffusion brazing (DFB). For some applications, diffusion brazed joints have better properties than do fusion welded joints. With diffusion brazing, the metallographic structure should reflect the diffusion of the brazing filler metal into the base metal. There is generally no significant reduction in mechanical properties of the base metal.

Other advantages of DFB, compared to fusion welding, are minimization of atmospheric contamination, little or no distortion, and potential weight savings. DFB is generally performed below grain growth or phase transformation temperature that could adversely affect the mechanical properties of the base metal.

Cleanliness of the surfaces to be joined has greater importance with this process than with conventional brazing because there is no melting nor flow of filler metal to flush entrapped impurities from the joint. Surface preparation for DFB is similar to that for conventional brazing. In addition, the surfaces to be joined must be flat and smooth so that intimate, uniform contact can be attained at the interface without the use of excessive pressure. The applied pressure should be just sufficient to hold the parts in contact. Mating surface finish requirements are not stringent.

The faying surfaces are electrolytically plated with a thin film of either pure copper or a series of elements, such as copper and nickel. When heated in an inert atmosphere to the brazing temperature of 1650 to 1700°F (900 to 927°C), the copper layer reacts with the titanium alloy to form a molten eutectic at the joint interface. The assembly is held at temperature for at least 1.5 hours, or is given a subsequent heat treatment at this temperature for several hours, to reduce the chemical composition gradient across the joint. A typical diffusion brazed joint between two Ti-8Al-1Mo-1V alloy sections is shown in Figure 27.1. In the illustration, a Widmanstatten structure formed at the joint as a result of copper diffusion into the titanium alloy.

Diffusion brazed joints made with a copper interlayer at 1700°F (927°C) for four hours exhibited tensile, shear, smooth fatigue, and stress-corrosion properties equal to those of the base metal. However, they had slightly lower notch fatigue and corrosion fatigue properties, and significantly lower fracture toughness.

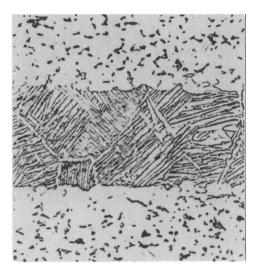

Figure 27.1 — Diffusion Brazed Joint in Ti-8Al-1Mo-1V Alloy Made With a Thin Copper Interlayer (500X)

Diffusion brazing is being used to fabricate light-weight cylindrical cases of titanium alloys for jet engines. In this application, the titanium core is plated with a very thin layer of selected metals that react with the titanium to form a eutectic. During the brazing cycle in a vacuum of 10^{-5} torr, a eutectic liquid forms at 1650°F (900°C). This liquid performs the function of a brazing alloy between the core and face sheets. The eutectic quickly solidifies due to rapid diffusion at the joints. The assemblies are held at temperature for one to four hours to reduce the composition gradient at the joint.

Zirconium And Zirconium Alloys. Zirconium and zirconium alloys, like titanium, can be diffusion brazed using a thin film plating of copper or other metal to form an in situ liquid phase. Joints can be diffusion brazed using copper at 1900°F (1038°C) for 0.5 to 2 hours at temperature under a pressure of about 30 psi. Molten metal forms in a joint after about 0.25 hours at 1550°F (843°C). During subsequent exposure at higher temperature, the copper diffuses into the zirconium and forms a solid joint. A distinct layer of filler metal does not exist.

Torch Brazing

Oxyacetyene torch brazing can be used for titanium, but a special flux and a skilled brazer are required. The flux is preplaced in the joint, and the filler metal is preplaced or manually fed into the joint during brazing. The use of preplaced filler metal is recommended with some fluxes. A slightly reducing flame is used. Brazing temperature is limited by the characteristics of the flux as well as the melting temperature of the filler metal.

Torch brazing is more economical than induction and furnace brazing. However, the problems of removal of flux residue and surface contamination after brazing and possible flux entrapment in the joint limit the use of torch brazing operations. For these reasons caution must be employed when torch brazing titanium. It is highly recommended that development tests, simulating the joint, be performed to demonstrate that the joint will perform as intended during service.

APPLICATIONS

Titanium

Titanium is used widely for aerospace hardware. Commercially brazed assemblies include jet engine inlet vanes, hydraulic tubing or fittings, honeycomb sandwich panels, and heat exchangers. Figure 27.2 shows a large front fan turbojet engine with fan blades of 6Al-4V titanium base metal. Each blade has a wrap-around skin and corrugated core member, which are brazed in vacuum with a Ti-15Cu-15Ni filler metal.

Titanium plumbing systems that provide additional weight savings were a logical outgrowth from brazed high-pressure fluid line joints and fittings of stainless steel used on the XB-70 supersonic airplane and the lunar excursion modules. Application of hinged induction heating tools and protective argon atmosphere permitted rapid, reliable brazes to be made in the production of Air Force and Navy high-performance

TITANIUM VANE BRAZEMENTS ⟶

Figure 27.2 — Large Fan Turbojet Engine With Brazed Sheet Metal Vanes

aircraft. One wide body jetliner utilizes over 250 brazed joints involving commercially pure titanium tubing and Ti-6Al-4V fittings. Tubing sizes include 1/4, 5/16, 3/8, and 1/2 in. (6.4, 7.9, 9.5, and 12.7 mm) and the brazing filler metal is 90Ag-10Pd alloy.

Titanium joints also brazed with filler metal, 82Ag-9Pd-9Ga, offer improved produceability and metallurgical compatibility. The results of impulse fatigue, burst, and corrosion tests were very satisfactory.

An American supersonic transport prototype design relied heavily on high-efficiency honeycomb structures. One very promising material for fuselage, wing, and control surfaces was metallurgically bonded titanium sandwich panels. A 24 ft (7.3 m) long demonstration wing test panel is shown during lay-up in Figure 27.3. Honeycomb core was commercially pure 0.002 in. (0.05 mm) thick titanium foil; facings were chemically milled Ti-6Al-4V alloy of 0.010 in. (0.25 mm) thickness. Type 3003 aluminum alloy foil was preplaced between all faying surfaces. This assembly was brazed successfully using ribbon heated ceramic tooling and an argon atmosphere. The U.S. Air Force F-15 fighter aircraft also had this aluminum brazed titanium alloy honeycomb

sandwich system. Engine cowls are in current production.

Honeycomb structures made with Ti-3Al-25V alloy core and Ti-6Al-4V alloy face sheets are in production using diffusion brazing techniques. One of these applications is a light-weight jet engine case shown in Figure 27.4. It is 75 in. (190.5 mm) in diameter and 22 in. (559 mm) long.

Successful joining of dissimilar metal combinations has been achieved with titanium, stainless steel, and copper. A brazed transition section located between a titanium tank and a stainless steel feed line was evaluated during the course of the space program. Ti-6Al-4V alloy was vacuum induction brazed to Type 304L stainless steel with the Au-18 Ni brazing filler metal. Formation of brittle intermetallic compounds was minimized by heating the joint rapidly and holding it at brazing temperature for a minimum of time. The same procedures have been used to make joints between titanium and mild steel, tool steel, and other metals.

Alloy development studies were conducted to develop a brazing filler metal for joining titanium to stainless steel for other uses in the space program. The filler metal

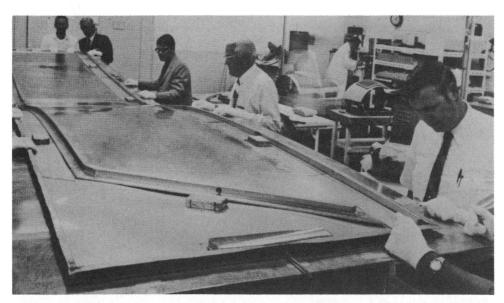

Figure 27.3 — Lightweight Titanium Honeycomb Sandwich Panel, 24 ft. (7.3 m) in Length, for Supersonic Plane Wing

Figure 27.4 — Titanium Alloy Honeycomb Rear Fan Case for a Jet Engine

Pd-9.0Ag-4.2Si was used to braze stainless steel to titanium in vacuum at 1360°F (738°C). Excellent flow properties were exhibited by this filler metal. Titanium alloys have been brazed successfully to carbon steel and austenitic stainless steel with the 48Ti-48Zr-4Be filler metal.

The alloy Ti-3Al-1.5Mn has been brazed to Cu-0.8Cr alloy using silver-base brazing filler metals. Three braze filler metals were evaluated during these programs: Ag-28Cu, Ag-40Cu-35Zn, and 68Ag-27Cu-5Sn. The joints were brazed in a vacuum at a temperature of 1520°F (827°C) for 5 minutes. Using these conditions, shear strengths of 28.4 to 38.3 ksi (196 to 264 MPa) were obtained. Heating rate and brazing temperature were critical. For maximum joint strength, the brazing temperature had to be controlled between 1518 and 1526°F (826 and 830°C).

A process for vacuum brazing of copper-plated titanium has been developed. Copper was electroplated on the surface of Ti-3Al-1.3Mn alloy after the surface had been hydrided in a sulfuric acid solution. Hydriding of the titanium surface removed hydrogen or hydrogen compounds, thereby creating an effective surface for the copper plate. Joints were made between the titanium alloy and (1) commercially pure copper, (2) stainless steel, and (3) a nickel-base alloy using 68Ag-27Cu-5Sn brazing filler metal. The joints were brazed at 1400 to 1500°F (760 to 816°C) for 15 to 20 minutes. The average shear strength was 28 ksi (193 MPa).

Zirconium

Applications for brazing of zirconium and zirconium alloys usually relate to nuclear requirements. One noteworthy use involved Zr-1.5Sn (UNS R60802) fuel tube cap assemblies 3 in. (76.2 mm) in diameter by 2 in. (50.8 mm) long. Approximately 500 of these 19 tube clusters were brazed at 1868°F (1020°C) in vacuum using Zr-5Be filler metal. Figure 27.5 shows a typical assembly. Excellent mechanical and thermal properties were achieved.

A recent application of electron beam brazing was for a Zr-1.5Sn alloy in-pile tube burst specimen. A stainless steel capillary tube was brazed into molybdenum adapter with copper, and then the adapter was brazed into the Zr-1.5Sn specimen with 48Ti-48Zr-4Be. Both brazes were made simultaneously by impinging a defocused electron beam on the molybdenum adapter. In this application, visual control of the brazing cycle was used.

The 48Ti-48Zr-4Be braze filler metal has also been used successfully to join alumina to Zr-1.5Sn alloys.

Beryllium

Brazing is used in the fabrication of complex, high-performance aerospace hardware of beryllium. Because of its remarkable modulus-to-weight ratio and commercial availability of large wrought sheet sizes, beryllium is used for spacecraft structure and instrument payloads. These are brazed by several filler metals in general use which give strong joints with good manufacturability.

An example of a small, complex brazement is a beryllium electronic-camera pressure housing, shown in Figure 27.6(A). This is 4 in. (102 mm) in diameter by 8 in. (203 mm) in height and weighs only 105 grams. Sleeves of Fe-29Ni-17Cu alloy were first brazed to the beryllium base using Ag-7Cu-0.2Li (BAg-19) filler metal. The electrical lead-throughs were then arc welded to the sleeves. Finally, the housing components, including the base, were brazed using aluminum alloy filler metal.

A primary structural brazement of beryllium was the main support ring for a satellite instrument. Figure 27.7 shows the ring that was assembled by furnace brazing with Al-12Si (AWS BAlSi-4) filler metal.

An example of a structural brazement is shown in Figure 27.8. It is an optical support structure of beryllium components and titanium inserts and is composed of 107 parts that were brazed using Type 1100 aluminum filler metal.

For nuclear applications, five Apollo scientific stations were powered by nuclear power generators. Figure 27.9 shows a critical beryllium casing brazement. Beryllium sheet fins were brazed to a machined beryllium case with 92.5Ag-7.5Cu-0.2Li (AWS BAg-19) filler metal. A stainless steel transition ring was brazed to each end of the beryllium housing using the same filler metal.

Numerous dissimilar metal combinations with beryllium have been brazed successfully. Stainless steel tubes have been brazed to beryllium end caps with 49Ti-49Cu-2Be filler metal. This filler metal readily wet the beryllium, and the joints exhibited adequate strength for the applications — provided the brazing time was kept sufficiently short to minimize the formation of inter-metallic compounds. Induction brazing in vacuum was used to make this assembly.

Beryllium window assemblies for X-ray tubes have been vacuum brazed into Monel retainers with silver-base filler metal BAg 8a. All assemblies were vacuum tight and suitable for use in X-ray tubes.

Figure 27.5 — Zirconium — 1.5 Percent Tin Vacuum Brazed End Ring

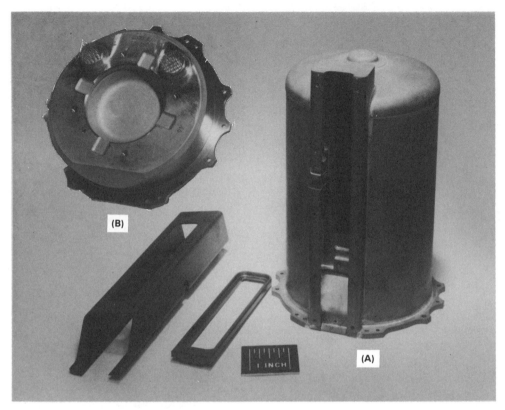

Figure 27.6 — Beryllium Housing For Spacecraft Lander (A) Brazed Beryllium Electronic Camera Pressure Housing, (B) Beryllium Base With Brazed Electrical Lead-Through Connections

Other vacuum brazing applications, currently being designed, developed, tested, or used, are multiwalled beryllium sandwich structures for antennas, space hulls, and interstage adapters.

Type 303 stainless steel pressure fittings were joined to a 0.080 in. (2.03 mm) thick beryllium sheet using pure silver filler metal. The joint was eutectic diffusion brazed at 1690°F (921°C) for 20 to 30 minutes in a vacuum. Brazing temperature and time at temperature must be kept to a mini-

mum to obtain high joint strengths when silver filler metal is used.

Beryllium was brazed to Type 304L stainless steel to demonstrate the feasibility of joining these metals for nuclear applications. The steel surface was coated with a thin film of Ag-28Cu filler metal by vapor deposition techniques. The brazing temperature was 1510°F (821°C). Lap shear strengths ranged from 14.2 to 20.5 ksi (98 to 141 MPa).

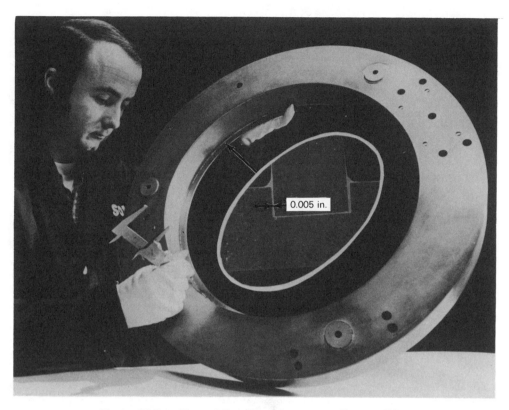

Figure 27.7 — Brazed Beryllium Instrument Support Ring

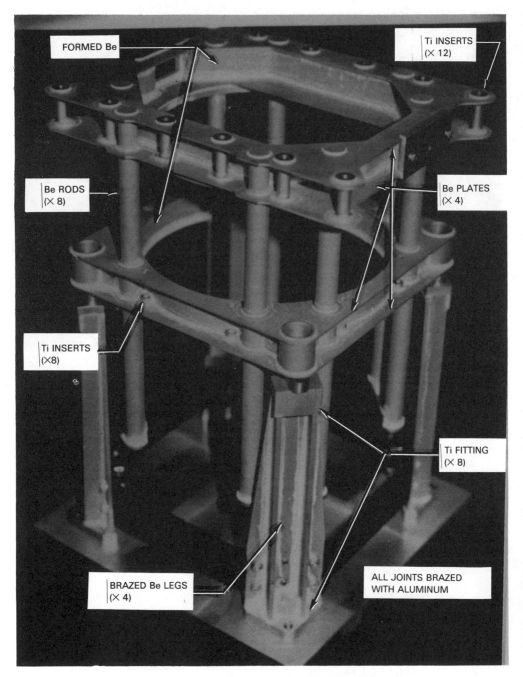

Figure 27.8 — Brazed Beryllium Optical Support Structure

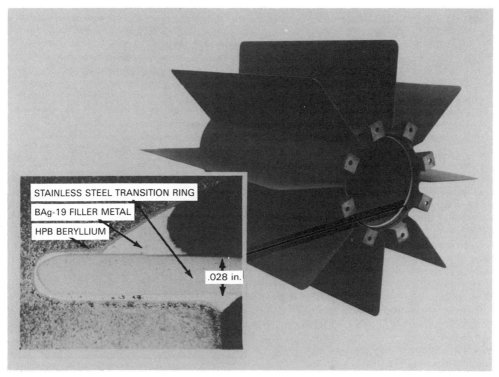

Figure 27.9 — Brazed Beryllium Nuclear Power Generator Casing

Supplementary Reading List

Beal, R. E. and Saperstein, Z. P. "Development of brazing filler metals for Zircaloy" *Welding Journal* 50(7): 275s-291s; July 1971.

Elrod, S. D., Lovell, D. T., and Davis, R. A. "Aluminum brazed titanium honeycomb-sandwich structure — a new system." *Welding Journal* 52(10): 425s-432s; October 1973.

Glenn, T. G., et. al. "Vacuum brazing beryllium to Monel." *Welding Journal* 61(10): 334s-338s; October 1982.

Howden, D. G. and Monroe, R. E. "Suitable alloys for brazing titanium heat exchangers." *Welding Journal* 51(1): 31-36; January 1972.

Kaarlela, W. T. and Margolis, W. S. "Development of the Ag-Al-Mn brazing filler metal for titanium." *Welding Journal* 53 (10): 629-636; October 1974.

Kimball, C. E. "Aluminum brazed titanium acoustic structures." *Welding Journal* 59(10): 26-30; October 1980.

Lam, S. W. "Laminating brazing filler metals for titanium assemblies." *Welding Journal* 61(10): 23-28; October 1982.

Schwartz, M. M. "Diffusion brazing titanium sandwich structures." *Welding Journal* 57(9) 35-43; September 1978.

Schwartz, M. M. "Brazing of reactive and refractory metals." *Metals Handbook*, Vol. 6, 9th Ed., 1049-1060. Miami: American Welding Society, 1983.

Wells, R. R. "Microstructural control of thin-film diffusion brazed titanium." *Welding Journal* 55(1): 20s-27s; January 1976.

Keywords — refractory metals, refractory alloys, niobium, molybdenumy, tantalum, tungsten, high melting point metlas, refractory metal melting points, recrystallization temperatures of refractory metals, brazing of refractory metals, brazes for high surface temperatures

Chapter 28

REFRACTORY METALS — NIOBIUM, MOLYBDENUM, TANTALUM, AND TUNGSTEN

The development of refractory alloys was influenced by the electric light and electronic tube industries through their use of tungsten and molybdenum. Today, the greatest impetus for development of the refractory metals niobium, molybdenum, tantalum, tungsten and their alloys has come from the aerospace industry's need for air frames and rocket motors. Experts predict that in the future the greatest manufacturing developments for refractory metals will be market driven. As a consequence, these materials will evolve through increased production and increases in demand for greater performance of high temperature structural components.

BACKGROUND

Mechanical strength is of primary importance to high temperature, structural component applications of refractory metals. To better understand the elements that comprise the refractory metals and their unique high temperature properties, it is possible to classify these metals according to their position in the Periodic Table. There are, however, other metals having high melting points. Figure 28.1 indicates the region of the Periodic Table containing metals having melting points approaching 3600°F (2000°C) or greater.

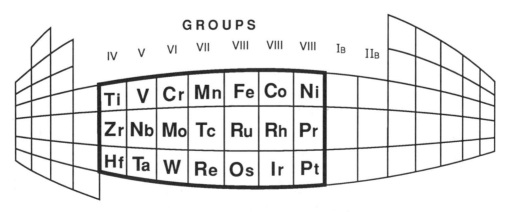

Figure 28.1 — High Melting Point Metals (Transition Elements) in the Periodic Table

Only a dozen metals have quite high melting points. Table 28.1 lists these metals in order of decreasing melting point with their crystal structure. This group can be further divided into two groups based on crystal structure. Six metals, W, Ta, Mo, Nb, V and Cr (i.e., the metals in Groups V and VI) have body center cubic (bcc) crystal structure. The remaining six elements (Rh, Os, Ru, Ir, Hf, Rh) have either face center cubic (fcc) or close packed hexagonal (cph) structures. The bcc metals are structurally superior for high-temperature applications compared to other metals with fcc or cph structures. Therefore, since refractory metals are chosen for structural applications when strength and good mechanical properties are required at high temperatures, those materials with cph or fcc structures will not be given further consideration in this chapter. In addition, commercial applications of these metals are limited due to their availability and exceptionally high cost for engineering functions. Furthermore, if the two lower temperature metals, vanadium and chromium, are deleted [the melting points of these metals are below 3600°F (2000°C)], then the remaining metals to be considered are niobium, molybdenum, tantalum, and tungsten.

Although the latter four metals have much more in common than presented in the refractory metal classification and organization scheme, there are differences that have a direct bearing in the manner in which these metals are used and joined. The metals niobium and tantalum occupy the period Group V and tend to have relatively high solubility for interstitials, low ductile-to-brittle transition temperature and are greatly affected by strain aging. In contrast, the metals molybdenum and tungsten found in period Group VI have low solubility for interstitials, a definite structure-sensitive transition temperature and are less affected by strain aging.

As a whole, niobium, molybdenum, tantalum, and tungsten are characterized by very high melting temperatures, high recrystallization temperatures, high to very high densities, low specific heats, and low coefficients of thermal expansion. The corrosion resistance of these materials in acid ranges from good to excellent, while the corrosion resistance of niobium, molybdenum, and tantalum is also good in liquid alkali metals. Niobium and molybdenum have very low capture cross sections for thermal neutrons; however, the cross sections for tantalum and tungsten are much higher.

Several characteristics of refractory metals related to mechanical properties are important because they may vary with service environment and must be considered when procedures are established to braze the refractory metals. These characteristics are ductile-to-brittle transition behavior, their recrystallization temperature, and their reactions with interstitial elements such as carbon and various interstitial gases.

Table 28.1
Refractory Metal Melting Points and Crystal Structures

Metal	Melting Point		Crystal Structure
	°F	°C	
Tungsten	6100	3380	b.c.c.
Rhenium	5700	3170	c.p.h.
Tantalum	5400	2996	b.c.c.
Osmium	4890	2700	c.p.h.
Molybdenum	4740	2620	b.c.c.
Niobium	4475	2468	b.c.c.
Ruthenium	4450	2450	c.p.h.
Iridium	4400	2443	f.c.c.
Hafnium	4025	2220	c.p.h.
Rhodium	3500	1960	f.c.c.
Vanadium	3380	1860	b.c.c.
Chromium	3350	1850	b.c.c.

As previously mentioned, niobium, molybdenum, tantalum, and tungsten all have a body centered cubic (bcc) crystal structure. A significant and characteristic behavior of the bcc crystal structure is the well defined ductile-to-brittle transition behavior. In notched bar impact testing, large decreases in energy absorption over a narrow temperature range are associated with distinct changes in fracture morphology from ductile tearing to brittle cleavage. The transition temperature ranges for the pure refractory metals are given in Table 28.2.

Factors that influence this transition temperature range for a particular metal are not a fixed property of the metal but are functions of strain rate, alloying additions, impurities, heat treatment, and method of fabrication. Thus, tungsten, compared to the other refractory metals being considered, has a relatively high ductile-to-brittle transition temperature. Therefore, care in brazing and handling tungsten must be exercised to insure that a stress-free condition is maintained to avoid damage. Although molybdenum is relatively tough at room temperature when in a pure form, it can become quite brittle due to impurities which raise the transition temperature. These impurities are oxygen, nitrogen, and carbon and can be present in commercial grades of molybdenum. The detrimental influence of these impurities on the transition temperature can be observed in Figure 28.2. Therefore, care must be taken when handling and brazing these refractory metals and their alloys, especially when impurities may be present in the parent material or service environment, or both.

Heating above the recrystallization temperature in a refractory metal is also an important problem. When the recrystallization temperature of the metal is exceeded during processing, the mechanical properties can be degraded. Reduction in strength of refractory metals occurs with microstruc-

Table 28.2
Transition Temperature Ranges for Pure Refractory Metals

Niobium	-328 to -103°F	(-200 to -75°C)
Molybdenum	-50 to 100°F	(-46 to 38°C)
Tantalum	<-320°F	(<-196°C)
Tungsten	150 to 450°F	(150 to 450°C)

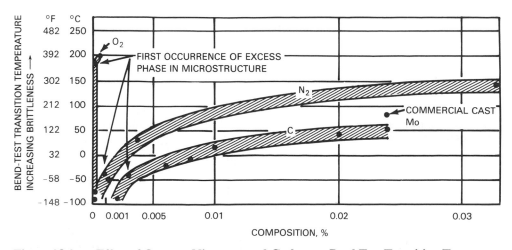

Figure 28.2 — Effect of Oxygen, Nitrogen, and Carbon on Bend Test Transition Temperature of Cast Molybdenum (Transverse Grain Orientation)

tural changes associated with recrystallization. During recrystallization, rapid diffusion of impurities to dislocations occurs. These impurities pin the dislocations and cause severe limitations in ductility. Table 28.3 provides the recrystallization temperature ranges for unalloyed refractory metals.

Several factors have been shown to alter recrystallization behavior and temperature. These recrystallization effects vary with alloying additions, interstitial content, fabrication method, and time at temperature.

Research has resulted in improved high-temperature properties of the refractory metals and has increased their respective recrystallization temperatures by alloying. For example, it has been found that titanium, zirconium, and hafnium can be used to strengthen molybdenum and increase its recrystallization temperature. The recrystallization temperature of unalloyed molybdenum is roughly 2100 to 2200°F (1149 to 1204°C). In contrast, the recrystallization temperature of TZM, (Mo-0.5Ti-0.07Zr molybdenum alloy) is about 2700°F (1482°C).

Deleterious effects of the recrystallization temperature can be avoided by designing the assembly to permit braze filler metals that melt below the recrystallization temperature range of the metal or alloy. When applications require the use of braze filler metals that melt above this temperature range, the joint must be designed to accommodate the loss in properties that is associated with recrystallization and diffusion of impurities.

As previously mentioned, reactions between the refractory metals and interstitial impurities in the form of gases or solids have an effect on the mechanical properties. These impurities can enter the base metal from the brazing filler metal, furnace atmosphere, as a contaminant on the base metal, or from fixtures made of graphite. Therefore, determining the environment in which

the refractory metals are brazed is an important consideration.

All refractory metals react with oxygen at moderately elevated temperatures, but different types of oxides are formed. The surface oxide films formed on niobium and tantalum at temperatures up to 626 F (330 C) are essentially amorphous and consist of Nb_2O_5 and Ta_2O_5, respectively. At higher temperatures [572 to 932°F (300 to 500°C)], niobium and tantalum form hard, pore-free, tenacious oxides. Oxides form on molybdenum and tungsten below 750°F (400°C). These oxides volatilize at temperatures above 750 to 950°F (400 to 510°C). The volatization of these oxides can influence the brazing process and metals. The surfaces of the refractory metals must be protected during brazing to insure wetting by the braze filler metal. If the refractory metals are expected to be exposed in air at elevated temperatures during brazing, protection can be achieved by an oxidation resistant material coating. Under this brazing condition, the brazing filler metal must be compatible with both the base metal and the coating.

Hydrogen influences the mechanical properties of the refractory metals in different ways. Niobium and tantalum are embrittled by the presence of hydrogen at relatively low temperatures. In contrast, molybdenum and tungsten can be brazed in a hydrogen atmosphere.

Gas-metal reactions involving nitrogen and the refractory metals niobium, molybdenum and tantalum commence at relatively low temperatures and embrittle these metals at high temperatures. For example, nitrogen reacts with molybdenum at temperatures as low as 1100°F (593°C), but severe embrittlement does not occur until the temperature exceeds about 2000°F (1093°C). In a similar way, tantalum behaves in much the same manner as molybdenum.

Table 28.3		

Table 28.3
Recrystallization Temperature Ranges for Unalloyed Refractory Metals

Niobium	1787 to 2102°F	(975 to 1150°C)
Molybdenum	2102 to 2192°F	(1150 to 1200°C)
Tantalum	2012 to 2552°F	(1100 to 1400°C)
Tungsten	2192 to 3000°F	(1200 to 1650°C)

All refractory metals react with carbon to form hard, brittle carbides. Failure in preventing this reaction has serious consequences in the degradation of properties in the base material adjacent to the joint. Therefore, the presence of carbon in any quantity or its compounds in solid, liquid and gaseous forms must be prevented.

Carbon can be introduced into the metal or alloy during brazing by hydrocarbon contamination on the surface of the metal, graphite brazing fixtures, or hydrocarbon contamination on the furnace walls and floor. Braze surfaces should be thoroughly clean and free of hydrocarbons. Brazing of refractory metals in reactive atmospheres is not recommended. The desired conditions are vacuum or dry inert (argon, helium) atmosphere.

To summarize the brazing of refractory metals the following applies:

(1) If maximum joint strength is required, the refractory metals must be brazed at temperatures below the recrystallization temperature.

(2) If brazing at higher temperatures is required by the service conditions, some loss of ductility in the material joint will occur and must be considered in joint design.

(3) All of the refractory metals can be brazed in a vacuum or in an argon or helium atmosphere with a very low dew point; molybdenum and tungsten can be brazed in a dry hydrogen atmosphere. When brazing in a vacuum, high vapor pressure braze filler metals are not used to avoid contamination of the vacuum system.

(4) Graphite fixturing should not be used to position refractory metal parts during brazing since these metals readily form carbides; graphite tooling may be acceptable if coated with a refractory material. Ceramics can be used for fixturing, but care must be exercised in their selection to avoid possible reactions and out-gassing under vacuum. Refractory metals with melting temperatures greater than the materials being brazed can be used for fixturing.

(5) Filler metals for brazing the refractory metals are selected on the basis of service conditions, the specific application, and compatibility with the base metal and coating (if a coating is used). The brazing of individual refractory metals is discussed in subsequent sections.

REFRACTORY METAL BRAZING

Refractory metals for low temperature service can be brazed with many silver and copper base braze filler metals. Refractory metals requiring high-temperature service can be brazed with braze filler metals based on nickel, gold, palladium and platinum based alloys. These braze filler metals and others will be presented in more detail in the following section. Also, specific examples of brazed components with general methodology and results will be discussed.

NIOBIUM AND ITS ALLOYS

Niobium is used mainly for nuclear and aerospace applications, so most braze filler metals developed for niobium are intended for these applications. The following section will present information concerning material preparation before brazing, brazing environment, corrosion resistance and compatibility of various filler braze metals with niobium.

Material Preparation

A primary step in brazing niobium, as with other materials, is starting with a clean surface free of grease and dirt. Niobium can be cleaned by both mechanical and chemical methods as shown in Table 28.4. Prior to brazing niobium, all gases, liquids, or solid residue must be removed from the surface to avoid undesirable reactions. The gases that must be eliminated are: oxygen, carbon monoxide, ammonia, hydrogen, nitrogen, and carbon dioxide. This is particularly important when brazing niobium at high temperatures. Protection from oxidation must be provided to ensure surface wetting and good bond formation.

Brazing Environments

If brazing under vacuum is not possible, or the removal of gases and contaminants is improbable, one method of preparing niobium prior to brazing is to electroplate either copper or nickel onto an acid-cleaned surface. The composite is then heated to a temperature of 2200°F (1204°C) under a vacuum of 0.001 Torr (0.15 Pa) to allow dif-

Table 28.4

Cleaning Methods for Refractory Metals

TUNGSTEN

Method 1	Immerse in 20% potassium hydroxide solution (boiling)
Method 2	Electrolytic etching in a 20% potassium hydroxide solution
Method 3	Chemical etching in a 50% vol. HNO_3 — 50 50% vol. HF solution
Method 4	Immerse in molten sodium hydroxide
Method 5	Immerse in molten sodium hydride

TANTALUM

Method 1	a.	Shot blast tantalum surface
	b.	Follow by immersion into HCl solution to remove embedded iron particles
	c.	Immerse in glass cleaning solution composed of 95% H_2SO_4, 4.5% HNO_2, 0.5% HF and Cr_2O_3 (Equivalent to 18.8 g/l)

MOLYBDENUM

Method 1	Immerse in a glass cleaning etch composed of 95% H_2SO_4, 4.5% HNO_3, 0.5% HF and Cr_2O_3 (Equivalent to 18.8 gm/l)
Method 2	a.	Immerse in alkaline bath of 10% NaOH, 5% $KMnO_4$ and 85% H_2O by weight operating at 150-180°F (66 - 82°C), 5-10 min.
	b.	Immerse 5-10 min. in second bath to remove smut formed during first treatment. Bath consists of 15% H_2SO_4 15% HCl, 70% H_2O, AND 6-10 WT. % PER UNIT volume of chromic acid.
Method 3		Mo-0.5 Ti alloy
	a.	Degrease 10 min. in trichloroethylene
	b.	Immerse in commercial alkaline cleaner 2-3 min.
	c.	Rinse in cold water.
	d.	Buff and vapor blast
	e.	Repeat step b
	f.	Rinse in cold water
	g.	Electropolish in 80% H_2SO_4 at 120°F (54°C) with 8-12 amp
	h.	Repeat step c

NOIBIUM

Method 1	a.	Degrease
	b.	Immerse in commercial alkaline cleaner 5-10 min.
	c.	Rinse with water
	d.	Immerse in 35-40% HNO_3 for 2-5 min. at room temperature
	e.	Rinse with tap water, followed by distilled water
	f.	Force air dry

fusion of the copper (nickel) plate with niobium. Brazing may then be accomplished by using the plated surfaces as a base. However, it is necessary for the braze filler metal to be compatible with the plating.

There is another technique for brazing niobium in air. This brazing technique can be accomplished through the use of a cover flux when inert gas (argon or helium) atmosphere or vacuum is not available. Although brazing niobium in air can be performed by using a suitable flux for a particular braze filler metal, niobium normally will require a protective coating, such as nickel or copper electroplate.

Braze Materials

When brazing niobium alloys without electroplated surfaces, several oxidation resistant braze filler metals are suggested and are available for high-temperature structural assemblies. Two of these brazing filler metals suited for a corrosive service environment are Ti-48Zr-4Be and Zr-19Nb-6Be. These braze filler metal joints are corrosion resistant to alkali metals and possess useful properties at temperatures of 1300 to 1500°F (704 to 816°C). In addition, these braze filler metals produce as-brazed shear strengths ranging from 20 to 33 ksi (138 to 228 MPa) at room temperature and from 12 to 20 ksi (80 to 138 MPa) at 1300°F (704°C). A decrease in the room temperature shear strength occurs with specimens aged for 100 hr at 1500°F (816°C) before testing. (Other alloys commercially available are Ag-Cu-Ti, Ag-Cu-In-Ti, Ag-Cu-Sn-Ti and can be brazed directly to niobium. Also, niobium to ceramic braze has been made with silver alloys.)

Niobium Braze Examples

The brazing of niobium can be accomplished with a variety of braze materials. Titanium and titanium-based braze filler metals produce joints with excellent ductility and with less base-metal reaction. In comparing the flow characteristics of these brazing filler metals, the flow characteristics of the B120 VCA (Ti-11Cr-13V-3Al) braze filler metal are more viscous than those of pure titanium. As a result of the high viscosity, the flow behavior of the filler metal is advantageous in bridging wide joint clearances.

Another brazing example for niobium uses pure titanium braze metal and titanium alloy braze metals. As an example, honeycomb sandwich panels were brazed with pure titanium braze metal at 3200°F (1760°C). Attached to this structure were heat shield panels that were brazed with B120 VCA braze filler metal at 3000°F (1649°C) in the same environment. Testing of the heat shields in a simulated re-entry environment indicated the panels could be used at temperatures as high as 2400°F (1316°C). Room and elevated temperature tests of the structural panels indicated retention of mechanical properties to 2300°F (1260°C).

Niobium-compatible braze filler metals in the Ti-V-Nb and Ta-V-Ti alloy systems have been investigated. The compositions of these and other filler metal alloys with their respective brazing temperatures are shown in Table 28.5. The Ti-V-Nb and Ta-V-Ti braze filler metal alloys are resistant to corrosion in alkali metals.

One example of their successful use was observed in joining a Nb-1Zr base metal assembly. Joining of this material was accomplished by vacuum brazing at 2280°F (1249°C) with a Ti-28V-4Be braze filler metal. Corrosion testing showed that the brazed joints were unaffected in an assembly exposed to potassium liquid and vapor up to 1000 hr at 1500°F (816°C).

Manufacturing procedures for fabricating flat and curved niobium alloy sandwich panels and heat shield panels have been established. Niobium alloy D36 (Nb-10Ti-5Zr) was used in producing a resistance welded honeycomb core with sheets of the same material brazed to the face. Among the filler metals evaluated for brazing the assembly were palladium, Pd-30Cu, titanium, and Ti-11Cr-13V-3Al (B120 VCA). Palladium and palladium-copper produced ductile joints with excellent filleting. However, during brazing considerable grain boundary attack occurs between the base metal and the palladium filler metal.

Other niobium alloys, such as D43 (10W-1Zr-0.1C-Nb), Nb-752 (10W-2.5Zr-Nb), and C-129Y (10W-11Hf-0.07Y-Nb), have been brazed successfully with two braze filler metals, B120VCA (Ti-11Cr-13V-3Al) and Ti-8.5Si [Braze temperature 2650°F (1454°C)]. A demonstra-

Table 28.5
Shear Strengths* of Ta-V-Nb and Ta-V-Ti Brazed Joints in Vacuum at 1093°C (2000°F)

Temperature Composition, Wt%	Brazing Temperature		Shear Strength+	
	°F	°C	ksi	MPa
25Ta-50V-25Nb	3400	1870	24.4	168
30Ta-40V-30Nb	3500	1925	22.4	154
5Ta-65V-30Nb	3300	1815	25.5	170
30Ta-65V-5Nb	3400	1870	23.4	161
30Ta-65V-5Ti	3350	1845	25.6	177
25Ta-55V-20Ti	3350	1845	18.9	130
20Ta-50V-30Ti	3200	1760	30.2	208
10Ta-40V-50Ti	3200	1760	20.5	141

* Base metal used for all tests was TZM (Mo-0.5Ti-0.07Zr Wt%)
+ Each data point for shear strength is an average of two tests.

tion of the variety of braze filler metals is the use of copper or zirconium base braze filler metals for brazing Nb-1Zr fins to Nb-1Zr tubes. On the basis of the test results, the Zr-28V-16Ti-0.1Be braze filler metal was used to braze full scale heat receivers.

Brazes for High Service Temperatures

Two new braze filler metals have been developed for D36 niobium. These braze filler metals and braze techniques have been designed to produce high remelt temperatures and retention of mechanical properties of the joint at high temperatures. The brazing techniques developed are based on diffusion brazing and reactive brazing concepts. The diffusion brazing technique involves movement and dispersion of the braze filler metal into the base metal under a controlled reaction condition. At this time in the joining process, dilution of the braze filler metal by the base metal increases the melting temperature of the joint. An example of a diffusion braze filler metal is Ti-33Cr. This braze filler metal requires a brazing temperature of 2650 to 2700°F (1454 to 1482°C).

The reactive brazing concept uses a braze filler metal containing a melting temperature depressant. The depressant is selected to react with the base material or powder additions to form a high-melting intermetallic compound during a post-braze diffusion treatment. The formation of this high-melting intermetallic compound increases the joint remelt temperature. Suc-

cessful application of this concept is dependent on controlling the intermetallic compound reaction to form discrete particles. A reactive braze filler metal is Ti-30V-4Be. This reactive braze filler metal requires a brazing temperature of 2350 to 2400°F (1288 to 1316°C).

A comparison between diffusion brazing and reactive brazing was conducted. Test results of Ti-33Cr diffusion brazed joints indicated an increase in lap shear strength from approximately 2.5 Ksi (17 MPa) to more than 4.5 Ksi (31 MPa) at 1500°F (1371°C) and from 0 to 1 Ksi (7 MPa) at 3000°F (1649°C). When the joint is given a diffusion treatment at 300°F (148°C) below the original brazing temperature, the joint remelt temperature is increased by 600°F (316°C), the lap shear strength is doubled at 2500°F (1371°C), and excellent strength is retained to 3000°F (1549°C).

Lap joints reactive brazed with Ti-30V-4Be exhibited substantially lower strength than similar joints diffusion brazed with Ti-33Cr, even though remelt temperatures were similar for both systems. Lap shear strength of Ti-30V-4Be joints was approximately one-half and one-third the strength of Ti-33Cr braze filler metal joints at 2500°F (1371°C) and 3000°F (1649°C), respectively. The decrease in properties observed in the lap shear strength of Ti-30V-4Be braze filler metal joints indicated potential for lower stress level applications.

MOLYBDENUM AND ITS ALLOYS

Because of its high melting point, there has been an emphasis in recent years on the use of molybdenum as a structural material for applications in the electronic, aircraft, missile, and nuclear reactor fields. The melting temperature of molybdenum is about 1800°F (982°C) above that of common high temperature iron base, cobalt base, or nickel base alloys. This melting point is exceeded only by tungsten and tantalum. In addition to its high melting temperature, high thermal conductivity, low specific heat, and low coefficient of thermal expansion, molybdenum provides good dimensional stability under rapidly changing thermal conditions.

Discussion on the brazing of molybdenum structures will focus on several areas of importance: oxide removal and material preparation before brazing, brazing environment, and compatibility of various filler braze metals with molybdenum.

Surface Preparation

In preparing molybdenum prior to brazing, removal of surface oxide is mandatory. The cleaning operation should be performed immediately before brazing to prevent contamination and reoxidation. Degreasing agents should be used to remove oil, fingerprints, and grease. Mechanical and chemical cleaning have been found satisfactory for oxide removal. Grit blasting, liquid abrasive cleaning, or abrasion may be used on simple parts to remove oxide films. However, chemical cleaning is more satisfactory, especially for complex assemblies. Electrolytic etchants may be used to remove oxides on simple parts but can result in severe grain boundary attack of the molybdenum. Chemical etchants are the most popular cleaning method. Heavy oxide films greater than several thousand angstroms thick have been removed successfully by using molten salt baths. Two baths which have been used are the following:

(1) A mixture of 70% sodium hydroxide and 30% sodium nitrite operating at 500 to 700°F (260 to 371°C) (Cleaning cycle must be controlled carefully to avoid excessive attack on the molybdenum).

(2) Commercial martempering salt (a mixture of sodium and potassium nitrates) operating at 700°F (371°C). No gross attack has been noted after this procedure. Light surface oxide films should be removed after the salt bath by an appropriate mechanical or chemical treatment.

Brazing Environments

Protection from oxidation is imperative during brazing. Purified, dry hydrogen and inert gas (helium and argon) atmospheres are suitable for brazing molybdenum. For brazing pure molybdenum, a dry hydrogen atmosphere is not critical. A dew point of 80°F (27°C) can be tolerated in hydrogen when reducing molybdenum oxide prior to brazing. However, a low dew point [-50°F (-46°C) or lower] at 2200°F (1204°C) is required when brazing titanium-bearing molybdenum alloys. Vacuum furnaces also have been used to braze molybdenum-titanium alloys. When brazing these molybdenum-titanium alloys, vacuums with less than .001 Torr (0.15 Pa) are desirable. When brazing in vacuum, precautions should be taken in selecting a proper braze filler metal. The material must not volatilize during the brazing cycle, which will result in contamination of the vacuum system.

Torch brazing of molybdenum has also been accomplished. When utilizing an oxyacetylene torch, protection must be obtained by using a flux. An acceptable commercial flux for torch brazing is a silver-borate base brazing flux accompanied by a high-temperature flux containing calcium fluoride. The temperature range over which the fluxes are active is between 1050 to 2600°F (566 to 1427°C). As heating occurs, the silver brazing flux is active in the lower end of the temperature range while the high temperature calcium fluoride flux is active in the upper temperature range to 2600°F (1427°C).

Braze Filler Metals

Many of the BAg series of brazing filler metals can be used to braze molybdenum or tungsten to a copper alloy base. For best results, it is often desirable to precoat the faying surfaces. Copper and silver base braze filler metals can be used to braze molybdenum for low temperature service. For high-temperature applications, molybdenum can be brazed with gold, palladium, and platinum filler metals, nickel filler metals, reactive metals, and refractory metals that melt at lower temperatures than molybdenum. However, evidence of liquid metal embrittlement has been observed in molybdenum and TZM (0.5Ti-0.07Zr-Mo) with gold-nickel braze alloy coupons at temperatures as low as 1832°F (1000°C). In addition to the liquid metal embrittlement, nickel base filler metals also have applicability for high temperature service, because nickel and molybdenum form a low melting eutectic at about 2400°F (1316°C).

There are three concerns when brazing molybdenum and its alloys for use above 1800°F (982°C). These three metallurgical concerns are the following:

(1) Embrittlement of the molybdenum resulting from the rapid diffusion of impurities near the recrystallization temperature of molybdenum.

(2) Formation of brittle intermetallics between refractory metal and braze filler metals.

(3) Relative weakness of braze filler materials at elevated temperatures as compared to the refractory base material.

Problems have been encountered, however, in making ductile, unalloyed molybdenum joints for high temperature service, because unalloyed molybdenum recrystallizes at about 2100°F (1149°C). Thus, when brazing is conducted above the recrystallization temperature, time must be kept at a minimum to avoid excessive grain growth and a subsequent increase in the transition temperature of molybdenum. This problem has been reduced by the development of molybdenum alloys, such as TZM which have increased recrystallization temperatures.

As in all metal systems, formation of continuous, intermetallic phases at the grain boundaries may occur and can be quite harmful. Therefore, proper selection between the braze filler metals and the refractory metal is very important due to possible formation of intermetallics which are brittle and may fracture easily at relatively low loads.

The relative weakness of most braze filler materials at elevated temperatures poses a basic limitation to the use of brazed molyb-

denum assemblies. Most of the nickel base braze filler materials for elevated temperature service melt in the range of 1800 to 2100°F (982 to 1149°C). Refractory metals are used in this temperature range due to their high temperature strength characteristics. There exist special high melting point braze filler metals for molybdenum and its alloys. These filler metals must provide adequate strength, not produce undesirable intermetallics, and must be stable in the operating environment.

Two binary braze filler metals appropriate for high- temperature applications are V-35Cb and Ti-30V. These braze filler metals have been evaluated for use with the Mo-0.5Ti molybdenum base metal. T-joints were brazed in a vacuum for five minutes at temperatures of 3000°F (1649°C) for the Ti-30V braze filler metal and 3400°F (1870°C) for the V-Cb braze filler metal. The braze filler metals had excellent metallurgical compatibility with the molybdenum base metal, and minimum erosion of the base metal occurred during brazing.

Several investigations have examined conventional brazing, reactive brazing and diffusion-sink brazing concepts for new refractory braze filler metals. Diffusion brazing with titanium and Ti-30V [braze temperatures used in this investigation were 3050 to 3200°F (1677 to 1760°C)] has produced remelt temperatures exceeding 3800°F (2093°C) in tee and lap joints. Diffusion brazing with 33Zr-34Ti-33V [braze temperature 2600°F (1427°C)] has produced remelt temperatures exceeding 3200°F (1760°C). The remelt temperature indicates that service temperature for the titanium and Ti-30V alloys could be 3000°F (1649°C).

Brazements of the diffusion-sink braze filler metal, Ti-8.5Si, which melts at approximately 2425°F (1330°C), exhibited excellent fillet forming and wettability characteristics resulting in crack-free ductile joints. In addition, specimens brazed at 2550°F (1400°C) with molybdenum powder added to the braze filler metal powder, increased the liquidus considerably and allowed greater service temperatures to be used.

The procedure described above has been utilized with the molybdenum alloy, TZM

(0.5Ti-0.07Zr-Mo). TZM has been brazed successfully at 2550°F (1399°C) with molybdenum powder added to the Ti-8.5Si braze filler metal. Additional work on TZM with powder additions have shown that other braze filler metals can be used. A Ti-25Cr-13Ni braze filler metal [braze temperature 2300°F (1260°C)] has produced the highest remelt temperature on TZM. Remelt temperature of brazes in tee and lap joints was about 3100°F (1710°C).

Because of its excellent high-temperature properties and compatibility with certain environments, sintered, isotropic molybdenum powder components brazed in assemblies are prime candidates for use in certain nuclear reactor components and chemical processing systems. As a result of extremely harsh conditions (liquid alkali metals effects on braze filler metal), a series of iron base brazing filler metals has been developed. The best filler metal based on overall performance is Fe-15Mo-5Ge-4C-1B.

Joint Design Considerations

The clearance requirements and joint design are similar to those of the other refractory metals. Molybdenum has an extremely low coefficient of thermal expansion that should be considered in the design of brazed joints, particularly when molybdenum is joined to other metals. Joint clearances in the range of 0.002-0.005 in. (0.05-0.13 mm) at the brazing temperature usually are suitable. The selection of a fixture material also should be made with care. Special consideration should be given to thermal expansion characteristics of both the braze filler metal and the base material.

Investigations of tube to socket joint configurations have been made between molybdenum and other metals with higher coefficients of thermal expansion, such as stainless steels. When the joint is configured such that the tube part is made from molybdenum and the socket made from stainless steel, unreliable joints result. In contrast, when the socket material is made from molybdenum and the tube made from stainless steel, a sound, reliable joint can be achieved.

A common application for molybdenum is resistance welding electrodes tipped with pure molybdenum or tungsten. These electrodes are especially useful for welding nonferrous, high conductivity materials. Welding of braided copper wire using the molybdenum tipped electrode is a typical use.

TANTALUM AND ITS ALLOYS

As a high temperature structural material, tantalum has outstanding resistance to chemical attack due to the production of a tough, self-healing oxide film. In addition to structural applications where high-temperature corrosion resistance is required, tantalum has several other advantageous properties. Tantalum has high ductility at low temperatures. Tantalum oxide has excellent dielectric properties which are utilized in tantalum foil condensers. Another property of tantalum is the ability to absorb gases, which makes it extremely useful as a getter in electronic equipment.

The joining of tantalum is very similar to the joining of other refractory metals. Many of the same precautions and procedures as previously discussed for niobium and molybdenum still apply. These include provisions for material cleaning and joint preparation, brazing methods, and braze filler metals.

Surface Preparation

Prior to brazing, preparation of the tantalum surface must be considered to ensure joint integrity. Tantalum can be cleaned by both mechanical and chemical methods (see Table 28.4). One chemical method utilizes hot chromic acid (glass cleaning solution) that cleans quite satisfactorily. However, hot caustic cleaning solutions are not recommended due to corrosive attack of the metal.

A more efficient means of cleaning tantalum is a combination of mechanical treatment followed by chemical cleaning. Prior to the chromic acid cleaning, tantalum can be grit-blast cleaned to remove surface contaminants and debris. This procedure should be followed by an immersion in a hydrochloric acid solution to dissolve any iron particles entrained in the tantalum during processing. Following this intermediate treatment, tantalum can be effectively cleaned in the hot chormic acid solution. Once cleaned, brazing of the tantalum should be conducted immediately to avoid thick oxide formation.

Tantalum has a tenacious oxide film that reforms immediately on exposure to air or vapor after any cleaning treatment. One method of preparing tantalum prior to brazing is to electroplate either copper or nickel onto an acid cleaned surface. The composite is then subjected to a diffusion treatment at a temperature of 2200°F (1204°C) under a vacuum of at least 0.15 Pa (.001 Torr). The deposits are bonded to the tantalum by diffusion and, in the case of copper, melting actually occurs. Once tantalum has obtained this required protective base coating of nickel or copper, brazing may be performed in air using fluxes suitable for the particular braze filler metal used.

When brazing unplated tantalum surfaces, several methods are available. These methods include inert gas (argon or helium) atmosphere and vacuum. When an inert gas atmosphere is used in brazing, gas purity must be carefully controlled. Constituents in the atmosphere, such as oxygen, nitrogen, and hydrogen, must be avoided or else a loss of ductility ensues due to the formation of oxides, nitrides and hydrides.

Since most uses of tantalum and its alloys are for elevated temperature applications [3000°F (1649°C) and above], there are few high-temperature braze filler metals available for tantalum other than the refractory metals discussed previously.

Currently, 90-95 percent of the available braze filler metals for tantalum are in powder form, which can be difficult to work with in some applications. New powder braze filler metals such as Hf-7Mo, Hf-40Ta, and Hf-19Ta-2.5Mo are examples of new braze filler metals that are being developed for tantalum. These braze filler metal powders are processed into foil-type forms by either direct rolling of the braze filler metal or pack diffusion heat treatment of the alloying elements with hafnium foil.

The term *pack diffusion* refers to a method of alloying the hafnium foil material by placing it in a container with alloying elements such as the tantalum and molybdenum and heating the material to a sufficiently high temperature. After an extended time, up to eight hours or more, at this elevated temperature, diffusion of the alloy elements into the hafnium foil is complete.

Brazing tantalum, as in brazing other refractory metals, requires selection of braze filler metals compatible with the application. A series of nickel base braze filler metals (such as the nickel-chromium-silicon filler metals) have been used to braze tantalum. In these alloys, nickel penetration into the base metal forms a homogeneous solution with tantalum up to a tantalum content of 36 percent by weight. As this reaction at the interface occurs, the liquidus is reduced from 2640 to 2460°F (1449 to 1349°C). These braze filler metals are satisfactory for service temperatures below 1349°F (982°C).

Copper-gold and silver alloy braze filler metals have also been used in brazing tantalum. Braze filler metals containing less than 40 percent gold have been successful in joining tantalum, but gold in amounts between 40 and 90 percent tends to form brittle, age-hardening compounds. Although silver base braze filler metals have been used to join tantalum, they are not recommended because of a tendency to embrittle the base metal. Other braze filler metals that have been used in brazing tantalum include copper-tin, gold-nickel, gold-copper, and copper-titanium.

TUNGSTEN

As stated previously, tungsten has the highest melting point of the materials reviewed. This unique property of tungsten has many applications in areas that require extremely high temperatures such as heat shield and plasma arc components. Many of the preparations and brazing techniques discussed for the other refractory metals are applicable in the brazing of tungsten.

Cleaning methods found acceptable for tungsten are given in Table 28.4. The most effective cleaning procedure will depend upon the tenacity of the oxide film, and whether it is pore free or is spalling. In cases where wrought tungsten sheet has been mill cleaned, degreasing is sometimes the only cleaning operation necessary prior to brazing. However, the optimum conditions for preparation of tungsten should be determined for each particular application. In some cases, electroplating of tungsten with nickel or other elements has been used satisfactorily to stop diffusion of elements that form brittle intermetallic compounds with the base metal. A furnace cleaning operation using a hydrogen reducing atmosphere at 1950°F (1065°C) for 15 minutes has been effective in chemically reducing light oxide films.

Several conditions in addition to cleaning must be considered when brazing tungsten. Care must be exercised in handling and fixturing tungsten parts because of the inherent brittleness due to the high ductile/brittle transition temperature. Because of the extreme brittleness at room temperature, parts should be assembled in a stress-free condition. Contact between graphite fixtures and tungsten must also be avoided to prevent the formation of brittle tungsten carbides.

Tungsten can be brazed in much the same manner as molybdenum and its alloys, using many of the same braze filler metals. When brazing tungsten, surface protection is required. There are several methods in which this can be achieved. Brazing of tungsten can be accomplished in either an inert gas atmosphere (dry helium or argon), a reducing atmosphere (hydrogen), or in a vacuum.

In vacuum, tungsten structures have been brazed very successfully with a wide variety of braze filler metals. Some of these braze filler metals are pure metals. The melting points of these metals range from 1200 to 3500°F (649 to 1927°C). Table 28.6 shows the variety of braze filler metals that have been used to braze not only tungsten, but also other refractory metals and their alloys.

Table 28.6
Braze Filler Metals for Refractory Metals

Braze Filler Metal*	Liquidus Temperature	
	°F	°C
Nb	4380	2416
Ta	5425	2997
Ag	1760	960
Cu	1980	1082
Ni	2650	1454
Ti	3300	1860
Pd-Mo	2860	1571
Pt-Mo	3225	1774
Pt-30W	4170	2299
Pt-50Rh	3720	2049
Ag-Cu-Zn-Cd-Mo	1145-1295	619-701
Ag-Cu-Zn-Mo	1325-1450	718-788
Ag-Cu-Mo	1435	780
Ag-Mn	1780	971
Ni-Cr-B	1950	1066
Ni-Cr-Fe-Si-C	1950	1066
Ni-Cr-Mo-Mn-Si	2100	1149
Ni-Ti	2350	1288
Ni-Cr-Mo-Fe-W	2380	1305
Ni-Cu	2460	1349
Ni-Cr-Fe	2600	1427
Ni-Cr-Si	2050	1121
Mn-Ni-Co	1870	1021
Co-Cr-Si-Ni	3450	1899
Co-Cr-W-Ni	2600	1427
Mo-Ru	3450	1899
Mo-B	3450	1899
Cu-Mn	1600	871
Nb-Ni	2175	1190

(Continued)

Table 28.6 (Continued)

Braze Filler Metal*	Liquidus Temperature	
	°F	°C
Pd-Ag-Mo	2400	1306
Pd-Al	2150	1177
Pd-Ni	2200	1205
Pd-Cu	2200	1205
Pd-Ag	2400	1306
Pd-Fe	2400	1306
Au-Cu	1625	885
Au-Ni	1740	949
Au-Ni-Cr	1900	1038
Ta-Ti-Zr	3800	2094
Ti-V-Cr-Al	3000	1649
Ti-Cr	2700	1481
Ti-Si	2600	1427
Ti-Zr-Be	1830+	999+
Zr-Nb-Be	1920+	1049+
Ti-V-Be	2280+	1249+
Ta-V-Nb	3300-3500+	1816-1927+
Ta-V-Ti	3200-3350+	1760-1843+

* Filler metals not always commercially available

\+ Varies with composition

While nickel-base braze filler metals have been used successfully to braze tungsten, a reaction between nickel and tungsten results in depression of the recrystallization temperature of the base metal at the bond line. This reaction can be minimized by short brazing cycles, minimum brazing temperatures and the use of a minimal amount of braze filler metal.

To some extent, the selection of the braze filler metal depends on the brazing atmosphere and conditions used in brazing. For example, braze filler metals that contain elements with high vapor pressures at the brazing temperature cannot be used effectively in a vacuum. When vacuum brazing, special caution should be observed concerning vapor pressures. Vapor pressures of the compositional elements in the braze filler metal must be low enough so that metal vapors are not emitted into the vacuum system. Failure to prevent this can result in vacuum system contamination and thereby poor vacuum.

In addition to the above methods, brazing tungsten through the use of fluxes is also acceptable. One common example where low temperature fluxes are used is in brazing tungsten for electrical contact applications. In this type of service application, silver-base and copper-base braze filler metals are often used.

Braze filler metals based on platinum-boron and iridium-boron systems were developed to braze tungsten for service at 3500°F (1927°C). They contained up to 4.5 percent boron and could be used for brazing below the recrystallization temperature of tungsten. Tungsten lap joints were brazed and diffusion treated in a vacuum at 2000°F (1093°C) for three hours. This cycle resulted in the production of joints with remelt temperatures of about 3700°F (2038°C). A slight increase in joint remelt temperature was noted when tungsten powder was added to the braze filler metal. The highest remelt temperature, 3940°F (2171°C), was obtained when joints were brazed with Pt-3.6B plus 11 percent tungsten powder.

Studies were conducted to develop and evaluate braze filler metals that could be used to braze tungsten for nuclear reactor service at 4530°F (2499°C) in hydrogen. Gas tungsten arc braze welded butt joints were achieved by using filler metals W-25, W-50Mo-3Re or Mo-5Re.

A common brazing application for tungsten is tungsten-copper tooling tips. In many cases the tungsten-copper tooling materials are not a wrought, solid-solution alloy but are a ratio of mixed powder alloys which are pressed and sintered into a tooling form. Analysis of a tungsten-copper tooling material commonly used for this application is 80W-20Cu. The tips produced from the powder form have two areas of application, resistance welding electrodes and electrical discharge machining (EDM).

Resistance welding materials are usually brazed to a copper-chromium alloy, which constitutes the main current carrying portion of the electrode. Any of the BAg series of braze filler metals are suitable for this application. The copper-tungsten tips are applied by induction or torch heating.

The same composition or a similar composition of tungsten-copper (W-Cu) is used for electrical discharge machining. In the finish machining electrode, only the end of the electrode is made from the tungsten-copper alloy. A base of phosphorized copper or oxygen-free-high-conductivity (OFHC) copper is used to attach the electrode to the current supply.

In most cases, brazing is accomplished in a neutral or reducing atmosphere to minimize oxidation. The most suitable braze filler metal is BAg-8. Alignment of the parts is facilitated by locating pins which are incorporated in the design. Braze filler metal is preplaced in the form of shim stock located between the mating parts.

Often when brazing tungsten-tipped resistance welding electrodes in a furnace at relatively low temperatures [1500°F (816°C)], precoating of the tungsten surface with copper or nickel aids in a uniform flow of the braze filler metal between the faying surface of the joint. Without pretreatment, poor wetting of the tungsten by the braze filler metal often results, causing premature joint failure under the high pressures and temperatures in resistance welding.

Mechanical property data on tungsten brazes is more limited than data on other refractory metal brazes. One example of joint shear strength and braze filler metal composition involves the development of braze filler metals for use in joining unalloyed tungsten. In this investigation, tungsten samples were vacuum brazed with two experimental niobium braze filler metals, Nb-2.2B and Nb-20Ti. The lap shear strength of joints brazed with Nb-2.2B was about 5 ksi (35 MPa) at 3000°F (1649°C) and 8 ksi (55.52 MPa) at 2500°F (1371°C). The strength of the joints brazed with Nb-20Ti was somewhat lower, about 3 ksi (21 MPa) at both test temperatures. A decrease in the shear strength of the Nb-20Ti braze filler metal resulted from poor wetting as a consequence of sluggish flow of filler metal compared to that obtained with Nb-2.2B.

DISSIMILAR REFRACTORY METAL JOINING

Molybdenum base metals have been brazed to other metals. As one example, silver-palladium braze filler metals have been used to braze TZM face sheets to honeycomb core fabricated from a nickel-base superalloy, AMS 5550. The assembly was brazed below 2200°F (1204°C) to prevent recrystallization of the molybdenum base metal.

TZM foil has been brazed to AMS 5536, 5545 and 5537 sheet stock. Each dissimilar metal combination was brazed with the following filler metals: Ni-7Si-5Fe-5Cr-4Co-3W-0.7B, Ni-15.5 Cr-16Mo-5Fe-4W-2.5Co-1Mn-1Si, and Pd-40Ni.

A tungsten nozzle has been brazed to a molybdenum plenum for a propulsion engine application. These components were brazed in argon with pure iron filler metal for service at temperatures up to 1800°F (980°C) and with Cr-25V for service in the 2700 to 3000°F (1480 to 1650°C) range. Also, molybdenum-to-tungsten joints have been brazed with Cu-35Mn.

Niobium and its alloys have been successfully joined to cobalt-base alloys with BAu-4 braze filler metal. Type 316 stainless steel has been brazed to the Nb-1Zr alloy using 21Ni-21Cr-8Si-50Co braze filler metal at 2150°F (1177°C); and niobium brazed to alumina using Zr-Ni as braze filler metal. B66 and D43 niobium alloys were joined to AMS 5536, 5545 and 5537 with the following braze filler metals: NX-77 (Ni-7Si-5Fe-5Cr-4Co-3W-0.7B-0.1Mn), AMS 5530, (Ni-15.5Cr-15Mo-5Fe-4W-2.5Co-1Mn-1Si), and Pd-40Ni.

Tantalum sheet has been vacuum brazed to oxygen-free high-conductivity (OFHC) copper plate for missile applications. This joint combined the high-temperature strength and abrasion resistance of tantalum with the heat-dissipation capabilities of copper. The bimetallic material was produced in plate form. The plate was then cold drawn into a nose cone configuration for evaluation in rapidly flowing oxidizing gases at 3600°F (1982°C). Joints that withstood this environment for one minute performance period were produced with the following braze filler metals: Cu-8Sn, BAu-4, and BAg-23.

Tantalum has been brazed to molybdenum in vacuum at 3244°F (2700°C) using pure zirconium as the braze filler metal. Gold-nickel and pure copper braze filler metals have produced excellent tensile test results in brazing Type 304 stainless steel to tantalum and Ta-10W.

SELECTED READING

Albom, M. J. "Diffusion bonding tungsten." *Welding Journal* (Research Supplement) 41(11): 4915 to 502s; 1962.

Barth, S. D. "The fabrication of tungsten." Defense Metals Information Center Report 115, 1955.

Bertossa, R. C., and Rau, S. "Heat extractive brazed bimetals show promise for missile and industrial applications." *Welding Journal* (Research Supplement) 38(7): 273s to 281s; 1959.

Briggs, J. Z. *A new look at joining molybdenum.* Climax Molybdenum Co.

Fox, C. W., Gilliland, R. G., and Slaughter, G. M. "Development of alloys for brazing columbium." *Welding Journal* (Research Supplement) 42(12): 535s to 540s; 1963.

Freedman, A. H., and Mikus, E. B., 1964 "Brazing of columbium D36 honeycomb structures." *Welding Journal* (Research Supplement) 43(9): 385s to 391s; 1966.

———. "High remelt temperature brazing of columbium honeycomb structures." *Welding Journal* (Research Supplement) 45 (6): 258s to 265s; 1964.

Gilliland, R. G. "Joining of tungsten." USAEC Report ORNL-TM-1606. Oak Ridge: Oak Ridge National Laboratory, 1966.

Gilliland, R. G. and Slaughter, G. M. "The development of brazing filler metals for high temperature service." *Welding Journal* (Research Supplement) 48(10): 463s to 468s; 1969.

Hendricks, J. W., Cole, N. C., and Slaughter G. M. "Compatibility of brazed joints with potassium and vacuum." *Welding Journal* (Research Supplement) 51(7): 329s to 336s; 1972.

McDonald, M. M., Keller, D. L., and Johns, W. L. "Wetting in molybdenum by commercial brazing filler metals." *Advances in Welding Science and Technology*, ASM Conference Proceedings. Metals Park, OH: ASM, 1986.

McDonald, M. M., Keller, D. L., Hieple, C. R., and Hofmann, W. "Comparison of wettability of braze alloys on molybdenum and TZM." *Welding Journal*, forthcoming.

McDonald, M. M. "Corrosion of brazed joints." *Metals Handbook*, 9th Edition, Volume 13. Metals Park, OH: ASM, 1987.

Monroe, R. E. "The joining of tungsten." Defense Metals Information Center Memo 74, 1960.

Pattee, H. E. and Evans, R. M. "Brazing for high temperature service." Defense Metals Information Center Report 149, 1961.

Robbins, W. P. "Brazing superalloys and refractory metals." AEC Research and Development Report No. MLM-1322. Miamisburg: Mount Laboratory, Monsanto Research Corporation, Contract no. AT-33-1-gen-53, 1967.

Scott, M. H. and Knowlson, P. M. "The welding and brazing of the refractory metals niobium, tantalum, molybdenum, and tungsten." *Journal of the Less Common Metals* (5): 205-44; 1963.

Stone, L. H., Freedman, A. H., and Milkus, E. B. "High-remelt temperature brazing of tantalum using a diffusion sink concept." *Metal Engineering Quarterly* 6 (1): 35-42; 1967.

———. "Brazing alloys and techniques for tantalum honeycomb structures." *Welding Journal* (Research Supplement) 46(8): 343s to 350s; 1966.

Keywords — brazing of components for electron tubes and vacuum devices, brazing filler metals for vacuum devices, brazing of refractory metals, brazing of ceramic to metal structures

Chapter 29

ELECTRON TUBE AND VACUUM EQUIPMENT BRAZING

INTRODUCTION

Brazing of structures for vacuum tubes and other high-vacuum devices has required the development of highly refined procedures, furnace equipment, and high purity, low vapor pressure brazing filler metals. Vacuum tubes, of necessity, are operated at very low pressures (10^{-6} to 10^{-8} torr [10^{-4} to 10^{-6} Pa]), and must maintain this very low pressure for the thousands of hours of their useful life. High-vacuum devices and specialized equipment cannot tolerate conditions which inhibit the ability to secure and maintain extremely low pressures. As a further complication, electron tubes and other vacuum devices are heated to temperatures beyond 932°F (500°C) for extended periods during their gas evacuation cycle ("bake-out") to drive out the gases entrapped in their metal structures. Vacuum tubes operate at elevated temperatures and are constructed of widely dissimilar materials; hence, the brazed joints must also withstand differential thermal expansion stresses. The foregoing indicates that the brazing of components for electron tubes and vacuum devices equipment should only be undertaken after a thorough analysis of the problem.

BASE MATERIALS AND FILLER METALS

Oxygen-free copper, nickel, stainless steels, copper-nickel alloys, iron-nickel-cobalt (fen-ico) alloys, molybdenum and tungsten, as well as alumina, fosterite, and sapphire ceramics, are commonly used materials for constructing electron tubes, special vacuum devices, and vacuum equipment. While the need to join such a wide range of materials has stimulated the development of many new brazing filler metal compositions, the number of possible filler metals available for brazing of vacuum devices is reduced considerably by the necessity of using only low vapor pressure elements in their composition. This limitation, for all practical purposes, restricts the selection of filler metals to essentially pure elements or alloys of platinum, copper, silver, gold, palladium, nickel, indium, tin, and gallium. All other elements have either too high a melting point or too high a vapor pressure. Brazing filler metals, therefore, must be completely free of zinc, cadmium, lithium, boron, and phosphorus. Vacuum system components (equipment) made of stainless steels are readily brazed in protective atmospheres using BNi and BAu series filler metals.

The necessity for completely vacuum-tight joints, rather than typically mechanically strong and reasonably air-tight or liquid-tight joints normally expected of a braze, complicates brazing procedures on vacuum devices and systems. Chemical fluxes normally cannot be used as their subsequent removal is questionable and their presence, even in trace amounts, in a vacuum device is very objectionable. Consequently, vacuum device components generally are brazed in hydrogen, argon, nitrogen, vacuum, or helium enriched environments. Helium should be a last choice if the component is to be later leak checked on a helium mass spectrometer for hermeticity.

Voids in brazed joints are to be avoided. Voids impair joint strength, and more importantly they may conceal gases or matter that if released during operation of a vacuum device could seriously impair its performance and in some instances, destroy the device.

In these type devices, materials to be joined by brazing and the brazing filler metal must be clean and essentially free of oxides. A light acid etch thoroughly rinsed will usually suffice. Brazing filler metals for vacuum devices must be produced with far more care than conventional filler metals to prevent inclusion of high vapor pressure elements, and to keep mechanically occluded impurities such as dirt and carbon to a minimum. Mechanically incorporated impurities generally are picked up during ingot casting or in the subsequent wire drawing or sheet rolling procedures as a result of the entrapment of organic lubricants used in these processes. The entrapped organic compound is converted to carbon during the numerous annealing treatments required to reduce the original ingot to its final wire or sheet form. Dirt, if present, is pressed into the metal during rolling or wire drawing reductions. On melting a contaminated filler metal, the carbon or dirt will generally rise to the surface of the molten filler metals. Any foreign materials, because of the remotest possibility of evolving gases when heated, are always to be avoided in vacuum devices and systems. Another danger is the possibility that the dirt and carbon may remain within the braze joint, form a gas pocket (a virtual leak), and possibly a leak to atmosphere.

Brazing filler metals for vacuum devices should be free of the aforementioned impurities, and have low oxygen and metal oxide contents. Oxides are particularly objectionable when brazing in a hydrogen atmosphere since they will react with the hydrogen to form water vapor. If such a reaction takes place while the filler metal is molten, the rapidly expanding water vapor splatters the molten filler metal into areas adjacent to the braze joint where their presence is often very undesirable.

The choice of a particular brazing filler metal depends on the maximum operating temperature of parts to be joined, the vapor pressure of the filler metal at this temperature, the number of subsequent brazes to be made on the assembly, and the properties of the materials to be joined. Brazing filler metals are available ranging from the high melting point platinum and platinum alloys for brazing of tungsten and molybdenum parts through pure copper down to a silver-copper-indium alloy having a liquidus of 1301°F (705°C).

Good copper-to-copper, copper-to-nickel, and nickel-to-nickel braze joints can be made with silver-copper eutectic, as well as with various gold-copper and gold-copper-nickel alloys. Economic considerations generally dictate the use of the silver-copper eutectic filler metal for these applications; however, since the gold-copper and gold-copper-nickel filler metal have considerably lower vapor pressures, vacuum system performance may require the use of the more costly filler metals. Since the liquidus and solidus of the gold filler metals are higher than those of the silver-copper, they also are often used for applications in which a component is best fabricated by two or more successive brazing steps. In such cases, the solidus of the first filler metal used must occur at a higher temperature than the brazing temperature of the next filler metal so that the first braze joint does not soften and deform during any subsequent brazing cycle. The choice between the various gold-copper filler metals available normally is determined by their brazing temperature range since they flow about equally well on copper, nickel, fenico-type alloys, and nickel plated stainless steels. The copper-gold-nickel and the BAu-4 nickel filler metal, however, generally are preferred for brazing of fenico and stainless alloys since they flow better on these iron-bearing materials and produce a stronger braze joint than do the binary gold-copper filler metals.

The fenico and nickel-iron alloys are widely used for sealing metal to glass and brazing to metallized alumina ceramic structures because they more closely match the thermal expansions of these insulating materials. While the silver-copper eutectic filler metal will wet and flow well on nickel-iron and fenico-type alloys, it is somewhat risky to use because of the its tendency to penetrate the grain boundaries of these alloys, when they are under stress, and cause them to crack (Figure 29.1). The stress corrosion cracking of nickel-iron and iron-nickel-cobalt alloys by the silver-copper filler metals can be minimized by stress relieving the formed parts prior to brazing. During stress relieving, care should be taken to avoid excessive grain growth. Electroplating these alloys with nickel or copper also appears to reduce stress corrosion by the silver filler metals. When these alloys are brazed to a higher expansion metal such as copper, care must be exercised in joint design to prevent restressing them during the brazing cycle. The addition of as little as 5% of palladium to the silver-copper eutectic filler metal will greatly inhibit the stress corrosion effect on fenico. Filler metals now are available with 5, 10, 15, and 25% palladium additions to silver-copper. Since they are not too costly and far safer to use, the 5% and 10% palladium-bearing silver-copper filler metals are being increasingly favored over the silver-copper eutectic for brazing of nickel-iron and iron-nickel-cobalt alloy structures.

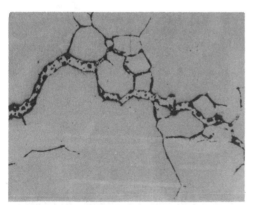

Figure 29.1 — Penetration of Silver into Grain Boundaries of Fenico Alloy Brazed With Copper Silver Eutectic. Such Penetration is Avoided by Adding Palladium to the Braze Filler Composition, or by Nickel-Plating the Penico Prior to Being Joined. Both Techniques Provide a Diffusion Barrier to the Copper Silver Filler. 100 x Mag. (Reduced by 35% on Reproduction)

When it becomes necessary to make a vacuum tight braze joint between two metals with widely different coefficients of thermal expansion, such as copper to fenico or molybdenum, a difficult brazing problem arises. The use of any braze filler metal with a relatively large difference between its solidus and liquidus temperatures should be avoided. Any differential movement between parts being joined during the period in which the filler metal is in a very weak liquid-solid state, will crack the body of the filler metal. To minimize cracking, the filler metal should have little or no temperature differential between its solidus and liquidus. Designing such joints to be in compression also will help.

BRAZING OF REFRACTORY METALS

Molybdenum, tungsten, and tantalum are used extensively in electron tube and other vacuum device construction and often are brazed to each other or to other metals (see Chapter 28). Tungsten and molybdenum are readily brazed with platinum and platinum alloys and palladium-gold-nickel filler metals. The choice between these filler metals generally is determined by the operating temperature of parts to be brazed, thus tungsten electron tube support rods can be brazed with either pure platinum or a platinum-4% tungsten alloy. Molybdenum vacuum tube "plate" structures also are commonly brazed with these filler metals. The tungsten contacts of vacuum switches are brazed to copper conductors with the BAu-4 filler metal. Tantalum is wet readily by copper-gold, gold-nickel, palladium-nickel, and palladium-gold-nickel filler metals. Tantalum reacts with hydrogen and thus is most often brazed in a vacuum.

AVAILABLE FORMS OF BRAZING FILLER METALS

Vacuum tube brazing filler metals (Table 29.1) are readily available from manufacturers in the form of wire of any diameter, any thickness of foil down to 0.0003 in. (0.008 mm), powder, and preforms such as wire rings, stampings, and formed wire shapes. Extrudables, pastes, and "tapes" made of filler metal powders mixed with appropriate organic vehicles also are available; however, they are seldom used in vacuum device brazing because of the undesirable residual carbon left by the decomposition of the organic vehicle. A serrated strip that can be bent around virtually any configuration enables the user to produce preforms of a wide variety of sizes and shapes. (Figure 29.2)

BRAZING OF CERAMIC TO METAL STRUCTURES

Vacuum-tight ceramic structures, generally of high-purity alumina, are widely used in electron tubes, vacuum switches, and vacuum equipment, having largely replaced

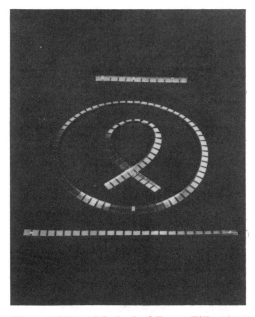

Figure 29.2 — Method of Braze Filler Application Used by Vacuum Equipment Manufacturers to Make Large Diameter Preforms Without the Need to Stamp Them Out of Sheet, Thereby Eliminating Tooling and Large Dropouts of Precious Metal Fillers

glass as an electrical insulator in these devices. Ceramic shapes formed to very close dimensional tolerances are far stronger than glass and can be obtained with a high bond strength metal coating in selected areas to which a metal part can be brazed. The ceramic component can also withstand higher operating temperature applications.

The more common method of brazing a ceramic to a metal is by the use of a nickel-plated moly-manganese metallization applied to the ceramic surface only where the braze is to be made. In that way, the flow of braze filler can be limited to the specific area desired. Control of braze flow is necessary as many applications involve a pin or tube being brazed into the ceramic and must be electrically isolated and able to pass voltage breakdown testing. Other uses of this type of configuration include a peripheral braze attaching a metal ring, so that a device may then be joined into a metal container.

The ceramics are metallized by applying a coating of molybdenum-manganese powder mixture, then firing at $2732°F$ ($1500°C$) in wet hydrogen with a dew point of approximately $+35°F$ ($2°C$). The sintered moly-manganese coating is then nickel plated to a thickness of 0.0001 in., and resintered at $1832°F$ ($1000°C$). The nickel coating enhances wetting and facilitates leak-tight joints.

Ceramic to metal joints must be designed carefully to avoid failure of the brittle ceramic as a result of differential thermal expansion stresses set up in the brazing process. Often the design is such that the metal of the braze joint leaves the ceramic in compression, its strongest condition. (Figure 29.3). Fenico-type alloys are preferred because their thermal expansion more closely matches that of alumina ceramics. When using these alloys, the metal section still must be designed to yield at a stress level below that which will rupture the ceramic. When for other considerations fenico-type alloys cannot be used, ductile metals such as copper and occasionally copper-nickel are used. While these metals have a considerable expansion mismatch with alumina, the section joined to the ceramic by brazing is designed to yield before the ceramic. A knife-edge contact accomplishes this quite readily. Ideally, the braz-

Table 29.1
Composition of Brazing Filler Metals Specially Produced for the Brazing of Vacuum Devices and Equipment

AWS Classification	Filler Metal	Liquidus		Solidus		Nominal Composition, %								
		°F	°C	°F	°C	Pd	Au	Ag	Cu	Ni	Sn	In	Co	Ti
	Pt	3210	1768	3210	1768									
	Pd	2826	1552	2826	1552	100								
	Pd-35Co	2255	1235	2246	1230	65							35	
	Pd-Au	2264	1240	2192	1200	8	92							
	Au-Pd-Ni	2050	1121	2019	1102	25	50			25				
BAu-3	Au-Cu-Ni	1886	1030	1832	1000		35		62	3				
	Au-Cu	1850	1010	1814	990		35		65					
BAu-1	Au-Cu	1841	1005	1805	985		37.5		62.5					
	Au-Cu	1778	970	1751	955		50		50					
BAu-4	Au-Ni	1742	950	1742	950		82			18				
	Ag-Pd-Cu	1742	950	1652	900	25		54	21					
	Au-Cu-Ni	1697	925	1670	910		81.5		16.5	2				
	Ag-Pd-Cu	1652	900	1562	850	15		65	20					
	Au-Ag-Cu	1643	895	1625	885		75	5	20					
	Ag-Pd-Cu	1566	852	1515	824	10		58	32					
	Au-Ag-Cu	1553	845	1535	835		60	20	20					
	Ag-Cu-clad	1562	850					68.8	26.7					4.5
	Ag-Pd-Cu	1490	810	1485	807	5		68	27					
	Ag-Cu-Ni	1463	795	1436	780			71.5	28.1	.75				
BAg-8	Ag-Cu	1436	780	1436	780			72	28					
	Ag-Cu-Sn	1406	760	1369	743			68	27		5			
BAg-18	Ag-Cu-Sn	1325	718	1116	602			60	30		10			
	Ag-Cu-In	1301	705	1166	630			61.5	24			14.5		

NOTE: Impurities are Zn<0.001%; Cd<0.001%; Pb<0.002%; C<0.005%. All other impurities not specified having a vapor pressure higher that 10^{-7} torr at 500°C, are limited to 0.002% each. Impurities having a vapor pressure lower than 10^{7} torr at 500°C are limited to a total of 0.05%.

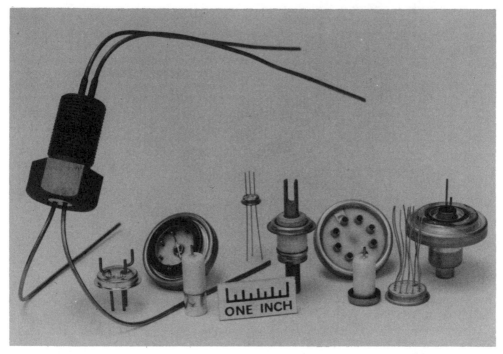

Figure 29.3 — Miscellaneous Metal-to-Ceramic Assemblies for Typical Vacuum Applications. Filler Metals Consist of Copper-Silver Eutectic, Nickel-Copper-Gold, Gold-Nickel, and OFHC Copper

ing filler metals used must also have a low yield point, the silver-copper eutectic and the 5% palladium-silver-copper filler metal preferred. Should higher melting point and lower vapor pressure filler metals become necessary, the gold-copper filler metals are indicated (Figure 29.4).

Ceramics are also brazed directly to metals, without metallizing the ceramics, using filler metals containing one or more of the active metals, such as zirconium or titanium. Some common filler metals for this application are silver-copper eutectic clad titanium and nickel-clad titanium. These clad filler metals contain about 5% titanium and must be brazed in high vacuum or inert atmospheres since they will react with hydrogen (Chapter 23).

Currently there are true alloy systems in the marketplace which include active metals greatly simplifying this technology.

The medical field has exploited the use of the active metal systems to join ceramics and graphites for miniature feed-throughs for use in pacemakers and nerve stimulation devices. In medical devices all materials must be bio-compatible to the human body. Materials such as nickel, manganese, silver, or copper must be avoided because of their possible reaction to the body's chemical or electro-chemical system. Successful metallizing applications of titanium are currently being accomplished by sputtering 1500-2000 angstroms of titanium to the ceramic substrate. Masking is suggested so that the coating is applied only to the area to be brazed. This is followed by sputtering 1500 angstroms of platinum to the same area of the substrate. To effect the braze, use a gold ring and braze in a hard vacuum (10^{-6} torr). This braze composition is totally bio-compatible with the human body.

Other unusual applications of joining ceramics to metal sections where high electrical charges are encountered for vacuum uses may involve the use of nonmetallized

Figure 29.4 — Operator Loading a Metal Section of a Large Vacuum Tube in a Bell Type Brazing Furnace

preparations when titanium is the metal to be joined to the ceramic. Highly successful joints have been made by the use of a thin aluminum foil between the ceramic and ti-tanium sections, resulting in devices that withstand up to 25 million volts, and severe thermal cycling far beyond other braze filler applications. See Figure 29.5 for an exam-

ple of a section of an electron accelerator
mounted in a brazing fixture.

**Figure 29.5 — Section of an Electron Accel-
erator Mounted in Brazing Fixture. White
Sections are Alumina; Braze Filler is Copper-Silver Eutectic**

Keywords — carbide tools, hard carbide composition, filler metals for brazing carbides

Chapter 30

CARBIDE TOOLS

INTRODUCTION

The materials commonly known as *carbides* are actually a wide variety of alloys that contain one or more types of hard carbide particles held together in a metallic binder.

Although the hard carbide alloys are best known for their use in metal cutting tool bits, they are found in many applications that require one or more of the outstanding mechanical properties of these materials (e.g., high compressive strength, rigidity and impact resistance, excellent corrosion-wear resistance). Typical applications include punches and dies, slitters and rolls, seal rings and valve trim sets. The large difference between the coefficient of thermal expansion of carbides and that of most mating structural alloys necessitates that in brazements the size of the carbide alloy piece be kept small to minimize thermal strains.

Several techniques for fastening the carbides have been developed. The major techniques are brazing and mechanical fastening. Brazing was one of the first successful methods of mounting carbides to steel or other base metals. Brazing is quite satisfactory for small area, short length joints. Through the use of certain design principles, the process can be applied satisfactorily to larger joints.

BASE METALS

The hard carbides vary widely in composition, but for brazing purposes the following general classifications can be made:

(1) Tungsten carbides with a cobalt binder. The cobalt binder may vary from 3 to 25 percent depending upon the use but normally will be on the order of 6 to 10 percent by weight.

(2) Tungsten carbides mixed with moderate percentages of carbides of titanium, tantalum, or columbium and a cobalt binder.

(3) Titanium or tantalum carbides mixed with tungsten carbides and nickel or cobalt binders.

(4) Chromium carbides mixed with nickel or cobalt binders.

(5) Various carbides impregnated with hardenable tool steels to combine the properties of the steels with those of the carbides.

(6) Hard carbides joined to support materials whose composition and properties are dictated by the expected application of the part. Generally, the metals are iron base, i.e., steel, martensitic and austenitic stainless steel, etc. In addition, high temperature applications utilize nickel or cobalt-base materials.

FILLER METALS

Although it is possible to use many of the BAg filler metals, those which contain nickel (e.g., BAg-3, BAg-4, and BAg-22) are generally recommended because nickel improves wettability. RBCuZn-D and BCu filler metals also have been used, particularly where a postbraze heat treatment is required. BAg-23 or 52.5 Cu — 38 Mn - 9.5 Ni filler metals may be used where the braze is to be subjected to elevated temperatures in service or additional wetting of titanium- or chromium-base carbides is required.

Some filler metals of the Cu-Mn-Ni series (e.g., 52.5 Cu — 38 Mn 9.5 Ni) wet carbides at temperatures compatible with steel heat-treatment temperatures. They also have the reported ability to fill wide-gap joints and provide good elevated temperature properties. Carbides brazed with the manganese-containing filler metals, coupled with higher brazing temperatures, require less critical surface preparation than when using the lower melting temperature filler metals.

Tungsten base carbides generally are readily wetted by BAg and BCu filler metals. However, titanium-base carbides are somewhat more difficult to wet. Where it is necessary to mount titanium carbides by brazing, the joint must be made in an inert or vacuum atmosphere. The surfaces may be specially treated by the carbide manufacturer, or be nickel plated, thus permitting the use of standard BAg and BCu filler metals.

In specifying a filler metal for a given job, first consideration should be given to the temperature range of anticipated application. This and the other considerations of corrosion and mechanical properties will dictate the filler metal composition selected, the brazing temperature required, the equipment to be used, and the brazing atmosphere. Generally, the filler metals mentioned previously (with the exception of the BCu alloys) are considered for the majority of applications requiring simple equipment, fluxes, and brazing temperatures below 1800°F (982°C). The most common filler metals used to join hard carbides to steel in this bonding temperature range are BAg-3 and BAg-4. The joints so produced have tensile strengths of up to 70 ksi (483 MPa), and shear strengths of 25 to 35 ksi (173 to 241 MPa). For high temperature applications pure copper (BCu) may be used as a filler metal, with good wetting results.

The carbides, in general, are not readily wet by most filler metals. That makes it preferable, when possible, to preplace the filler metal in the joint rather than to face feed the filler metal in the form of wire. For larger surfaces, braze filler composites that have a core of copper with a coating of filler metal on both surfaces are frequently used. In sandwich-type joints, the soft core of the composite yields to, and thereby relieves, the stresses set up by differences in thermal expansion between the carbide and the base metal. Core of the shim is generally about 50 percent of the total thickness with 25 percent thickness of filler metal on each side, as shown in Figure 30.1. Braze filler composites normally used may vary in thickness from 0.010 to 0.025 in. (0.25 to 0.64 mm). The thickness of the core of the composite should be increased as the area of the joint is increased.

If operation of the part involves heavy loading or high impact, a ductile-metal shim may be displaced locally from the joint. Thus, it would not be available to provide the uniform support required to prevent breakage of the carbide. For this reason, a braze filler composite with a copper shim is useful only for a light or medium duty application.

JOINT DESIGN

Carbides have thermal expansion coefficients approximately one-third to one-half those of steel and most other base metals to which they are attached. This factor must be considered in the design of carbide joints. A carbide core can be brazed readily into a steel ring or cylinder, although the steel expands away from the carbide on heating. On cooling, the joint will be under compression and crack resistant. If the combination is reversed, the result will be insufficient clearance combined with differences in thermal

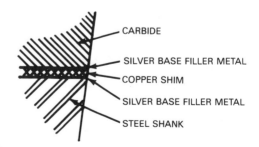

Figure 30.1 — Sandwich Braze Assembly Method

expansion at brazing temperature. This will cause a tensile stress buildup during heating, possibly resulting in cracking of the carbide. Sufficient clearance at room temperature to permit the steel to expand at the brazing temperature without stressing the carbide will put the joint in tension, possibly cracking the filler metal or carbide during cooling.

Small carbide pieces usually are brazed directly to an outer surface or in a groove machined to fit the piece. In the latter case, the carbide may be brazed on several surfaces as shown in Figures 30.2 and 30.3. The size and shape of the carbide to be brazed must be considered when selecting the filler metal form. With carbides over 0.5 in. (12.7 mm) long, cored shims are used to absorb the contracting stresses (Figure 30.1). With pieces longer than 3 in. (76.2 mm), it may be necessary to cut the carbide into segments as shown in Figure 30.4. The filler metal joining the ends of the segments will serve to relieve the excessive stresses that would occur in a single length. Stress may often be relieved or reduced on a poorly proportioned assembly by brazing only on one surface as shown in Figures 30.5 and 30.6. Using a stopoff paint or a relief gap will reduce flow and prevent wetting on the other surfaces.

Compensation for thermal stresses can be made by prestressing the assembly in a jig

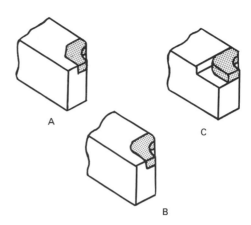

Figure 30.3 — Tip Cannot Fit All Three Shoulder Surfaces at A. The Redesign to Nonconflicting Shoulders at B Permits Proper Braze. Removing the Steel Shoulder as at C and Brazing Only on Bottom and End of Tip Improves Strength of Assembly

and allowing differential expansion and contraction to straighten the assembly as shown in Figure 30.7(A). Another technique used to obtain a similar effect is to peen the steel surface opposite the braze joint as shown in Figure 30.7(B). Simultaneously brazing an opposing carbide on the other side of the holder to balance stresses as shown in Figure 30.7(C) is another method of brazing a long length of carbide. This method of overcoming curvature provides a counter strained assembly, with two wear faces that can be indexed to double the life of the part.

The overall effect of joint thickness or clearance has been discussed in Chapter 17. As mentioned earlier, the thicker the braze layer, the more readily it can absorb thermal strain and shock loading. This aspect of improving tensile strength is especially im-

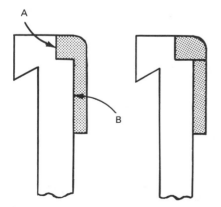

Figure 30.2 — Single Piece of Carbide Cannot be Brazed Properly to Both A and B Surfaces. Two-Piece Construction Permits Each to Seat Properly

Figure 30.4 — Use of Multiple Inserts Prevents Braze Strain Cracks

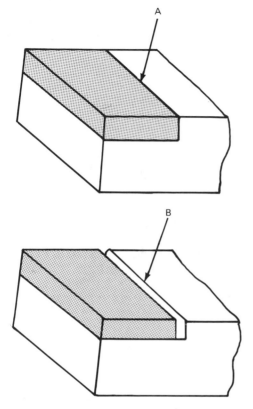

Figure 30.5 — Use of Stopoff Paint to Inhibit Braze, A, or Relief Slot, b, Reduces Brazing Strain

creases; however, the increasing joint clearance suggests improved loading and thermal strain capacities at larger joint spacings.

EQUIPMENT

Torch brazing and induction brazing are commonly used with the BAg, BCu, copper-manganese and copper-manganese-nickel filler metals. These filler metals are also used in controlled atmosphere furnaces, vacuum, or, less frequently, in induction brazing apparatus arranged for use in a protective atmosphere.

PRECLEANING AND SURFACE PREPARATION

Much of the difficulty in brazing carbides is the result of improper cleaning. The carbide surfaces should be grit blasted or ground on a silicon carbide or diamond wheel to remove any surface carbon enrichment. Such a surface is not wet readily by the filler

portant when large carbide pieces are being brazed.

Stressed brazed joints tend to elongate prior to failure. The reduction in area normally accompanying elongations is restrained in the joint by the closely spaced hard faces of the steel and carbide on either side of the filler metal. Because of this restraint, elongation is minimized — resulting in an apparent tensile strength several times that of the tensile strength of the cast filler metal as the joint thickness is reduced. Table 30.1 illustrates these effects for the shear testing of steel (4340/tungsten carbide joints brazed with pure copper). Note the significant fracture shear elongation in these 0.5 x 0.25 in. (12.7 x 6.4 mm) joints. These data exhibit the behavior of reduced fracture shear strength as joint clearance in-

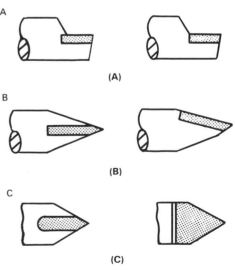

Figure 30.6 — Avoid Enclosing Carbide in a Steel Slot; Redesign For Single Braze Surface

(A)

PRESTRESSING IN JIG
DURING BRAZING AND
COOLING...

THEN RELEASE FOR
STRAIGHT BUT STRESSED
ASSEMBLY.

(B)

BRAZE, ALLOW TO
WARP IN COOLING,
THEN...

PEEN OPPOSITE FACE
UNTIL STRAIGHT.

(C)

BRAZE CARBIDE TO
OPPOSITE FACES
SIMULTANEOUSLY.

Figure 30.7 — Three Methods of Counter-straining to Overcome Curvature of Assembly Due to Brazing Strains

metal. Therefore, the usual precaution of degreasing the surface prior to brazing also should be taken. Occasionally some of the more difficult-to-wet carbides, such as titanium carbide, are prepared by plating or painting on a paste. The paste is made from copper oxide or nickel oxide fired in a reducing atmosphere to fuse the copper or nickel onto the surface. This surface is wet readily by the common filler metals. Normally when these surfaces are required, the carbide supplier will prepare the carbide for brazing.

FLUXES AND ATMOSPHERES

Most carbide brazing using BCu and copper-manganese filler metals is done in a controlled atmosphere furnace, usually with a thin flux coating. Flux is always used with the other filler metals. The AWS flux Type FB3C is used most widely. Some of the difficult-to-wet carbides or higher melting filler metals may require the use of Type FB3D or FB4A flux. The recommendations of the supplier of the filler metal should be followed in these cases.

BRAZING TECHNIQUES

The carbide insert is usually the lightest part of the assembly, and for this reason the mass differential must be considered. If this is not done, the carbide may be overheated and oxidized before the holder is hot enough to be wet by the filler metal. Once the entire assembly reaches brazing temperature, it is advisable, if possible, to rub the carbide back and forth slightly. The rubbing action aids in the wetting of the carbide surface. The added movement squeezes out excess filler metal and flux and reduces inclusions and gas porosity.

Table 30.1
Properties of Tungsten Carbide/Steel Joints Brazed with Copper (BCu), Tested in Shear

Joint Clearance		Ultimate Engineering Shear Strength		Fracture Engineering Shear Strength		Fracture Shear Elongation (Total)	
in.	mm	psi	MPa	psi	MPa	in.	mm
0.000	0.00	26 700	184	26 700	184	0.010	0.25
0.005	0.13	28 800	199	25 750	178	0.021	0.53
0.010	0.25	33 625	232	22 800	157	0.049	1.24
0.015	0.38	31 300	216	23 500	162	0.060	1.52
0.020	0.51	33 200	229	19 400	134	0.066	1.68
0.025	0.64	29 700	205	18 450	127	0.082	2.08

PRE AND POSTBRAZE OPERATIONS

Cooling or quenching should be carefully considered for each specific circumstance.

Although carbides are resistant to thermal shock, they will not withstand stresses caused by severe thermal gradients when brazed to a material (such as steel) with a significantly different coefficient of thermal expansion. Failure can result from such stresses during cooling. However, slow cooling affords some reduction in brazing stress during cool-down. Consequently, if it is desirable to braze a hard carbide to a steel support that must operate in the hardened condition, one of several methods must be used:

(1) Use an air hardening steel. It should be brought to recommended temperatures, soaked long enough for uniform heating, and cooled in air to achieve hardness prior to any brazing operation. With the hardness properly established, the material can be reheated to the brazing temperature for a short time, brazed with a suitable brazing filler metal for the particular temperature involved, and air cooled. The assembly can then be tempered to required hardness of the steel. A single heating generally is not satisfactory, except on small parts. The time required at temperature for proper heat treating is excessively long for a suitable braze operation.

Many of the air hardening steels contain substantial percentages of chromium. This complicates brazing since it is difficult to flux away the oxide of chromium which forms on the steel. A nickel-containing air-hardening steel rather than a chromium-base steel should be used.

(2) Harden the steel support member by quenching, and then braze at a temperature below about 1300°F (704°C).

(3) Immerse the support itself (but not the carbide and braze area) in a liquid quenching medium immediately after brazing.

(4) Residual flux may be removed from the assembly by washing in hot water or commercial flux-removal compounds, followed by pickling in a solution suitable for removing the oxide coating from the base metal holder. Nitric acid solutions should be avoided because they rapidly attack many brazing filler metals.

INSPECTION

Inspection usually consists of close scrutiny of the carbide and the joint for cracks. This inspection should be made before and after any final grinding operations, as cracking of the carbide may be caused by local overheating and overstressing during the finishing operations. A test for proper wetting of the carbide may be made by rapping the holder sharply. A poorly brazed carbide insert will pop off. Sound joints usually result if proper wetting and flow of filler metal within the joint is observed during the brazing operation.

Keywords — ceramic brazing, traditional ceramics, structural ceramics, aluminum oxide, zirconium oxide, silicon carbide, silicon nitride, sialons, experimental active braze filler metals, vapor coating process ceramic-to-metal joining

Chapter 31

CERAMIC BRAZING[1,2]

INTRODUCTION

Ceramics are inorganic nonmetallic materials that can be separated into two broad categories: traditional ceramics and structural ceramics. A common characteristic of ceramic materials is that they are manufactured from powders which are formed to a desired shape, and then heated to high temperature with or without the application of external pressure to achieve a final densified part. Traditional ceramics include clay products and refractories. These materials typically have low densities (or relatively high porosity contents), and normally are not used in applications where joining by techniques such as brazing is practical. The term *structural ceramics* may be taken to include monolithic materials such as aluminum oxide (Al_2O_3), zirconium oxide (ZrO_2), silicon carbide (SiC), aluminum nitride (AlN), silicon nitride (Si_3N_4), and silicon-aluminum oxynitrides (SiAlON's), as well as composites made entirely of ceramics like Al_2O_3 containing SiC whiskers or SiC containing titanium diboride particles. Generally, care is taken during the manufacture of structural ceramics to ensure that chemical composition is controlled and that high densities (or relatively low porosity contents) are achieved. Brazing of structural ceramics is possible and widely practiced.

The technological interest in structural ceramics is directly related to their unique properties when they are compared to metals. Many ceramics are characterized by high strength, not only at room temperature but at elevated temperatures as well. Silicon carbide, for example, can maintain a tensile strength in excess of 29 Ksi (200 MPa) at 2786°F (1530°C), the melting point of iron. Other ceramics like Si_3N_4 and certain ceramic composites also maintain high strength at high temperatures. Besides high strength, other properties that make ceramics attractive candidates for applications usually reserved for metallic alloys include excellent wear resistance, high hardness, excellent corrosion and oxidation resistance, low thermal expansion, and high electrical resistivity. Structural ceramics are being used or considered for use as cutting tools, bearings, machine-tool parts, dies, pump seals, high-temperature heat exchangers, and a variety of internal combustion and turbine engine parts. Typical properties for some ceramic materials and some metallic alloys are given in Table 31.1.

1. Sponsored by the U.S. Department of Energy, Assistant Secretary for Conservation and Renewable Energy, Office of Transportation Technologies, as part of the Ceramic Technology for Advanced Heat Engines Project of the Advanced Materials Development Program, under contact DE-ACO5-840R21400 with Martin Marietta Energy Systems, Inc.
2. The submitted manuscript has been authored by a contractor of the U.S. Government under contract No. DE-ACO5-840R21400. Accordingly, the U.S. Government retains a nonexclusive, royalty-free license to publish or reproduce the published form of this contribution, or allow others to do so, for U.S. Government purposes.

Table 31.1
Selected Properties of Some Pure Metals and Structural Ceramics[1]

Material	Strength[2] MPa	Modulus of Elasticity GPa	Coefficient of Linear Thermal Expansion, μm/m/°C	Electrical Resistivity, μΩ · cm	Thermal Conductivity, W/m · K
Al	34	62	23.6	2.6548	221.75
Cu	69	110	16.5	1.6730	393.71
Fe	130	196	11.7	9.71	75.31
Mo	345	324	4.9	5.2	142.26
Ni	152	207	13.3	6.84	92.05
Ti	207	116	8.41	42	21.9
Al_2O_3	300	380	6.8	$>10^{20}$	27.2
SiC	500	480	4.2	10^7	62.8
Si_3N_4	1000	304	3.2	$>10^{20}$	10
ZrO_2	700	205	9.7	10^6	2

1. Values given are typical values for each material at or near room temperature. The property values for the strength, electrical resistivity, and thermal conductivity of metals vary significantly with composition.

2. Yield strengths are given for metals; modulus of rupture strengths are given for ceramics.

Even though there is keen interest in the development of structural ceramics and their use in new and unusual engineering applications, it is in the electronics industry where the largest fraction of ceramics is actually being used. Likewise, while the development of materials like ZrO_2, Si_3N_4, and SiC has been vigorously pursued in recent years, Al_2O_3 is still the most widely used structural ceramic with a sizeable commercial market at the present time. With this in mind, the brazing of Al_2O_3 using commercially available practices will be emphasized for the remainder of this chapter.

(A) Insulated Electrical Leads

CERAMIC MATERIALS

Aluminum Oxide

Aluminum oxide (or alumina or Al_2O_3) is generally extracted from the mineral ore bauxite with world production around 35 million metric tons annually. About 90% of this alumina is reduced to produce aluminum metal, and about 3% is used for alumina specialty products including structural ceramic grades of Al_2O_3. The ceramics market for Al_2O_3 encompasses a wide range of applications from glass and chinaware to spark plugs, biomedical ceramics, and integrated circuitry.

In the United States, applications for Al_2O_3 are dominated by the electronics field. These applications generally require an insulating material with some or all of the following properties: high mechanical strength (e.g., for rectifier housings), high volume resistivity, low dielectric losses (e.g., for transmitter tubes), high density, and translucency (e.g., for sodium vapor discharge lamps). Most of these requirements can be met only by high-purity structural grades of Al_2O_3. There are numerous specialty electronic applications for Al_2O_3, including spark plugs, high-voltage insulators, vacuum-tube envelopes, rf windows, rectifier housings, integrated circuit packages, and thick- and thin-film substrates. Two examples of electronic components that contain Al_2O_3 braze joints are shown in Figure 31.1.

The compositions of structural aluminas can range from 90 to 99.99% Al_2O_3 with the

(B) Ceramic Vacuum Tube

Figure 31.1 – Examples of Ceramic-To-Metal Braze Joints

balance being impurities which are intentionally added to aid in the processing and densification of the materials. For example, alumina made especially for metallization may be doped with fluxing additives to form a microstructure that is optimum for ceramic-to-metal bonding.

Zirconium Oxide

Zirconium oxide (or zirconia or ZrO_2) occurs in nature most commonly as the mineral zircon, $ZrSiO_4$, and less frequently as free zirconia. Annual production of zirconia ore reportedly has been as high as 650 000 tons, with most of it being used as refractory material and only a relatively small portion in structural ceramics. Much of the recent technological interest in ZrO_2-based ceramics stems from the fact that it can exist in three different crystal structures: cubic, tetragonal, and monoclinic. The structure or combination of structures that exists in ZrO_2 ceramics can be controlled by alloying and heat treatment to produce materials with high strength and relatively high toughness. The good mechanical properties are a result of stress-induced phase changes that can occur between the different crystal structures. Several distinct types of toughened ZrO_2 have been developed: partially stabilized zirconia, PSZ; tetragonal polycrystalline zirconia, TZP; and composites of zirconia with other ceramics such as Al_2O_3 containing ZrO_2 particles, a material which is referred to as zirconia-toughened Al_2O_3, or ZTA.

Excellent resistance to thermal shock and mechanical impact damage has lead to the use of PSZ materials as extrusion die materials. Superior wear properties and low susceptibility to stress corrosion has led to the use of PSZ as a structural biomedical material. Wear properties, high strength, low thermal conductivity, and relatively high thermal expansion have resulted in the application of PSZ in a variety of internal combustion engine components. An important use of zirconia-toughened Al_2O_3 is for metal cutting tool bits where dramatic improvements in bit life have been achieved.

Silicon Carbide

Silicon carbide (SiC) occurs naturally, but not in quantities large enough to be of significant commercial value. In the western countries, about 500 000 tons/year of SiC are produced by reacting silica, SiO_2 with carbon at high temperatures (3992°F or 2200°C). Major uses of SiC are for abrasives where it is commonly used as loose or bonded grits, as a siliconizing agent in steel making, and as a structural ceramic material. Structural grades of SiC are most commonly made by two processes. In one case, a porous body of SiC and carbon are infiltrated with liquid silicon metal, and then

heat treated (reaction sintered) to produce further reaction of the carbon and silicon to form more SiC. SiC parts produced by this technique are known as *reaction-bonded SiC*, and typically have very high densities and contain about 10-15% of free, unreacted silicon. The other process commonly used to produce SiC parts is pressureless sintering. In this case, small amounts (about 1-3 wt%) of boron and carbon are mixed with fine SiC powder to promote densification, and then the parts are sintered in inert atmosphere at high temperature (3632°F or 2000°C). SiC produced by the method is known as sintered SiC, and it is routinely densified to near its theoretical value.

One of first important engineering applications of SiC was for rotating mechanical seals where reaction-bonded SiC was found to outperform hard metals and Al_2O_3 by a wide margin. Sintered SiC has become an important material for the fabrication of sliding seals in hermetically sealed pumps particularly where hazardous, corrosive, or abrasive media must be pumped. The corrosion resistance and mechanical properties of sintered SiC make it an attractive candidate material for many high-temperature applications. Automotive applications are an area of intense interest in SiC, where this material is being used to make critical components of advanced design turbine engines.

Silicon Nitride

Silicon nitride (Si_3N_4) does not occur in nature, and the high-quality Si_3N_4 powders required for making structural material must be derived. Common processes for producing Si_3N_4 powders include the reaction of silicon metal with nitrogen gas, the decomposition of silicon diimide, the reaction of silicon chloride and ammonia, and by the carbothermic reaction of silica, carbon and nitrogen gas. Si_3N_4 parts also can be made by a technique referred to as *reaction bonding*. For Si_3N_4, this process involves producing a shape from silicon powder, and then heating it to near 2552°F (1400°C) in nitrogen atmosphere. Reaction bonded Si_3N_4 is relatively porous and of low strength. The production of high-density, high-strength Si_3N_4 requires the use of hot-temperature processing, as well as the use of small quantities of liquid phase "sintering aids". Yttrium oxide, magnesium oxide, or aluminum oxide singly or in combination at concentrations of 4-15 wt% are commonly added to Si_3N_4 to achieve consolidation.

Widespread use of Si_3N_4 parts has been hampered by the difficulties of producing high-quality, low-cost Si_3N_4 powders, and densifying them into flaw-free articles. Nevertheless, Si_3N_4 figures prominently in all of the advanced engine concepts currently under development. Silicon nitride is considered to be one of the toughest and strongest of ceramics at temperatures above 1832°F (1000°C). It is also a prime candidate for lightweight engine components requiring good wear resistance. Potential applications for Si_3N_4 include valves, valve train parts, and turbocharger rotors for internal combustion engines, and power turbine rotors and shafts for gas turbine engines.

Sialons

Silicon-aluminum oxynitrides (or sialons) do not occur in nature, but because the elements used to produce sialons are among the most abundant on earth there is no possibility of a raw materials shortage for their production. Sialon powders can be produced by a number of techniques including the heating of clay and coal mixtures, or sand and aluminum mixtures, in nitrogen. Sialons are a solid solution of Si-Al-O-N and are described by the chemical formula, $Si_{x-6}Al_xO_xN_{X-8}$, where X can vary from 0-4. The most common form of sialon used in industry is β'-sialon. There are two major variations of the β'-sialons: one which contains a small amount of glassy phase used to promote densification during sintering, and another in which heat treatment causes the glassy phase to crystallize. The material containing the glassy phase has very high strength at low temperatures, but its strength quickly decreases with increasing temperature. The heat treated version has lower strength but retains it to higher temperatures, and it has better creep resistance than the glass-containing materials.

One of the most successful applications of sialons has been in cutting tools for machining cast irons, hardened steels and nickel-based alloys where because of superior strength, wear resistance, and thermal shock resistance, it outperforms both tungsten carbide and alumina tools. Sialons are also being used as extrusion die inserts in the production of ferrous and nonferrous metals, and as die inserts and mandrels in the production of stainless and high-alloy steel tubing. The excellent strength and thermal shock resistance of sialons also makes them preferred materials for gas shrouds for automatic welding operations. Sialons are also well suited for many seal and bearing applications.

CERAMIC BRAZING

Both direct methods like diffusion bonding as well as indirect methods like brazing can be used to bond ceramics to ceramics or to metals, and many process variations exist in both categories of bonding techniques. For example, in diffusion bonding many options exist for applying pressure and heat to the joint during the process. Also, in some cases, diffusion bonding may employ an intermediate layer, which may or may not melt, as a bonding agent. A technologically important variation of indirect bonding involves the use of glasses as bonding materials. A full discussion of ceramic joining techniques, however, is beyond the scope of this handbook, and only brazing techniques will be described in the following paragraphs. Also, as mentioned in the introduction, ceramic joining is an area of a considerable amount of developmental work. Discussion of this work will be minimal, and the focus will be on brazing techniques which are well established and used commercially. Since Al_2O_3 is by far the most widely used ceramic in applications that require brazing, most of the following discussion will deal with techniques developed for and used on this ceramic material.

Brazing Techniques

Structural ceramics are among the most stable compounds known, and a result of their chemical stability is that they are difficult for liquid metals to wet. Because wetting is necessary for producing useable braze joints, an essential consideration in ceramic brazing is the need to promote wetting of the ceramic surfaces by the braze filler metals. Wetting of ceramic surfaces can be obtained by two general methods: (1) applying coatings that promote wetting to the ceramic surfaces prior to brazing; and, (2) alloying braze filler metals with elements that activate wetting.

Sintered-Metal-Powder Process

The most widely used approach for brazing Al_2O_3 involves the sintered-metal-powder or moly-manganese (Mo-Mn) process. The Mo-Mn process actually refers to a method of metallizing Al_2O_3 surfaces by coating them with a mixture of metal and oxide powders followed by a sintering treatment at high temperature. The basic features of the process actually date back to the 1930's; they were refined in the 1950's into the metallizing treatment used today in the 1950's. In the Mo-Mn process, a slurry consisting of powders of Mo and MoO_3, Mn and MnO_2 and various glass-forming compounds is applied to an Al_2O_3 surface in the form of a paint. The coated ceramic is then fired in wet hydrogen at a temperature near 2732°F (1500°C) which causes the glassy material to densify the metallic layer and to bond it to the ceramic surface. Successful completion of the process depends on factors such as the amount and viscosity of the glassy phases, the ratio of grain size in the ceramic to the pore size in the Mo-Mn powder, and the processing temperature and atmosphere.

The Mo-Mn process has been widely accepted by industry as a standard method for metallizing Al_2O_3 surfaces, and numerous variations have been developed to extend the usefulness of the process. For instance, the metallizing mixture can be applied by brush for small production runs or unique or oddly shaped parts, or it can be applied by mass production techniques such as spraying or silk screening when necessary.

Alumina surfaces metallized by the Mo-Mn process can be brazed with a number of standard braze filler metals including BAg-8 (72Ag-28Cu), BAu-1 (64Cu-36Au), and BAu-4 (82Au-18Ni). Prior to actual brazing, nearly all metallized parts are nickel coated either by painting with a nickel oxide paint followed by sintering in hydrogen, or they are nickel plated. The nickel plating greatly improves wetting of the Mo-Mn coating by alloys such as those mentioned above. One version of the Mo-Mn process is shown schematically in Figure 31.2 where it is compared to the active filler metal process. Even though the Mo-Mn process requires many more steps than other ceramic brazing techniques it is still widely used and often preferred for several reasons: there is wide familiarity with the process; it is easily automated; minor deviations in the process variables do not significantly affect the quality of the braze joints; and brazing can be accomplished in wet hydrogen, vacuum, or inert atmosphere giving the manufacturer a variety of processing options.

Active Filler Metal Process

Early experimental work on reactions between liquid metals and oxides showed that when a liquid metal contains an element which forms a more stable oxide than the solid oxide on which the liquid metal is held, then wetting or spreading of the liquid occurs. Titanium and zirconium were shown to be particularly effective at wetting oxides, and it has been known since the 1950's that small additions of Ti to other metals promoted their wetting of oxides. Today the ability and effectiveness of Ti additions to promote wetting of Al_2O_3 and other ceramics like SiC, Si_3N_4, and sialons is well documented. Filler metals specially alloyed to promote wetting on ceramics are often referred to as active braze filler metals.

Commercial Active Filler Metals. The selection of commercially available active braze filler metals is limited, particularly for high temperature applications. Some of the active braze filler metals available today include: Ag-27Cu-4.5Ti, Ag-35Cu-2Ti, Ag-27.5Cu-2.0Ti, Ag-23.5Cu-14.5In-1.25Ti, and Ag-1Ti.

Experimental Active Braze Filler Metals. A number of researchers have identified filler metals that will wet and adhere to ceramics, but because of low industrial demand these filler metals have not been commercialized. Nevertheless, the usefulness of some of these filler metals in severe conditions has been demonstrated. Some filler metals of particular note include the following:

(1) Alloys of 48Ti-48Zr-4Be and 49Ti-49Cu-2Be with brazing temperatures of 1922°F (1050°C) and 1832°F (1000°C) respectively were used for joining Al_2O_3.

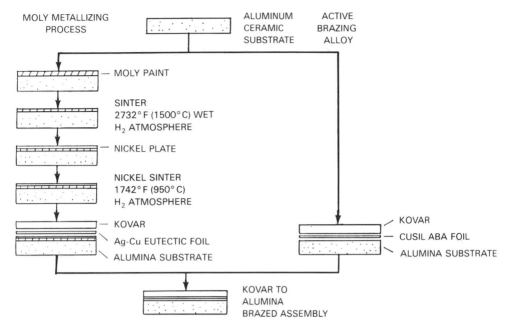

Figure 31.2 — Schematic Representation of Two Techniques For Brazing Ceramics: the Mo-Mn (Moly-Metallizing) Process, and Active Filler Metal Brazing. (Cusil ABA is a Trademark of GTE-WESGO; Kovar is a Trademark of Westinghouse Corporation.)

Al_2O_3 joints made with the Ti-Cu-Be filler metals withstood repeated exposure to high-temperature steam and severe thermal transients.

(2) A filler metal with the composition of Ti-25Cr-21V and a brazing temperature of 2912°F (1600°C) is reported to work well for brazing Al_2O_3.

(3) A Cu-Ag-Ti filler metal was found to produce excellent wetting on a variety of oxide ceramics. Alumina joints made with this alloy had strengths above 24 ksi (165 MPa) at 752°F (400°C); ZrO_2 joints had strengths above 119 ksi (130 MPa).

(4) The alloy Cu-Au-Ti was reported to give excellent braze joints of both Al_2O_3 and ZrO_2. With this alloy, Al_2O_3 joint strengths up to 32 ksi (220 MPa) were achieved at 752°F (400°C). Room temperature strengths of 37 ksi (260 MPa) were obtained for ZrO_2 joints.

(5) Morphus filler metal with compositions of 57Cu-43Ti and 70Ti-30Ni have been reported to produce strong braze joints of both SiC and Si_3N_4.

Vapor Coating Process

Another variation of the metallizing approach is the use of vapor deposited coatings of metals to promote the wetting of ceramics. The ability of metallic vapor coatings to enhance the brazing characteristics of ceramics has been known for some time, but has only been exploited for practical purposes in recent years. The key element of the process is the deposition of a thin coating, generally of a reactive metal such as titanium, onto ceramic surfaces prior to brazing. These coatings either isolate the ceramic from directly contacting the liquid filler metals, or they react with the ceramic during the brazing process. The coatings typically are made by the sputter coating process or the electron beam vapor coating process, and coating thicknesses of a few microns or less appear to be satisfactory for promoting wetting. Some examples of the use of this technique include:

(1) Sputter-deposited coatings of titanium were used to enhance the wetting of

partially stabilized zirconia and alumina by Ag-30Cu-10Sn (BAg-18) filler metal. For brazing PSZ to cast iron or titanium, joint shear strengths in the range of 30 Ksi (200 MPa) were obtained.

(2) Vapor-deposited coatings of hafnium, tantalum, titanium, and zirconium have been used to promote the wetting of SiC and Si_3N_4 by a wide variety of filler metals including Ag-28Cu (BAg-8), Au-18Ni (BAu-4), and Au-25Ni-25Pd (BVAu-7). A shaft consisting of Si_3N_4 (PY6) brazed to Incoloy 909 is shown in Figure 31.3.

JOINT DESIGN

Brazing a ceramic material to itself represents no special problems in joint design.

However, a major concern when brazing ceramics to metals is the accommodation of residual stresses that result from the differences in thermal expansion coefficients between the two materials. Ceramics have low coefficients of thermal expansion, generally lower than those of metals to which they might commonly be joined. Strains produced by the differences in thermal expansion coefficients can be large enough to cause spontaneous fracture of ceramic components. The average mismatch strain, em, in a ceramic-to-metal joint can roughly be estimated by the expression:

$$em = \Delta a \Delta T \qquad \text{(Eq. 1)}$$

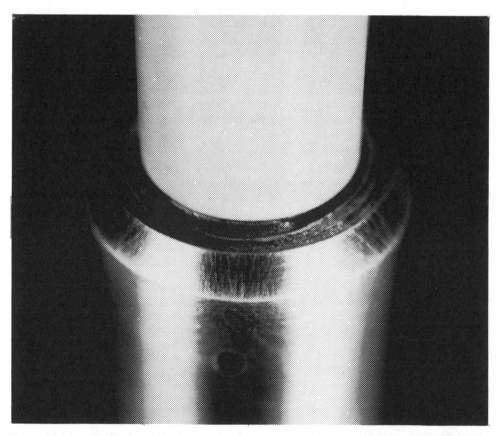

Figure 31.3 — Shaft of PY6 S_3N_4 brazed to Incoloy 909 with a Filler Metal of AU-5Pd-2Ni. The Si_3N Was Vapor Coated With Titanium Prior to Brazing. Photo Courtesy of GTE Laboratories, Inc., Waltham, MA. (Py6 Is a Trademark of GTE-Wesgo. Incoloy 909 is a Trademark of Inco Alloys International)

where Δa is the difference in the linear thermal expansion coefficient between the ceramic and the metal used in the joint, and ΔT is the difference between the temperature at which the braze filler metal solidifies and room temperature. This expression considerably oversimplifies the residual stress state of ceramic-to-metal joints but is instructive because it illustrates the effect of two important variables. Low strains, and therefore low residual stresses are promoted by minimizing differences in thermal expansion coefficients, and by brazing at the lowest temperature possible based on service conditions of the joint and good brazing techniques. In reality, there is a distribution of residual stress in ceramic-to-metal joints and the stresses will vary from compressive to tensile depending on factors such as the geometry of the component, and the materials used for the joint. For example, Figure 31.4 shows the results of subjecting a Si_3N_4-to-metal shaft joint like that shown in Figure 31.3 to a rigorous stress analysis. A complex distribution of varying residual stress is produced in the brazed component.

In practice, there are several ways to reduce residual stresses in ceramic-to-metal joints to acceptable levels. The first is by selecting a ceramic and metal with the same or similar thermal expansion coefficients. This approach is effective, but is often not practical because the components of a joint are usually selected for other reasons like strength or corrosion or oxidation resistance. Another method for reducing residual stresses is by using compliant interlayers in making the joints. In this case, a thin sheet of a highly ductile material or one with a thermal expansion coefficient near that of the ceramic is inserted between the ceramic and metal parts making the desired joint. The interlayer material accommodates the residual stresses by itself deforming, or by isolating the ceramic from the most intense stresses. Graded interlayers are yet another method for reducing resid-

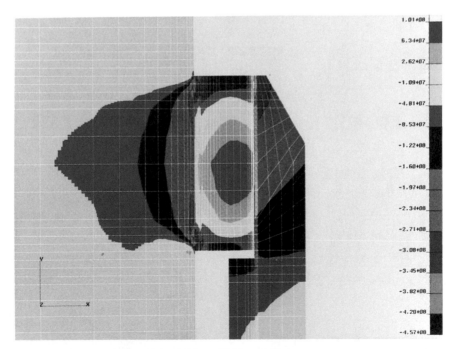

Figure 31.4 — Results of a Residual Stress Analysis For a Si_3N_4-to-Metal Joint Like That Shown in Figure 31.3. Contour Plot of Maximum Principal Stress (in Pa) at 68°F (20°C) in the Vicinity of the Joint. (Courtesy of GTE Laboratories, Inc., Waltham, MA)

ual stresses. In this approach, a composite material with varying composition is placed between the ceramic and metal parts. The thermal expansion characteristics of the composite vary with its composition and within wide limits the mismatch strains in ceramic-to-metal joints can be reasonably controlled. Joint design can also be used to minimize residual stress and some recommendations for several types of joints are given in Figure 31.5.

Processes and Equipment

Many ceramics, particularly oxides like Al_2O_3 and ZrO_2, have low thermal conductivities compared to metals. The result of this characteristic is that ceramics take longer to heat to a certain temperature than most metals and alloys would, and they are difficult to heat evenly by localized heating techniques. Also, most ceramics are electrical insulators especially at low temperatures (1832°F or 1000°C) and therefore cannot be directly heated by induction. Ceramics can be induction heated by using a metal or graphite susceptor, but this approach is dif-

ficult to control and not recommended. Because of their unique electrical and thermal properties ceramics are best suited to furnace brazing. For most applications vacuum or controlled atmosphere brazing in inert gas are the processes most often used for brazing ceramics.

Since ceramics do not heat as rapidly as metals, care should be taken to control heating rates during brazing so that both materials are maintained at the same temperature. Once melted, braze filler metals will be drawn to regions where the temperature is highest. If a metal component of a ceramic-to-metal joint reaches the brazing temperature faster than the ceramic component, then the filler metal may be drawn away from the joint, and a defective joint or no joint at all may be produced. One technique for minimizing this problem is to equilibrate the temperature of assemblies just below the solidus temperature of the filler metal, followed by slow controlled heating to the final brazing temperature.

The rapid cooling of ceramic-to-metal joints should also be avoided. If the cooling

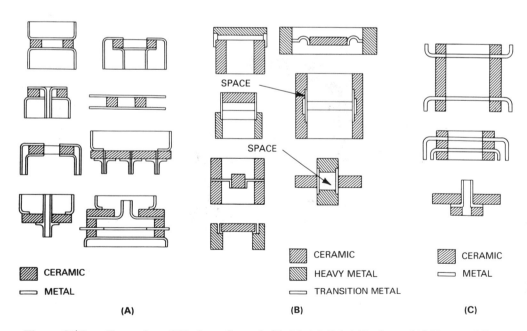

(A)   (B)   (C)

Figure 31.5 — Examples of Various Ceramic-To-Metal Joint Designs: (A) Butt and Lap Seal Joints; (B) Joints for Transitions to Thick Metal Parts; (C) Backup of Metal Seal With Blank Ceramic

rate is too fast, the metal component will cool more rapidly than the ceramic and any residual stress problem will be magnified. In addition, slow controlled cooling will permit some plastic deformation of ductile braze filler metals under the stresses due to thermal expansion coefficient mismatches, and thereby reduce the residual stress level retained in the joint at room temperture.

Surface Preparation and Cleaning

As with most brazing operations, cleanliness is an essential step in obtaining acceptable ceramic and ceramic-to-metal joints. Ceramic surfaces may contain the same kind of contaminants normally found on metal surfaces: traces of oil or grease, fingerprints, and metallic or abrasive particles, to name some of the more common ones. During Mo-Mn processing, or active filler metal brazing, most oily substances will be pyrolized to carbonaceous residues and metallic particles may be oxidized. These residues will likely result in inconsistent sintering or brazing, however, and should be removed. Surface contaminants can also adversely affect the adhesion of vapor-deposited coatings, as well as the brazing characteristics of vapor-coated ceramics.

Ultrasonic cleaning or scrubbing with a mild detergent is effective for removal of loose particles and some organic oils. Metallic contaminants can be removed by immersion in dilute acids followed by rinsing in a neutralizing solution. Other possible treatments include rinsing in acetone and alcohol followed by air drying at a low temperature [212 to 392°F (100-200°C)], or heating in air to the range of 1472 to 1832°F (800 to 1000°C).

For brazing processes other than the Mo-Mn process it may be desirable to polish previously ground ceramic surfaces, or to heat treat them at a very high temperature (resintering) prior to the brazing operation. The reason for this is that the grinding operations often used to shape ceramic components typically produce a high degree of near-surface microcracking. This surface damage can increase the probability that residual stresses will produce catastrophic cracks during the cooling of ceramic-to-metal joints, and can reduce the load-carrying capability of both ceramic and ceramic-to-metal joints. To obtain the highest reliability in a ceramic-containing joint, surface damage on the ceramic surfaces should be removed to hatever extent possible.

REFERENCES

Brush, E. F., Jr. and Adams, C. M., Jr. "Vapor-coated surfaces for brazing ceramics." *Welding Journal* 47(3): 106-s-114-s; 1968.

Fox, C. W. and Slaughter, G. M. "Brazing of ceramics." *Welding Journal* 43(7): 591-597; 1964.

Hammond, J. P., David, S. A., and Santella, M. L. "Brazing ceramic oxides to metals at low temperatures." *Welding Journal* 67(10): 227-s-232-s; 1988.

Hart, L. D., Editor. *Alumina chemicals*, Westerville, Ohio: The American Ceramic Society, Inc., 1990.

Iseki, T., Yamashita, K. and Suzuki, H. "Joining of self-bonded silicon carbide by germanium metal." *Proceedings of the British Ceramic Society* (31): 1-8; 1981.

Kang, S., Dunn, E. M., Selverian, J. H., and Kim, H. J. "Issuesin ceramic-to-metal joining: an investigation of brazing a silicon nitride-based ceramic to a low-expansion superalloy." *Ceramic Bulletin* 68(9): 1608-1617; 1989.

Mizuhara, H. and Huebel, E. "Joining of ceramic to metal with ductile active filler metal." *Welding Journal* 65(10): 43-51; 1986.

Mizuhara, H. and Mally, K. "Ceramic-to-metal joining with active brazing metal." *Welding Journal* 64(10): 27-32; 1985.

Mizuhara, H., Huebel, E., and Oyama, T. "High-reliability joining of ceramic to metal." *Ceramic Bulletin* 68(9): 1591-1599; 1989.

Moore, T. J. "Feasibility study of the welding of SiC." *Journal of the American Ceramic Society* 68: C-151-C-153; 1985.

Moorhead, A. J. "Direct brazing of alumina ceramics." *Advanced Ceramic Materials* 2(4): 159-166; 1987.

Moorhead, A. J. and Keating, H. "Direct brazing of ceramics for advanced heavy-duty diesels." *Welding Journal* 65(10): 17-31; 1986.

Naka, M., Tanaka, T., Okamoto, I., and Arata, Y. "Non-oxide ceramic joint made with amorphous $Cu_{50}Ti_{50}$ and $Ni_{24.5}Ti_{75.5}$ filler metals." *Transactions of the Japanese Welding Research Institute* (12): 337-340; 1983.

Pattee, H. E. "Joining ceramics to metals and other materials." *WRC Bulletin 178.* New York: Welding Research Council, 1972.

Oyez Scientific and Technical Services Ltd. *Proceedings of the first European Symposium on engineering ceramics.* Bath House, London: Oyez Scientific and Technical Services Ltd., 1985.

Santella, M. L. "Brazing of titanium-vapor-coated silicon nitride." *Advanced Ceramic Materials* 3(5): 457-462; 1988.

Sheppard, L. M. "Sialon, another super structural ceramic." *Advanced Materials and Processes* (1): 35-39; 1986.

Twentyman, M. E. "High-temperature metallizing, part 1. The mechanism of glass migration in the production of metal-ceramic seals." *Journal of Materials Science* (10): 765-776; 1975.

Twentyman, M. E. and Popper, P. "High-temperature metallizing, part 2. The effect of experimental variables on the structure of seals to debased aluminas." *Journal of Materials Science* (10): 777-790; 1975.

———. "High-temperature metallizing, part 3. The use of metalizing paints containing glass or other inorganic bonding agents." *Journal of Materials Science* (10): 791-798, 1975.

Weiss, S. and Adams, C. M., Jr. "The promotion of wetting and brazing." *Welding Journal* 46(2): 49-s-57-s; 1967.

Keywords — carbon, graphite, carbon materials, wettability of carbons and graphite, the brazing of graphite

Chapter 32

CARBON AND GRAPHITE

INTRODUCTION

Carbon, graphite, and composites based on these materials, i.e., carbon/carbon composites, are somewhat similar to structural ceramics and ceramic-matrix composites. However, although both types of material have high melting points and relatively low coefficients of thermal expansion, structural ceramics are compatible with oxidizing environments while the carbon based materials typically begin oxidizing at temperatures ranging from 750° to 1200°F (400 to 650°C) (depending on the perfection of the carbon structure and its purity). Also, ceramics generally have little or no open porosity, while many carbon materials have significant amounts of open porosity. Such porosity complicates the brazing process by wicking filler materials away from a joint. On the other hand, carbon materials have higher thermal conductivity, and coupled with their low coefficients of thermal expansion and high strengths, are much less sensitive to thermal shock. Finally, the carbon materials are also more creep resistant than structural ceramics.

At present there is significantly less research being conducted on the joining of the carbon based materials than there is on brazing of structural ceramics. However, there are several important applications or potential applications that utilize some of the unique properties of these materials. For example, in recent years there has been a significant interest in graphite for several critical components of experimental fusion power reactors. Graphite is favored for such applications because of its excellent thermal shock resistance, high melting point (both properties critical when the hot plasma occasionally touches a surface), and low atomic number (minimizes disruption of the plasma if sputtered from plasma-facing surfaces). Also, there is recent interest in the fabrication of components of carbon-carbon composite material for a variety of aerospace programs, and even more recently in the technology, including brazing, for joining these components into structures.

Carbon materials vary widely in the degree of crystallinity, in the degree of orientation of the crystals, and in the size, quantity, and distribution of porosity in the microstructure. These factors are strongly dependent on the precursor materials and on processing and, in turn, govern the physical and mechanical properties of these products. This potential for variation in properties can be advantageous because properties that are significant from a brazing standpoint (such as coefficient of thermal expansion) can be readily modified by processing. For example, coefficients of thermal expansion can be varied from essentially zero in the fiber direction in fiber reinforced composites to 8 x 10^{-6}/°C for isostatically molded isotropic graphites.

MATERIAL PRODUCTION

Carbons and graphites can be manufactured by several processes that yield materials with a wide range of crystalline perfection and properties. In the most widely used process, polycrystalline graphites are made from cokes produced from the residue remaining from petroleum after all volatile materials have been removed, or from natural pitch sources. The coke product is broken up and then calcined at temperatures from 1650 to 2550°F (900 to 1400°C) to further reduce the volatile content and create a filler coke that will not shrink excessively during subsequent heat treatments.

The calcined coke is crushed, milled, and sized through screens into various fractions. The size of the particles in the coarse fraction may exceed 0.04 in. (1 mm) in large bodies and may be as small as .004 in. (0.1 mm) in smaller blocks, depending on the desired properties. The shape and properties of the crushed coke particles depend largely on the coke source. Some cokes naturally break up into highly anisotropic particles with platelike or needlelike shapes, while others produce rounder isotropic particles.

The size fractions are mixed according to the properties desired in the final material, and blended with a hot coal tar pitch. The pitch creates a plastic mix that can be shaped by extrusion or molding. The shape of particles is important in that forming processes tend to preferentially align the anisotropic particles. Particles that are aligned in this manner as the result of extrusion or molding processes produce bodies with highly anisotropic properties.

The formed body is baked to pyrolize the binder pitch at temperatures from 470-1650°F (800 to 900°C), usually in large gas-fired floor furnaces. During baking, the binder experiences approximately 50 to 60% weight loss and an even greater volume loss. The effect is a reduction in overall density and subsequently increased porosity. Density can be increased by rebaking following impregnation with low-melting-point, high-viscosity pitches. Special impregnants, such as thermal setting resins or mixtures of resins and pitches, can be used to control porosity. Final graphitization of carbon bodies is achieved at temperatures ranging from 4170-5430°F (2300 to 3000°C) in an electric furnace similar to ones used primarily for the production of silicon carbide grains.

Carbon-carbon composites consist of carbon or graphite fibers in a carbon or graphite matrix. The fibers are made into yarns that are woven into two-dimensional fabrics or in three-dimensional bodies of the desired shape. The preforms so produced are fabricated into composites by either of two processes — liquid impregnation or chemical vapor infiltration (Ref. 1). In the former process, a preform is impregnated with liquid phenolic resin or pitch, baked to reduce porosity, and graphitized. The process of impregnation under pressure, carbonization, and high-temperature heat treatment is typically repeated five or more times in order to produce a composite having higher density and stronger matrix. In the chemical vapor infiltration (CVI) process, gaseous hydrocarbons such as methane are used to deposit an internal carbon matrix within a carbon fiber preform. Several specialized procedures have been developed to overcome a major problem with the CVI process which is to achieve a uniform deposition of the carbon matrix throughout a thick preform.

The carbon-carbon composite bodies resulting from either process generally have very high tensile strength, as well as low coefficients of thermal expansion, which provides for excellent resistance to thermal shock or stresses. However, these materials tend to be anisotropic, with much lower mechanical strength in directions that are not reinforced with fibers, i.e., between plies in two dimensional bodies. In addition, for applications where oxidation resistance of the composite is of concern, the materials are given a primary oxygen barrier consisting of an external coating of a ceramic such as SiC or Si_3N_4, and internal inhibitors such as compounds of boron, titanium, and silicon.

At high temperatures the latter materials form glassy oxide phases within the composite that seal any thermal expansion mismatch cracks formed in the coating and restrict passage of oxygen through the porous matrix. Of course such oxidation protection methods can significantly affect the brazing of these materials. One consideration, for example, would be whether the external coating is applied before or after brazing as the coatings are applied at 1830-2550°F (1000° to 1400°C), depending on the process. The glass-forming inhibitors may also affect wetting or adherence of a brazing filler metal.

APPLICATIONS

Carbon and graphite find widespread use as electrodes in metallurgical applications and as moderator materials in nuclear applications. Specialized uses include rocket nozzles, guide vanes, nose cones, electric motor brushes and switches, bushings and bearings, high-temperature heat exchangers, and plumbing, as well as heart valves, posts for synthetic teeth, air frame composites, and high-performance brake linings.

BRAZING CHARACTERISTICS

Wettability

The wetting characteristics of all the carbons and graphites are strongly influenced by impurities, such as oxygen or water, that are either absorbed on the surface or absorbed in the bulk material. Moisture absorption always occurs to some extent, with levels as high as 0.25 percent by weight.

Brazeability also depends on the size and distribution of pores, which can vary significantly from one grade to another. For example, some graphites are so porous that all available filler metal is drawn into them, resulting in filler metal-starved joints. Others are so dense and impervious that adherence of the filler metal is poor.

Thermal Expansion

A major consideration when brazing carbon and graphite is the effect of the coefficient of thermal expansion of these materials. This can range from zero in the fiber direction in carbon-carbon composites to 8 x $10^{-6}/°C$ depending on the type and grade of product, as well as within a given piece, depending on the degree of anisotropy. In these materials, expansion coefficients may be less than, equal to, or greater than those of the reactive or re-

fractory metals. However, they are always less than the more common structural materials such as iron and nickel. Before brazing graphite, the type and grade of carbon or graphite must be established to ascertain the expansion characteristics of the particular material. This information is also important when brazing carbon or graphite to itself. Joint failure, particularly during thermal cycling, may occur if too great a difference exists between the coefficients of thermal expansion of the graphite and the brazing filler metal.

Brazing to a Dissimilar Material

If the braze gap increases significantly on heating because of a large mismatch in coefficient, the brazing filler metal may not be drawn into the joint by capillary flow. However, if the materials and joint design cause the gap to become too small, the filler metal may not be able to penetrate the joint. In conventional brazing of dissimilar materials, the material having the greater coefficient of expansion is made the outer member of the joint. Joint tolerances are used that do not allow the gap between the surfaces to become too great for capillary flow.

Additional problems occur in brazing dissimilar materials when one part of the joint is a carbon or graphite. Carbons and graphites have little or no ductility and are relatively weak under tensile loading. These adverse conditions are usually compensated for in graphite-to-metal joints by brazing the graphite to a transition piece of a metal, such as molybdenum, tantalum, or zirconium, with a coefficient of expansion near that of the graphite. This transition piece can be subsequently brazed to a structural metal if required. This minimizes shear cracking in the graphite by transposing the stresses resulting from the large difference in thermal expansion to the metallic components. Thin sections of metals, such as copper or nickel, that deformed easily when stressed have also been successfully used for brazing dissimilar metals.

A special graded transition piece was developed by Hammond and Slaughter (Ref. 3) to accommodate the mismatch between the coefficients of expansion of graphite and of a nickel-based structural alloy AMS 5607. Using powder metallurgy techniques, a series of seven W-Ni-Fe rings having different tungsten contents were fabricated. By varying the tungsten content from 97.5% in the first ring down to 40% in the seventh, they were able to fabricate a graded seal with a thermal coefficient of expansion near that of graphite on one end and of Hastelloy N on the other. All of the W-Ni-Fe compacts and the graphite and AMS 5605 terminal pieces were then brazed in a single operation using pure copper as the filler metal. Brazing of the heavy alloys to the graphite was made possible by a prior metallization of the graphite with chromium.

FILLER METALS

Graphite is inherently difficult to wet with the more common brazing filler metals. Most merely ball up at the joint, with little or no wetting action. Two techniques are used to overcome this wetting deficiency: the graphite is coated with a more readily wettable layer, or brazing filler metals containing strong carbide-forming elements are used. Several researchers have developed techniques for coating graphite with either a metallic or intermetallic layer so that brazing can be accomplished with a conventional filler metal.

Example 1. Chemical Vapor Deposition of Graphite with Molybdenum or Tungsten

Graphite was coated with molybdenum or tungsten (0.008 to 0.31 mils thick) by a chemical vapor deposition (CVD) process (Ref. 4). This was accomplished by passing a mixture of hydrogen and the appropriate hexafluoride gas (molybdenum or tungsten) over the graphite at temperatures of 1470 and 1110°F (799 and 599°C) respectively. These metals were selected because of their low coefficients of thermal expansion, which are approximately that of the particular graphite being used, not because of carbide formation at the interface. Carbide formation does not occur at these low temperatures so the coating-to-graphite bond is essentially mechanical. In this instance, the coated graphite parts were brazed to molybdenum with copper (BCu-1), but other filler metals are equally acceptable. Although the CVD process is not complicated, it is not widely used. There are commercial compa-

nies, however, that do utilize this process to apply coatings.

Example 2. Formation of a Chromium Carbide Substrate on Graphite by Chromium Vapor Plating

Hammond and Slaughter (Ref. 3) developed a process for treating graphite that produces a metallurgically attached chromium carbide substrate. This treatment is applied by a novel procedure involving vapor plating chromium on the graphite in a partial vacuum at 2550°F (1400°C). The chromium carbide forms by chemical reaction as chromium deposits on the graphite. The chromium vapor is supplied by reacting a mixture of fine carbon and chromium oxide powder, spread over the hearth of a graphite crucible in which the plating is carried out.

Push-pin shear tests were conducted on graphite specimens coated by this technique and then brazed with pure copper filler metal. Reported shear strengths were 20 ksi (138 MPa) both at room temperature and at 1290°F (700°C). Metallographic evaluation of failed test pieces showed the specimens coated with chromium carbide to have failed by shear in the graphite pin just inside the bond region.

FILLER-METAL COMPOSITIONS

A number of experimental brazing filler metals have been developed for brazing of graphite either to itself or to refractory metals. These filler metals typically contain one or more of the strong carbide-forming elements such as titanium, zirconium, or chromium. For example, Donnelly and Slaughter (Ref. 5) reported on the successful brazing of graphite using filler metals of composition 48%Ti-48%Zr-4%Be, 35%Au-35%Ni-30%Mo, and 70%Au-20%Ni-10%Ta. In addition, Fox and Slaughter (Ref. 6) recommended the use of a filler metal with composition 49%Ti-49%Cu-2%Be for brazing of graphite as well as oxide ceramics. These filler metals wet graphite and most metals well in either a vacuum or inert atmosphere (pure argon or helium) and span a fairly wide range in brazing temperatures [from 1830°F (999°C) for 49%Ti-49%Cu-2%Be to 2460°F (1349°C) for 35%Au-35%N-

30%Mo]. However, they have not been evaluated for oxidation resistance or mechanical properties. These materials are not available commercially, and this presents a problem for a potential user who does not have access to arc-melting services.

At least two commercially available brazing filler metals reportedly wet carbon or graphite, as well as a number of metals. One is a modified version of the silver-copper filler metal with a small titanium addition to promote wetting of oxide ceramics and graphite. This filler metal has the composition of 68.8%Ag-26.7%Cu-4.5%Ti, with a solidus of 1525°F (829°C) and a liquidus of 1560°F (849°C). This filler metal is suitable for low-to-medium temperature applications but appears to have only moderate oxidation resistance. The second commercially available filler metal for graphite brazing has the composition of 70%Ti-15%Cu-15%Ni. It has a somewhat higher melting range (1670°F [910°C] solidus and 1760°F [960°C] liquidus) than the first and, by virtue of its greater titanium content, has better oxidation resistance than the silver-bearing filler metal.

A considerable amount of work has been done by researchers in the USSR on joining of graphite. For example, a Russian paper reported that graphite was brazed to steel at 2100°F (1149°C) using a filler metal of 80%Cu-10%Ti-10%Sn (Ref. 7). In another technique, known as diffusion brazing, a metallic interlayer was placed between the graphite components: the components were pressed together with a specific pressure and heated to the temperature of formation of a carbon-bearing melt or a eutectic (Ref. 8 and 9). On heating to higher temperature, the melt dissociated with the precipitation of finely divided crystalline deposits of carbon that interacted with the graphite base material to form a strong joint. Depending on the physical nature of the metal of the interlayer and on the type of carbon-metal phase diagram, carbon is formed in the joint either during the thermal disassociation of a carbon-bearing melt or a carbide-carbon eutectic. Iron, nickel, and aluminum are typical metals that form carbon-bearing melts when heated at high temperature in contact with graphite. Molybdenum is capable of forming a thermally dissociating carbide-carbon eutectic.

For *in situ* formation of a liquid film of brazing material that is subsequently dissociated at a graphite interface, a specific compressive force of 0.5 kgf/mm^2 was used, and argon pressure of 0.3 to 0.5 atm was supplied. The optimum temperature range for joining of graphite using an intermediate nickel layer was 3810°F (2099°C) to 3990°F (2199°C); for iron, 3990°F (2199°C) to 4350°F (2399°C); and for molybdenum, 4350°F (2299°C) to 4710°F (2399°C).

Metallographic examination of joints made using this technique showed that with increasing temperature the amount of metallic or carbide phase in the joint decreased, but the amount of the graphite phase increased. This increase in graphite content resulted in a marked increase in joint strength as compared to those samples brazed at lower temperatures and having significant metal or carbide-containing microstructures.

Amato et al. (Ref. 10) developed a procedure for brazing a special grade of graphite to a ferritic stainless steel for a seal in a rotary heat exchanger. It seems apparent that the selection of type 430 stainless steel was based at least partly on its lower coefficient of thermal expansion (13.1 x 10^{-6}/°C) as compared to that of a typical austenitic stainless steel (18.7 x 10^{-6}/°C). In addition, a joint geometry was developed that minimized the area of the braze joint, thereby reducing thermally induced stresses to acceptable levels. Specimens of graphite brazed in a vacuum furnace to thin type 430 stainless steel sheet with either Ni-20%Cr-10%Si or Ni-18%Cr-8%Si-9%Ti at 2060 to 2150°F (1144 to 1194°C) performed well in tests at 1200°F (649°C).

APPLICABLE HEATING METHODS

Since the carbon based materials begin to oxidize at about 750°F (400°C) (depending on the grade), brazing operations must be conducted in environments that exclude oxygen. This can be accomplished either in a vacuum at a pressure of 1 x 10^{-4} torr, or less or through the use of high-purity inert gas (argon or helium) protection. Typically, carbon based components are brazed in a furnace, but induction heating also has been used.

REFERENCES

(1) Diefendorf, Russell J. "Continuous carbon fiber reinforced carbon matrix composites." *Engineered Materials Handbook,* Vol. 1, 911-914 Cyril a Dostal, et al. Ed. ASM International; 1987.

(2) King, J. F. et al. "Materials tests and analyses of Faraday shield tubes for ICRF antennas." *Fusion Technology* (2B): 1093-1096; March 1989.

(3) Hammond, J. P. and Slaughter, G. M. "Bonding graphite to metals with transition pieces." Welding Journal, 50(1) 33-40; 1971.

(4) Werner, W. J. and Slaughter, G.M. "Brazing graphite to Hastelloy N for nuclear reactors." *Welding Engineer* (65) March 1968.

(5) Donnelly, R. G. and Slaughter, G.M. "The brazing of graphite." *Welding Journal* 41(5): 461-469; 1962.

(6) Fox, C. W. and Slaughter, G. M. "Brazing of ceramics." *Welding Journal* 43 (7): 591-597; 1964.

(7) Kochetov, D. V., et al. "Investigation of heat conditions in the brazing of graphite to steel." *Weld. Prod.* (English translation) 24(1): 39-41; 1974.

(8) Anikin, L. T. et al. "The high temperature brazing of graphite." *Weld. Prod.* (English translation) 24(7): 23-25; 1977.

(9) ———. "The high temperature brazing of graphite using an aluminum brazing alloy." *Weld. Prod.,* 24(7): 23-25 (English translation, 1977).

(10) Amato, I., et al. "Brazing of special grade graphite to metal substrates." *Welding Journal* 53(10): 623-628, 1974.

Keywords — honeybomb structure brazing, honeycomb structure manufacturing, brazed sandwich construction design, open face honeycomb structures, honeycomb sandwich brazing, acoustic structures

Chapter 33

HONEYCOMB STRUCTURE BRAZING

Among the foremost achievements of modern brazing technology is the manufacture of sophisticated honeycomb sandwich structures. These remarkably light-weight, temperature and corrosion resistant structures have been used successfully since the early fifties in aircraft, space, and naval applications. They have replaced riveted structures in those areas where high strength-to-weight ratio is of critical importance. Brazed honeycomb cylindrical structures have also found wide use recently in the automotive industry.

The all-metal honeycomb integrated structure may be a flat sheet assembly as well as a cylindrical core enclosed in, and bonded to, a cylindrical shell. Specifically, the sheet sandwich panel comprises a central honeycomb core and top and bottom closure panels, or face sheets (Figure 33.1). The central core consists of a plurality of cells which have four- or six-sided polygonal cross-sections. The standard procedures for fabricating the honeycomb core is to resistance weld or braze-join corrugated strips of metal.

The face sheets are the prime load-bearing members. Complete stabilization of the facing surfaces by proper panel design permits ultimate compressive strength to be attained, even in thin gage materials. The core performs the vital function of providing essentially continuous support to the face sheets by preventing buckling, while at the same time transmitting shear forces. Furthermore, excellent stiffness, vibration dampening, thermal, acoustic, and insulation properties are inherent, while the moment of inertia can be very large for a given structural weight. The favorable properties of the brazed sandwich are maintained even at elevated temperatures.

Although the basic honeycomb panel consists of foil gage metal core and thin sheet face sheets, it is necessary to incorporate edge close-outs, ribs, densified core, massive inserts, or doublers. An optimum design can, therefore, lead to a complex structure which may be difficult to manufacture, assemble, and braze.

There are a number of ways to preplace the brazing filler metal, which may be in a tape, powder or foil form. When powder is used, it is applied either as tape (powder in a solid acrylic matrix) pressed into the back side of the honeycomb or spun into the cells and sealed in place with a binder. In the case of a foil filler, thin sheets of brazing material (i.e., the foil) are placed between the core honeycomb and the faying surface of the panels. Braze filler metal foil also may be placed in the core, bisecting the core cells. This usually is done by spot welding the foil to the cell nodes. Pressure, directed downward, and heat are then applied to the sandwich structure, causing the filler metal to melt and to flow by capillary action to the cell nodes and edges adjacent to the face sheets.

Diffusion brazing, another quality method of joining, also has taken its place in manufacturing honeycomb sandwich con-

Figure 33.1 — Basic Honeycomb Sandwich Elements

struction. In this process, a small amount of selected metal is plated over the faying surfaces. When heat is applied to the parts, the specially selected plated metal reacts with the base metal to form a liquid eutectic intermediate layer. This layer rapidly dissolves into the base metal by diffusion mechanisms. Because the plated layer usually is thin, the result is a very narrow joint zone containing transition or eutectic brittle intermetallic phases, or both.

MATERIALS AND PROCESSING

Base Metals

A wide variety of alloys has been used in brazed honeycomb panels. The most predominant are materials which withstand moderate and high temperatures, and also have high ductility and strength. Among the most notable are precipitation-hardening stainless steels, such as AMS 5604, 5529, 5520 and 5601, which are used in fuselage

and wing aircraft panels. AMS 5540, 5599, 5596, 5536 and 5545 have found use in various parts of aircraft turbine engines, working under high temperature conditions. A family of titanium- and zirconium-base alloys, and the aluminum alloys 2024, 3003, 5052, and 5056, are also widely applied in aircraft and naval airfoil structures. Other base metals include beryllium, various special steels, and refractory metals.

Filler Metals

Nickel- and cobalt-base eutectic alloys are considered to be the best suited filler materials for joining honeycomb structures for use under highly oxidizing conditions. These alloys contain silicon or boron or both as melting point depressants, while some of them have additions of palladium, chromium, iron, or molybdenum to enhance their resistance to corrosion and oxidation. Most of these alloys correspond to the AWS BNi group of compositions, such as BNi-1, -2, -4, -5 and -8. They were applied first in a powder form, but it was found in the late seventies that, in certain applications, BNi alloys may perform better in the form of a rapidly solidified amorphous 100 percent metal foil. Improper use of organic binders may cause voids and contaminating residues which weaken joints. Amorphous metals melt more uniformly, and wet and flow into gaps much better than conventional powder filler metals. That is because amorphous metals have a more homogeneous elemental distribution and lower oxygen concentration than powder and conventional surface coating materials.

Honeycomb structures working at intermediate temperatures are brazed with nickel-manganese-copper, nickel-phosphorus, nickel-manganese-copper-silicon (BNi-6, -7, and -8) alloys, nickel-palladium-silver, silver-copper-palladium, and gold filler metals such as BAu-1, -2, -3, and -4. For service temperatures below about 1000°F (538°C), silver-base filler metals of the AWS BAg-19 and BAg-8 types are widely used.

Aluminum 3003 alloy finds successful application as a filler metal in large and complex titanium structures. Brazements made with this material can withstand a continuous service at 600°F (316°C) temperature and up to 900°F (477°C) for a short time.

When selecting a proper filler metal and brazing conditions, one should note an important specific characteristic of honeycomb structures: namely, the cores are made of rather thin gauge sheets of base metal, normally no thicker than 0.02 in. (0.5 mm), and often no thicker than about 0.004 to 0.005 in. 0.10 to 0.13 (100-200 mm). Therefore, excessive erosion or dissolution of this thin core material in the liquid filler may undercut the core strength if the parts are exposed to excessive temperature, time, or quantity of filler metal during brazing. This is particularly important when nickel-boron-silicon fillers are used, due to their strong capacity to dissolve various base metals.

Critical factors to be considered in choosing a filler metal while planning the brazing operation are base metal characteristics, strength level required, operational environment, assembly size, filler metal form (powder, foil, plating, or plastic-bonded tape),

and quantity. Because braze panels often have very large dimensions such as aircraft wing panels, the choice of brazing facility (i.e., type of furnace, possible heating and cooling rates, uniformity of furnace temperature field etc.) is of utmost importance.

MANUFACTURING AND PREBRAZE ASSEMBLY

It is critical for honeycomb structure manufacturing to satisfy stringent dimensional criteria that are typical, for example, of the aircraft industry. These specifications have to be maintained for parts which are probably the largest ever produced by brazing technology.

The heating, melting, and solidifying of filler metals causes substantial changes in part dimensions after the brazing cycle is completed. Therefore, these changes must be considered when choosing part dimensions, setting brazing gaps, and assembling the parts before the brazing cycle. A typical case may involve machining and forming individual components — or even an entire core — which can be 12 ft (4 m) or more in length with a dimensional tolerance of a few thousandths of an inch. The same accuracy is required for mating edge members, inserts, internal doublers, etc. Electrical discharge machining and electrochemical machining are favored for finishing stainless steel foil cores.

Normally, preliminary assembly of all components is made on special supporting tables to check out tolerance accumulations. Necessary adjustments are then made to obtain proper fits. When preliminary assembly is satisfactory, the components are dismantled, cleaned, and reassembled. Final assembly generally includes fixturing components into position by tack welding. Core, edge members, and inserts also are joined by minute resistance tack welds.

Face sheets frequently are held in place by small foil tabs during assembly. When brazing titanium panels, the face sheets are sometimes fusion welded or seam welded directly to the close-out or edge members. This approach essentially converts the panel into an in situ hermetically tight retort, while ensuring panel integrity.

Such a design also allows evacuation and urging of the internal volume of the sandwich, and ensures brazing conditions with accurate control of internal gas pressure. (See Figure 33.2.)

BRAZING

Brazing techniques used for the fabrication of honeycomb structures must provide a nonoxidizing environment. This involves either a purified inert gas atmosphere, vacuum, or a reducing atmosphere. For small and moderately large parts, batch-type sealed or open furnaces are used. These furnaces heat each work load separately. Sealed batch-type vacuum furnaces have internal electrical heaters and cold wall construction. Vacuum furnaces presently in use can accept panels up to 15 ft (4.6 m) in diameter — for example, a complete acoustic honeycomb shroud for a large aircraft turbine engine.

Another widespread method utilizes a sealed retort which can accommodate a very large honeycomb panel. The sealed retort technique provides an economical method for applying a rather large load over large face sheet surfaces in order to keep the face sheet and core in contact during braz-

ing. The sealed retorts can be heated by placing them inside an open furnace and using a large number of infrared lamps, electric blankets, or strip heaters as the external source of heat.

Honeycomb panels are highly sensitive to erosion processes. This characteristic makes it necessary that precise time-temperature conditions are maintained during brazing or heat treating. Careful control is required to ensure uniform temperature distribution. This condition is achieved by placing many thermocouples in direct contact with the panel, and regulating the multi-zone heaters accordingly.

Gravity is another specific and peculiar effect to be avoided when brazing large honeycomb panels. Gravity can affect filler metal distribution across the honeycomb core nodes, resulting in poor attachment of filler metal to vertical braze surfaces. At the same time it contributes to excessive accumulation of liquid metal on upward face sheet horizontal surfaces. These gravity effects are eliminated by brazing the items either in a rotating vacuum sealed retort, which is being rotated on a central support, or in a stationary vacuum furnace in which the parts being brazed are rotated. The following is an example of the latter case. One manufacturer brazed a 36 in. (91 cm) I. D.

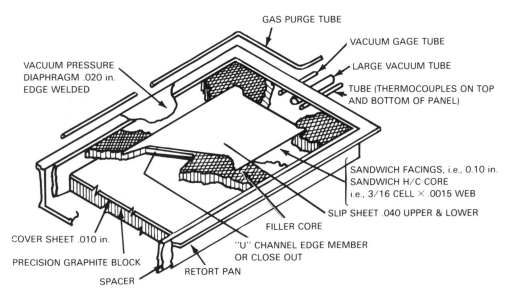

GAS PURGE TUBE

VACUUM GAGE TUBE

VACUUM PRESSURE DIAPHRAGM .020 in. EDGE WELDED

LARGE VACUUM TUBE

TUBE (THERMOCOUPLES ON TOP AND BOTTOM OF PANEL)

SANDWICH FACINGS, i.e., 0.10 in. SANDWICH H/C CORE i.e., 3/16 CELL × .0015 WEB

SLIP SHEET .040 UPPER & LOWER

FILLER CORE

COVER SHEET .010 in.

"U" CHANNEL EDGE MEMBER OR CLOSE OUT

PRECISION GRAPHITE BLOCK

RETORT PAN

SPACER

Figure 33.2 — Retort Assembly for Brazing Precision Honeycomb Sandwich

titanium alloy cylinder with a 0.5 in. (12.7 mm) core and 0.02 in. (0.51 mm) thick inner and outer skins at 1255°F (679°C). The assembly was rotated in a stationary vacuum furnace. A stainless steel mandrel was selected, based on its thermal expansion coefficient. The vacuum furnace had a seal which accepted the rotating shaft, and provided a pressure level better than 5 x 10⁻⁵ torr during brazing.

NONDESTRUCTIVE INSPECTION

Appropriate nondestructive testing techniques are normally used to inspect and evaluate brazed panels and ascertain that quality is acceptable. Each panel is examined for continuity, proper size of fillets, and lap joint integrity. A control sample, brazed at the same time as the production panel is mechanically tested in addition to nondestructive testing of the production panel.

The sequence of the various nondestructive tests is typically as follows: (1) visual inspection and dimensional measurement; (2) eddy current scan of 100 percent of core nodes to check distribution and flow of filler metal; (3) radiographic testing in standard areas to determine node flow, core crush, and shear tie; (4) ultrasonic testing for 100 percent of panel brazements to determine fillet size and facing sheet-to-core voids. Each panel is accompanied by the results of these tests as a permanent record.

Advances are being made in the use of holography, which may prove to be a highly sensitive and practical tool. Automated inspection also appears feasible with this method, while acoustic emission research is yet another emerging possibility.

APPLICATIONS

The 1950s supersonic B-58 bomber was the first production in which a brazed all-metal honeycomb sandwich was used. Approximately 1200 ft² (111.5 m²) of engine nacelle and wing paneling were fabricated from thin gage AMS 5529 stainless steel. Over 100 ship sets resulted from this major development and production program.

Command Modules (CM) incorporating brazed stainless steel honeycomb for the outer cone shell and aft heat shield were produced for the Apollo program. One manufacturer fabricated building-block panels; another welded these panels into the CM vehicle. Figure 33.3 shows various brazed sandwich elements that combine to form the high strength, lightweight truncated cone structure approximately 13 ft (4 m) in diameter by 10 ft (3 m) high. The CM cone structure typically comprises brazed honeycomb panels 0.5 in. (12.7 mm) in thickness; the aft heat shield's four quadrant-shaped panels utilize 2 in. (50.8 mm) deep, relatively high-density core.

The aft shield was welded to the core (Figure 33.4). Cores, face sheets, and attachments for the CM panels are of AMS 5601 stainless steel. Chemical milling techniques were used extensively to thin the panel face sheets to 0.008 in. (0.1032 mm), while maintaining a "picture frame" of full sheet thickness to permit high strengths as butt welded. Ingenious weld joint designs made integration of the sandwich subassemblies practical and reliable. This was accomplished by using two opposing gas tungsten arc welding heads operating simultaneously on front and back honeycomb sandwich face sheets.

Several filler metals were utilized in the CM brazing operations. A silver base, filler metal, Ag-7.5Cu-5.5In-2.2Pd-0.2Li, was used in a nickel matrix composite (80/20) for core-to-facing joints. This material has a solidus of 1400°F (760°C), and a liquidus of 1615°F (879°C). A second silver-base filler metal, 62.5Ag-32.5Cu-5Ni, was specified for flat faying surfaces, such as internal and external doublers and inserts. This filler metal is very fluid, whereas the nickel matrix or wick-containing composite is sluggish.

Panels were brazed at 1685 ± 25°F (918 ± 14°C) for 15-20 min in argon atmosphere. The AMS 5601 brazements were cooled under conditions which were favorable for the base metal phase transformations needed to provide the desired hardening and strengths. In fact, parts were cooled to 1525°F (829°C), held for two hours, and then cooled to -140°F (-96°C) for four hours to increase the amount

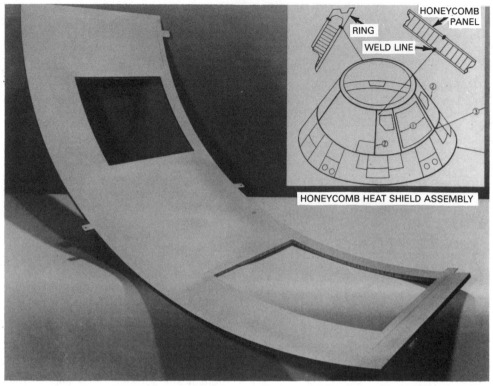

Figure 33.3 — Sketch of Command Module Used in the Apollo Program Showing the Honeycomb Heat Shield Assembly

of martensite phase. After aging at 1072 +15°F (578 ± 8°C) for one hour, the following mechanical properties were obtained (in accordance with specifications):

Ultimate tensile strength	(190,000 psi) 1310 MPa
0.2% Yield strength	(175,000 psi) 1210 MPa
Elongation	3%

Brazed sandwich construction design of the U.S. supersonic transport was largely based on manufacturing technology established during the previously mentioned programs. Brazed titanium alloy honeycomb·sandwich was also incorporated as a high efficiency structure as a result of NASA-sponsored development.

Today's advanced commercial jetliners and military aircraft incorporate honeycomb brazed structures in many places. The most widespread application involves structures working at low and high temperatures for sound attenuation. Fan ducts, exhaust ducts, turbine shrouds and seals, splitters, and center body structures represent important areas where honeycomb sandwich with perforated facing sheet or open faced honeycomb is being used. Proper choice of face sheet and honeycomb core cell materials, as well as parts design, is required to achieve optimum results.

Figure 33.5 is an example of an acoustic sandwich structural component. This 9 ft (2.7 m) long jet engine exhaust plug is used on each power plant for a jumbo commercial jet. Hot gases up to 1400°F (760°C) flow over the contoured AMS 5599 plug which supports the thrust reverser linkage and controls the nozzle flow area. Weight savings of 100 lb. (45.4 kg) per unit were

Figure 33.4 — Joining the Apollo Outer Cone Shell and Aft Heat Shield. Stainless Steel Cores and Face Sheets used to make the Honeycomb Panels.

achieved with the brazed honeycomb sandwich construction. Thousands of manufactured exhaust nozzle plugs attest to the efficiency of these extraordinary brazements.

Open-face honeycomb structures — or so-called air seals — can be brazed for use in high temperature heat-resistant applications. The development of these efficient airseal structures has made a major contribution to the evolution of large, lightweight turbojet and turbofan engines. Figure 33.6 shows the configuration of two different types of seals. The unique attribute of such honeycomb shrouds is that the core structure provides an effective seal for the stream of hot gases. At the same time honeycomb structures tolerate actual interference or rub between rotating and stationary components without damaging the turbine/compressor rotors. Consequently, engines can be

assembled tighter and cases made lighter because distortion and rubbing can be accommodated.

A small segmented rub-ring seal is shown in Figure 33.7. Here, both 0.125 in. (3.2 mm) core cell and backing are made of AMS 5540, and they are brazed with a nickel-silicon-boron filler metal (BNi-4). The enlarged inset illustrates a severe interference rub caused by the turbine rotor blades operating at 1750°F (954°C) gas temperature. The large contact area, formed on core edges as a result of abrading, is evident, as well as the efficiency of such a deformable/abradable structure. In this particular case, the core nodes were welded to form closed cells. The cell dimensions vary from 0.031 in. (0.8 mm) to 0.125 in. (3.2 mm). The brazing operation includes extended annealing at temperatures substantially higher than the service tempera-

Figure 33.5 — Sound Attenuation Plug for Commerical Jumbo Jet

ture. That makes it possible to achieve high strength and fatigue resistance in the seal brazements.

An example of the microstructural changes occurring during this stage is shown in Figure 33.8. Here, dissolution of eutectic phases from the fillet and brazement zone, and formation of a more or less uniform microstructure across the node area, are clearly seen. This occurs due to extensive interstitual diffusion of silicon and boron in the solid state out of the filler metal into the base metal. Such diffusion plays an important role in manufacturing honeycomb structures. Several new processes, such as diffusion brazing, brazing with transient liquid phase, etc., have been successfully applied to satisfy stringent specification requirements.

Aircraft gas turbines rely heavily on a variety of brazing processes. Honeycomb-type seals cover the temperature range up to 2100°F (1150°C); at these temperatures,

abradable, porous fillers are used in conjunction with an open face honeycomb sandwich supporting structure. Developmental seals are being tested with gas temperatures up to 3000°F (1650°C).

There is another concept of honeycomb manufacturing which utilizes integral amorphous brazing foil introduced at the time of core manufacture. The amorphous foil is sandwiched between the nodes of the core, and is then resistance-welded in place to form the core blanket. The advantages of this type of core are as follows:

(1) No binder required

(2) Controlled amount of braze alloy

(3) Structural configurations have high reliability and design integrity

(4) Improved strength/stiffness-to-weight ratio because of the reduced amount of filler metal

One problem eliminated in acoustic shields, the largest braze foil application area, was clogging of perforations in the

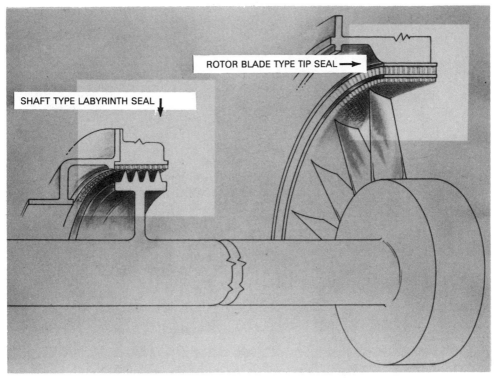

SHAFT TYPE LABYRINTH SEAL ↓

ROTOR BLADE TYPE TIP SEAL →

Figure 33.6 — Two Types of Basic Honeycomb Airseals for Gas Turbine Engines

duct internal surface by filler metal powder. These perforations are needed to absorb the engine noise. Capillary action draws the melted brazing foil into brazing gaps, thus keeping it out of the perforations.

The advantage of amorphous foil application versus powder can be demonstrated in the case of solving a tailpipe braze problem encountered on the Boing 727. Changing from powder to foil on these tailpipes resulted in a dramatic decrease of the part reject rate and brazing cycle time. It also improved the lifetime of brazing furnaces due to the absence of contaminating binders. Most importantly, it produced stronger parts with lower weight. The lower weight is a general feature of parts brazed with foil instead of the powder filler metal. Indeed, a *controlled* amount of the foil is placed in the brazing gap vicinity. Upon melting, it selectively fills brazing gaps through capillary action, whereas powder is normally applied over all part surfaces. A substantial portion

of the powder becomes attached to sites other than brazing gaps.

Titanium honeycomb panels were developed successfully for use in supersonic aircraft. An impressive example of titanium honeycomb structure is a very thin, rear fan duct used for acoustic attenuation of a military aircraft engine. This case is a one-piece structure 7.5 in. O.D. x 22 in. long x 1/2 in. thick (19 cm O.D. x 55.9 cm long x 1.27 cm thick). Face sheets are made of Ti-6Al-4V 0.02 in. (500 μm) thick sheets while the 3/8 in. (9 mm) cell core is manufactured from Ti-3Al-2.5V 0.003 in. (75 μm) thick foil. The outer facing is chemmilled from a 0.063 in. (1.575 mm) thick sheet. The internal face sheet is perforated. Copper-base filler metal alloy is plated on the parts, and diffusion brazing is then performed.

Honeycomb panels, particularly titanium ones, are used extensively in modern military planes (about 16 000 sq ft (1486 m²) per aircraft). A wide variety of brazing filler metals,

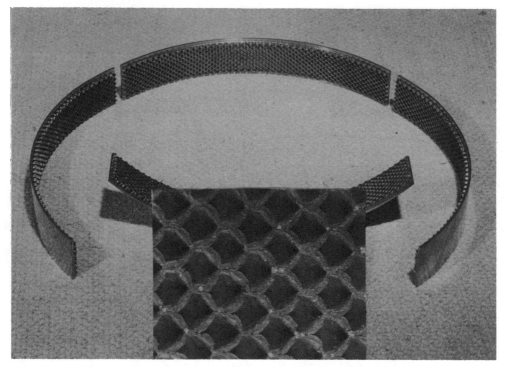

Figure 33.7 — Segmented Rub-Ring Seal

such as silver-base, aluminum-base, tita-nium-nickel, titanium-zirconium-beryllium, silver-palladium, etc., was introduced, sub-jected to very rigorous testing, and imple-mented. As a result, aluminum brazed tita-nium panels are being used for engine fairings for the F-15 fighter airplane, inlet ramps for the F-14, and engine shrouds for B-1B bomb-ers. Lightweight secondary structures, such as leading and trailing edge wedges and control surfaces, acoustic sandwiches for jet engine noise suppression, high-load structural mem-bers, etc., are all manufactured using this ex-ceptional material.

Another new and large application using brazed honeycomb core is for the mono-lithic metal supports used in exhaust gas catalytic converters. Millions of cars have already been equipped with these new parts, replacing previously used ceramic ones. Here again, joining a corrugated and wrapped core in a monolithic body, and at-taching it to the supporting case, is accom-plished by using amorphous brazing foil.

SELECTED READINGS

Davis, R. A., Elred, S., and Lovell, D. T. "Titanium honeycomb structure." Proc. Symposium on Welding, Bonding, and Fastening, 1-5, 1972.

Filippi, F. J. "Qualitative analysis of brazed sandwich." Paper read at meeting of Society for Nondestructive Testing. Cleve-land: October 27, 1958.

Kramer, Bruce E. "Brazed titanium hon-eycomb core sandwich." Paper read at Third Air force Metalworking Conference, WES-TEC, March 14, 1972.

Long, John V. and Cremer, G. D. "Braz-ing bonanza." Paper read at ASM National Metal Congress, October 17, 1967.

Schwarz, M. *Brazing.* 361-365, 375-381, Metals Park, OH: ASM International; 1987.

———. "Diffusion brazing titanium sandwich structures." *Welding Journal* (9): 35-38; 1987.

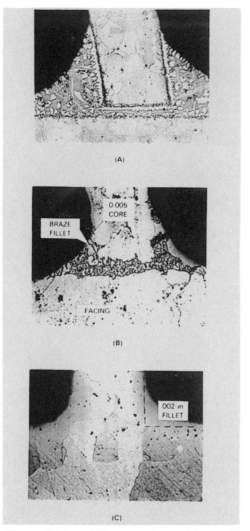

Kimball, Ch. "Acoustic structures." *Welding Journal* (10): 26-30; 1980.

Blair, W. "Procedure development for brazing Inconel 718 honeycomb sandwich structures." Ibid. (11): 433s-440s; 1973.

McNeill, W. "Nickel brazing honeycomb for aircraft gas turbine engines." Ibid. (11): 32-36; 1978.

Yeaple, F. "Metallic glass foil solves jet engine brazing problem." *Design News* (11): 1986.

De Cristofaro, N., and Datta, A. "Rapidly solidified filler metals in brazing and soldering applications." Proc. of Fifth Int. Conf. on Rapidly Quenched Metals, Wurzburg, W. Germany, September 3-7, 1984. North-Holland, 1985.

Figure 33.8 — Microstructure of Inconel[1] Honeycomb After Various Heat Treatments. Core is 0.13 mm (0.005 in.) Inconel; Facing is 2.5 mm (0.1 in.) Inconel; Filler Metal is BNi-4, High Temperature, Corrosion-Resistant. Magnification Approximately 275 x, Reduced by 30 Percent on Reproduction (A) Brazed for 1 Minute at 1950°F (1066°C); (B) Brazed for 10 Minutes at 1950°F (1066°C); (C) Soaked for 5 Hours at 1950°F (1066°C). Extended Solution Treatment (C) Yielded a Homogeneous Structure Which Has No Trace of Brittle Eutectic Phases [1]Trademark of Inco Alloys International.

Appendix A

Definitions of Terms

(This Appendix is not a part of the *Brazing Handbook*, but is included for information purposes only.)

This Appendix was written to provide the *Brazing Handbook* reader with a dictionary of the technical terms that are unique to the brazing industry. This compilation of terms includes some entries from ANSI/AWS A3.0 — *Standard Welding Terms and Definitions* and the *Metals Handbook* (Ninth Ed., Vol. 6, pages 1-20).

-A-

arc brazing. A brazing process using heat from an electric arc. **See carbon arc brazing.**

as-brazed. Pertaining to the condition of brazements after brazing, prior to any subsequent thermal, mechanical, or chemical treatments.

automatic brazing. Brazing with equipment that requires only occasional or no observation of the brazing, and no manual adjustment of the equipment controls.

-B-

balling up. The formation of globules of molten brazing filler metal or flux due to lack of wetting of the base metal.

base material. The material that is welded, brazed, soldered, or cut. **See also base metal.**

base metal. The metal or alloy that is welded, brazed, soldered, or cut.

blanket brazing. A brazing process using heat from flexible, resistance heated blankets. Brazing filler metal preforms are placed in the joint before arranging the workpieces and heating blanket(s). Blanket brazing is commonly used for large honeycomb panels.

blind joint. A joint, no portion of which is visible.

block brazing (BB). A brazing process that uses heat from heated blocks applied to the joint. This is an obsolete or seldom used process.

blowpipe. A device used to obtain a small, accurately directed flame for fine work. A portion of any flame is blown to the desired location by the blowpipe which is usually mouth operated.

bond. A uniting force that holds things together. **See bonding force, covalent bond, mechanical bond, and metallic bond.**

bonding force. The force that holds two atoms together; it results from a decrease in energy as two atoms are brought closer to one another.

braze. A weld produced by heating an assembly to the brazing temperature using a filler metal having a liquidus above 840°F (450°C) and below the solidus of the base metals. The filler metal is distributed between the faying surfaces of the joint by capillary action.

braze interface. The interface between filler metal and base material in a brazed joint.

braze welding. A welding process variation in which a filler metal, having a liquidus above 840°F (450°C) and below the solidus of the base metals, is used. Unlike

441

brazing, in braze welding the filler metal is not distributed in the joint by capillary action.

brazeability. The capacity of a material to be brazed under the imposed fabrication conditions imposed into a specific, suitably designed structure, and to perform satisfactorily in the intended service.

brazement. An assembly whose component parts are joined by brazing.

brazer. One who performs manual brazing.

brazing (B). A group of welding processes that produces coalescence of materials by heating them to the brazing temperature in the presence of a filler metal having a liquidus above 450°C (840°F) and below the solidus of the base metal. The filler metal is distributed between the closely fitted faying surfaces of the joint by capillary action. See also Master Chart of Welding and Allied Processes.

brazing alloy. A nonstandard term for brazing filler metal.

brazing filler metal. The metal that fills the capillary joint clearance and has a liquidus above 840°F (450°C) and below the solidus of the base materials.

brazing foil. A nonstandard terms for brazing filler metal in strip form.

brazing operator. One who operates automatic, furnace, or mechanized brazing equipment.

brazing powder. A nonstandard term for brazing filler metal powder.

brazing procedure. The detailed methods and practices involved in the production of a brazement.

brazing procedure qualification record (BPQR). A record of brazing variables used to produce an acceptable test brazement and the results of tests conducted on the brazement to qualify a brazing procedure specification.

brazing rod. A nonstandard term for brazing filler metal rod.

brazing rope. A nonstandard term for brazing filler metal in the form of fine wires woven together.

brazing sheet. Brazing filler metal in sheet form.

brazing shim. A nonstandard term for brazing filler metal in strip form.

brazing tape. A nonstandard term for brazing filler metal in strip form.

brazing technique. The details of a brazing operation which, within the limitations of the prescribed brazing procedure, are controlled by the brazer or brazing operator.

brazing temperature. The temperature to which the base material is heated to enable the filler metal to wet the base material and form a brazed joint.

brazing temperature range. The temperature range within which brazing can be performed.

butt joint. A joint between two members aligned approximately in the same plane.

-C-

capillary action. The force by which liquid, in contact with a solid, is distributed between closely fitted faying surfaces of the joint to be brazed or soldered.

carbon arc brazing. A nonstandard term for twin carbon arc brazing.

carburizing flame. A reducign oxyfuel gas flame in which there is an excess of fuel gas, resulting in a carbon-rich zone extending around and beyond the cone.

chemical-bath dip brazing. A dip brazing process variation.

clad brazing sheet. Metal sheet having one or both sides clad with brazing filler metal.

coil (inductor). An electrical device connected to an induction generator designed to provide induction heating of a workpiece.

cold braze joint. A joint with incomplete coalescence caused by insufficient application of heat to the base material during brazing.

complete braze fusion. Molten brazing filler metal over the entire base material surface intended for brazing and between all adjoining brazes.

complete joint penetration. Brazing filler metal penetration for the full extent of the intended joint.

complete penetration. A nonstandard term for complete joint penetration.

copper brazing. A nonstandard term for brazing with a copper filler metal.

corner joint. A joint between two members located approximately at right angles to each other.

covalent bond. A primary bond arising from the reduction in energy associated with overlapping half-filled orbitals of two atoms.

crack. A fracture type discontinuity characterized by a sharp tip and high ratio of length and width to opening displacement.

-D-

defect. A discontinuity or discontinuities that by nature or accumulated effect (for example, total crack length) render a part or product unable to meet minimum applicable acceptance standards or specifications. This term designates rejectability. **See also discontinuity and flaw.**

differential thermal expansion. The difference between the dimensional changes of two (or more) materials having different expansion coefficients, which is caused by temperature changes at constant pressure.

diffusion bonding. A nonstandard term for diffusion brazing and diffusion welding.

diffusion brazing (DFB). A brazing process that forms liquid braze metal by diffusion between dissimilar base metals or between base metal and filler metal preplaced at the faying surfaces. The process is used with the application of pressure.

dilution. The change in chemical composition of a brazing filler metal caused by the admixture of the base metal or previous brazing filler metal. It is measured by the percentage of base metal or previous brazing filler metal in the braze joint.

dip brazing (DB). A brazing process using heat from a molten chemical or metal bath. When a molten chemical bath is used, the bath may act as a flux. When a molten metal bath is used, the bath provides the filler metal.

discontinuity. An interruption of the typical structure of a brazement, such as a lack of homogeneity in the mechanical, metallurgical, or physical characteristics of the filler metal, base material, or brazement. A discontinuity is not necessarily a defect. **See also defect and flaw.**

dissolution. The dissolving of brazing filler metal in one or more of the base materials of a joint.

duty cycle. The percentage of time during an arbitrary test period that a power supply or its accessories can be operated at rated output without overheating.

-E-

edge joint. A joint between the edges of two or more parallel or nearly parallel members.

electric brazing. A nonstandard term for arc brazing and resistance brazing.

electromagnetic field. The field created when alternating current passes through as inductor.

electron beam brazing. A brazing process using heat from a slightly defocussed electron beam. Normally, electron beam brazing is used in special applications where the joint is small and the workpieces can be heated only in a vacuum.

erosion. A condition caused by dissolution of the base metal by molten filler metal resulting in a reduction of base metal thickness.

exothermic brazing (EXB). A brazing process using an exothermic chemical reaction between a metal oxide and a metal or inorganic nonmetal as the heat source, with filler metal preplaced int he joint.

-F-

face feed. The application of filler metal to the joint, usually by hand, during brazing or soldering.

faying surface. That mating surface of a member that is in contact with or in close proximity to another member to which it is to be bonded.

filler metal. The metal or alloy to be added in making a welded, brazed, or soldered joint.

fillet. A radiussed region of brazing filler metal where workpieces are joined.

fillet joint. A nonstandard term for a brazed joint that is designed to have a fillet.

fit. A nonstandard term for joint clearance.

flash coat. In brazing, a thin coating usually less than 0.0002 inches (0.005 mm) thick.

flat position. The brazing position used to braze from the upper side of the joint; the face of the braze is approximately horizontal.

flaw. An undesirable **discontinuity**.

flowability. The ability of molten brazing filler metal to flow or spread over a surface.

flow brazing (FLB). A brazing process that bonds metals by heating them with molten nonferrous filler metal poured over the joint until brazing temperature is attained. The filler metal is distributed in the joint by capillary action.

flux. A material used to hinder or prevent the formation of oxides and other undesirable substances in molten metal and on solid metal surfaces, and to dissolve or otherwise facilitate the removal of such substances. See also *active flux* and *neutral flux*.

flux coated rod. Brazing filler metal in rod form that is coated with flux.

focal spot. In an energy beam the location having the smallest cross-sectional area and, consequently, the highest energy density.

freezing point. A nonstandard term for liquidus and solidus.

fuel gas. A gas usually used with oxygen for heating; examples include acetylene, natural gas, hydrogen, propane, methylacetylene propadiene stabilized, and other synthetic fuels and hydrocarbons.

furnace brazing (FB). A brazing process using a heated furnace.

-G-

gap. A nonstandard term for joint clearance.

gas brazing. A nonstandard term for torch brazing.

gas generator. Equipment that produces a furnace atmosphere.

gas pocket. A nonstandard term for porosity.

getter. A material that is used to purify low pressure gases (usually vacuum furnace atmospheres) by chemically combining with impurities.

-H-

hard solder. A nonstandard term for silver base brazing filler metals.

hazardous material. A substance that can harm humans.

heat-affected zone. The portion of the base metal whose mechanical properties or microstructure have been altered by the heat of wleding, brazing, soldering, or thermal cutting. See Figure 20G.

heat pattern. The area heated by the coil in induction brazing.

hot crack. A crack that develops during solidification.

hydrogen brazing. A nonstandard term for any brazing process that takes place in pure hydrogen or a hydrogen-containing atmosphere.

-I-

impedence. A combination of electrical resistance, inductance, and capacitance.

imcomplete fusion. A condition in which some of the brazing filler metal in a joint did not melt.

incomplete joint penetration. Joint penetration that is unintentionally less than the thickness of the weld joint.

indistinct fillet. A condition in which the brazing filler metal did not result in a fully formed fillet.

induced current. Circulating currents produced in the workpiece when placed in an electromagnetic field.

induction brazing (IB). A brazing process using heat from the resistance of the workpieces to induced electric current.

induction generator. An electrical device used in induction heating to convert line frequency into higher frequency.

inert gas. A gas that normally does not react chemically with materials.

infrared brazing (IRB). A brazing process using heat from infrared radiation.

intergranular penetration. The penetration of a brazing filler metal along the grain boundaries of a base metal.

ionic bond. A primary bond arising from the electrostatic attraction between two oppositely charged ions.

-J-

joint. The junction of members or the edges of members which are to be bonded or have been bonded.

joint brazing procedure. The materials, detailed methods, and practices employed in brazing a particular joint.

joint clearance. The distance between the faying surfaces of a joint. In brazing, due to thermal expansion of the workpieces, joint clearance may vary as the workpieces are heated and cooled.

joint design. The joint geometry together with the required dimensions.

joint efficiency. The ratio of the strength of a joint to the strength of the base metal, expressed in percent.

-K-

karat. In gold base alloys, one karat is one twenty-fourth gold by weight.

-L-

lack of fill. A nonstandard term for imcomplete penetration.

lap joint. A joint between two overlapping members in parallel planes.

laser. A device that produces a concentrated coherent light beam by stimulated electronic or molecular transitions to lower energy levels.

laser brazing (LB). A brazing process using heat from a focussed laser beam. Normally, laser brazing is used where the joint is extremely small.

liquation. The separation of a low melting constituent of an alloy from the remaining constituents, usually apparent in alloys having a wide melting range.

liquidus. The lowest temperature at which a metal or alloy is completely liquid.

longitudinal crack. A crack with its major axis approximately parallel to the joint axis.

-M-

machine brazing. A nonstandard term for mechanized brazing.

manual brazing. A brazing operation performed and controlled completely by hand.

mechanical bond (thermal spraying). The adherence of a thermal spray deposit to a roughened surface by the mechanism of particle interlocking.

mechanized brazing. Brazing with equipment which performs the brazing operation under the constant observation and control of a brazing operator. The equipment may or may not load or unload workpieces. **See also automatic brazing.**

metal-bath dip brazing. A dip brazing process variation.

metallic bond. The principal bond that holds metals together and is formed between base metals and filler metals in all brazing processes. This is a primary bond arising from the increased spatial extension of the valence electron wave functions when an aggregate of metal atoms is brought close together. **See also bonding force and covalent bond.**

metallizing. A nonstandard term for thermal spraying; or applying (by silk screening, chemical vapor deposition, spraying, etc.) material (usually molybdenum manganese) to ceramic surfaces in preparation for brazing.

metallurgical bond. A nonstandard term for metallic bond.

micron. A unit of pressure equal to one thousandth of a torr; i.e., one milli-torr. **See also torr.**

-N-

neutral flame. An oxyfuel gas flame which is neither oxidizing nor reducing. **See also oxidizing flame and reducing flame.**

noncorrosive flux. Brazing flux which in neither its original form nor its residual form chemically attacks the base metal.

-O-

oxidizing flame. An oxyfuel gas flame in which there is an excess oxygen, resulting in an oxygen-rich zone extending around and beyond the core. **See also neutral flame and reducing flame.**

-P-

parent metal. A nonstandard term for base metal.

partial joint penetration. Joint penetration that is intentionally less than complete.

partial pressure. Pressure, usually of a furnace atmosphere or constituent of a furnace atmosphere, that is below 15 pounds per square inch; or the pressure of any constituent in a gas mixture at any pressure.

paste brazing filler metal. A mixture of finely divided brazing filler metal with an organic or inorganic flux or neutral vehicle or carrier.

peel test. A destructive method of inspection which mechanically separates a lap joint by peeling.

penetration. A nonstandard term for joint penetration.

porosity. Cavity type discontinuities formed by gas entrapment during solidification.

postheating. The application of heat to an assembly after welding, brazing, soldering, thermal spraying, or thermal cutting. **See also postbraze heat treatment.**

postbraze heat treatment. Any heat treatment after brazing.

power density. The kilowatts per square inch within an induction brazing coil.

precoating. Coating the base material in the joint by dipping, electroplating, or other means before brazing.

preform. Brazing filler metal fabricated in a shape or form for a pecific application.

preheat. The heat applied to the base metal or substrate to attain and maintain preheat temperature.

preheating. The application of heat to the base material immediately before brazing or soldering.

preheat temperature. The temperature of the base metal or substrate in the welding, brazing, soldering, thermal spraying, or thermal cutting area immediately before these operations are performed. In a multipass operation, it is also the temperature in the area immediately before the second and subsequent passes are started.

procedure. The detailed elements of a process used to produce a specified result.

procedure qualification. The demonstration that a brazed joint or a soldered joint made by a specific procedure can meet prescribed standards.

process. A grouping of basic operational elements used in welding brazing, soldering, thermal cutting, or thermal spraying.

protective atmosphere. A gas or vacuum envelope surrounding the workpieces, used to prevent or reduce the formation of oxides and other detrimental surface substances, and to facilitate their removal.

-Q-

quench. Accelerated cooling, frequently in liquid (oil, water).

-R-

reaction flux (soldering). A flux composition in which one or more of the ingredients reacts with a base metal upon heating to deposit one or more metals.

reaction stress. A stress that cannot exist in a member if the member is isolated as a free body without connection to other parts of the structure

reducing atmosphere. A chemically active protective atmosphere, which at elevated temperature will reduce metal oxides to their metallic state.

reducing flame. An oxyfuel gas flame with an excess of fuel gas. **See also neutral flame and oxidizing flame.**

remelt temperature. The temperature necessary to melt a brazing filler metal in a completed joint.

repair brazing. The process of rebrazing a joint that exhibited repairable defects.

residual stress. Stress present in a joint member or material that is free of external forces or thermal gradients.

resistance brazing (RB). A brazing process that uses heat from the resistance to electric current flow in a circuit of which the workpieces are a part.

-S-

salt-bath dip brazing. A dip brazing process variation.

sandwich brazing. A brazed assembly of dissimilar materials using a preplaced shim, other than the filler metals, as a transition layer to minimize thermal stresses.

semiautomatic brazing. Manual brazing with equipment which automatically controls one or more of the brazing conditions.

semiblind joint. A joint in which one extremity of the joint is not visible.

shielding gas. Protective gas used to prevent or reduce atmospheric contamination.

shrinkage stress. A nonstandard term for residual stress.

shrink crack. A nonstandard term for a hot crack or a shrinkage void.

shrinkage void. A cavity type discontinuity normally formed by shrinkage during solidification.

silver alloy brazing. A nonstandard term for brazing with a silver-containing filler metal.

silver soldering. A nonstandard term for brazing with a silver-base filler metal.

simultaneous brazing. A nonstandard term for producing several brazed joints at the same time.

skull. The unmelted residue from a liquated filler metal.

slag inclusion. Nonmetallic solid material entrapped in filler metal or between filler metal and base material.

soldering (S). A group of welding processes that produces coalescence of materials by heating them to the soldering temperature and by using a filler metal having a liquidus not exceeding 450 (840°F) and below the solidus of the base metals. The filler metal is distributed between closely fitted faying surfaces of the joint by capillary action. See also the Master Chart of Welding and Allied Processes.

solidus. The highest temperature at which a material is completely solid.

spool. A filler metal package consisting of a continuous length of welding wire in coil form wound on a cylinder (called a barrel), which is flanged at both ends. The flange contains a spindle hole of smaller diameter than the inside diameter of the barrel.

step brazing. The brazing of successive joints on a given part with filler metals of successively lower brazing temperatures so as to accomplish the joining without disturbing the joints brazed previously. A similar result can be achieved at a single brazing temperature if the remelt temperature of prior joints is increased by metallurgical interaction.

stopoff. A material used on the surfaces adjacent to the joint to limit the spread of brazing filler metal or flux.

stress relief cracking. Intergranular cracking in the heat-affected zone or filler metal as a result of the combined action of residual stresses and postbraze exposure to an elevated temperature.

stress relief heat treatment. Uniform heating of a structure or a portion thereof to a sufficient temperature to relieve the major portion of the residual stresses, followed by uniform cooling.

strike. A nonstandard term for flash coat.

susceptor. An electrically conductive material heated by induction and used to assist in heating a workpiece by radiation.

-T-

thermal expansion. The dimensional change exhibited by solids, liquids, and gases, which is caused by temperature changes at constant pressure.

thermal expansion coefficient. The fractional change in length or volume of a material for a unit change in temperature at constant pressure.

thermal stress. Stress resulting from non-uniform temperature distribution or differential thermal expansion.

torch brazing (TB). A brazing process using heat from a fuel gas flame.

torr. A unit of pressure normally used to describe very low pressures; one torr exerts the same force as one millimeter of mercury.

twin carbon arc brazing (TCAB). A brazing process that uses heat from an arc between two carbon electrodes. This is an obsolete or seldom used process.

-U-

undercut. A groove melted into the base metal adjacent to the braze and left unfilled by filler metal.

-V-

vacuum brazing. A nonstandard term for various brazing process that take place in a chamber or retort below atmospheric pressure.

-W-

weld brazing. A joining method that combines resistance welding with brazing.

wetting. The phenomenon whereby a liquid filler metal or flux spreads and adheres in a thin continuous layer on a solid base metal.

workpiece. A part that is brazed.

Appendix B

Properties of Brazeable Metals and Alloys

NOTE: Yield strengths are reported for the 0.2% offset method unless otherwise specified. Values suffixed YP, PL, or 0.5% signify yield point, proportional limit, or 0.5% extension under load.

Hardness values are Brinell Hardness Numbers unless suffixed R_b, R_c, or R_f (Rockwell B, Rockwell C, or Rockwell F).

[1]Hard temper
[2]As-cast
[3]Age-hardened
[4]Annealed
[5]Half hard temper
[6]Solution heat treated
[7]Artificially aged only
[8]Solution heat treated, then artificially aged
[9]Monel 326 same as Monel except nickel content 55-60%
[10]Heat treated
[11]Water quenched, drawn at 200 °F (93 °C)
[12]Water quenched, drawn at 1200 °F (649 °C)
[13]Normalized at 1200 °F (649 °C)
[14]Water quenched, drawn at 1250 °F (667 °C)
[15]Normalized at 1150 °F (621 °C)
[16]Water quenched, drawn at 1150 °F (621 °C)
[17]Rare earths

Metal	Nominal Composition (%)	UNS Number	Condition and Temper	Yield Strength ksi	Yield Strength MPa	Tensile Strength ksi	Tensile Strength MPa	Elongation in 2 in. (50 mm) (%)	Hardness Bhn 500 Kg Load (70 mm Ball)	Density lb/in³	Density 10³kg/m³
Aluminum 1100	Al 99+	A91100	Annealed -0	5	34	13	90	45	23		
			Cold worked[5] - H14	17	110	18	121	20	32	0.098	2.71
			Cold worked[1] - H18	22	152	24	165	15	44		
Aluminum 3003	Al Bal. / Mn 1.2	A93003	Annealed -0	6	41	16	110	40	28		
			Cold worked[5] - H14	19	131	21.5	148	16	40	0.099	2.74
			Cold worked[1] - H18	26	179	29	200	10	55		
Aluminum 3004	Al Bal. / Mn 1.2 / Mg 1.0	A93004	Annealed -0	10	69	26	179	25	45		
			Cold worked[5] - H34	27	186	34	234	12	63	0.098	2.71
			Cold worked[1] - H38	34	234	40	276	6	77		
Aluminum 5050	Al Bal. / Mg 1.2	A95050	Annealed -0	8	55	21	145	24	36		
			Cold worked[5] - H34	24	165	27.5	190	8	50	0.097	2.68
			Cold worked[1] - H38	28	193	31	214	6	57		
Aluminum 5052	Al Bal. / Mg 2.5 / Cr 0.25	A95052	Annealed -0	13	28	30	45				
			Cold worked[5] - H34	31	38	38		14	68	0.097	2.68
			Cold worked[1] - H38	37	42	8	85				
Aluminum 5053	Al Bal. / Si 0.7 / Mg 1.3 / Cr 0.25	A95053	Annealed -0	8	55	16	110	35	26		
			Heat treated[6] - T4	20	138	30	138	21	62		
			Heat treated[7] - T5	27	186	21	186	15	60	0.097	2.68
			Heat treated[8] - T6	32	221	37	221	13	80		
Aluminum 6061	Al Bal. / Cu 0.25 / Mg 1.0 / Cr 0.25 / Si 0.6	A96061	Annealed -0	8	55	18	124	30	30		
			Heat treated[6] - T4	21	145	35	241	25	65	0.098	2.71
			Heat treated[8] - T6	40	276	45	310	17	95		
Aluminum 6062	Al Bal. / Cu 0.25 / Si 0.6 / Mg 1.0	A96062	Annealed -0	6.5	45	17	117	30	28		
			Heat treated[6] - T4	21	145	35	241	25	65	0.098	2.71
			Heat treated[8] - T6	40	276	45	310	17	95		
Aluminum 6063	Al Bal. / Si 0.4 / Mg 0.7	A96063	Heat treated[6] - T42	13	90	22	152	20	42		
			Heat treated[7] - T5	21		21	145	12	60	0.098	2.71
			Heat treated[8] - T6	31	214	35	241	12	73		
Aluminum 4043	Al Bal. / Si 5	A94043	Sand cast[2]	8	55	19	131	8	40		
			Perm. mold cast	9	62	23	159	10	45	0.097	2.68
Aluminum A612	Al Bal. / Cu 0.5 / Mg 0.7 / Zn 6.5		Sand cast[2]	25	172	35	241	5	75	0.102	2.82
Aluminum C612	Al Bal. / Cu 0.5 / Mg 0.35 / Zn 6.5		Perm. mold cast	18	124	35	241	8	70	0.103	2.85
Copper, oxygen free (OF)	Cu 99.95 / Min	C10200	0.050 mm anneal	10 (.5%)	69	32	221	45-55	40R$_f$		8.89
			Cold rolled, hard	45 (.5%)	310	50	345	6-30	55R$_f$		to
			Hot rolled	10 (.5%)	69	34	234	40-45	40-45R$_f$		8.94
Copper, (a) electrolytic tough pitch (ETP)	Cu 99.9+ / O 0.04	C11000	0.050 mm anneal	10 (.5%)	69	32	221	45-55	40R$_f$	0.321	8.89
			Cold rolled, hard	45 (.5%)	310	50	345	6-20	85-95R$_f$	to	to
			Hot rolled	10 (.5%)	69	34	234	45-55	45R$_f$	0.323	8.94
Copper deoxidized (DLP)	Cu 99.9+ / P 0.02	C10800	0.050 mm anneal	10 (.5%)	69	32	221	45-55	40R$_f$	0.323	8.94
Commercial bronze, 90%	Cu 90 / Zn 10	C22000	0.035 mm anneal	12 (.5%)	83	38-40		45-50	57R$_f$		8.80
			1/2 Hard	45 (.5%)	310	52-60		6-15	58R$_b$	0.318	
			Hot rolled	10 (.5%)	69	37	255	45	53-60R$_f$		
Red brass, 85%	Cu 85	C23030	0.035 mm anneal	14 (.5%)	97	41	283	46	63R$_f$		8.75
			1/2 Hard	49 (.5%)					65R$_b$	0.316	
Low Brass, 80%	Cu 80 / Zn 20	C24000	0.035 mm anneal	15 (.5%)	103	46	317	48	66R$_f$		8.66
				50 (.5%)	345	61	421	18	70R$_b$	0.313	
Cartridge brass, 70%	Cu 70 / Zn 30	C26000	0.025 mm anneal	19 (.5%)	131	51	352	55	72-75R$_f$		8.55
			1/4 hard	40 (.5%)	276	54-70		20-43	55R$_b$	0.308	
			1/2 hard	52 (.5%)	359	62-70		25-30	70-80R$_b$		
Yellow brass	Cu 65 / Zn 35	C27000	0.025 mm anneal	19-21 (.5%)		51	352	55	72R$_f$		8.47
			1/2 hard	50-52 (.5%)		61-88		15-30	70R$_b$	0.306	
			Hard	64-66 (.5%)		74-110		8	80-82R$_b$		
Muntz metal	Cu 60 / Zn 40	C28000	Hot rolled	21 (.5%)	145	54	372	45	85R$_f$		8.39
			1/2 hard	50 (.5%)	345	70	483	10	75R$_b$	0.303	
			Cold rolled								
Low-leaded brass	Cu 65 / Pb 0.5 / Zn 34.5	C33000	1/2 hard	50 (.5%)	345	61	421	23	70R$_b$		8.47
			Hard	60 (.5%)		74		8	80R$_b$	0.306	

(Continued)

Melting Range °F	°C	Specific Heat (32-212°F) Btu/lb·°F	(0-100°C) J/kg·K	Coeff. of Thermal Exp. (8-212°F) 10⁻⁶ in/in·°F	(21-100°C) 10⁻⁶ mm/mm·°C	Thermal Conductivity (Room Temp.) Btu·in/h·ft²·°F	W/m·K	G Shear Modulus ksi	GPa	E Tensile Modulus of Elasticity 10³ ksi	10³ GPa
1190-1215	643-657	0.22	921	13.1	7.3 1510	1540 218	222	3.8	25.9	10	69
1190-1210	643-654	0.22	921	12.9	7.2	1335 1105 1075	192 159 155	3.8	25.9	10	69
1165-1205	629-652	0.22	921	12.9	7.2	1130	169	3.8	25.9	10	69
1160-1205	627-652	0.22	921	13.2	7.3	1130	192			10	69
1100-1200	593-649	0.22	921	13.2	7.3	958	138	3.8	25.9	10.2	70
1075-1205	579-652	0.22	921	13.1	7.3	1190 1075 1190 1075	171 155 171 155	3.8	25.9	10	69
1080-1205	582-652	0.22	921	13.1	7.3	1190 1075 1075	171 155 155	3.8	25.9	10	69
1080-1205	582-652	0.22	921	13.0		1190 1070 1070	171 154 154			10	69
1140-1205	616-652	0.22	921	13.0	7.2	1335 1450 1450	192 209 209	3.8	25.9	10	69
1065-1170	574-632	0.22	921	12.3	6.8	987		3.9	26.9	10.3	71
1105-1195	596-646	0.22	921	13.4	7.4	957	138	3.7	25.2	9.7	67
1120-1190	604-643	0.22	921	13.1	7.3	1103	159			10.3	71
1981	1083	0.092	385	9.4		2712	391	6.4	44.1	17.0	117
1949-1981	1065-1083	0.092	385	9.4		2712		6.4	44.1	17.0	117
1981	1083	0.092	385	9.4		2676		6.4	44.1	17.0	117
1870-1910	1021-1043	0.09	377	10.2	5.7	1306		6.4	44.1	17.0	117
1810-1880	988-1027	0.09	377	10.4	5.8	1104	159	6.4	44.1	17.0	117
1770-1830	966-999	0.09	377	10.6	5.9	972	140	6.0	41.4	16	110
1680-1750	916-954	0.09	377	11.1	6.2	840	121	6.0	41.4	16.0	110
1660-1710	904-932	0.09	377	11.3	6.3	804	116	5.6	38.6	15.0	103
1650-1660	899-904	0.09	377	11.6	6.4	852	123	5.6	38.6	15.0	103
1650-1700	899-927	0.09	377	11.3	6.3	804	116	5.6	38.6	15.0	103

(Continued)

Metal	Nominal Composition (%)		UNS Number	Condition and Temper	Yield Strength		Tensile Strength		Elongation in 2 in. (50 mm) (%)	Hardness Bhn 500 Kg Load (70 mm Ball)	Density	
					ksi	MPa	ksi	MPa			lb/in³	10³kg/m³
Medium-leaded brass	Cu Pb Zn	65 1.0 34	C33100	1/2 hard Hard	48-50 (.5%) 60 (.5%)	345	61-88 74		7-30 7	70R$_b$ 80R$_b$	0.306	8.47
High-leaded brass	Cu Pb Zn	65 2.0 33	C33200	1/2 hard Hard	50 (.5%) 60 (.5%)	345	61-65 74		20-25 7	70R$_b$ 70-75R$_b$	0.306	8.47
Leaded Muntz metal	Cu Pb Zn	60 0.6 39.4	C36500	Hot rolled	20 (.5%)	138	54	372	45	80R$_f$	0.304	8.41
Inhibited admiralty	Cu Zn Sn	71 28 1	C44300	Hot rolled	18-25 (.5%)		48-55		50-65	70-88R$_f$	0.308	8.55
Aluminum brass	Cu Zn Al	76 22 2	C68700	0.025 mm annealed	27 (.5%)	186	60	414	55	77R$_f$	0.301	8.33
Naval brass	Cu Zn Sn	60 39.25 0.75	C46700	Light anneal 1/2 hard Hot rolled	28-30 (.5%) 53-57 (.5%) 25 (.5%)	172	63 75-80 55	434 379	40-43 20 50	60R$_b$ 82-85R$_b$ 55R$_b$	0.304	8.41
Free-cutting brass	Cu Zn Pb	61.5 35.5 3	C36000	1/4 hard (11%)	45 (.5%)	310	56	386	20	62R$_b$	0.307	8.50
Phosphor bronze, 5% (A)	Cu Sn	95 5	C51000	1/2 hard Hard	55-80 (.5%) 75 (.5%)	517	68-85 81	558	8-28 5-10	78-81R$_b$ 87-90R$_b$	0.320	8.86
Phosphor bronze, 8% (C)	Cu Sn	92 8	C52100	1/2 hard Hard	55-65 (.5%) 72 (.5%)	496	71-105 93-96		27-40 10-12	81-89R$_b$ 93-96R$_b$	0.318	8.80
Phosphor bronze, 10% (D)	Cu Sn	90 10	C52400	1/2 hard Hard	65-72 (.2%) 91-94 (.2%)		82-118 99-147		32-35 12-17	87-92R$_b$ 94-96R$_b$	0.317	8.77
Phosphor bronze 1.25% (E)	Cu Sn P	98.75 1.25 Trace	C50500	1/2 hard Hard	51 (.2%) 60-63 (.2%)		53-55 61-65		12-16 5-8	59-64R$_b$ 67-75R$_b$	0.321	8.89
Cupro-nickel 30%	Cu Ni	70 30	C71590	Hot rolled Cold rolled strip	51 (.2%) 60-63 (.2%)	138	55 84	379	45 3	35R$_b$ 86R$_b$	0.323	8.94
Cupro-nickel, 10%	Cu Ni Fe	89 10 1	C70600	0.25 mm anneal Light drawn	20 (.5%) 79 (.5%)	393	44 60	414	42 12	15R$_b$ 72R$_b$	0.323	8.94
Nickel silver, 65-18	Cu Zn Ni	65 17 18	C75200	0.035 mm anneal 1/2 hard Hard	16 (.5%) 57 (.5%)	172	58 86 70-103	400	40 8 3	40R$_b$ 83R$_b$ 87R$_b$	0.316	8.75
Nickel silver 55-18	Cu Zn Ni	55 27 18	C77000	Hard Spring	85 (.5%) 93 (.5%)	586	100 115-145	690	3 2	91R$_b$ 99R$_b$	0.314	8.69
Nickel silver 65-15	Cu Zn Ni	65 20 15	C75400	0.050 mm anneal 1/2 hard Hard	19 (.5%) 62 (.5%) 75 (.5%)	131 427 517	55 74 85	379 510 586	42 10 3	73R$_f$ 80R$_b$ 87R$_b$	0.314	8.69
Nickel silver 65-12	Cu Zn Ni	65 23 12	C75700	0.050 mm anneal 1/2 hard Hard	19 (.5%) 60 (.5%) 75 (.5%)	131 414 517	54 73 85	372 503 586	45 11 4	30R$_b$ 80R$_b$ 89R$_b$	0.314	8.69
Nickel silver 65-10	Cu Zn Ni	65 25 10	C79500	0.050 mm anneal 1/2 hard Hard	19 (.5%) 60 (.5%) 75 (.5%)	131 414 517	51 73-85 86-105	352	46 7-12 4	28R$_b$ 80R$_b$ 89R$_b$	0.314	8.69
High-silicon bronze, (A)	Cu Si	94.8 min 3	C65500	0.035 mm anneal 1/2 hard hard	25 (.5%) 45-59 (.5%) 58-85 (.5%)	172	60 78-98 93-135	414	60 17 5-8	62R$_b$ 87R$_b$ 93R$_b$	0.308	8.53
Aluminum bronze, 5% nickel	Cu Al; Ni Fe Mn	81.5 10 5 2.5 1	C63000	Extruded 1/2 hard (10%) Note 20	50-69 (.5%) 60-75 (.5%)		90-100 100-118		15 15-20	96R$_b$ 97R$_b$	0.274	
Aluminum bronze, 9.5%	Cu Al Fe	87.5 9.5 3	C62300	Extruded	35 (.5%)		75		30	80R$_b$	0.276	7.64
Aluminum bronze, 8%	Cu Al Fe	90.5 7 2.5	C61400	Soft Hard	33-45 (.5%) 35-60		76-82 78-89		40-45 32-40	81-84R$_b$ 84-91R$_b$	0.285	
Magnesium AZ31B	Al Zn Mn	3.0 1.0 0.4	M11311	Cold-worked -H24 Annealed - O As extruded - F	18-24 15 20		34-39 32-42 34		3.0 12.0 7.0	73Bhn 58Bhn 49Bhn	0.064	1.77

(Continued)

Melting Range		Specific Heat (32-212°F) (0-100°C)		Coeff. of Thermal Exp. (8-212°F) 10⁶ (21-100°C) 10⁶		Thermal Conductivity (Room Temp.)		G Shear Modulus		E Tensile Modulus of Elasticity	
°F	°C	Btu/lb · °F	J/kg · K	in/in · °F	mm/mm · °C	Btu · in/h · ft² · °F	W/m · K	ksi	GPa	10³ ksi	10³ GPa
1630-1700	888-927	0.09	377	11.3	6.3	804	116	5.6	38.6	15.0	103
1630-1670	888-910	0.09	377	11.3	6.3	804	116	5.6	38.6	15.0	103
1630-1650	888-899	0.09	377	11.6	6.4	852	123	5.6	38.6	15.0	103
1650-1720	899-938	0.09	377	11.2	6.2	768	111	6.0	41.4	16.0	110
1710-1780	932-971	0.09	377	10.3	5.7	696	100	6.0	41.4	16	110
1630-1650	888-899	0.09	377	11.8	0.66	804	116	5.6	38.6	15.0	103
1630-1650	888-899	0.09	377	11.4	6.3	804	116	5.3	36.5	14	97
1750-1920	954-1049	0.09	377	9.9	5.5	480		6.0	41.4	16	110
1620-1880	822-1027	0.09	377	10.1	5.6	432	62	6.0	41.4	16	110
1550-1830	843-999	0.09	377	10.2	5.7	348	50	6.0	41.4	16	110
1900-1970	1038-1077	0.09	377	9.9	5.5	1440	208	6.4	44.1	17	117
2140-2260	1171-1238	0.09	377	9.0	5.0	204	29	8.3	57.2	22	152
2010-2100	1099-1149	0.09	377	9.5	5.2	312	45	6.8	46.9	18	124
1960-2030	1071-1110	0.09	377	9.0	5.0	228	33	6.8	46.9	18	124
1930	1054	0.09	377	9.3	5.2	204	29	6.8	46.9	18	124
1900-1970	1038-1077	0.09	377	9.0	5.0	252	36	6.8	46.9	18	124
1900	1038	0.09	377	9.0	5.0	276	40	6.8	46.9	18	124
1870	1021	0.09	377	9.1	5.1	312	45	6.8	46.9	17.5	121
1780-1880	971-1027	0.09	377	9.0	5.6	252	36	5.6	38.6	15	103
1895-1930	1035-1054	0.09	377	9.0		271		6.4	44.1	17.5	
1905-1915	1040-1046	0.09	377	9.0	5.0	372		6.4	44.1	17	117
1905-1915	1040-1046	0.09	377	9.0	5.0	468		6.4	44.1	17	
1120-1170	604-632	0.25	372	14.5	8.1	552		2.4	16.5	6.5	45

(Continued)

Metal	Nominal Composition (%)		UNS Number	Condition and Temper	Yield Strength		Tensile Strength		Elongation in 2 in. (50 mm) (%)	Hardness Bhn 500 Kg Load (70 mm Ball)	Density	
					ksi	MPa	ksi	MPa			lb/in³	10³kg/m³
Magnesium M1A	Mn	1.2	M15100	Cold-worked - H24	26		35		7.0	48Bhn	0.063	1.74
				Annealed - O	18		33		17.0	48	0.063	1.74
Magnesium	Zn	1.2	M16100	Cold-worked - H2419	28		34-38		8-12.0	After cold	0.063	1.74
ZE10A	RE¹⁷	0.17		Annealed - O	26	179	33	228	23.0	work only	0.063	1.74
Nickel (pure)	Ni	99.9	N02200	Annealed	8.5		46	317	30	94	0.322	8.91
Nickel (wrought)	Ni	99.9 min	N02201	Annealed	10-30		50-80		50-35	90-120	0.321	8.89
				Hot rolled	15-45		55-80		65-30	90-120		
				Cold drawn	40-90		65-95		35-15	125-230		
				Cold rolled ¹	75-105		90-115		15-2	92Rᵦ		
Low-carbon nickel	Ni	99.0 min	N02270	Annealed	15	103	60	414	50	90	0.321	8.89
	C	0.02 max										
Nickel (cast)	Ni	Rem.		Sand cast ²	25	172	50	345	25	100	0.301	8.33
	Si	2.0										
	Fe	1.25										
"D" Nickel	Ni	95	N02211	Annealed	35	241	75	517	40	140	0.317	8.77
	Mn	4.5		Hot rolled	34				40	147		
				Cold drawn	80	552	100	690	25	190		
"E" Nickel	Ni	98		Annealed	35	241	75	517	40	140	0.32	8.86
	Mn	2		Hot rolled	50	345	90	621	35	150		
				Cold drawn	80	552	100	690	25	190		
"Dura-nickel•	Ni	94	N03301	Hot rolled	35-90		90-130		55-30	140-240		
	Al	4.4		Hot rolled ³	115-150		110	160	30-15	300-375	0.298	8.25
				Cold drawn	60-130		210	200	35-15	185-300		
				Cold drawn³	125-175		170-210	150	25-15	300-380		
Monel (wrought)	Ni	67	N04400	Annealed	25-45		70-85		50-35	110-140	0.319	8.83
	Cu	30		Hot rolled	40-90		80-110		45-20	140-220		
				Cold drawn	85-125		55-120		35-10	160-250		
				Cold drawn ¹	100-120		90-110		15-2	204		
Monel (cast)	Ni	63		Sand cast	32-40		65-90		45-25	125-150	0.312	8.64
	Cu	32										
	Si	1.25										
"R" monel	Ni	67	N04405	Hot rolled	35-60							
	Cu	30		Cold drawn	75	517	75-90	621	45-30	130-170	0.319	8.83
	S	0.035					90			180		
"K" monel	Ni	66	N05500	Hot rolled	45-110		95-155		45-20	140-315	0.306	8.47
	Cu	29		Hot rolled ³	110-150		140-190		30-20	265-346		
	Al	2.75		Cold drawn	70-125		100-140		35-13	175-260		
				Cold drawn ³	95-160		135-185		30-15	255-370		
"H" monel	Ni	65		Sand cast ²	60-80		100-130		25-10	240-290	0.308	
	Cu	29.5										
	Si	3.0										
"S" monel	Ni	63	N04019	Sand cast ⁴	70	483	90	621	3	275	0.302	8.36
	Cu	30		Sand cast ²	80-115		110-145		4-1	273-350		
	Si	4		Sand cast ³	100	690	110-145		4-1	275-375		
Inconel "X•	Ni	73	N07750	Notes 10, 23	100-135		160-185		15-30	302-263	0.298	8.25
	Cr	15		Notes 10, 24	115-140		162-193		15-30	300-390		
	Fe	7.0										
	Ti	2.5										
	Al	0.7										
	Cb	1.0										
80 Ni 20 Cr alloy (wrought)	Ni	78	N06009				100-200	1138	25-35	165-184	0.303	8.39
	Cr	19					165		1	241-287		
60 Ni 15 Cr alloy (wrought)	Ni	61		Annealed			105		30	157	0.298	8.25
	Fe	Bal.										
	Cr	15.5										
60 Ni 15 Cr alloy (cast)	Ni	60		Cast ²			65	448	2	184	0.298	8.25
	Fe	Bal.										
	Cr	14										
"Hastelloy" Alloy A	Ni	Bal.		Sand cast ⁴	40	331	73	503	12	168		
	Mo	21		Hot rolled ⁴	44	303	110	758	50	175	0.318	8.8
	Fe	19		Investment cast ²	45	310	76	524	21	173		
"Hastelloy" Alloy B	Ni	Bal.	N10001	Sand cast ⁴	58	400	80	552	8	207		
	Mo	28		Hot rolled ⁴	60	414	135	931	50	201	0.4	9.25
	Fe	5		Investment cast ²	61	421	85	586	14	216		

(Continued)

Melting Range		Specific Heat (32-212°F) (0-100°C)		Coeff. of Thermal Exp. (8-212°F) 10^{-6} (21-100°C) 10^{-6}		Thermal Conductivity (Room Temp.)		G Shear Modulus		E Tensile Modulus of Elasticity	
°F	°C	Btu/lb·°F	J/kg·K	in/in·°F	mm/mm·°C	Btu·in/h·ft²·°F	W/m·K	ksi	GPa	10^3 ksi	10^3 GPa
1198-1200	648-649	0.25	372	14.5	8.1	958	125	2.4	16.5	6.5	45
1198-1200	648-649	0.25	372	14.5	8.1	870				6.5	45
1100-1195	593-646	0.25	372	14.5	8.1	920		2.4	16.5	6.5	45
1100-1195	593-646	0.25	372	14.5	8.1	956				6.5	45
2650	1454	0.11	164	7.4	4.1	575	83			30	207
2615-2615	1435-1446	0.11	164	7.2	4.0	420	61	1.2	8.2	30	207
2615-2635	1435-1446	0.11	164	7.2	4.0	420	61			30	207
2540-2600	1393-1427	0.11	164	7.2	4.0	410	59			21.5	148
2500-2600	1371-1427	0.11	164	7.4	4.1	335	48			30	207
2570-2600	1410-1427	0.11	164	7.4	4.1	335	48			30	207
2615-2635	1435-1446	0.104	435	7.2		165		1.1	7.9	30	207
2370-2460	1299-1349	0.13	544	7.8	4.3	180	26	9.5	65.5	26	179
2400-2450	1316-1343	0.13	544	7.2		180	26			19	131
2370-2460	1299-1349	.010		7.8	4.3	180	26			26	179
2400-2460	1316-1349	0.11		7.8	4.3	130	19	9.5	65.5	26	138
2350-2400	1288-1316	0.13	544	6.8	3.8	180	26			20	138
2300-2350	1260-1288	0.13	544	6.8	3.8	180	26			21	145
2540-2600	1393-1427	0.103-0.107		6.8		83		1.2	8.3	31	214
2550	1371	0.107	448			93				31	214
2600	1427	0.107	448	7.5		93	14			31	214
2600	1427	0.107	448	7.5	3.4	93	14			30.5	210
										24	165
2370-2425	1299-1329	0.094	394	6.1	3.4	116	17			27	186
										26.5	183
2410-2460	1321-1349	0.091	381	5.6	3.1	78.5	11			30.8	212
										28.5	197

(Continued)

Metal	Nominal Composition (%)		UNS Number	Condition and Temper	Yield Strength		Tensile Strength		Elongation in 2 in. (50 mm) (%)	Hardness Bhn 500 Kg Load (70 mm Ball)	Density	
					ksi	MPa	ksi	MPa			lb/in³	10³kg/m³
"Hastelloy" Alloy C	Ni Mo Cr Fe W	Bal. 16 16 5 4	N10002	Investment cast [2] Sand cast [4] Hot rolled	60 58 55	414 400 379	80 78 130	552 538 896	7 8 45	225 218 211	0.323	8.94
"Hastelloy" Alloy D	Ni Si Cu	Bal. 10 3		Sand cast [4]	118	814	118	814	0-2 (in. 1 in.) (25 mm)	390	0.282	7.81
"Hastelloy" Alloy X	Cr Mo Fe Ni	22 9 20 Bal.	N06002	Cast, solution treated	43	296	77	531	23.0		0.3	8.3
Rene 41	Cr Co Mo Ti Al Fe Ni	19 11 10 3 1.5 2.5 Bal.	N07041		150	1034	206	1420	14.0	32R_c	0.298	8.25
Kovar	Ni Co Fe	29 17 Bal.	K94610	Annealed	50	345	75	517	25.0	150	0.3	8.3
Multimet (N-155)	Cr Ni Co Mo W Cb	21 20 20 3 3 1	R30155	Cr 20% Annealed Investment cast	135 58 58	931 400 400	145 118 101	1000 814 696	30.0 49.0 31.0	194 180	0.294	8.17
Stellite 21	Ni Cr Mo Fe Co	2.5 27 5.5 2 Bal.	R30021	As cast Aged	82 110	565 758	101 124	696 855	8.2 1.7	30R_c 34R_c	0.3	8.30
Stellite 25 (L-605)	Ni Cr W Mn Si Co	10 20 15 1 1 Bal.	R30605	Sheet, annealed Sheet, CR 20%	68 163	469 1124	148 192	1020 1324	61.0 27.0	250 43R_c	0.33	9.13
Tungsten	W		R07005	Sintered Drawn 15 min.			18.5 285	128 1965	1-4	225 375	0.697	19.29
Tungsten alloy	Mo	15			88	608	108	745	22.0	265	0.615	17.02
Molybdenum	Mo		R03600	Drawn 14 min.	57-85	393-586	215-240	1482-1655	2-5	155-250	0.368	10.19
Molybdenum	Ti	0.5	R03670	Stress relieved 1600 F	85	586	110	758	35.0	275	0.37	10.24
MIA ("Dow-metal" M)	Mg Mn	Bal. 1.2	M15100	Hot rolled Cold rolled Annealed	28 18	193 124	33 37 33	228 255 228	12 7 16	56 48	0.064	1.77
Silver (pure)	Ag	99.9 +	P07010	Annealed Cold rolled [1]	12 38	83 262	23 43	159 296	45 6	30 90	0.379	10.49
Gold (pure)	Au	99.9	P00015	Annealed Cold rolled [1]	0 PL. 30 PL.	0 PL. 207 PL.	17 32	117 221	45 4	25 60	0.698	19.32
Platinum (pure)	Pt	99.99	P04980	Annealed Cold rolled [1]	10 PL. 27 PL.	69 PL. 186 PL.	24 36	165 248	24 2.5	40 100	0.772	21.37
Platinum (commercial)	Pt	99+	P04995	Annealed Cold rolled [1]	6.5 PL. 27 PL.	45 PL. 186 PL.	23 37	159 255	29 3.5	57 113	0.772	21.37
Iridium-Platinum 10%	Pt Ir	Bal. 10	P04840	Annealed Cold rolled [1]	30 PL. 54 PL.	207 PL. 372 PL.	53 82	365 565	27 2.5	110 170	0.778	21.54
Rhodium Platinum 10%	Pt Rh	Bal. 10		Annealed Cold rolled [1]	17 PL. 56 PL.	117 PL. 386 PL.	47 84	324 579	37 3	65 160	0.742	20.54
Palladium (refined)	Pd	99.5	P03980	Annealed Cold rolled [1]	5 PL. 32 PL.	34 Pl. 221 PL.	30 47	207 324	40 1.5	44 106	0.432	11.96
Palladium (hard)	Pd Ru Rh	Bal. 4 1		Annealed Cold rolled [1]	28 PL. 46 PL.	193 PL. 317 PL.	59 72	407 496	23 3		0.432	11.96
Tantalum	Ta	99.9+	R05210	Annealed or unannealed sheet			42 to 178	290 to (in 1227	45-25 1 in.) (25 mm)	55 to 125	0.60	16.60

(Continued)

Melting Range		Specific Heat (32-212°F)	(0-100°C)	Coeff. of Thermal Exp. (8-212°F) 10⁻⁶	(21-100°C) 10⁻⁶	Thermal Conductivity (Room Temp.)		G Shear Modulus		E Tensile Modulus of Elasticity	
°F	°C	Btu/lb · °F	J/kg · K	in/in · °F	mm/mm · °C	Btu · in/h · ft² · °F	W/m · K	ksi	GPa	10³ ksi	10³ GPa
2320-2380	1271-1121	0.092	385	6.3	3.5	87	13			24.5	169
										26	179
2030-2050	1110-1121	0.108	452	6.1	3.4	145	21			28.9	199
2350	1288	0.11	461	7.7	4.3	750	108				
		0.108	452	7.5	4.2	85	12			25.0	172
2715	1491	0.16	670	2.3	1.3	133	19			20.0	138
2350-2475	1288-1357	0.104	435	7.9	4.4	101	15			30.0	207
2465	1352			7.8	4.3	100	14			16.0	110
2425-2570	1329-1410	0.09	377	7.2	4.0	90	13			34.0	234
6170 +35	+20	0.034	142	2.2	1.2	1160	167			50	345
4760	2627	0.065	272							48.0	331
4750	3621	0.06	251	3.0	1.7	1010	146			42	290
1198-1200	648-649	0.25	1047	3.4	1.9	820	118			46.0	317
				14.4	8.0	870	125			6.5	45
1760	960	0.056	234	10.9	6.1	2900	418			11	76
1945	1063	0.031	130	7.8	4.3	2000	288			12	83
3225	1774	0.032	134	4.9	2.7	480	69			21	145
3225	1774	0.032	134	4.9	2.7	480	69			22.6	156
3230-3270	1777-1799			4.93	2.74						
3345	1841										
2830	1554	0.058	243	6.5	3.6	490	71			17	117
2860	1571										
5425	2996	0.036	151	3.57	1.98	375	54			27	186

(Continued)

Metal	Nominal Composition (%)		UNS Number	Condition and Temper	Yield Strength		Tensile Strength		Elongation in 2 in. (50 mm) (%)	Hardness Bhn 500 Kg Load (70 mm Ball)	Density	
					ksi	MPa	ksi	MPa			lb/in³	10³kg/m³
Wrought iron	Fe	Bal.		Hot rolled	30	207	48	331	30 (in 8 8 in.) (200 mm)	100	0.278	7.70
	Slag	2.5										
Ingot iron	Fe	99.9+		Hot rolled	29	200	45	310	26	90		
				Annealed	19	131	38	262	45	67	0.284	7.86
Gray cast iron	C	3.1	F10007	Cast ²	None	None	32	221	0.5 max.	180	0.260	7.20
	Si	1.8										
	Mn	0.7										
	Fe	Bal.										
Malleable iron	C	2.5		Cast ²	36.5 YP	252 YP	55	379	20	130	0.264	7.31
	Si	1										
	Mn	0.55										
"Ni-Tensyl-iron"	C	2.7		Cast ²	30 PL.	207 PL.	40	276		220		
	Si	1.5		Cast ¹⁰	40 PL.	276 PL.	100	690		350	0.260	7.20
	Mn	0.8										
	Ni	2.3										
	Cr	0.3										
	Fe	Bal.										
	Mo	0.4										
"Ni-Resist" Type 1	C	2.8	F47004	Cast ²			30	207	2	150	0.270	7.47
	Si	1.5										
	Mn	1.0										
	Ni	14.0										
	Cr	2.0										
	Cu	7.0										
	Fe	Bal.										
"Ni-Resist" Type 2	C	2.8		Cast ²			30	207	2	150	0.270	7.47
	Si	1.8										
	Mn	1.3										
	Ni	20										
	Cr	2.5										
	Fe	Bal.										
Ductile iron (Mg containing)	C	3.4		Cast²	70	483	100	690	3	245		
	Si	2.5		Cast⁴	55	379	70	483	20	160	0.26	7.2
	Mn	0.7										
	P	0.1 max										
	Ni	1.5										
	Mg	0.06										
	Fe	Bal.										
Ductile "Ni-Resist" (Mg containing)	C	2.8		Cast ⁴	35	241	60	414	10	175	0.268	7.42
	Si	2.5										
	Mn	1										
	P	0.01 max										
	Ni	20										
	Cr	2										
	Mg	0.1										
	Fe	Bal.										
"Ni-Hard" low-carbon	C	2.7		Sand cast²					55	550		
	Si	0.6		Chill cast²					75	625	0.275	7.61
	Mn	0.5										
	Ni	4.5										
	Cr	1.5										
	Fe	Bal.										
"Ni-Hard" high-carbon	C	3.0		Sand cast ²					30	600		
	Si	0.6		Chill cast²					80	700	0.275	7.61
	Mn	0.5										
	Ni	4.0										
	Cr	1.5										
	Fe	Bal.										
Carbon steel SAE 1020	Fe	Bal.	G10230	Annealed	40	276	60	414	35	130		
	Mn	0.45		Hot rolled	45	310	69	476	31	146	0.284	7.86
	Si	0.25		Hardened¹¹	80	552	104	717	6	205		
	C	0.20		Hardened¹²	62	427	90	621	22	175		
Cast carbon steel	Fe	Bal.	J03006	Cast⁴	40	276	72	496	26	140		
	Mn	0.70		Cast¹³	45	310	80	552	30	160	0.281	7.83
	Si	0.40		Cast¹⁴	60	414	90	621	25	185		
	C	0.30										
Cast alloy steel	Fe	Bal.		Cast⁴	60	414	105	724	20	225		
	Ni	1.75		Cast¹⁵	95	655	120	827	17	260	0.284	7.86
	Mn	0.80		Cast¹⁶	135	931	150	1034	13.5	325		
	Cr	0.75										
	C	0.30										
	Mo	0.25										

(Continued)

Melting Range		Specific Heat (32-212°F)	(0-100°C)	Coeff. of Thermal Exp. (8-212°F) 10^{-6} (21-100°C) 10^{-6}		Thermal Conductivity (Room Temp.)		G Shear Modulus		E Tensile Modulus of Elasticity	
°F	°C	Btu/lb·°F	J/kg·K	in/in·°F	mm/mm·°C	Btu·in/h·ft²·°F	W/m·K	ksi	GPa	10^3 ksi	10^3 GPa
2750	1510	0.11	461	6.7	3.7	418	60			29	200
2795	1535	0.108	452	6.5	3.6	494	71			30.1	208
2075	1135			6.0	3.3	310	45			13+ 1.5	90+ 10
2250	1232	0.122	511	6.6	3.7	435	63			25	172
2150	1177			6.5	1.6	320	46			20+ 1.5	138+ 10+
2150	1177			10.6	5.9	275	40			14.5	100
2150	1177			10.3	5.7	275	40			14.5+ 1.5	100+ 10
2100	1149			7.5	4.2					25	172
2250	1232			10.4	5.8					18.5	128
2150	1177			4.8	2.7						
2150	1177			4.8	2.7						
2760	1516	0.116	486	6.5	3.6	357	51			30	207
2745	1507	0.107	448	6.7	3.7	400	58			30	207
2745	1507	0.107	448	6.7	3.7					30	207

(Continued)

Metal	Nominal Composition (%)		UNS Number	Condition and Temper	Yield Strength		Tensile Strength		Elongation in 2 in. (50 mm) (%)	Hardness Bhn 500 Kg Load (70 mm Ball)	Density	
					ksi	MPa	ksi	MPa			lb/in³	10³kg/m³
4340	C	0.4	G43400	As rolled	147	1014	180	1241	9.0	380		
	Mn	0.7		Annealed	99	683	120	827	17.0	240	0.283	7.83
	Cr	0.8										
	Mo	0.25										
	Fe	Bal.										
Vascojet 1000 (H-11)	Cr	5	T20811	Hardened and tempered	215	1482	268	1848	5.0		0.280	7.75
	Mo	1.3										
	V	0.5										
	C	0.4										
	Fe	Bal.										
Stainless steel Type 202	Cr	18	S20200	Annealed	55	379	105	724	55.0	90R_b	0.28	7.75
	Ni	5										
	Mn	9										
	C	0.1										
Stainless steel Type 301	Fe	Bal.	S30100	Annealed	35	241	100	690	65	160		
	Cr	17		C. R. 52%	164	1131	190	1310	32	385	0.29	8.03
	Ni	7										
	C	0.11										
Stainless steel Type 302	Fe	Bal.	S30200	Annealed	30	207	90	621	55			
	Cr	18		C.R. 51%	156	1076	177	1220	5.5	160	0.29	8.03
	Ni	9										
	C	0.10										
Stainless steel Type 304	Fe	Bal.	S30400	Annealed	30	207	85	586	50			
	Cr	19		C.R. 50%	152	1048	167	1151	8	160	0.29	8.03
	Ni	9.0										
	C	0.08										
Stainless steel Type 304L	Cr	19	S30403	Annealed	30	207	80	552	55.0	76R_b	0.29	8.03
	Ni	9										
	C 0.03 max											
Cast stainless steel Type 304	Fe	Bal.	S30400	Cast[a]	35	241	79	545	54	131	0.286	7.92
	Cr	19										
	Ni	9										
	C	0.1										
Stainless steel Type 309	Fe	Bal.	S30900	Annealed	30	207	82	565	50			
	Cr	23		C.R. 45%	132	910	144	993	6	165	0.29	8.03
	Ni	13										
	C	0.20										
Stainless steel Type 310	Fe	Bal.	S31000	Annealed	40	276	100	690	50			
	Cr	25		C.R. 50%	135	931	145	1000	3	165	0.29	8.03
	Ni	20										
	C	0.25										
Stainless steel Type 316	Fe	Bal.	S31600	Annealed	40	276	90	621	50	165		
	Cr	17		C.R. 49%	136	938	148	1020	6	275	0.292	8.08
	Ni	12										
	Mo	2.5										
	C 0.08 max											
Cast stainless steel Type 316	Fe	Bal.	J92810	Cast[a]	35	241	83	572	60	161	0.286	7.92
	Mo	9										
	Mo	2.0										
	C	0.05										
Stainless steel Type 321	Fe	Bal.	S32100	Annealed	35	241	85	586	50	160		
	Cr	18		C.R. 50%	141	972	160	1103	4.0	300	0.292	8.08
	Ni	10										
	C	0.08										
	Ti	5 X C										
Stainless steel Type 347	Fe	Bal.	S34700	Annealed	40	276	90	621	50			
	Cr	18		C.R. 52%	144	991	169	1165	4.0	160	0.292	8.08
	Ni	11										
	C	0.08										
	Cb 10 X C											
AM 350	Fe	Bal.	S35000	Sol. treated and hardened	165	1138	190	1310	13.0	41R_c	0.290	8.03
	Cr	16										
	Ni	4.3										
	Mo	2.8										
	Mn	0.8										
	Si	0.3										
15-7-Mo	Cr	15	S15700	TH 1050	200	1379	210	1448	7.0	45R_c		
	Ni	7		RH 950	215	1482	240	1655	6.0	48R_c	0.277	7.68
	Mo	2.5										
	Al	1.0										
	Mn	0.7										
	Fe	Bal.										

(Continued)

Melting Range		Specific Heat (32-212°F) (0-100°C)		Coeff. of Thermal Exp. (8-212°F) 10⁶ (21-100°C) 10⁶		Thermal Conductivity (Room Temp.)		G Shear Modulus		E Tensile Modulus of Elasticity	
°F	°C	Btu/lb · °F	J/kg · K	in/in · °F	mm/mm · °C	Btu · in/h · ft² · °F	W/m · K	ksi	GPa	10³ ksi	10³ GPa
2740	1504	0.107	448	6.3	3.5	260	37			29.5	203
2550-2590	1399-1421	0.12	502	6.1	3.4	198	29			31.0	214
				9.8	5.4						
				8.0	4.4	113	16			29	200
2550-2590	1399-1421	0.12	502	8.0	4.4	113	16			28	193
2550-2650	1399-1454	0.12	502	8.0	4.4	113	16			28	193
2550-2650	1399-1454	0.12	502	8.0	4.4	113	16			28.0	193
2550-2590	1399-1421	0.12	502	8.0	4.4	110	16			29	200
2550-2650	1399-1454	0.12	502	8.0	4.4	108	16			29	200
2550-2650	1399-1454	0.12	502	8.0	4.4	96	14			30	207
2500-2550	1371-1399	0.12	502	8.8	4.9	114	16			29	200
2550-2590	1399-1421	0.12	502	8.0	4.4	110	16			29	200
2550-2600	1399-1427	0.12	502	8.3	4.6	113	16			28	193
2550-2600	1399-1427	0.12	502	8.3	4.6	113	16			28	193
2500-2550	1371-1399	0.120	502	6.8	3.8	114	16			30.4	210
		0.110	461	61.	3.4					28.0	193

(Continued)

Metal	Nominal Composition (%)	UNS Number	Condition and Temper	Yield Strength		Tensile Strength		Elongation in 2 in. (50 mm)	Hardness Bhn 500 Kg Load	Density	
				ksi	MPa	ksi	MPa	(%)	(70 mm Ball)	lb/in³	10³kg/m³
17-7 PH	Cr 17 Ni 7 Al 1.3 Mn 0.7 Si 0.7 Fe Bal.	S17700	TH 1050 RH 950	150 190	1034 1310	180 210	1241 1448	6.0 6.0	38R$_c$ 44R$_c$	0.267	7.39
Stainless steel Type 405	Cr 13 Al 0.2 C 0.08 max	S40500	Annealed	40	276	70	483	30.0	80R$_b$	0.28	7.75
Stainless steel Type 410	Fe Bal. Cr 12.5 C 0.15	S41000	Annealed Heat treated	40 115	276 793	75 180	517 1034	30 15	150 300	0.28	7.75
Stainless steel Type 414	Fe Bal. Cr 12.5 Ni 1.75 C 0.15	S41400	Annealed Heat treated	80 150	552 1034	100 200	690 1379	22 17	217 387	0.28	7.75
Stainless steel Type 420	Fe Bal. Cr 13 C 0.30	S42000	Annealed Heat treated	60 200	414 1379	98 250	676 1724	28 8	180 480	0.28	7.75
Stainless steel Type 430	Fe Bal. Cr 16 C 0.12	S43000	Annealed C.R. 45%	40 111	310 765	70 115	483 793	30 2.5	150 217	0.28	7.75
Stainless steel Type 431	Fe Bal. Cr 16 Ni 1.75 C 0.20	S43100	Annealed Heat treated	80 150	586 1034	120 195	827 1345	25 20	250 400	0.280	7.75
Stainless steel Type 440A	Cr 17 C 0.7	S44002	Hardened and tempered	240	1655	260	1793	5.0	51R$_c$	0.28	7.75
Stainless steel Type 446	Fe Bal. Cr 25 C 0.35	S44600	Annealed	50	345	80	552	30	165	0.27	7.47
Invar	Fe Bal. Ni 36		Annealed Hot rolled	42 50	290 345	70 73	483 517	41 37	130 140	0.289	8.00
Stainless steel Type 312	Fe Bal. Cr 29 Ni 9	S31200	Hot rolled	45	310	90	621	30	170		
Cast 28 Cr 10 Ni alloy ACI Type HE	Fe Bal. Cr 29 Mo 9	S31200	Cast²	45 (.5%)	310	85	586	10	200	0.277	7.67
Stainless steel Type 330	Fe Bal. Ni 36 Cr 16	N08332	Hot rolled Cold drawn¹ 55 Cold drawn³	 379 	 80 	100 552 150	690 35 1034	 200 	 0.284 	7.86	2515
A286	Ni 26 Cr 16 Mn 1.4 Mo 1.3 Ti 2 Al 0.2 V 0.3 Fe Bal.	S66286	Sol. treated quenched, and aged	105	724	153	1055	20.0	24R$_c$	0.286	7.92
Cast 35 Ni 15 Cr alloy ACI Type HT	Fe Bal. Ni 36 Cr 16		Cast²	38 (.5%)	262	69	476	10	192	0.290	8.03
Iron-silicon alloy	Fe Bal. Si 14.5 C 0.8 Mn 0.35		Cast²	17	117	17	117	0	500	0.253	7.00
"Durichlor"	Fe Bal. Si 14.5 Mo 3 C 0.8		Cast²	17	117	17	117	0	500	0.254	7.03
"Durime. T"	Fe Bal. Ni 22 Cr 19 Mo 2.5 Si 1 Cu 1		Cast⁴ Hot rolled⁴	35 45	241 310	70 85	483 586	40 50	140 140	0.283	7.83
Beryllium			Hot pressed Extruded	32 40	221 276	45 82	310 565	2.0 16.0	90	0.067	1.85

(Continued)

Melting Range		Specific Heat (32-212°F) (0-100°C)		Coeff. of Thermal Exp. (8-212°F) 10⁻⁶ (21-100°C) 10⁻⁶		Thermal Conductivity (Room Temp.)		G Shear Modulus		E Tensile Modulus of Elasticity	
°F	°C	Btu/lb · °F	J/kg · K	in/in · °F	mm/mm · °C	Btu · in/h · ft² · °F	W/m · K	ksi	GPa	10³ ksi	10³ GPa
2700-2790	1482-1532	0.11	461	6.13.4	117	117	473				200
				6.0	3.3	150	22			29.0	200
2700-2790	1482-1532	0.11	461	5.1	2.8	173	25			29	200
2600-2700	1427-1482	0.11	461	5.5	3.1	173	25			29	200
2650-2750	1454-1510	0.11	461	5.5	3.1	173	25			29	200
2600-2750	1454-1510	0.11	461	5.1	2.8	165	24			29	200
2600-2700	1427-1482	0.11	461	6.5	3.6	140	20			29	200
2500-2750	1371-1510	0.11	461	5.6	3.1	140	20			29.0	200
2550-2700	1399-1482	0.12	502	5.7	3.2	145	21			29	200
2600	1427	0.123	515	0.6	0.3	73	11			21	145
		0.14	586	10.5	5.8					22	152
1379		0.11	461	6.3	3.5	90	13				
2500-2600	1371-1427	0.110	461	9.4	5.2	105	15			29.1	201
		0.135	565	9.7	5.4					23	159
2300	1260	0.12	502	3.6	2.0	360	52				
2350	1288	0.12	502	3.6	2.0	360	52				
2650	1254	0.12	502	7.8	4.3	145	21			22.7 27.8	157 192
2345	1285	0.473	1989	6.9	3.8	1090	157			40.0	276

(Continued)

Metal	Nominal Composition (%)	UNS Number	Condition and Temper	Yield Strength		Tensile Strength		Elongation in 2 in. (50 mm) (%)	Hardness Bhn 500 Kg Load (70 mm Ball)	Density	
				ksi	MPa	ksi	MPa			lb/in³	10³kg/m³
Columbium			Annealed sheet	30	207	50	345	35.0	75	0.309	8.55
Titanium			High purity	20	138	35	241	50.0	90		
			Commercial pure	75	517	90	621	22.0	225	0.163	4.51
Zirconium			Extruded, air			65	448	24.0	80R_b	0.235	6.50

Notes:

(a) Lake copper (silver bearing, fire refined) has essentially the same properties as electrolytic tough pitch copper.

Footnotes:

(1) Hard temper
(2) As-cast
(3) Age Hardened
(4) Annealed
(5) Half hard temper
(6) Solution heat treated
(8) Solution heat treated, then artifically aged
(9) 326 Monel same as Monel except Ni content 55-60%
(10) Heat treated
(11) Water quenched, drawn at 200°F
(12) Water quenched, drawn at 1200°F

Melting Range		Specific Heat (32-212°F)	(0-100°C)	Coeff. of Thermal Exp. (8-212°F) 10⁻⁶ (21-100°C) 10⁻⁶		Thermal Conductivity (Room Temp.)		G Shear Modulus		E Tensile Modulus of Elasticity	
°F	°C	Btu/lb · °F	J/kg · K	in/in · °F	mm/mm · °C	Btu · in/h · ft² · °F	W/m · K	ksi	GPa	10³ ksi	10³ GPa
4380	2416	0.065	272	4.0	2.2	360	52			15.0	103
3075	1691	0.139	582	4.7	2.6	1190	171			15.5	107
3325	1829	0.067	281	2.9	1.6	116	17			11.3	78

(13) Normalized at 1200°F
(14) Water quenched, drawn at 1250°F
(15) Normalized at 1150°F
(16) Water quenched, drawn at 1150°F
(17) Rare earths
(18) Unclad product
(19) Elongation samples for aluminum are 1/2 in. diameter round bars
(20) Parenthesis number is % reduction of test coupon after last annealing step. Strength is very dependent on grain size and distribution.
(21) Unable to verify data; retained for historical interest only. Alloy designations have changed as companies have changed proprietary product chemistries.
(22) No UNS Number available
(23) H.F.B., EQ. and P-T: 1625°F/24 hr, A.C. and 1300 °F/20 hrs A.C. (AMS 5667)
(24) H.F.B., S-T and P-T: 1800°F/1 hr, A.C., 1350°F/8 hrs F.C. to 1150°F and hold for 18 hrs (AMS 5670)

Appendix C

Thermal Expansion Data

The following thermal expansion data has been compiled from numerous sources. It is presented for reference only because its accuracy cannot be assured. These curves are useful in calculating braze joint clearances when joining dissimilar metals. The method for calculating this information in presented in Chapter 2, Figure 2.3.

Materials Listed	Chart No.
Alumina 95%	C-3
Alumina 97.5%	C-3
Aluminum 1100	C-2
Aluminum 2024	C-8
Aluminum 2618	C-6
Aluminum 6061	C-2
Beryllium Copper 255	C-2
Beryllium-33 Al	C-7
Ceramvar	C-10
Chromium	C-1
Columbium-10 Ti-10 Mo	C-7
Copper OFHC	C-11
Gold	C-9
Graphite CDJ-97	C-5
Graphite P-03-XHT	C-5
Graphite 692	C-5
Hastelloy A	C-8
Hastelloy B	C-9
Hastelloy X	C-4
Inconel X	C-9
Inconel 600	C-4
Inconel 625	C-11
Inconel 718	C-4
Inconel 750	C-11
Invar 36	C-4
Iron	C-8
Kovar	C-10
Kovar 22	C-6
Molybdenum	C-1
Molybdenum-0.5 Ti-0.08 Zr	C-7
Monel K500	C-4
Monel 400	C-11
Ni Fe 42	C-3

(Continued)

THERMAL EXPANSION VS. TEMPERATURE

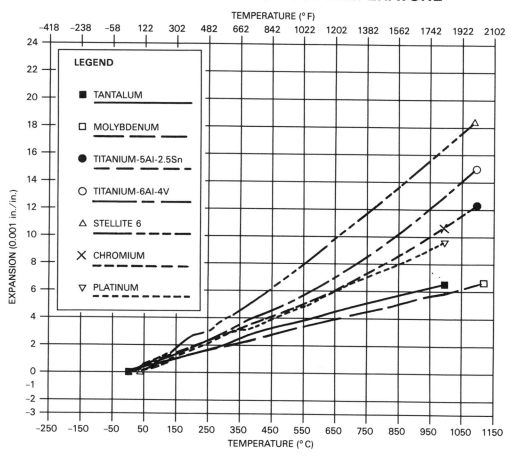

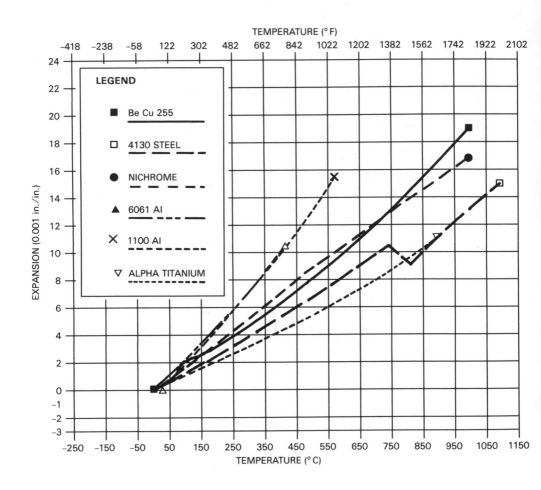

THERMAL EXPANSION VS. TEMPERATURE

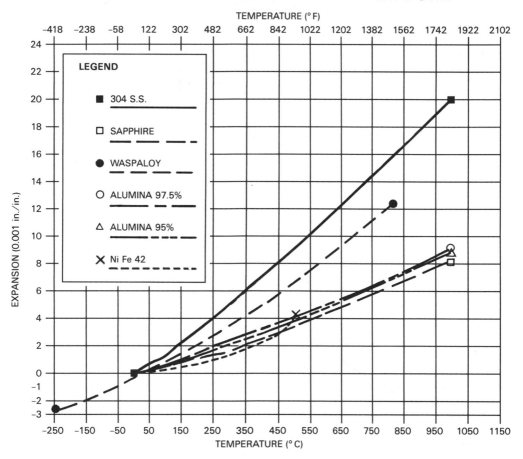

THERMAL EXPANSION VS. TEMPERATURE

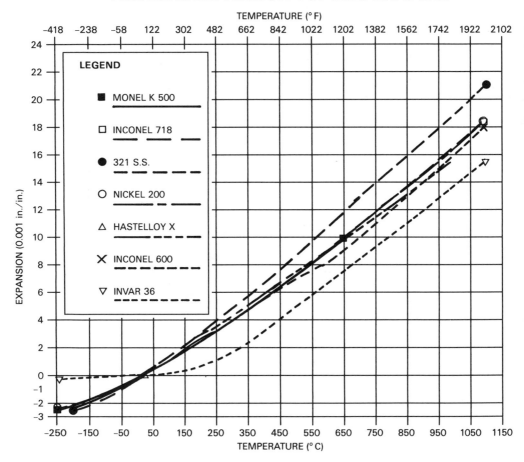

THERMAL EXPANSION VS. TEMPERATURE

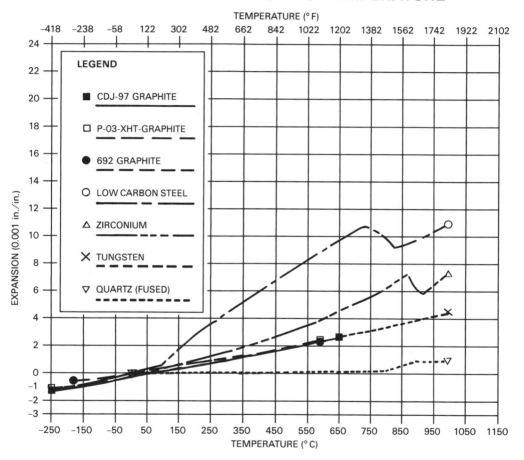

THERMAL EXPANSION VS. TEMPERATURE

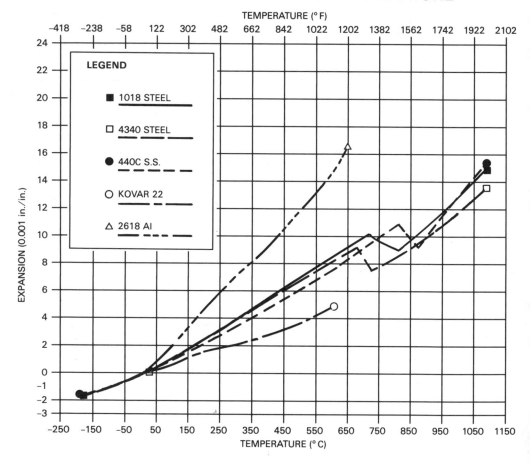

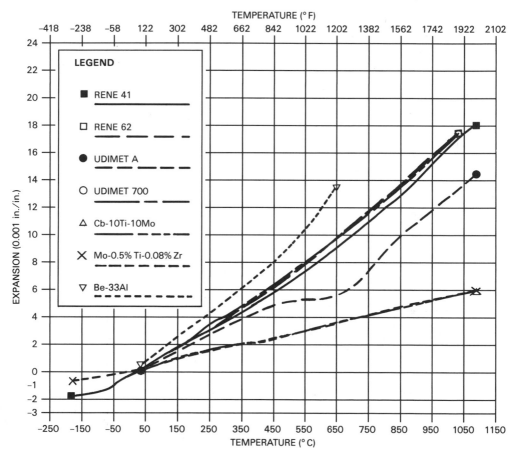

THERMAL EXPANSION VS. TEMPERATURE

THERMAL EXPANSION VS. TEMPERATURE

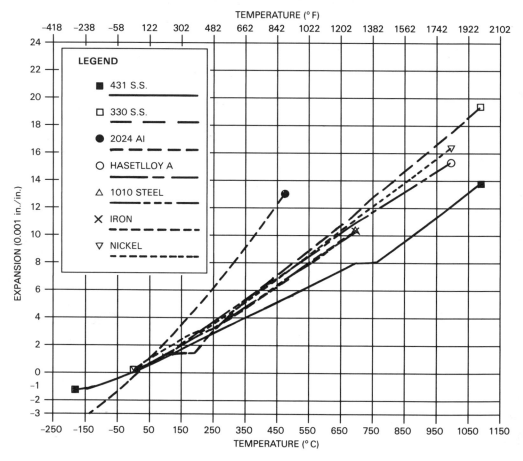

THERMAL EXPANSION VS. TEMPERATURE

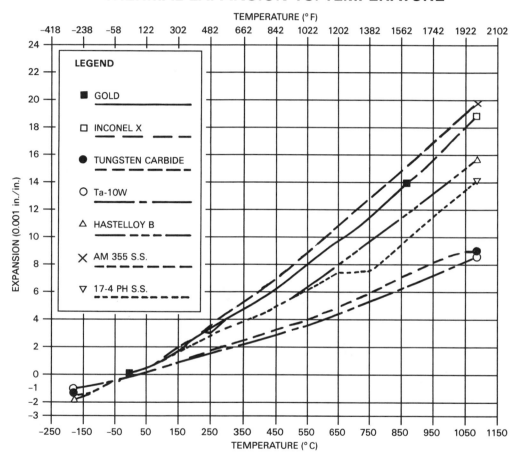

THERMAL EXPANSION VS. TEMPERATURE

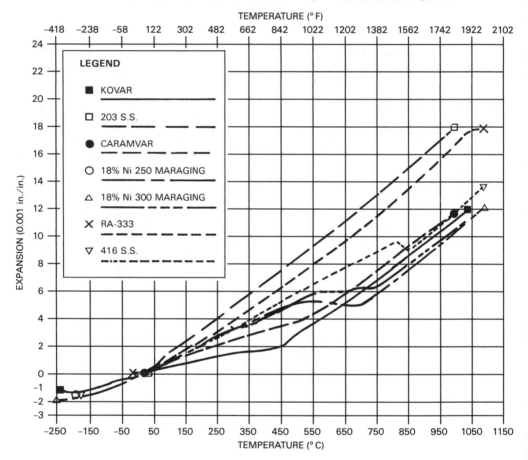

THERMAL EXPANSION VS. TEMPERATURE

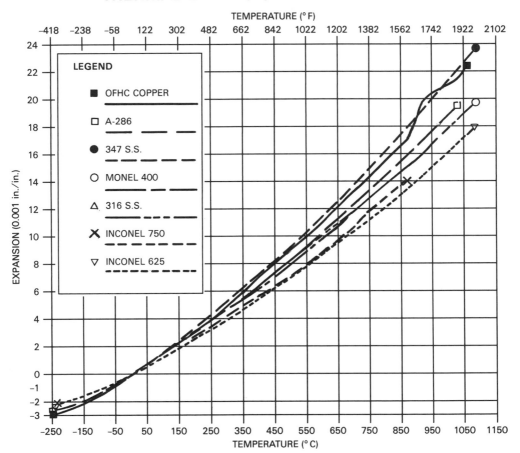

INDEX